TECHNOLOGY AND ETHICS
A European Quest for Responsible Engineering

Technology and Ethics
A European Quest for Responsible Engineering

Editors
Philippe Goujon
Bertrand Hériard Dubreuil

PEETERS
2001

ISBN 90-429-0950-1
D. 2000/0602/155

© 2001 – Peeters – Bondgenotenlaan 153 – B-3000 Leuven – Belgium

No part of this book may be reproduced in any form or by any electronic or mechanical means, including information storage or retrieval devices or systems, without prior written permission from the publisher, except the quotation of brief passages for review purposes.

International Editorial Team

José-Angel CEBALLOS-AMANDI
ICAI-ICADE
Universidad Pontificia Comillas de Madrid

Göran COLLSTE
THE CENTRE FOR APPLIED ETHICS
Linköping University

Gérard FOUREZ
EMSTES
Facultés Universitaires Notre-Dame de la Paix

Philippe GOUJON, coordinator
DÉPARTEMENT D'ETHIQUE
Université Catholique de Lille

Bertrand HÉRIARD DUBREUIL, chairman
ICAM-DÉPARTEMENT D'ETHIQUE
Université Catholique de Lille

Christiaan HOGENHUIS
Institute for Church and Society

Martin MEGANCK
Katholieke Hogeschool Sint-Lieven, Gent

Sally WYATT
DEPARTMENT OF INFORMATION STUDIES
University of East London
SOCIAALWETENSCHAPPELIJKE INFORMATICA
Universiteit van Amsterdam

Lille Operative Committe

Centre d'Ethique Technologique
Université Catholique de Lille

Jean-Marc Assié
Polytechnicum de Lille

Christelle Didier
Département d'Éthique

Annie Gireau-Geneaux
Institut Supérieur d'Électronique du Nord

Philippe Goujon
Département d'Éthique

Bertrand Hériard Dubreuil
Institut Catholique des Arts et Métiers

Jean-Marie Lhôte

Emmanuel Thévenin
Association Régionale pour la Promotion Pédagogique
et Professionelle dans l'Enseignement Catholique

Michel Veys
Institut Supérieur d'Agriculture

Address:

Centre d'Éthique Technologique
Université Catholique de Lille
B.P. 109-Boulevard Vauban, 60
59016 Lille cedex France
Tel: + 33.320134011
Fax: + 33.320134143
Email: bertrand.heriard@icam.fr

European Ethics Network
Board of Directors

Michel FALISE (Université Catholique de Lille): President ad interim

Johan VERSTRAETEN (K.U.Leuven): Chairman of the Board

Guillaume DE STEXHE (FUSL, Brussels)

Maurice DE WACHTER (Brussels): Representative of the *European Association of Centres of Medical Ethics* (EACME)
José Luis FERNANDEZ-FERNANDEZ (U.P. Comillas, Madrid)
Heidi Hövik-VON WELTZIEN (Norwegian School of Management, Oslo): President *European Business Ethics Network* (Eben)
Walter LESCH, (U.C. Louvain-la-Neuve)
D. LIPS (Gent): *European Ethics Network of Institutes of Higher Learning*, (this network will be fully integrated into the EEN)
Lars REUTER (Aarhus University): Secretary General of *Societas Ethica*
Jean-Pierre WILS (K.U. Nijmegen)

Address:

European Ethics Network Coordination Centre
Deberiotstraat 26
B-3000 Leuven
Tel: + 32.16.32.37.95/ 45.34
Fax: + 32.16.32.45.34
Website: http://www.kuleuven.ac.be/een
Email: johan.verstraeten@theo.kuleuven.ac.be

TABLE OF CONTENTS

Foreword: Professional Ethics and Ethics Education: Vision of the Core Materials Project XIII
Johan Verstraeten, Chairman of the Board of Directors, European Ethics Network.

General Introduction 1
Christelle Didier, Philippe Goujon, Bertrand Hériard Dubreuil, Christiaan Hogenhuis
0.1 Engineers Facing Ethical Debates 15
Gerard M. Fourez
0.2 Ethics in Engineering: Foundations of Ethical Practice 31
Brenda Almond
0.3 The Ethics of Technology 45
Günter Ropohl

1. Engineers within Technical Institutions 67

Introduction 69
Martin Meganck
1.1 Description 73
  1.1.1 Emergence and Growth of the Engineering Profession in Europe in the 19th and Early 20th Centuries 75
  André Grelon
  1.1.2 Today's Engineer Must Be More than a Technician 101
  Helmuth Lange and André Städler
1.2 Examples 119
  1.2.1 Virtual Games Inviting Real Ethical Questions 121
  Bernard Reber
  1.2.2 Workers and Engineers: Two Different Worlds? 133
  Stanislas Dembour
  1.2.3 The Safety of Risk-Prone Sociotechnical Systems: Engineers Faced with Ethical Questions 137
  Michel Llory
  1.2.4 From Accusations to Causes: Integrating Controversies and Conflicts into the Innovation Process 157
  Madeleine Akrich
  1.2.5 Interview with Professor Sir Joseph Rotblat 169
  Sally Wyatt and Joseph Rotblat

1.3 Reflection 183
   1.3.1 Handling Ethical Dilemmas in Everyday Engineering Work   185
      Boel Berner
   1.3.2 Engineers' Tools for Inclusive Technological Development . 207
      Christiaan T. Hogenhuis and Dick G.A. Koelega

2. The Development of Technical Systems 231

Introduction 233
   Sally Wyatt
2.1 Description 237
   2.1.1. Constructive Technology Assessment as Reflexive Technology
      Politics 239
      Johan Schot
   2.1.2 The State as Constructor of the Living Environment:
      Its Abilities and Limitations 251
      André Talmant and Pierre Calame
2.2 Examples 269
   2.2.1 Swissmetro: Engineers' Fancy or Major Project? 271
      Pierre Rossel
   2.2.2 The Superphénix Controversy: An Ethical Perspective 285
      Bertrand Hériard Dubreuil
   2.2.3 A Practical Perspective of Information Ethics 305
      Simon Rogerson
   2.2.4 Codes of Ethics: Conduct for Computer Societies:
      The Experience of IFIP 327
      Jacques Berleur and Marie d'Udekem-Gevers
2.3 Reflection 351
   2.3.1 Taking Risks and the Value of Human Life 353
      Göran Möller
   2.3.2 Who Will Bear Moral Responsibility: Engineers or
      Corporations? 369
      José Luis Fernández Fernández

3. Technological Development as a Societal Issue 387

Introduction 389
   Göran Collste
3.1 Description 395
   3.1.1. The Role of Technological Development in the Economic
      Process 397
      Luc Soete and Berit Schneider
   3.1.2 Globalisation and Ethical Commitment: The Challenge of
      the 21st Century 413
      Riccardo Petrella
   3.1.3 Globalisation from an Ethical Perspective 423
      Göran Collste

3.2 Examples 439
   3.2.1 The Hemacard Project: Applying the Constructive
        Technology Assessment Method to Computerised Health
        Cards.. 441
        Dominique Dieng
   3.2.2 The Case of rDNA Techniques: Ethics and the Complexities
        of Decision-Making 455
        Philippe Goujon
   3.2.3 Objective Science? The Case of Climate Change Models ... 485
        H.J.M. De Vries
   3.2.4 Sustainability: A Vision for a New Technical Society? 511
        Raoul Weiler
3.3 Reflection 525
   3.3.1 Engineers and the Dialogue on Extending their Horizon
        of Action: Awareness of Responsibility as a Claim to
        Competence and as Moral Behaviour 527
        Eva Senghaas-Knobloch
   3.3.2 Managing Technology: Some Ethical Considerations for
        Professional Engineers 543
        Peter W.F. Davies

4. Epilogue... 555
     Philippe Goujon, Bertrand Hériard Dubreuil, Jean Marie Lhôte,
     Emmanuel Thévenin, Michel Veys

Postscript: The Achievement of Technology and Ethics: A Perspective
   from the United States 565
   Carl Mitcham

Bibliography 583
Contributors 605

FOREWORD

PROFESSIONAL ETHICS AND ETHICS EDUCATION

Vision of the Core Materials Project

Johan Verstraeten
Chairman of the Board of Directors, European Ethics Network

This book is a result of the core materials project for the development of courses in professional ethics initiated by the *European Ethics Network* and subsidised by the European Commission DG XXII. The target group of this project are not students, but lecturers and professors. The texts are conceived as a source of inspiration for the development of courses adapted to local situations and personal interpretations. In the first stage the dissemination of the project is realised via the publication of four books:

- Professional Ethics: General Perspectives
- Business Ethics
- Engineering Ethics
- Media Ethics

The project, in which ethicists from all over Europe have participated, is a response to a real need: a better ethical formation and education of students via (1) an improvement of courses and (2) the development of an integral life-enabling education project of universities and institutions of higher education.

The project starts from the assumption that the students of today are the professionals of tomorrow. In a world dominated by the power of knowledge, professional experts such as scientists, (bio)engineers, physicians, lawyers, public servants, media experts, economists and business administrators exercise a crucial influence on the lives and the quality of life of millions of citizens. In the future this will increase under the influence of new (bio)technological, biomedical and managerial developments. Their implementation by professionals will affect the human and natural environment, the solution of problems with regard to life and death, employment and the quality of information and public office.

Very often the actual and future professionals are not sufficiently prepared to deal with the ethical aspects of their professional decisions and with the social consequences of their work. They need a broader education in which their professional knowledge and expertise is completed with the ability to resolve ethical dilemmas and with the capacity to discern the values that are at stake in every professional decision.

Providing students with greater ethical expertise is necessary but not sufficient, since the tendency to hyperspecialisation in already quite specialised disciplines goes together with a loss of the ability to integrate everything into a larger and meaningful whole. In such a context, there is need for a broad education in which space is created for an integral interpretation of reality, for initiation into traditions of thought, for the development of the student's civic sense and the configuration of persons as moral subjects.

Learning to Interpret

Before one can ethically reflect about the solution to a problem, one must clarify what the problem means, what meanings are connected with the problem and how the problem fits into the wider social context.[1] It would make little sense to pose questions of business ethics if one does not understand what business as a human activity means. It makes little sense to have a technical discussion about euthanasia without asking oneself about the meaning of human life and death. This is confirmed by the European Commission's white paper on *Teaching and Learning: Towards the Learning Society*, where an argument is made, precisely in the context of policy proposals, for better technical and scientific education. According to this document professional ethics requires more than simply a transfer of ideas from a differentiated ethical discipline or of models of ethical argumentation. It is rather a matter of opening minds and improving the capacity to interpret reality by way of additional education

---

[1] P. VAN TONGEREN, 'Ethiek en traditie' in *Tijdschrift voor Filosofie* 58(1996) pp. 84-102.

in literature and philosophy. The white paper considers these subjects to be important because, in a world where knowledge is quantitatively increasing, they *'arm the individual with powers of discernment and critical sense. This can provide the best protection against manipulation, enabling people to interpret and understand the information they receive.'*

That a wider and especially a literary education is one of the conditions for acquiring ethical competence, has been suggested by Martha Nussbaum, among others. In her book *Poetic Justice* she shows that it is not enough to initiate students into rational ethical argumentation.[2] One must also educate their emotional intelligence and their capacity to put themselves in the place of others. One of the conditions for this is an initiation into the reading of literary texts. Without an activation of their imaginative powers, according to Nussbaum, future professionals will be unable to put themselves as unprejudiced spectators in the situation of the people about whom they will make decisions in their professional life. Initiation into the reading of literary texts is also a way to liberate future professionals from limited or enclosed circles of interpretation. Their situation is sometimes comparable to that of the cave dwellers in Plato's *Republic*, particularly if they consider a severely limited approach to reality to be the only true one. Sometimes they see only those dimensions of a problem that are considered to be relevant according to the premises of a certain scientific discipline. Especially the applied sciences are apt to get stuck in what Weber called the steel cage of bureaucratic or technical rationality, and what MacIntyre described as a kind of thinking permeated by instrumental rationality and a kind of action reduced to manipulative expertise.

By teaching future scientists and professionals how to deal with new, different, even poetic possibilities for interpreting reality, through an introduction to literary and philosophical texts, one provides them with the means of breaking out of their closed or limited hermeneutic circles. This is not to say that scientific interpretations of reality are meaningless or without value. I rather mean that it is necessary, through initiation into literature and philosophy, to give students access to a horizon of interpretation which is different than

---

[2] M.C. NUSSBAUM, *Poetic Justice: The Literary Imagination and Public Life*, Boston, Beacon Press, 1995, pp. 72-78.

the scientific one in which they have been trained. Through a creative confrontation between various possibilities of interpretation, their field of vision expands. One might echo Marcel Proust and say that the *true voyage of discovery is not to seek out new territory, but to learn to see with new eyes*. The most important change that a person can undergo is the change in the way they view reality. One can change studies, job, neighbourhood, country or even continent and yet remain the same. If the fundamental hermeneutic perspective changes, however, then the way in which one experiences reality also changes.

In addition to a specialised course in ethics, something more is needed: the embedding of technical, scientific education within a *universitas* education, a broadly literary, philosophical, and cultural education that provides future professionals with the capacity to 'meaning-fully' interpret the reality in which they live and act.

A Purely Neutral Ethics Education Is an Illusion

The basic philosophy behind the core materials project is not based on the idea of a value neutral ethical expertise. The project also differs from theories in which it has become almost a dogma that ethics should take distance from particular moral or philosophical convictions or religious beliefs if it is to remain meaningful in a pluralistic society. In these theories one starts from a *method of avoidance* or in the best case from the *thinning out of the conflictuous thickness of moral concepts*. The project texts do not deny that overcoming philosophical and religious differences and achieving a reflexive equilibrium or an overlapping consensus is necessary if we are to achieve a minimum consensus regarding difficult questions. But there are doubts about whether a pedagogical process can work with ethical concepts and models that have been *totally* cut off from their original philosophical or religious frame of reference. People do not merely act on the basis of abstract principles or de-natured rational arguments. In making moral decisions, they make use of a scale of values and this is influenced in part by what Charles Taylor would call fundamental frames of reference, i.e., fundamental principles which are not as such rationally justified, but which determine the perspectives on the basis of which one attaches importance to specific values.³ Ultimate frames of reference and meaning, then, have an enormous influence on action. Like it or not, one always belongs to a tradition of thought or belief or, in a fragmented culture, to various traditions from which one draws inspiration. Even when one tries *a priori* to put the influence of traditions out of play, one belongs to a tradition, namely the tradition that uses this conception. This is the reason why it remains of crucial importance to the educational mission of a university to initiate students into tradition(s) of thought and to show them how different particular convictions have ethical implication. According to some scholars such as Alasdair MacIntyre, to be initiated into a tradition is even a precondition for meeting and understanding other cultures.⁴ There exists no Archimedean or neutral point from which one can approach these traditions. On the basis of a particular rationality, one can also recognise and clarify the reasonableness of other traditions and, where necessary, even critically integrate their incommensurable aspects into one's own tradition.

Ethics Education Requires Training in Civic Sense

'Par leurs activités de formation et de recherche, les universités sont des acteurs importants de la transformation sociale. Elles peuvent contribuer de façon substantielle à l'évolution des mentalités, des comportements, des structures qui favorisent le renouveau de la citoyenneté...'

Colloque 'L'université européenne acteur de citoyenneté'

As a standard-bearer for the knowledge industry, the university is often too one-sidedly oriented towards producing *employable* individuals in the service of the economic system (R. Bellah, *The Good Society*). One then runs the risk of losing sight of an important aspect of education: training in civic sense. Students must also learn to think about their social responsibility and take account of the social consequences of exercising professional power or knowledge power. Training in civic sense requires not only the transfer of knowledge about political and social ethics, but also the formation of social attitudes such as attention for the least advantaged in society. A univer-

---

³ C. TAYLOR, *Sources of the Self: The Making of Modernity*, Cambridge, MA, Harvard University Press, 1989, p. 26.

⁴ We adopt here a standpoint based on A. MACINTYRE, *Whose Justice, Which Rationality?*, Notre Dame, University of Notre Dame Press, 1989, p. 26.

sity as centre of excellence should not only focus on high points of culture and science, but also the depths of suffering into which a society and its citizens can fall.[5] Every society exhibits the face of its victims, and it is certainly not asking too much of students that they learn as professionals how to recognise that face.

There Is No Ethical Responsibility without Personal Development

The essential mission of education is to help everyone to develop their own potential and become a complete human being, as opposed to a tool at the service of the economy. The acquisition of knowledge and skills should go hand in hand with building up character (...) and accepting one's responsibility in society.

White Paper, p. 26

Ethics requires more than just knowledge acquisition. An ethical education implies more than providing future professionals with the means to make a critical judgement about problems through rational argumentation. Traditionally ethics has also been viewed as practical wisdom, aimed at the moral development of the acting person. Such a person is not only responsible for what he or she does in specific situations of choice, but also for the moral quality of his or her life, in other words, for the integration of moral choices and actions into a meaningful life that is configured as a narrative unity.

This development of moral identity has become exceptionally problematic nowadays. In fact, we currently find ourselves in a tension between a modern, liberal illusion of a completely autonomous and self-affirming subject and the factual break-up of the subject by a fragmented existence.[6] The latter implies not merely that the value and meaning systems outside us are breaking up (what the sociologists refer to as the disappearance of plausibility structures), but also that the inner life of people is falling apart into a multiplicity of

---

[5] D. HOLLENBACH, *Intellectual and Social Solidarity: Comment on J.M. Buckley's The Catholic University and the Promise Inherent in its Identity*, in J.P. LANGAN, L.J. O'DONNOVAN, *Catholic Universities in Church and Society*, Washington, Georgetown University Press, 1993, pp. 90-94.

[6] J. MORNY, 'Reflections on Paul Ricoeur's *Soi-même comme un autre*', in R.C. CULLEY and W. KLEMPA (eds.), *The Three Loves: Theology, Philosophy and World Religions: Essays in Honour of Joseph C. McLelland*, McGill Studies in Religion, Vol. 2, Atlanta, Scholars Press, 1994, p. 85.

experiences and possibilities. Fragmentation is embedded inside us as a cognitive possibility. In order to get around this impasse, it is necessary to once again offer the conditions of possibility for a new personal configuration. According to Ricoeur, this is primarily a matter of integrating the autonomous ego-identity that maintains itself through the course of time (*idem* identity) and the identity that is built up through the encounter with the other (*ipse* identity).[7] This other is not only the concrete other or the community, but also the texts which offer models in which a person can imaginatively recognise his/her conditions of existence as one step in the direction of a reconfiguration of himself (herself). For Ricoeur, understanding oneself is always a *se comprendre devant le texte*. If this conception is correct, then it means, once again, that initiating students into literary, philosophical and religious texts is a *sine qua non* condition for their moral education. Admittedly these texts have no direct influence on the concrete moral choices that a person must make, but they do open up an entire world of meaning that can stimulate the moral imagination with which moral subjects can recognise new ways of acting and being.

In light of these four points the steering committee of the European Ethics Network holds the opinion that ethical education requires a global pedagogical project. A university is more than an institution devoted to scientific research and education, more than a place where knowledge is acquired and transmitted. It is also a community of professors and students in which life-promoting learning processes are inculcated. As a school of life, the university can make a contribution to the education of citizens with a sense of responsibility.[8]

It is in this broader framework that professional ethics gets its necessary place. The participants in the core materials project do not want to relativise the fact that, in an academic environment, characterized by an increasing number of specialisations and hyperspecialisations, the rapidly evolving scientific and technological innovations require adequate ethical answers and a permanent adaptation of the law.

Before we can even think of excellent courses in professional ethics, we need to consider that first a two-fold expertise is required.

---

[7] Cf. P. RICŒUR, *Soi-même comme un autre*, Paris, Seuil, 1990, pp. 137-198.
[8] J. VERSTRAETEN, 'De spirituele bronnen van burgerzin en de taak van de katholieke universiteit' in *Ethische Perspectieven* 8(1998)2, pp. 59-64.

On the one hand, there is a need for ethicists with highly specialised knowledge who can function as discussion partners with experts; on the other hand, these experts themselves must learn to discover ethical problems and moral dilemmas in their own field. Ethics is not the application of abstract norms that would be imposed on a discipline like a *deus ex machina*. It begins by uncovering the problems and dilemmas as they present themselves in a specialised domain of knowledge. It is only after an adequate and expert analysis that a reasonably legitimate judgement can be passed. For instance, one cannot say anything meaningful about the ethics of international financial markets without an expert knowledge of derivatives and speculative techniques. It is impossible to pass a legitimate ethical judgement on prenatal diagnosis without a thorough knowledge of the medical aspects of the problem, and the techniques involved. This is why a specialised course in *ethics* for licentiate and postgraduate students is necessary, in addition to training the ethics instructors to be experts in a specific domain. Education in ethics, however, requires more than the introduction of a course in *applied* ethics and a pedagogic relationship between expert ethicists and students. At least as important is dialogue with specialists themselves, not only because ethicists should acquire a better knowledge, but also because the instructors and experts from various disciplines who are not ethicists, should themselves learn to pose ethical questions about their own specialised domain. Whenever *ethics* is restricted to a separate discipline, it is easily perceived by the students to be something standing somewhere outside their own discipline or professional training.

In other words, training in the students' ethical competence demands a simultaneous training in the ethical competence of all the instructors. This is the reason why the texts of the *Core Materials Project* are both destined to ethicists and experts in different fields.

GENERAL INTRODUCTION

TECHNOLOGY AND ETHICS

A European Quest for Responsible Engineering

*Christelle Didier, Philippe Goujon,
Bertrand Hériard Dubreuil, Christiaan Hogenhuis*

What are the reasons for reflecting about technology and ethics? Is modern technology getting out of control?

In his article 'The Risks of the *"Risikogesellschaft"'* the Polish-British sociologist Zygmunt Bauman states that in current technological development means are 'liberated' from their ends. What he thereby means to say is that means are created without predefined ends. Every technologist knows however that this is not true in the literal sense. Technologies are developed for ends like enlarging comfort, personal autonomy, safety, health and wealth and of course for having a satisfying job and making a good living. What is true, however, is that modern techniques and modern technological products and processes often turn out to be used for many more and quite different ends than they were originally developed for. Think of examples like nylon, Teflon, Internet, satellite communication, virtual reality software, the laser etc. What is true also is that modern technologies like transportation systems, information and communication technology, medical technologies and biotechnology by giving us personal autonomy, safety and wealth, also seem to contribute to undermining (the need for) interpersonal caretaking, the beauty of landscapes, cultural diversity, a sense of belonging to a tradition, community and environment and the view that life is a gift instead of our own making. And – making things more complicated – modern technology in some ways limits personal freedom and societal safety as well. Widespread introduction of new technologies leaves individuals often no choice but to make use of the technology themselves too. And it needs no extensive discussion here that the use of some technologies creates great social, economical and ecological dangers.

Obviously this calls for reflection on the ends that technological development should meet. In other words this calls for an 'ethics of technology'. But at the same time Bauman's statement points to the difficulty of such reflection. If there is no unequivocal connection between means and ends in modern technology, how then should we reflect on the goals of technological development and what is the sense of it?

These observations form the background of the underlying book. The book gives ample material for reflection on the role, meaning and ends of technology. But the emphasis is on showing what such reflection can consist of and how it can be made fruitful for actual processes of technological development. Therefore it is necessary to have a good idea of what ethics is about and what it can do. That will be the main topic of the first chapters on which later chapters will build. At the same time it is necessary to have a clear insight into what technology encompasses, what role it plays in society and how it comes about, within technological institutions and in society as a whole. The book is mainly devoted to giving a thorough introduction to the relation between technology and technological development on the one hand and individuals, (technological) institutions and society on the other hand, making use of knowledge from disciplines like history, sociology, economy and philosophy and ethics and the experience of engineers, businesspersons and politicians.

Engineering Ethics as Part of an 'Ethics of Technology'

This work aims at technical agents in general and in particular at engineers, as well as at those who, at school or within the firm, have as their mission to train these technical agents, to accompany them in their professional evolution or to advise them in their decisions. Therefore engineers are also the main focus of the book, especially in the first part. This is not because we think that engineers are the only 'agents of technology'. We all play a part in the making of technology, be it as consumer, businessperson, politician, citizen and voter, member of special interest groups like that for labour, environment, peace, development etc. What is more, this list of social roles makes clear we inevitably play *several* parts in the processes of development, implementation and use of technology. Yet engineers – even in their diversity – play a unique and crucial part, as will be clarified in several articles in the book. This legitimises the choice giving the engineers' reflection on their role and responsibility special attention. A further and maybe even more convincing reason is that engineers are criticised for the character and impact of modern technological developments more than other persons, although this is in many respects unfair. Anyway, the result is that engineers themselves show more and more the desire to reflect on their role and responsibility and to be prepared and equipped for that.

The objective of the book is therefore to give engineers and their trainers and advisors the means to reflect in a critical way on the engineers' contribution to the development of technologies. It is also to find out the 'loci', that is the moments and possibilities, of their social and moral responsibility, both within the institutions where they work and at their other positions in society. In this way the book can be seen as a working tool, a handbook designed for the initial or continuing formation of engineers and for professional's reflection. In doing so, it wishes to bring its contribution to the building of the discipline of engineering ethics; a discipline that is still immature, as we will clarify.

The foregoing presents engineering ethics as a part of or rather a special focus on an 'ethics of technology', just as there can be other ways of focussing; for instance on the role of economists, politicians, consumers etc. Clearly for a fruitful and useful reflection on the meaning and ends of technological development these other roles and the accompanying 'loci of responsibility' have to be discussed also in fora adequate for that, such as international committees for economic policy, political parties, employers' and labour organisations, consumer interest groups, churches etc.

Engineering Ethics as Part of Professional Ethics

Engineering ethics can and must not only be related to the 'ethics of technology', but also to professional ethics. There has been much talk in the field of professional ethics, particularly in the United States where the discipline is old, of the salaried status of engineers within large firms. The discussion has been based around whether to deny them the prestige reserved for 'genuine' professions and discredit any attempt to establish a professional ethics of their own, or to point out the narrow limits of their autonomy and clear them of any individual responsibility. To be sure, the status of salaried employee of a private firm – which is that of the great majority of engineers – is not negligible, because it almost inevitably makes them face a conflict of interests between loyalty to their employer and an obligation of service and/or protection to the public. But this salaried/non-salaried distinction and its moral implications are unsatisfactory in several ways. On the one hand, because they proceed from an illusion concerning the 'traditional' professions and their autonomy. For to take but one example, no one would consider that the restriction of the physician's autonomy in his practice today, often working in teams and tied to economic, technical, and legal constraints that have become more and more onerous, brings into question the very possibility of a reflection on medical ethics. On the other hand, the idea that the engineer is but a link in a chain of decisions and actions beyond his power, whose stakes are beyond his scope and whose impact is distant in space or in time, must not be an alibi for refraining from moral reflection. On the contrary, these considerations stress the importance of such reflection and only show that such reflection should not be limited to (professional organisations of) engineers, as we discussed before. Maybe nowhere else is the gap between our common intellectual development and our moral development so obvious as in the way that we manage the utilisation and the impact of new technologies in our societies. In the same way that the medical profession cannot do without a reflection on what disease means in our society and on the economic and political stakes of public decisions touching on health, so engineering cannot do without a reflection on the relations between technology and humans. That the need for a special professional ethics for physicians is seldom questioned may reflect the fact that the medical profession has preserved its autonomy surprisingly well during the 20th century.

Of course, if the engineer is under some conditions answerable for his or her actions to the employer, the clients and other commercial partners, to colleagues and subordinates, to the neighbours of the institution and the other persons that directly or indirectly experience the impact of the engineer's actions, (as we think is the case), he or she also answers to mankind as a fellow human being. In this, the engineer is no different from other professionals. Also, the potential conflicts of interest and of values that an engineer may encounter may not all be of a completely different nature compared to that of other professionals. Still there are differences and special problems, connected to the complex and widespread character of modern technologies, which means that an engineer with his or her colleagues in the same firm and in competing firms have a 'collective relationship' with many indirect 'clients' at once, scattered in place and time. This special situation is a further legitimisation for engineering ethics as a special branch of professional ethics. No reflection on the morality at play in complex systems where many agents interact, where any decision is a sequence of micro-decisions, where responsibilities are diluted, can forego meticulous, precise observation and analysis of *just* these systems, of their mode of functioning, and of their regulating mechanisms.

The Need for this Book

There remains one question to be answered in order to give a sufficient explanation for the publication of this book: is there a need for a new book of this kind? Of course, we are convinced there is. Surely several introductions and textbooks on the *philosophy* of technology exist.[1] Philosophy of technology is a discipline with some history,[2]

---

[1] For textbooks see Mitcham and Mackey (1972), Sachsse (1974-76), Durbin and Mitcham (1980) and Durbin and Rapp (1983). For introductions see for instance Ihde (1979), Hottois (1984), Rapp (1994) and Achterhuis (1992, 1997).

[2] Generally the German Ernst Kapp is considered to be the first philosopher that wrote about technology, in his book *Grundlinien einer Philosophie der Technik* of 1877 From then on a growing number of authors in Europe contributes to the discipline: among the first were in Germany Spengler (1931), Heidegger (1954), among many others; in Russia Berdyaev (1934); in Spain Ortega y Gasset (1939); in Brittain Gill (1940); in the Netherlands Van Riessen (1949); in France Ellul (1954). Ellul deserves special attention because he was the first to study technology from a sociological perspective. In the same time the discipline developed in the United States. Already in 1934 Lewis Mumford grounded a line of study in which the relation between technology and culture was analysed with his book *Technics and Civilization*. German immigrants like Arendt (1958), Horkheimer and Adorno (1947, 1971), Marcuse (1967) and Jonas (1979) introduced the European philosophy of technology in the United States. By philosophers like Mitcham (1972) and Winner (1977) both traditions were brought together.

although in comparison to the whole body of western philosophy it is still young and immature. Although ethical questions are raised in some of the early works – of philosophers like Kapp, Dessauer, Mumford, Ortega y Gasset, Van Riessen and Heidegger – a real ethics of technology does not exist. Later philosophers continued the disengaged approach, but in a more focussed way: on technology as separate kind of knowledge (epistemological approach), on the relation between technology and man (anthropological) or between technology and society (sociological). Only a few philosophers offer early examples of the study of technology from a moral (and religious) point of view.[3]

What characterises these early ethics of technology studies, however, is their general nature. They offer rather a culture critique from a moral point of view rather than detailed ethical studies offering starting points for specific moral choices. This can be explained by the way in which technology was interpreted by most of the philosophers of technology at that time. Either they studied technology in general as a specific type of human activity, or they interpreted technology simply as a comprehensive word for technical products in general. This changed in the 1980s with what is now commonly called the 'empirical revolution' in philosophy of technology.[4] This led to the so called 'constructivistic' approach to technology in which the way specific technologies developed out of competing influences in society was studied. This approach is closely connected to the Technology movement, which started around 1970 in the USA. The constructivistic approach contributed to the insight that technologies are complex and intricate networks of products, processes, institutions, organisational patterns, people with different social roles, cultural values etc. The first author that took this approach was Thomas Hughes. European authors like Callon, Latour and Bijker continued this approach. They are characterised by a strong emphasis on sociology of technology, or maybe social philosophy of technology. What is striking is that the moral point of view is absent in this kind of studies. This is one of the reasons why Winner characterised these studies with the phrase 'opening the black box and finding it empty'.

[3] H. JONAS, *Das Prinzip Verantwortung*, Frankfurt a.M., Suhrkamp, 1979.
G. PICHT, *Mut zur Utopie*, München, Piper, 1969.
H. SACHSSE, *Technik und Verantwortung*, Freiburg, Rombach, 1972.
[4] H. ACHTERHUIS et al., *Van stoommachine tot cyborg. Denken over techniek in de nieuwe wereld*, Amsterdam, Ambo/Anthos, 1997.

The above seems to be a sufficient legitimisation for the statement that we still cannot speak of the *ethics of technology* as a discipline. On the other hand there is something resembling an *engineering ethics* as part of a broader discipline of professional ethics for which medical ethics has prepared the ground. Several textbooks and introductions have appeared (Unger; Martin and Schinzinger; Schaub and Pavlovic; Weil). Professional ethics, and especially engineering ethics, has mainly been developed in the USA. In Europe, Lenk and Ropohl are some of the few that have contributed to developing an engineering ethics.[5] The American professional ethics is connected to the many and strong professional organisations in the USA. Organisations of engineers had already started paying attention to professional ethics already in 1911, mainly by developing and discussing professional codes of ethics

However, it is noteworthy that in this line of engineering ethics almost no use is made of the results of the philosophy of technology studies, with the exception of the work of Lenk and Ropohl. In the USA Florman combines the two disciplines to a certain extent. But he denounces most of the philosophy of technology as mere 'antitechnology' and declares the tendency to pay special attention to the moral and social responsibility of engineers irrelevant or even dangerous. Obviously the two disciplines of engineering ethics and the philosophy of technology seem to be almost separate domains.

In about the last decade, germs of an engineering ethics as part of an ethics of technology and therefore building on the philosophy of technology tradition have developed, mainly in Europe as it seems.[6] Also the domain is developing into more specific branches like computer ethics and ethics of biotechnology.

---

[5] H. LENK and G. ROPOHL (eds.), *Technik und Ethik*. Stuttgart, Reclam, 1987, 2nd ed. 1993.

[6] K. OTT, 'Technik und Ethik' in *Angewandte Ethik*, Stuttgart, Kröner, 1996, pp. 650-717.

Ch. HUBIG, *Technik- und Wissenschaftsethik*, 2nd ed., Berlin/Heidelberg/New York, Springer, 1995.

P. VAN DIJK, *Op de grens van twee werelden. Een onderzoek naar het ethisch denken van de natuurwetenschapper C.J. Dippel*, Den Haag, Boekencentrum, 1985.

C.T. HOGENHUIS and D.G.A. KOELEGA (eds.), *Technologie als levenskunst. Visies op instrumenten voor inclusieve technologie-ontwikkeling*, Kampen, Kok, 1996.

Peter W.F. DAVIES, *The Contribution of the Philosophy of Technology to the Management of Technology*, Ph.D diss., Brunel University with Henley Management College, 1992.

I. BARBOUR, *Ethics in an Age of Technology*, San Francisco, Harper, 1993.

Engineering ethics as we wish to promote it here is perhaps first of all an attempt to strengthen the meeting and dialogue between these two disciplines. On the one hand stands engineering ethics as a branch of professional ethics and on the other, a philosophy of technology, including its historical, social philosophical and sociological approaches. It is deliberate that this work, bringing together the contributions of 35 authors hailing from 10 different European countries, does not fit into a single intellectual tradition, but gives a voice to various critical currents, elaborated from without as well as from within technical practice. The reader will discover, for instance, some theories dear to professional ethics as it has developed above all in Anglo-Saxon countries; elsewhere, it is more the current of technology assessment, renewed by the constructive evaluation of technologies, that will be in the limelight. Obviously, all the traditions and schools of thought existing in Europe are not represented. But we hope that our choices will allow the reader to get acquainted with the state of the question as it has appeared to us over the course of our three years of collective labour.

Overview of the Contents

To put these materials with their rich diversity in order, the handbook is made up of four parts. After an introductory section, the handbook separates methodologically three levels of analysis: the micro level developing the ethical problems met by engineers in their institutions; the meso level where technical systems and institutions are in competition; the macro level representing technical developments as societal problems. Each part has the same structure: specialists of the human sciences, mostly history and social sciences, give a first description of technical praxis; engineers, sustained by some social scientists and philosophers, show problematic examples; philosophers reflect on these materials. This double division into parts and sections gives the handbook a systematic structure that should allow the reader to better perceive the complexity of the problems treated.

The introductory section proposes a reflection on the concepts of ethics and of technology, so that the engineer may situate the process which he is invited to enter into within a broader theoretical framework. It first includes a reflection on what is ethics, as a philosopher/scientist and an ethicist see it. It then takes care to unravel the different approaches that an engineer interested in ethics should be able to distinguish and to articulate: the technical, the ethical, the political, the scientific. This leads us to ask in what sense it may be said that an engineering ethics exists.

After the introductory part, the first part effects an analysis of the place occupied by engineers in societies and their institutions, and more specifically in technical institutions. It is first a matter of situating the engineering profession in a historical context: what is the genealogy of the professional institutions that we know today. This historical study then gives way to a sociological analysis highlighting the roles, functions, tasks, status, organisations, and ideologies of engineers today. This analysis should enable them to become aware that their profession occupies a particular social space in society, that it has its 'natural' alliances, and that it meets with constraints that are not only technical, but also social and economic. It highlights the societal conflicts and the controversies of which engineers are a part. These questions are approached at once directly as themes of reflection, and through specific narratives displaying the place of engineers in society. This part concludes with reflections on the engineers' autonomy and a presentation of the means at their disposal to take up their responsibility in an effective and feasible way.

The second part centres on the development of socio-technical systems and the role of technical institutions (firms, professional bodies, national regulatory bodies), in relation to the place that engineers have in it. It starts with presenting Constructive Technology Assessment, a branch of technology assessment that emphasises the possibility of ceaselessly fashioning and refashioning technologies at every stage of their development and in function of their predictable impact on society and on individuals. This kind of Technology Assessment is carried through in close relation with the relevant technical research and development institutes. After this the role of the national state in processes of technological development is discussed. Numerous examples (computer ethics, new transportation systems, nuclear energy) illustrate this part which closes with reflections on risk management and on the moral responsibility of technical institutions.

Finally, the third part deals with the questions raised for society by technological development, as well as with the analytic tools available on this subject. It begins by presenting a macro-economic analysis of the role of technical development. Then the ethical aspects of the trends of globalisation are discussed. Then this part examines some societal and partly global problems raised by the evolution of techniques, including those of sustainable development and climatic change and of the development of biotechnologies. It ends with a critical reflection on the ideology of engineers and on the whole body of beliefs spread by modernity about technologies.

Such a book cannot be concluded in a real sense. An epilogue, however, permits us to take up the whole process again and to underline its systematic character. Finally, a postscript by Carl Mitcham develops the cultural differences relating to ethical debate, between North America and Europe in particular. Although each contribution has been written by recognised specialists of each question, they have been designed to be readable by non-specialists. Nevertheless, a pedagogical use could not do without a teacher who would build his course around these materials, giving his students the interpretative framework to enable them to make up their own personal idea and opening the space for discussion without which no ethical judgement can be founded.

Terminological Warnings

Some warnings are necessary here. Because of the diversity of traditions and original languages some concepts will have a slightly different meaning in the various contributions. Where Ropohl defines *ethics* as the theory of acting in a morally right way, Fourez interprets ethics in a more liberal or even voluntaristic way as answering the question 'what do I want to do?', whereas Almond gives a broader and more Aristotelian definition of ethics as the study of the best way to live. Furthermore, one should be aware that *Ethics of Responsibility* means something different to Ropohl as compared to Almond. Whereas the first refers to the concept '*Verantwortungsethik*' of the German sociologist Weber, which has come to be identified with teleological ethics or consequentialism or even utilitarianism, the second refers to the approaches of ethics developed by people like Levinas, Jonas and feministic philosophers, which she rightly distinguishes sharply from utilitarianism and consequentialism. Then some authors use the concept of *deontology*, mainly as a translation of the French '*déontologie*', meaning something like the set of norms and values within a certain group. In many other countries this will be called morals or morality (German: *Moral*; Dutch: *moraal*), where deontology refers to a non-consequential justification of moral choices. Something similar occurs around the concept of *technology*. In French speaking countries *'technologie'* often means a theory of technical means. In other countries often 'technics' and 'technology' are used interchangeably, but sometimes a difference is made between technics and technology (in German: *Technik* and *Technologie*; in Dutch: *techniek* and *technologie*). The first word then sometimes refers to ancient or simple technical means, sometimes to specific, more or less isolated technical artefacts. The second word then means modern, complex, sciencebased technics and sometimes the complex and widespread system or network that connects specific technical artefacts, the institutions where they are developed, marketed or used, the users, the regulations that constrain the use, etc. Furthermore in the English and the French there is the separate word *'technique'* for a specific way of processing materials – like a new steel production technique – or a specific skill, like a good piano technique or a good skating technique, but also new welding technique or a management technique. In German and Dutch however this is the same word *'Technik'* and *'techniek'* respectively.

Acknowledgements

This work was initiated by Johan Verstraeten as director of the European Ethics Network. This network was founded in order to develop a practical ethics for and with professionals. It began with the help of FUCE and Coïmbra universitas as well important networks and associations: such as the European Business Ethics Network, the European Association of Centers for Medical Ethics, and the Societas Ethica. Over a hundred universities and institutions of higher learning, including engineering schools, joined it. It has been recognised as a thematic network of the Socrates programme. In this capacity, it has been supported by the European Commission to realise four handbooks: one on professional ethics, one on media ethics, one on business ethics, and one on engineering ethics.

This last one has been put together by a team of eight researchers from six different countries (see the list of the international editorial committee) and assisted by a team of teachers of the Polytechnicum of Lille (see the list of the Lille committee). The Département d'éthique of the Université Catholique de Lille, which has assumed operational responsibility for the work, was financially supported by the Fondation Leopold Mayer pour le Progrès de l'Homme.

The editors of this book, wish to thank all the authors of this work (see the list of contributors), the referees for each article (whose names cannot be given in order to preserve their anonymity), the translators whose names are cited in a note on the first page, and Loveday Pollard, Olivier Alglave, Julien Dolidon, Edmund Guzman who have in the first stage of the editorial work reviewed all the texts and all the translations to make them more readable. The translations of the texts into English were coordinated and financially sponsored by the European Ethics Network Coordination Centre, K.U.Leuven which for that purpose obtained subsidies from the European Commission. The final editorial work was coordinated by Johan Verstraeten, John Ries and Maria Duffy (proof-reading, harmonisation of the texts).

References

ACHTERHUIS, H. et al., *Van stoommachine tot cyborg. Denken over techniek in de nieuwe wereld*, Amsterdam, Ambo/Anthos, 1997.
ACHTERHUIS, H., (ed.), *De maat van de techniek*, Baarn, Ambo, 1992.
BARBOUR, I., *Ethics in an Age of Technology*, San Francisco, Harper, 1993.
BAUMAN, Z., 'The Risks of the *"Risikogesellschaft"'*, lecture delivered on April 2, 1996, University for Agriculture, Wageningen, The Netherlands (published in Dutch in Z. BAUMAN, *Leven met veranderlijkheid, verscheidenheid en onzekerheid*, Amsterdam, 1998).
BIJKER, W., 'Life after Constructivism' in *Science, Technology and Human Values*, London, 1993, pp. 113-138.
DAVIES, P.W.F., *The Contribution of the Philosophy of Technology to the Management of Technology*, Ph.D diss., Brunel University with Henley Management College, 1992.
DURBIN, P.T., 'Introduction: Some Questions for Philosophy of Technology' in DURBIN and RAPP 1983, *Philosophy and Technology*, Dordrecht, Reidel, 1983, pp. 1-14.
DURBIN, P.T. and C. MITCHAM, *Research in Philosophy and Technology: An Annual Compilation of Research*, London, 1980.
DURBIN, P.T., and F. RAPP, *Philosophy and Technology*, Dordrecht, Reidel, 1983.
FLORMAN, S.C., *The Existential Pleasures of Engineering*, New York, St. Martin's Press, 1976.
HOGENHUIS, C.T., *Beroepscodes en morele verantwoordelijkheid in technische en natuurwetenschappelijke beroepen. Een inventariserend onderzoek*, Zoetermeer/Driebergen, MCKS, 1993.
HOGENHUIS, C.T., and D.G.A. KOELEGA, *Technologie als levenskunst. Visies op instrumenten voor inclusieve technologieontwikkeling*, Kampen, Kok, 1996.
HUBIG, C., *Technik- und Wissenschaftsethik*, Berlin, Springer, 1995.
IHDE, D., *Technics and Praxis*, Dordrecht, Reidel, 1979.
KOELEGA, D.G.A., (ed.), *De ingenieur buitenspel? Over maatschappelijke verantwoordelijkheid in technische en natuurwetenschappelijke beroepen*, Den Haag, Boekencentrum, 1989.
LENK, H., and G. ROPOHL (eds.), *Technik und Ethik*, Stuttgart, Reclam, 1987.
LENK, H., and G. ROPOHL, 'Toward an Interdisciplinary and Pragmatic Philosophy of Technology' in P. DURBIN (ed.), *Research in Philosophy and Technology*, vol. 2, Greenwich, CT, 1979.
MARTIN, M.W. and R. SCHINZINGER, *Ethics in Engineering*, New York, Mc Graw Hill, 1983.
MITCHAM, C. and R. MACKEY (eds.), *Philosophy and Technology. Readings in the Philosophical Problems of Technology*, New York, Free Press, 1972.

OTT, K., 'Technik und Ethik' in J. NIDA-RÜMELIN (ed.), *Angewandte Ethik*, Stuttgart, 1996.
PICHT, G., *Mut zur Utopie*, München, Pipert, 1969.
RAPP, F., *Die Dynamik der modernen Welt. Eine Einführung in die Technikphilosophie*, Hamburg, Junius, 1994.
SCHAUB, J.H. and K.R. PAVLOVIC, *Engineering Professionalism and Ethics*, New York, Wiley, 1983.
UNGER, S.H., *Controlling Technology: Ethics and the Responible Engineer*, Chicago, 1982; 2nd enlarged edition, New York, Wiley, 1994.
VAN DER POT, J.J.H., *Die Bewehrtung des technischen Fortschritts. Eine systhematische Übersicht der Theorien*, Assen, Van Gorcum, 1985 (English edition: *Steward or Sorcerer's Apprentice? The Evaluation of Technical Progress: a Systematic Overview of Theories and Opinions*, Delft, 1994)
VAN DIJK, P., *Op de grens van twee werelden. Een onderzoek naar het ethisch denken van de natuurwetenschapper C.J. Dippel*, Den Haag, Boekencentrum, 1985.
VAN DIJK, P., 'Op zoek naar een verantwoorde technologie-ethiek. Een bijdrage vanuit oecumenisch gezichtspunt' in KOELEGA 1989, pp. 127-48.
VAN RIESSEN, H., *Filosofie en techniek*, Kampen, 1949.
WEARE, K.M., Engineering Ethics: History, Professionalism and Contemporary Cases, in *Louvain Studies*, 13 (1988), 252-271.
WEIL, V., *Beyond Whistleblowing: Defining Engineers' Responsibilities*, Chicago, Illinois Institute of Technology, 1983.

0.1

ENGINEERS FACING ETHICAL DEBATES

Gerard M. Fourez

This introduction to engineering ethics starts with the existence of an ethical debate between engineers who have to decide what, all considered, they do (or do not) want to do. Engineers, like other human beings have their own self-esteem and do not want to do just anything. Conflicts of interests are often ideologically concealed behind these ethical discussions, but it is a fact that beyond technical issues there are ethical debates. Their standardisation provides a foundation for training in ethics. The concepts of ethical and/or moral paradigms, attitudes, principles, maxims, norms, deontology, responsibility, values, etc., are thus explored as well as the distinctions between individual and structural ethics or between heteronomous and autonomous ethics. The difference between an ethical debate and a political debate is introduced as well as several concepts related to the societal evaluation of technologies, e.g. technology assessment and technology as a societal organisation. To conclude, the distinction between an ethics of technology and an engineering ethics is conceptualised, which brings to the fore what it means to discuss ethical issues from the perspective of an engineer who refuses to be a mere technocrat and wants to be a responsible person.

Many engineers like neither ethics[1] nor the jargon of ethicists. They feel that, too often, ethical discourses speak of norms that could restrain creativity. The engineering profession is already difficult

---

[1] We will speak of ethical situations or debates when what is at stake are some fundamental options for our lives. Or, in other words, when our self-esteem or the consideration that others have for us is involved. And we will speak of technical situations or debates when what is at stake are only means towards an end, without any fundamental option being involved. For a clarification of the terms used here, see G. FOUREZ, *Nos savoirs sur nos savoirs, un lexique d'épistémologie*, Brussels, Éd. De Boeck Univ., 1997, pp. 104-118, especially the entries dealing with the term 'ethics'.

enough without meddling by moralists who would restrict the freedom to carry out new ideas and research, often without really knowing what they are talking about.

Moreover, the most common way of referring to ethics implies *heteronomy*, i.e. a moral obligation imposed from the outside. It tells us what we may do, what we ought to do, or what we should avoid doing. Ethical debate thus seems to be loaded with obligations coming from somewhere outside the realm of professional practice. Traditionally, these moral imperatives are based on legitimating authorities with names such as God, nature, conscience, duty, conviction, or even some mysterious entity called 'Ethics'[2] or 'Morality'. It is understandable that engineers – who already have their hands full simply coping with the constraints imposed by nature, the economy, institutions and the law – mistrust those who would confine them within an ethical framework.

In reality, however, ethical debate is not always carried on under the rubric of such external obligations. The ethical question 'What ought to be done?' can be approached in a different way. Indeed, from the point of view of an *autonomous* ethics, human beings can also debate about what, all things considered, they *want* to do, while recognising that, in such situations, their own self-respect is at stake. (When Ricoeur suggest self-esteem as the quality specifically at stake when Ethics is concerned,[3] he does not refer to the narcissistic self but to the person who commits him or herself, and may be looked upon as someone 'other'.[4])

When the emphasis is given to autonomy, ethical questioning becomes: 'What, ultimately, do I want to do?', 'What do we want to do?', 'What am I going to do?' or 'What are we going to do?' 'What, all things considered do I (we) take the risk of deciding?'.[5] In fact, we

---

(This book will be referred to as NSSNS).

[2] As when this word comes into expressions such as 'Ethics says this about that topic', or 'What does Ethics have to say about this?'. One could wonder about the ideological meaning of such a general term in the absence of any enunciator. It might be more appropriate to specify the latter by expressions such as 'the ethics of this person (or of this community) says this about this topic.' Specified in such a way, the word 'ethics' designates a set of results from an ethical debate, in relation to specific individuals or communities.

[3] According to Paul Ricoeur 'self-esteem [is] the dignity attached to the ethical quality of human beings' (in *Le juste*, Paris, Éd. Esprit, 1995, p. 199).

[4] Cf. P. Ricoeur, *Soi-même comme un autre*, Paris, Seuil, 1990.

[5] The 'decision', here, is not only what, empirically speaking, we decide or want.

often have only a vague idea of what we really want to do. For example, we do not know if we really want to assume the responsibility of placing an ailing parent in a nursing home. We do not know if we want to construct a society where everyone is running around with a cellular telephone. We do not know if euthanasia should or should not be left to the discretion of physicians. We do not know how much pollution we are prepared to accept in our pursuit of economic growth. With all of these questions, and beyond all the technical issues involved, there is room for an ethical debate aiming at clarifying our decisions, in the full knowledge that these decisions will irreversibly determine our personal and collective future.[6] Since we know that some solutions will cause suffering,[7] ethical debates inevitably begin, confronted by others and with the reality of evil on earth.[8]

It is thus incorrect to claim that giving up the classical perspective of heteronomy eliminates ethical debate. Many things remain to be discussed simply to clarify what exactly it is people want. From this angle, ethical reflection's starting point can be seen as anchored in three assertions: people do not want to do just whatever appears possible; ethical debates *do* indeed exist and ethical debate conceals conflicts.

Ethical Debates Exist, and We Take Part in Them

There is an initial, basic empirical reality: *human beings do not want to do just anything*. True cynics, the kind who can agree to whatever kind of behaviour, are quite rare, if they exist at all. Whether it be a

---

It is the kind of future we dare to assume. It is the guiding principle that we decide to allow to govern our life. As Ricoeur emphasises (cf. *Le juste*, p. 24), beyond the debates related to the 'good' or 'duty', it is the tragic dimension of life that is at stake. This is where ethical conscience has to decide, despite uncertainty and conflicts.

[6] This way of presenting the issue, involving the imperative to make choices, refers both to the point of view of modernity, which stresses the human construction of history, and to Kant's definition of moral choice in the categorical imperative: 'act as if the maxim of your action could become, by your own will, a universal law of nature'.

[7] An ethical debate most often arises because of the cries of suffering stemming from those crushed by some specific action or by the organisation of society.

[8] The future towards which we strive when we commit ourselves is often symbolised by the notion of 'good'. While the confrontation with others when we debate

matter of placing one's parents in a rest home, building a factory that would spoil the landscape or destroy the environment, or even launching a telecommunications network that would structure global relations, people ask themselves questions. They quickly discover that, apart from the norms that some would impose in the name of the law or of moral heteronomy, they are not indifferent to their own action and its consequences. Their choices are tied up with the respect they have for themselves and the respect they expect from others. They thus venture the very meaning of their life, in the sight of themselves, before others, and for some before God.

To come to the second empirical feature: in every society, *people debate about what they want to do*. Ethical debate began before we arrived, and continues to take place whether we like it or not. It questions our actions, especially by relaying the cries of distress of those who are crushed or the voice of those with whom we are in conflict. Contrary to what some may think, it is not because moral decisions are left to the conscience of each individual that there is not a debate – think of the hours of discussion which take place before making the decision to place a parent in a retirement home, or before deciding to end a friendship. Whether the subject is abortion or human rights, honesty in business, pollution of the seas or anti-personnel mines, ethical discussions take place all around us and influence our own positions regarding what we want to do. We are continually discussing in order to clarify what we want to do. In addition, ethical reflection – like language and culture – always precedes us. We take part in it, but we do not entirely initiate it. Sometimes we are able to make the debate a rational exchange of ideas.

Thirdly: *underlying many ethical debates are conflicts*, sometimes ideologically disguised conflicts.[9] Behind the abstract norm 'do not steal', conflicts can be identified between those with nothing and the owners who defend their position. Behind the issue of pollution, it is

---

can be referred to the structures of human personality, e.g. to the superego.

[9] By 'ideological discourses' we are referring to discourses intended primarily to motivate and mobilize people, to legitimate practices and to form cohesive groups, while concealing the place from which they speak. On closer examination, every discourse has an ideological component which, as Ricoeur has shown, is not necessarily negative, but which deserves an ethical analysis. (Cf. P. RICOEUR, 'Science et Idéolologie' in *Revue Philosophique de Louvain* 72(1974)14, pp. 328-356.) If people agree with Ricoeur that ideology is a constitutive component of community life and does not have to be viewed negatively, it becomes obvious that ethical discourses are ideolog-

easy to see conflicts of economic interests. In a similar way the morality of emotional life conceals – sometimes barely – conflicts of a relational nature (think of jealousy), or conflicts of a communal nature (think of the status of women). It is sometimes quite illuminating to wonder what interests and conflicts lie hidden beneath ethical pronouncements. What interests are being supported by such or such an ethical discourse? What interests are behind positions which claim to be purely technical?[10] What are the conflicts concealed by the harmonious ethical discourses that, apparently, only seek the common good? Ethical discourses reflect but also often conceal agreements and conflicts related to life and its meaning.

Contrary to what technocrats might believe,[11] the deliberations about what we want to do are not purely technical. It is not simply a matter of finding adequate means to reach already accepted goals. To the contrary, it is more a matter of determining what it is that we want. And to do so, we have to choose between incommensurable values. What is ultimately at stake is the meaning we give to our existence as well as our self-respect and the esteem others have for us. All of this is part of the ethical debate.

This situation gives a foundation to the affirmation that some people, more than others, have an *'ethical attitude'* (which also means having a 'responsible attitude'). To have an ethical attitude need not amount to adhering to certain specific moral norms or to defending specific ethical positions. It would rather amount to making room, in one's own deliberations about what to do, for the ethical discussions that are taking place within our society. For instance, at the moment of carrying out an organ transplant, a doctor with an ethical attitude might find it useful to get informed about the debates that others had when they were themselves in a similar situation. An engineer

ical.

[10] It is precisely this ideological concealment of conflicts that often gives ethical discourse an allure of unreality. Even the claim of ethicists who state that it is important to think about ethical norms may conceal a conflict: the conflict between these ethicists and people who do not like to submit themselves to their norms. Cf. G. FOUREZ, *Liberation Ethics*, Philadelphia, Temple University Press, 1982.

[11] The notion of 'technocracy' deals with the attitude of those who claim that professional choices are only technical and that, in these, no fundamental option related to the meaning of life is enacted. A technocrat believes that there is no need of any ethical debate or political negotiation because he or she only deals with science and technology. A technocrat is not aware that technological choices themselves involve

with an ethical attitude will not go rushing into some project without first examining what others are saying or have said about what he or she intends to do. The person possessing an ethical attitude knows that, before he or she has examined the issue at stake, others have already done so, and that there is a difference between the well thought-out position of someone who is familiar with the whole problematic and the mere opinion of a person who simply summarises superficial judgements. In other words, such an ethical attitude is related to being concerned with respect for others and for their opinion in respect of the issue. Such an attitude points to the concept of responsibility, defined not by reference to legal demands but by the way in which we decide to use our abilities and knowledge. And in this case, not to decide is to decide.[12] The opposite of an ethical attitude would be the behaviour of a technocrat for whom every problem is exclusively technical. For such an individual, there is no point in examining the fundamental options of life which are expressed in professional choices.

Standardisation Making Possible an Educational Project

A large part of ethical debates – just like scientific debates – refer to paradigms[13] which have become standardised over the course of history. Historians of science and epistemologists have shown the set of standardised presuppositions, problematics, traditions and instruments which gave a structure to disciplines such as physics, chemistry or computer science. Similarly history has also led to a standardisation of the ethical issues and of the methodologies used for analysing them (for instance, ethical debates related to human rights,

---

most of the time societal choices as well as societal organisations.

[12] Nowadays the concept of responsibility, especially in line with the philosophy of H. Jonas, is often used when referring to the human ability of responding to a situation. From the moment that we comprehend an issue and have the possibility of influencing its course, one can speak of responsibility, i.e. of the possibility, or more precisely, the necessity of responding. In this case not to decide is to decide. Hans Jonas used this principle of responsibility to analyse the position of people who foresee what the consequences might be of various ecological policies, and who find themselves in a position of having to respond and thus assume their responsibility.

[13] The concept of a paradigm is defined, or at least explained, in T.S. KUHN, *The Structure of Scientific Revolutions*. Cf. also G. FOUREZ, NSSNS, pp. 75 and 115. To put it briefly, a paradigm is the set of presuppositions; goals and methodologies that make

organ transplants or respect for the environment, have a specific standardised format in our Western culture).[14] Whether it is a matter of scientific debates aiming at producing adequate representations of the world, or of ethical debates about the question 'What will we do?', some degree of normalisation (standardisation) of the issues is indispensable in order for exchanges to have a minimum of rationality,[15] and to prevent them from turning into a dialogue of the deaf. In both ethical and scientific contexts, paradigms make communication possible by permitting the production of commensurable discourses. Yet they also have the inconvenience of confining and closing the debate. Ethical paradigms, like scientific paradigms, are both useful and confining. This is why it helps if the standardised problematics that they promote are regularly subjected to revision by paradigmatic revolutions or by the rise of new paradigms. It is important that they function more like benchmarks than boundaries.

In science, paradigms create disciplines, i.e. scientific approaches that can be taught. It is the same with ethical paradigms: they make ethical education possible. Such an education aims to teach traditions related to some ethical debates as they are dealt with in a group or a given society.[16] This kind of education does not intend to prescribe what should be done but it takes the risk of presenting some ways – classical ways – of selecting and dealing with the issues. In this way, it becomes possible to enter the ethical debate and to go beyond what one might call a haphazard approach to the questions. For example, if the problem is to analyse the various possible scenarios relating to an organ transplant, or in the case of the greenhouse effect, certain paradigms are established which will permit rational discussion of the various aspects of the issue: this standardisation is quite useful, even though in some cases it can also paralyse the debate.

possible the construction of a discipline.

[14] Ethical issues – even some as basic as human rights – do not have an universal shape. They are the outcome of a process of standardisation and normalisation through which they have been given a specific format (and this format bears the mark of the culture in which they have been developed). On the development of paradigms and ethical problematics, we can also refer to the concept of 'city' as used in L. BOLTANSKI and L. THÉVENOT, *De la justification, les économies de la grandeur*, Paris, Gallimard, 1991.

[15] We speak of rational discussions when they are carried out with enough agreement about the meanings of the words and the experiences to which they refer. This presumes an adequate level of standardisation.

Ethical paradigms, and the historical perspective they introduce, can clarify the role and place of ethical principles and moral norms. Ethical principles can be viewed as standardised forms, within a culture,[17] of certain key ideas that are frequently at work in ethical debates. This is true of the principle of the right to privacy, or of the principle stating that we should care for the environment. Ethical principles ultimately refer to the various forms of utopia that are part of a society.[18] As for moral norms – and the code[19] that they implicitly or explicitly make up – one can view them as simplified summaries of long ethical debates. This is the case with rules such as: 'do not steal' or 'do not throw all your factory's waste into the river', etc. Some claim that there are ethical principles or moral norms that are absolute and/or universal, even if their expression is always approximate and contingent.[20] Others, on the other hand, think that neither ethical principles nor moral norms are absolute. These maxims would be the result of human constructions having their particular history and bearing the signature of the culture and the ideologies that contributed to their construction.[21] Their function is to facilitate exchanges in which a certain understanding is reached, at least up to a point. Their concise formulation also makes

---

[16] Cf. NSSNS, pp. 112-113.

[17] An emphasis on the cultural context of ethical debates implies an awareness of their relative character, but does not at all imply that one adopts a relativist stance (according to which all norms and principles would be equal). This distinction is easy for engineers: they know that every technology is relative to the circumstances under which it was created and to which it responded, but this does not mean that engineers are relativists. They know quite well that technologies are neither equal, nor equivalent.

[18] A utopia is an imaginary construction symbolising a community's hope for a better world, and mobilizing its members in such a direction. A utopia is conditioned by the circumstances of life and organization of the society or the communities that generated it. Typical examples would be 'universal brotherhood', 'the kingdom of God', 'universal peace', etc.

[19] There is no clear standard way of drawing the distinction between ethics and morality. We speak here of ethics when people reflect on human action so as to give it meaning. We speak of morality or of a moral code when referring to the norms that are implicitly or explicitly accepted by some community.

[20] From the same perspective, there is meaning in distinguishing the historical processes that have led to the expression of norms from the rational justification of these norms.

[21] When people think that ethical maxims are a product of human history, this does not mean that they believe they are meaningless or arbitrary. Maxims are like technologies; they are the product of human history but some are more appropriate,

easy fast discussion (admittedly with the risk that too much rapidity will overlook the nuances). In order to participate meaningfully in ethical debate, a person has to master sufficient knowledge of the existing state of the discussion as it is effectively practised in the culture where he or she lives.

In connection with moral norms, one could mention deontology. A deontological code brings together a series of norms relating to professions or particular social roles. Here again, they are summaries of long ethical debates. But deontological rules are to be distinguished from legal rules, even though some of them might be enshrined in legal obligations. Because the law deals less with the question 'What are we to do?', than with the way in which, following various confrontation and strength ratios,[22] rules are imposed on everyone (with some public power to ensure its enforcement).

This leads to a conceptualisation of the distinction between ethical debate and *political debate*. Before what is generally called 'modernity', political debate was almost simply one aspect of ethical debate. It was occupied with searching for the necessary conditions to ensure collective existence. And it was believed that it was possible to reach a consensus with respect to that issue, at least between people of good will. After modernity, the ethical debate went on with some hope of a universal answer, i.e. some consensus around what 'the good' meant. People tend to look on the ethical debate as taking place in a community where it is assumed that a consensus (as broad as possible – even hopefully a universal one) can be reached about what should be done.[23] Political debate, on the other hand, tends to presuppose some irreducible pluralism, in connection with a variety of interests and ethical values. It tries to respond to the following question: 'given that there is ethical pluralism in a society – i.e., that its components do not share the same values and interests – what are the acceptable compromises[24] in view, for example, of

at least in some contexts, than others.

[22] Even when laws have been promulgated by force/the stronger over the weaker this does not mean that they do not bear any relation with ethical debates. Those who promulgate laws often do so because these laws fit their ethical project (or in other words because it seems they will lead towards what they believe to be 'good').

[23] Ethical debate is thus generally practiced under the banner of a possible universalisation, which goes along the Habermasian (or Kantian) goal of the universalisation of the debate.

[24] About this notion of 'compromise' when commensurable benchmarks are no more available, see BOLTANSKI and THÉVENOT, *De la justification*, p. 408. See also G. FOUREZ, 'Constructivism and Ethical Justification' in M. LAROCHELLE, N. BEDNARZ and

establishing public peace?' Political debate does make reference to ethics, but without there being any consensus on ethical issues.

It is customary to distinguish between *two dimensions of ethics: the individual and the collective*. Individual ethics involves an ethical debate about what an individual would do, assuming that the individual's action will not bring a broad change in the general structures and conditions of the action's context. It concerns, for instance, an engineer who submits an expert report on the risks surrounding certain production conditions. Or it deals with relieving the misery being suffered by a city's poor people. Collective or structural ethics, on the other hand, concerns action undertaken in order to modify the structural conditions of existence. This would be the case in a situation where it is a matter of promoting legislation that would modify the powers of engineers with respect to the shareholders or the employees. Or when it is a matter of not just improving the lot of the poor, but of altering the economic conditions in order to reduce unemployment and poverty. The point of this distinction is to show that there are some questions which cannot be effectively addressed by considering spontaneous individual action alone, but that the solution must come about by way of a complex (and interdisciplinary) analysis of structures and an effective organisation of the agents involved (with a necessarily political dimension).

An Education towards Participation in the Ethical Debate?

In this context, an ethical education implies – among other things –: learning about ethical issues that have historically been debated, mastering conceptual tools useful for analysing these issues, understanding various interests involved (and often in conflict), and being able to undertake a clarification of the implicit values. All of this requires the acquisition of knowledge, competence and know-how. Competence in interdisciplinary practice,[25] is especially helpful

---

J. GARRISON (eds.), *Constructivism and Education*, Cambridge, Cambridge University Press, 1998, pp. 123-15.

[25] Many interdisciplinary methodologies have been developed, although few of them have been standardised. Probably the most established methods are those related to technology assessment, especially to so-called 'constructive technology assessment'. This approach can be seen as one way of integrating ethical debate and technological construction. (Cf. G. FOUREZ, 'Le Technology Assessment, nouveau par-

since, in the majority of cases, it is impossible to come to an adequate representation of a situation[26] without drawing on several disciplines (ranging from the so-called pure/hard sciences to the social sciences and humanities). In addition, to remain in touch with reality the analyses producing an overall representation of a situation must take into account the conflicts, tensions and controversies that might be concealed by too abstract a discourse.

Such a programme of education could proceed by lectures and theoretical explanation, but most often it is done by case studies the results of which can be transferred – *mutatis mutandis*, and this is sometimes a problem – to other situations. An examination of the debate about the placement of an elderly person in a nursing home, or about the consequences of the modification of the ozone layer through the greenhouse effect, can provide patterns which can be transferred to many other situations.

Ethics Linked to Society: The Engineering Profession

The handbook which this text is introducing is not about ethics in general. There is not just one ethical debate. In a certain sense, each individual carries out his or her own debate, both within oneself and with others (and confronted with oneself and with others). Moreover, every community must articulate its own discourse and paradigm about the issues that particularly concern its life. Faced with the question 'what are we to do?', human beings must look to who they are and to their place within society. Even though it might imply an attempt to situate themselves in an impartial manner,[27]

---

adigme éthique' in J. PLANTIER (ed.), *La démocratie à l'épreuve du changement technique*, Paris, l'Harmattan, 1996, pp. 249-278.

[26] Such a representation demands some appropriate answer to the question 'What is it about?'. Such an answer always has to deal with many approaches, from technical ones to more social ones. Such representations have been labeled 'Interdisciplinary rationality islands' (cf. NSSNS, p. 90, or G. FOUREZ, 'Scientific and Technological Literacy as a Social Practice' in *Social Studies of Science* 27(1997), pp. 903-936). This metaphor shows that what is conceptualised emerges from an ocean of ignorance, and that the emerging theoretical representation makes possible a discussion of the situation.

[27] Some moral philosophies give a privileged place to such a view, e.g. such is the 'veil of ignorance' and its role in the philosophy of John Rawls. Moreover, there is a debate among philosophers who disagree about deciding whether it is meaningful or not to pretend to an absolutely universal discourse on ethics.

many believe the questions they ask, their ways of analysing these questions, their sensibilities, responsibilities and values are influenced by their place in society. According to this point of view anyone claiming a universal position, a God's eye view, is either mistaken or deluded (such a person would be enclosed in his or her ideological perspective). Each profession's ethical debate would then be related to that group's interests as well as to their social power. For instance, the points of view held by teachers and students, parents and children, workers and managers, would be irremediably different: they would not generate equivalent ethical discourses.[28] Even if these various points of view maintain a dialogue (or sometimes come into conflict), it would be a lie to claim that they are the same. Ethical education for engineers, then, requires that they be able, from the very beginning, to analyse the power controlled by engineers as well as their particular interests and/or 'natural' loyalties.

For the ethical debate to go beyond the level of mere general discourse, the analyses must reach the level where technology interacts with social and economic aspects, and where the power relations governing our societies are exposed. This is what Michel Serres suggests: *'we should give priority to the kind of association between the human sciences, the political sciences and the applied sciences that is taking shape right in front of us ... [because] while the intellectuals talk about a return to Kant, those who are really making the decisions are telecom engineers, bioengineers, physicians, etc.'*[29]

As engineers' typical practice mainly deals with technologies, engineering ethics has to be supported by a neat, relevant and operational definition of technology. But there does not exist a well standardised definition of this word: its meaning changes according to the surrounding linguistic culture (e.g. 'technology' in English does not have the same meaning as *'technologie'* in French).

The way we choose to construct the concept of technology has much ideological weight. If indeed technology is reduced to mean-

---

[28] Emphasising the particular social place of every ethical position does not imply locking oneself into this position. On the contrary, if an ethical attitude starts with the specific concerns of a profession, it tends to get confronted with the perspectives of others and leads to seeing oneself 'as another'. An ethics that would be confined to a group, let us think of certain research ethics or of certain military doctrines, could only amount to the defence of a specific interest group.

[29] In *Le nouvel observateur*, nr. 1435, 1992.

ing technical objects that are put together, the work of engineers will appear in a technocratic light. They will look like producers of technical devices that would be socially and ethically neutral. Ethics would only be concerned with the decision of implementing technology. But if as contends the *American Academy of Science* technology has simultaneously material, mental and social dimensions, it carries intrinsic ethical values.

To give a concrete example: can railroad technology be reduced to rails, locomotives, wagons, etc., or does it comprise within itself, in its very structure, some social organisation of the personnel and of users, social conditions, and even a cultural way of considering time.

Depending on whether people opt for one or other of this perspective, engineers' profession will raise quite different issues. In the first case, engineers would be reduced to their role as technicians and their specific ethical problems would only amount to some professional honesty. But, in the second case technological creativity is looked upon as one of the ways through which people organise their society and build their future. Technological creativity thus appears as one of the most important components of ethical and political projects. And, correlatively, the social and ethical responsibility takes another dimension. Negotiating various interests and values is seen as a constitutive part of the very construction and stabilisation of a technology. And the profession of engineers gains, by the same token, much intrinsic societal value.

Finally, the decision to publish a handbook in engineering ethics leads to a distinction between the ethical debate specific to the profession, and the debate that is concerned with society at large. Engineers are clearly concerned by both. First by the latter, i.e. about issues dealing with the societal impact of technology, e.g. problems of communication, industrial pollution, genetic engineering, information technology, nuclear power, etc. Engineers must take part in the deliberations leading to an evaluation of the social impact of these technologies ('technology assessment'). And these evaluations go beyond the merely technical – they refer to organising society and thus to ethics and politics. Technology assessment has to go as far as pointing to the fact that technologies are not neutral but implement the results of strength ratios and negotiations between various interests.[30] Yet it would be too restrictive to reduce the ethical concern of

---

[30] Negotiations between social, financial, political, cultural and ethical interests do not only occur after the construction of a technology. On the contrary, technology

the engineering profession to technology assessment. This would not supply an adequate answer to some basic questions such as: 'given our powers as engineers, our knowledge, competencies, interests and values, what are we to do?' Issues like these have to be considered when dealing with specific situations whilst avoiding restricting oneself to general ethical issues. A handbook for engineering ethics, when proceeding to an analysis of the big issues mentioned above, should investigate how engineers – as individuals and as a social group – can step in.

One presupposition of this handbook is that engineers, like every social group, own a certain space of freedom regarding their actions. They are neither omnipotent nor powerless. This may seem obvious in theory, but it is often ignored in practice. For instance, technocrats tend to extend their freedom to the extreme by claiming – in practice if not in theory – that they may decide what is best for everyone, without taking account of all the conflicts, negotiations and debates implied by ethics and politics. On the other hand, many technicians deny that they have the slightest power, since they are hindered by technical, social and economic constraints. This handbook is based on the idea that engineers – individually and collectively – do indeed have some power, and are responsible[31] for their actions.

Appendix I
Listing of Relevant Qualifications for Engineers Wanting to Get an Understanding of the Ethical Issues Related to their Profession

This handbook intends to promote a specific profile of an engineer able to take into account the ethical dimension of the profession. Such engineers ought to be proficient in the following (this list is open for discussion and development):

is, at least partially, the outcome of such negotiations. It is false to think that engineers would first construct an 'ethically neutral' technology, while social agents would later proceed to negotiate its implementation (which would imply ethical stands). Negotiating between various interests is a constitutive part of the construction and the stabilisation of a technology.

[31] This responsibility should be looked upon neither in a legal nor in a guilt inducing sense. It is more a matter of engineers' ability to respond to a situation thanks to their know-how, their knowledge and their social status. Yet, 'being responsible for one's actions' does not imply that there are specific, predetermined things which have to be done. Rather, taking responsibility is to confront the particularity of a situation as well as the ethical and political debate surrounding it.

- Able to identify the technical, ethical, political and scientific dimensions of their work, and understand what ethics and ethical debate involve.
- Capable of recognising how, beyond what is technical, ethical issues originate in everyday life.
- Capable of analysing an ethical problem, clarifying what values are at stake, and debating opposing stands and arguments.
- Be familiar with, and know how to analyse the formal or informal deontological codes governing the profession.
- Have a historical perspective on the development of their profession.
- Be able to analyse engineers' contemporary roles, functions, status, alliances and social position.
- Understand how various paradigms regulate engineering science and affect the representation engineers have of the situations in which they are involved.
- Understand what an interdisciplinary approach to a problem includes, grasping epistemological foundations of interdisciplinarity, and mastering some methodologies dealing with interdisciplinary work.
- Be familiar with the professional culture of engineers and able to deal critically with their spontaneous ideologies, as well as with the kind of social conflicts in which they are a part.
- Understand and be able to analyse engineers' status and their social institutions, and specifically the socio-technical systems they are involved in.
- Be able to analyse the relative autonomy as well as the conflicting responsibilities of engineers regarding consumers, employers, workers, unions and colleagues.
- Understand issues related to security, risk and ecology.
- Have a sensible model of what a technological system is, especially a large technical system.
- Be able to analyse the impact of household technology, and other technologies affecting daily and domestic life.
- Be familiar with technology assessment (TA) as a social project and as a method, with a particular focus on constructive technology assessment (CTA).
- Know how to locate technological development within a wider epistemological, historical, social, economic and ecological framework.

- Know how to analyse ideological discourses about the 'needs' involved in technological development and understand what a technocratic attitude is.
- Understand contemporary debates concerning technology and gender within our patriarchal society.
- Understand issues related to the effects of technology on unemployment and on the concept of work within society.
- Understand the issues and debate concerning sustainable development.
- Understand what is meant by modernity, postmodernity, the ideology of progress, their relationship with the engineers' professional culture, as well as contemporary discussions of these notions.

References

BOLTANSKI, L. and L. THÉVENOT, *De la justification, les économies de la grandeur,* Paris, Gallimard, 1991.

FOUREZ, G., 'Constructivism and Ethical Justification' in M. LAROCHELLE, N. BEDNARZ and J. GARRISON (eds.), *Constructivism and Education,* Cambridge, Cambridge University Press, 1998, pp. 123-115.

FOUREZ, G., 'Le Technology Assessment, nouveau paradigme éthique' in J. PLANTIER (ed.), *La démocratie à l'épreuve du changement technique,* Paris, l'Harmattan, 1996.

FOUREZ, G., 'Scientific and Technological Literacy as a Social Practice' in *Social Studies of Science* 27(1997), pp. 903-936.

FOUREZ, G., *Liberation Ethics,* Philadelphia: Temple University Press, 1982.

FOUREZ, G., *Nos savoirs sur nos savoirs, un lexique d'épistémologie,* Brussels, Éd. De Boeck Univ., 1997.

GARRISON, (ed.), *Constructivism and Education,* Cambridge, Cambridge University Press, 1998.

KUHN, T.S., *The Structure of Scientific Revolutions,* 3rd ed., Chicago, University of Chicago Press, 1996.

RICOEUR, P., *Le juste,* Paris, Éd. Esprit, 1995.

RICOEUR, P., *Soi-même comme un autre,* Paris, Seuil, 1990.

RICOEUR, P., Science et Idéologie' in *Revue Philosophique de Louvain* 72(1974)14, pp. 328-356.

0.2

ETHICS IN ENGINEERING

Foundations of Ethical Practice

Brenda Almond

The questions this chapter seeks to address are: first, what kind of ethical questions arise in the field of engineering? When, and in what contexts do they arise? Then, secondly, how should these questions be addressed? Are there underlying ethical principles that can provide a foundation for good practice? The common view that the job of an engineer is simply the technical task of carrying out a brief provided by others is shown, by the use of some practical examples, to be too narrow, and the underlying ethical aspects of technical choices are explored. These involve issues of rights, of respect for life, the willingness to accept risks, and the balancing of competing interests, including those of the wider community and of future people.

Introduction

The questions this chapter seeks to address are: first, what kind of ethical questions arise in the field of engineering? When, and in what contexts do they arise? Then, secondly, how should these questions be addressed? Are there underlying ethical principles that can provide a foundation for good practice? In asking these questions, it is worth remembering that there is a common view that the job of an engineer is simply to carry out instructions that have been provided by other people – to fulfil a brief, whether the task is to install a bridge or a tunnel, to redesign a road layout, or to set up a telecommunications system. But codes of practice for engineers may include elements that are not simply technical or practical; they may, for example, call for attention to environmental considerations, or the need to have concern for safety or health; while in those rare cases when financial shortcuts taken in the building of apartment blocks or the design of bridges lead to deaths and injury, this may be regarded as not only ethically, but criminally blameworthy. There may also be more detailed aspects of projects or assignments which require ethical judgements of a distinctive nature. Many examples could be found of situations which offer a challenge to the 'facts first and last' view of the matter, but for purposes of the present chapter, the following three hypothetical scenarios can be used to illustrate this point:

1. After a serious rail-crash apparently linked to a design fault in the train concerned, all trains of a similar type are immediately withdrawn from service – a decision which creates substantial disruption and inconvenience for travellers and loss of revenue to the rail company.

2. A local authority decides to change the road layout at a notorious accident black-spot. A number of possibilities are explored, but the most promising proposal is ruled out on grounds of cost. A scheme is introduced instead which it is estimated will save lives, but is still not as effective as the more expensive proposal.

3. A government sets up an inquiry, as a result of which it is decided to commission a new nuclear power-station. Options such as coal- or oil-fired stations are also evaluated, but on a five-year basis the nuclear option is judged to be cheaper and cleaner, and this consideration is held to outweigh the very long-term disadvantages of nuclear energy and the miniscule risk of accident.

A first response in such situations may be to say that these are technical matters and that decisions can and should be based on straightforward empirical findings, usually of a statistical or cost-benefit nature, in which actuaries, experts in decision theory or game theory, in economics or in law may play a part, as well as scientific and technical experts. But closer inspection suggests that it is not as simple as that. In the case of the new road layout, engineers are working to a design in which the question: How many fatalities are *acceptable*? must be answered, for it may well be the case that there is no possibility of reducing these to zero. In the case of the rail safety-check, the issue is different, since the alternative outcomes are either that there will be no accident and hence no casualties at all, or that, should there be a crash, there may be many. And, as well as the possibility of withdrawing all trains immediately, there is also the option of withdrawing only a few trains at a time, thus taking a week or more to check for faults. The chances of a further accident in a matter of days must indeed be slight, so someone must have had to answer the question: what *degree of risk* is acceptable?

A judgement about the acceptability or otherwise of risk is also involved in the case of the power-station. But here the question is how to evaluate a risk, in itself minute, but which is nevertheless extremely widespread and enduring in its effects. The Chernobyl disaster, for example, was near-global in its impact, and residual contamination from that incident may stretch into the indefinite future. How is one to set the sheer *magnitude* and *persistence* of adverse consequences against their extreme improbability? This involves judging risk in the context of results spread widely in space and time and asking the question, 'Does a risk which is infinitesimal or vanishingly small become significant when its scope is unprecedentedly extensive?'

This issue also raises questions of a rather different kind: do future people, those as yet unborn, whose well-being may be affected by our present decisions, have a right to our consideration? And are we entitled to bequeath to them problems with significant political and security implications? Does it indeed make sense to speak of the rights of people who not only do not yet exist, but may indeed *never* exist? Or may we perhaps speak of duties and obligations independently of the issue of rights?

None of these questions can be answered directly by appeal to facts. They involve considerations which are not purely technical. And central amongst the non-technical considerations, there are, it seems, assumptions that are essentially ethical. It is, for example, the underlying presumption that everyone has a right to life and a right not to be knowingly or deliberately harmed by others that makes risks to life or health matters of ethical concern. It could be said, too, that a principle famously formulated by the German philosopher Immanuel Kant (1724-1804) is involved here: that each individual should be treated as a person, as an end in himself or herself, not simply as a means to someone else's ends.[1] So one aspect of what is

---

[1] I. KANT, *The Moral Law: Kant's Groundwerk of the Metaphysics of Morals*, ed. H. PATON, London, Hutchinson, 1948, p. 91.

wrong about disregarding the strong possibility of deaths or harms resulting from a procedure is that such indifference is a breach of the right of some person or persons as yet unknown to be treated with respect, and not to be treated as pawns in someone else's game.

The idea of not having one's important interests determined by others is sometimes described as the principle of autonomy, and it means, in practical terms, that where risk is concerned, it is pertinent to ask: 'Who is entitled to choose to take that risk?' It may be pointed out, for example, that, in the case of the train safety-checks, it is probably more dangerous for the would-be travellers to drive to the station in their own cars to catch the train than to travel on it. There is, however, an important difference in the two cases. This is that individuals accept the risk of the car-journey for themselves. In the case of the decision whether or not to withdraw trains for checking, and whether to do so immediately or gradually over a longer period, an engineer, or a manager, or a committee, has made that choice for the passengers. One factor involved in this case, then, is the right to make one's own decisions, including decisions about what risks one is prepared to take – and in this case that right has, whether justifiably or not, been pre-empted by others.

It is a further aspect of such situations that when someone else, whether an individual or a committee, does make a decision, it often turns out that they are taking it on behalf of not just one person, but many. This raises a further question about the relation between the simple notion of costs and benefits on the one hand and the more complex issue of their quantity and distribution, i.e. 'How serious are the harms or benefits involved? And how many people do they affect?' Such questions are familiar in ethical theory, particularly in relation to the ethical theory known as utilitarianism, but they may arise, too, in connection with other ethical theories. It would be useful, then, at this point, if these issues are to be properly considered, to say a little about the main ethical theories influential today and to set out briefly their background and origin.

Background to Ethics

Although today, especially in an academic context, ethics is often treated as an abstract and remote theoretical study, it was historically regarded as a study which was directed to answering practical questions about the way human beings should behave – about, in other words, the question, 'What is the best way to live?' Indeed, the roots of the terms 'ethics' and 'morals' are both to be found in the idea of customs or patterns of behaviour. But the way people behave is not just a personal matter; it affects the structure and texture of society as a whole. This is not a new insight, for when moral philosophy in the Western tradition was first recorded, in Athens in the fifth century B.C., both Plato (427-347 B.C.) and Aristotle (384-322 B.C.) took it for granted that this was not merely a matter for the individual, but also for society. They believed, that is, that the questions of the good person and the good society were inextricably linked.

To varying degrees, this link is maintained in the main ethical theories prevalent today. These include: utilitarianism; an ethics of rights and justice which is sometimes, but by no means inevitably, associated with the idea of a social contract; virtue ethics; a more personal ethic of care and responsibility for others derived from feminist philosophy; and positions such as relativism and postmodernism. These last two positions are characterised by a rejection of the rule of reason, usually associated with the Enlightenment, and an emphasis on the way in which ideas about right and wrong, good and bad, vary from culture to culture and even from person to person. Hence they deny that it is possible by rational reflection to reach conclusions on ethical matters which are valid for anyone and everyone. Since the search for ethical codes of practice in engineering is based on the assumption that there are approaches to ethical issues which have general validity, relativistic approaches, including postmodernism, will not be discussed here; indeed they are perhaps best regarded as a form of *anti*-ethics.

It is also possible to construe ethical debate along a political spectrum, in which case a contrast may be drawn between liberal approaches and conservative or traditionalist approaches, with communitarianism today being presented as a 'third way' between these contrasts. In the contemporary period, theoretical movements like Marxism and existentialism have also had significant practical implications. However, Marxism rejects traditional ethics entirely, dismissing it as the imposition of a self-serving bourgeois elite, while existentialism breaks with the key ethical notion of *pattern*, by making ethics an arbitrary matter of choosing specific actions on particular occasions as opposed to seeking to follow universally binding principles of behaviour. Again, within the limits of the present chapter, these political questions will not be further explored, but may be regarded as providing an interesting background for reflection.

However, in considering these more ideological approaches, it is worth noting that philosophy itself has influenced public attitudes to ethics. For the twentieth century saw a movement, centred at one time in Vienna, to place ethics, together with religion and metaphysics, on the margins of philosophy as a pseudo-study. This was based on a more general programme of seeking to bring philosophy into relation with science. Science and technology have advanced at a dramatic pace and there has been a popular welcome for their achievements, particularly in the early part of the twentieth century when logical positivism, a strongly empiricist philosophy, flourished. Logical positivism assigned to empirical knowledge and scientific method a privileged position, and applied this philosophically by rejecting as unintelligible substantive claims that were not subject to empirical verification. Thus, for the greater part of the twentieth century, professional philosophers in the positivist and empiricist tradition, if they chose to treat ethics at all, largely occupied themselves with the language of morals, often saying explicitly that practical issues were outside their brief.

The late twentieth century has brought a reappraisal of science and technology at the same time as a reappraisal of ethics itself. Technological developments that have affected the waging of war have been amongst the first to be seen as matters of ethical concern, and some scientists involved in weapons development have themselves sought curbs on the use of new weapons. Increasingly, too, the environmental impact of developments in science and technology has generated ethical debate. Along with today's interest in environmental issues has come, for example, a strong claim that other species, too, deserve consideration from human beings, and a sense that, after all, it was never right to see human beings, despite their technological brilliance, as supreme controllers of the earth, entitled to exploit it without regard for the interrelatedness of all forms of life. Developments in the field of medicine and medical and biomedical technology – reproductive medicine, end-of-life situations, the human genome project – are also the subject of intense ethical debate and are acknowledged to have generated important ethical dilemmas. Concurrent with these changing perceptions of science and technology, there have been changes in the way the scope and nature of ethics has been construed. The preoccupation with language has largely gone, and has been replaced with a more active concern for issues in the real world, especially in relation to these developments in science and technology. There is a certain competition between theories, but despite differences of emphasis, these theories are now more often seen as having substantive normative implications.

Contemporary Theories of Ethics

Probably the most currently influential of ethical theories is utilitarianism. Its main feature is its claim that what one ought to do is to maximise utility, interpreted variously as happiness, preference-satisfaction, or welfare. In the words of its first main exponent Jeremy Bentham (1748-1832), it aims to promote pleasure and minimise pain. The distinctive feature of this approach is that it makes the issues of right and wrong, duty and obligation, dependent on consequences. Thus it is often set in contrast to approaches which posit principles, and use a language of rights and justice. Critics of the position often ask whether utilitarians believe it would be right to commit some enormous injustice, some gross violation of human rights – for example, the execution of an innocent man or the use of torture – to avoid a great disaster. It is usually thought that utilitarians will have difficulty in showing that this would be wrong, since they are committed to evaluating an action entirely by its results.

However, those who adopt a principle-based approach to ethics may also have difficulty with this dilemma, but that is because they seem to be committed to allowing wholly unacceptable and avoidable disasters to result from their commitment to following justice and disregarding consequences. While the concept of commitment to a simple set of clear principles of behaviour is most often associated with religion and tradition, it is generally associated in philosophy with the views of Kant, who is generally described as a moral absolutist, since he argued that lying and breaking promises were wrong in all circumstances, and that principles should never be qualified by appeal to the facts of particular situations, still less their consequences for the happiness of those involved.

Kant's reasoning was logical rather than empirical. That is to say, he did not argue that following principles produced good results. Instead, he argued that the idea of an ethical 'ought' is *necessarily* universal in its application, and does not permit of exceptions. If we seek to adopt a 'let-out' clause for promising, for example, we end up defeating our own purpose by destroying the very conception of a promise as something the promisor will stick to whatever future circumstances apply. Some support for Kant's insight is today provided by developments in the institution of marriage. Interpreted as a promise to stay with a person for life, the widespread availability of the legal 'let-out' of divorce has in fact created a situation in which the institution of marriage in its original sense no longer exists – it is simply not legally *possible* to take on a permanently binding commitment.

Kant's approach is often described as deontological, in contrast to a consequence-based ethics like utilitarianism. The rigourism of a deontological approach is, however, often considered to be too stringent in the circumstances of contemporary life, even if the resort to the sheer expediency of utilitarianism is recognised to be morally unsatisfactory. One compromise is to view principles and the rights to which they give rise as the result of agreement, whether tacit or explicit. And if humans are the authors of rules, this may seem to give them a degree of flexibility, for what has been generated by agreement can be changed by agreement. This, however, may be to introduce too *much* flexibility into an area where it is the overriding nature of moral constraints and requirements that are their strength.

So some seek to avoid the problems of these approaches by focusing on the issue of character rather than on behaviour. The theory involved here is usually described as virtue theory. Virtue theory centres on the notions of function and flourishing. Aristotle, in particular, was interested in the question, 'What is a thing *for*? What is it designed to do?' And although this teleological approach is widely rejected today in favour of a more Darwinian approach, especially where human beings are concerned, it can be appreciated metaphorically. The function of a knife is to cut, and a good knife, as Aristotle pointed out, is one that fulfils that essential function well. Similarly, to know what a good *person* is, one must ask the question, what is the essential function of a human being? What distinguishes humanity from other creatures, or indeed from artifacts? Aristotle's own answer was that it was man's rational capacity that was his distinguishing feature, and he went on to offer a rational exposition of the virtues as a mean between extremes of vice.

In more general contemporary terms, one might find it easier to understand virtue theory *negatively* by considering what it is for a human to be *failing* to live up to the human potential. If we conceive of a human being not so differently from Aristotle, as a self-determining rational being, then we can see that various kinds of deprivation on the one hand, or of self-indulgence on the other, can both deprive a human life of its full potential. This applies whether we consider someone in a developing country who is forced by poverty to survive at a level below subsistence, or a drug-addict or alcoholic in a wealthy city in the Western world. Each in a different way is failing to flourish. That is to say, they could be *more* than what they currently are; they have a potential for physical and intellectual development which remains unfulfilled. One might add that they have a potential for moral and spiritual development too, which is also less likely to be fulfilled, and in this case their ideal potential – what they could ideally be – might well involve reference to the traditional virtues of temperance, courage, justice, and wisdom.

The virtues approach, which focuses on persons and their situation, has something in common with recent feminist approaches to ethics which have criticised the more legalistic 'justice and rights' tradition, and are also unsympathetic to the absolutism and impersonality of utilitarianism. The key feature of a feminist ethics is its emphasis on care and on responsibility for particular others. The theory arose from research into moral development, stemming originally from the work of the Swiss structuralist psychologist Jean Piaget (1896-1980) and from more recent developments in that work led by Lawrence Kohlberg (1928-87).[2] This research presented moral growth as culminating in a person's appreciation and acceptance of abstract and impersonal principles, particularly the principle of justice. It was criticised, however, for its focus on male rather than female responses, and when attention finally turned to female respondents, it was noticed that there were significant differences in male and female approaches to morality. Indeed, the psychologist Carol Gilligan identified a different 'voice' in the responses of female

---

[2] See J. PIAGET, *The Moral Judgement of the Child*, trans. M. GABAIN, London, Routledge and Kegan Paul, 1960, first pub. in England, 1932; and L. KOHLBERG, *Essays on Moral Development*, vol. 1: The Philosophy of Moral Development, San Francisco, Harper and Row, 1981.

subjects to moral questions.[3] What her research drew attention to was the way in which women were more responsive to context and to their own responsibilities to particular others – often a result of their position as carers – where men were more willing to take abstract principles of justice as moral determinants.

Applying the Theories

So how do these theories appear in relation to the examples set out above? Clearly, there will be a primary contrast between utilitarian approaches and a rights and justice approach, for there is often a conflict between the interests of the majority and the rights of the few. But the cases described at the beginning could be said to involve appeal, too, to issues of responsibility that evoke aspects of virtue ethics and even of the feminist 'ethic of care'.

Taking, first, the utilitarian approach, one major problem for utilitarians has already been mentioned. This is the potential conflict between being guided by utilitarian consequential considerations and having regard for justice and rights, allowing these to act as side-constraints or limitations on what it is permissible to do. But another consideration is relevant to the example described at the beginning. This is the issue of risk assessment. Utilitarianism may seem at first sight to have no place for an ethics of risk since it offers straightforward measurement of the utility of consequences as the criterion of right and wrong. So we may ask, were people better off as a result of the introduction of the new road layout? Was overall welfare increased by the new power-station? But, of course, the point is that such decisions have to be made in *advance* of knowledge of the outcome. Hence the utilitarian may retreat to the position that in deciding what to do one must seek to maximise *expected* utility. Assessing and comparing known probabilities is, of course, feasible under utilitarianism. Indeed, Bentham himself set out a 'felicific' ('hedonic') calculus showing how seven 'dimensions' of pleasure and pain might be fed into any decision. But in some cases, e.g., the nuclear power issue, probabilities are strictly unknown. This, however, immediately introduces the notion of risk into the picture. In

---

[3] C. GILLIGAN, *In a Different Voice: Psychological Theory and Women's Development*, Cambridge, MA, Harvard University Press, 1982, 2nd rev. ed. 1993.

general, then, trivial gains at the cost of dangerous risks will be ruled out; enormous gains for minor risks possibly ruled in. However, difficult ethical evaluations are bound to be involved here, to which straightforward utilitarian calculation is unable to provide the answers.

Rights theory might seem to offer clearer guidance here. The idea of natural rights, as has already been said, includes a right not to be killed or harmed. Thus, to take an example from the broader area of environmental impact, the discharge of toxic wastes into rivers and seas could be said to be clearly ruled out if it affects the health of any human being. Indeed, even if no harm is in fact done, a right may be violated if it was *possible* that harm might have resulted. But as another writer points out, a rule that nobody must ever do anything which just *might* harm somebody else 'would paralyse human life.'[4] Indeed, various codes of practice for engineers and others seem to make just such a concession to practical requirements, for example, by ruling against causing *unnecessary* environmental damage. So rights theory, too, would appear to involve at least an element of risk assessment. If this means that an absolute regard for rights must be judged unrealistic, it seems that would be a serious limitation of the theory. But if, on the other hand, that regard is qualified, then how is the scope of the flexibility required to be determined? In particular, how small does a risk have to be to be ignored? And how important does a goal have to be to justify overriding the wrong of imposing of a risk on someone without their consent? However, while these theoretical difficulties are far from trivial, for present purposes this element of judgement may simply have to be accepted as an aspect of the ragged borders of any human commitment.[5]

As far as contract theory is concerned, it is worth noting the asymmetry, which is prominent in the train example, between accepting risk for oneself and imposing it on others. Examples illustrating this particular human weakness might also be taken from the health field. People may be prepared to take risks themselves with sexual hazards or recreational drugs who would be incensed to discover that their routine hospital treatment had exposed them to risks

---

[4] J.E.J. ALTHAM, 'The Ethics of Risk' in *Proceedings of the Aristotelian Society* 84(1984), pp. 15-29, p. 18.

[5] These matters are dealt with more fully in my *Exploring Ethics: a Traveller's Tale*, Oxford, Blackwell, 1998. See in particular ch. 5, 'The Resort to Rights'.

of contamination. It is a further problem, too, of appealing to a social contract – common prior agreement about what may be permitted, and what risks to take – that some people are more risk-averse than others.

All these theories, however, despite the specific problems they may have to confront, have a role to play in an ethically sensitive approach to particular practical issues. Virtue theory, in particular, could have a special appeal to the engineer. Function and flourishing are notions that can be well understood by engineers, who know what it is to create an ideal example of its kind and what it is to supply a substandard one. Indeed, it could be said that the original conception of a virtue is an engineering one.

The ethics of care, too, is relevant to the engineering profession in the sense that engineers operate in particular contexts which involve a high degree of personal responsibility which cannot readily be transferred to others. When accidents occur, for instance, human error is often cited as the cause. But it is sometimes the case that the possibility of error could have been anticipated and its effects mitigated by good design. Indeed, after the event, this often happens. For example, the 'dead man's handle' on trains was a valuable design feature which only emerged after the experience of serious rail disasters resulting from the sudden death of a train-driver. So, to the extent that the care ethic points to the importance of retaining a sense of context and of individual responsibility, it, too, forms part of a morally sensitive approach to practical affairs.

Conclusion

While the various ethical theories are often presented as rivals, they are better seen, then, as different aspects of the truth, different facets, and all of them have something to say to engineers and practitioners.

Utilitarian calculation is relevant because the wholly praiseworthy objective of most engineering projects is to enhance human welfare. But rights theory reminds us that this calculation must be set within the side-constraints of rights and justice, and should not ride rough-shod over anybody's particular interests. Where individuals are likely to be adversely affected, consultation and compensation are clearly ethical requirements. But some appeal to virtue theory may be required, first to set the ideal of excellence as an ethical goal, but secondly, also to highlight the question of how welfare or happiness is to be construed – and it provides a reminder that this may not be entirely in crude materialistic terms. Changed attitudes to the environment illustrate this point, as environment is now seen in a much more holistic way, affecting approaches to areas as diverse as architecture and factory-farming.

At the same time, the insights of feminist ethics, like those of virtue ethics, may encourage us not to neglect the personal in pursuit of such objectives and they also locate the engineer in a context of personal responsibility for the well-being of others in a context which has made their well-being his or her particular responsibility.

Finally, one might add the further consideration that a broad ethical perspective, not confined within narrow academic schools of thought, must draw on a broad understanding of humans and their needs. These needs are first and foremost for food, shelter and the conditions of a materially good life. It is a salient task of engineering and technology to contribute to the supply of these needs. The enterprise, then, is ethical, and it is wholly appropriate that it should be bound, in its execution, by more general ethical guidelines: that the primary concern to promote human welfare should be constrained by regard for individual human rights; guided by ideals of excellence; and carried out by individuals aware of their own position as responsible for the well-being of others in the particular context in which they are involved.

References

ALMOND, B. (ed.), *Introducing Applied Ethics*, Oxford, Blackwell, 1995.
ALMOND, B., *Exploring Ethics: A Traveler's Tale*, Oxford, Blackwell, 1998.
ALMOND, B., *Exploring Philosophy: The Philosophical Quest*, Second edition, Oxford, Blackwell, 1995.
ALTHAM, J.E.J., 'The Ethics of Risk' in *Proceedings of the Aristotelian Society*, 84(1984), pp. 15-29.
GILLIGAN, C., *In a Different Voice: Psychological Theory and Women's Development*, Cambridge, MA, Harvard University Press, 1982. 2nd rev. ed. 1993.
KANT, I. *The Moral Law: Kant's Groundwerk of the Metaphysic of Morals*, ed. H. Paton, London, Hutchinson, 1948, p. 91.

KOHLBERG, L., *Essays on Moral Development*, Vol. 1: The Philosophy of Moral Development, San Francisco, Harper and Row, 1981.
PIAGET, J., *The Moral Judgement of the Child*, trans. M. Gabain, London, Routledge and Kegan Paul, 1960. First pub. in England, 1932.

0.3

THE ETHICS OF TECHNOLOGY[1]

Günter Ropohl

Ethics applies to technology, as (i) producing technical objects is a certain kind of human action, and (ii) the technical object, incorporating certain patterns of utilisation, will shape human action. The conceptions known in general ethics are discussed in engineering ethics as well. Consequential ethics (or ethics of responsibility), however, play an eminent role and will be analyzed in more detail, including types of responsibility, peculiarities of engineering practice, kinds of consequences, and normative arguments. Finally, it is suggested that engineering ethics ought to be integrated into social and political technology assessment.

Beginnings

Ever since there has been a *philosophy of technology* describing itself expressly as such, attention has also been called to the ethical problems of technology. E. Kapp mentions, although in passing, the ethical responsibility that persons bear together with their technical work for the community.[2] 'The perfection of technology does not run parallel with a moral perfection of human beings, although, in fact, increased technology means that more exacting demands have to be made of the moral strength of humans' notes the unjustly little-known J. Goldstein as early as the beginning of our century,[3] thus characterising a basic problem that comes up again and again in the philosophy of technology.[4]

---

[1] Original text translated by B. DECEUNINCK.
[2] KAPP 1877, p. 347.
[3] GOLDSTEIN 1912, pp. 12f; in essence this idea was already uttered by Sophocles 440 BC, in a choral song of 'Antigone', and JONAS 1979, p. 17f., forewords his book with this text.

Admittedly, the adopted viewpoints long tended to be rather wholesale, particularly since even the concept of technology itself was often not adequately differentiated. In their opposite corners, there were the extreme positions whereby 'Technology' was either seen as an intrinsic evil,[5] or as ethically valuable in itself.[6] On the other hand, there was also the widespread position that 'Technology' was ethically neutral, and ethical meaning was attached only to the ways and means in which technology was used.[7] Since such generalisations have all been discredited by more precise analysis and contradictory experience, the balanced assessment gained acceptance, during the last third of the century, that technology and technical action had a largely ambivalent character, that is, both negative and positive sides.

Previously, professional moral philosophers hardly ever occupied themselves with technology, and the philosophers of technology and philosophising engineers were, without exception, devoid of any deeper moral-philosophical competence. Shortly after World War II, for instance, under the impression of war technology in general and of atomic weapons in particular, an international discussion was held around the professional moral obligations of scientists and engineers. This was at the time when Bertolt Brecht rounded off his play *The Life of Galileo* with the now-famous passage on the 'Hippocratic Oath' for natural scientists. Inspired by this discussion, the Verein Deutscher Ingenieure (VDI) published a *Bekenntnis des Ingenieurs*[8] ('Profession of the Engineer') in 1950. However, beyond well-intentioned empty phrases – *'for the benefit of mankind'* (Brecht) or *'in the service of humanity'* (VDI) – such recommendations could not offer ethical bases, nor substantial and practice-oriented rules.

An ethics of technology in the true sense thus does not actually begin until around 1970, more specifically in Germany, with subsequently published radio lectures by philosopher of morals and religion G. Picht, in which he developed an ethics-of-responsibility solution to the question as to how 'science and technology come to reason' and also made institutional proposals that clearly outbid the

---

[4] See the summary discussions in HUBIG, HUNING and ROPOHL 1999.
[5] For instance, F.G. JÜNGER 1946 (written in 1939).
[6] For instance DESSAUER 1927, esp. pp. 143f.
[7] According to GOLDSTEIN 1912, p. 71; this view is apparently anticipated by MARX, 1867, esp. p. 465, who does not condemn 'Machinery' as such, but only its 'capitalist application'.
[8] Reprint in LENK and ROPOHL 1993, p. 314.

book by H. Jonas, published ten years later, in concreteness.[9] Shortly afterwards, H. Sachsse produced a collection of essays, which he placed, explicitly, under the title *Technology and Responsibility*; in the last contribution in the book, he is one of the first to treat, by the example of information technology, the fraught interchange between general moral rules, technology-specific applications and legal standards, refuting in advance some of the later rhetoric about an allegedly required 'new ethics'.[10] The explorations of ethics of technology since grew more numerous, not least in collaboration between philosophers of technology and the VDI in which Sachsse himself was involved.[11] Finally, the book by H. Jonas – which, however, written as it was in America, does not consider the previously mentioned publications – has made the 'Responsibility Principle' a watchword of public debate since the 80s.

I do not wish to go over the further development in chronological detail, particularly since two instructive research reports are on hand.[12] I would rather move on to a systematic representation and, firstly, explain certain conceptual and theoretical links between ethics and technology. This will be followed by a brief summary of the conceptions currently discussed in the ethics of technology. Because this will be, predominantly, a matter of consequentialist positions, I then wish to discuss the ethics of technological responsibility and its problems in some depth. Finally, I will be taking a short look at technology assessment, which I portray as a social-ethical form of ethics of consequence, and I will be arguing the case on behalf of a synthesis of ethics and technology assessment, which is also to be understood as a new unit of Practical Philosophy.

Ethics and Technology: Concepts and Correlations

'Why technology is an object for ethics' does not seem to be quite clear for all contemporaries even today; H. Jonas, in any case, in the

---

[9] PICHT 1969; JONAS 1979; Picht, however, banks on a 'concrete utopia' à la Bloch, whereas Jonas, widely dismissing Bloch, sees maintaining as the only chance for the future.

[10] SACHSSE 1972; reprint of the Chapter '*Ethical problems of technical progress*' in LENK and ROPOHL 1993, pp. 49-80.

[11] HUNING 1974; MOSER and HUNING 1975 and 1976.

[12] GRUNWALD 1996; OTT 1996.

80s, considered a systematic foundation necessary.[13] In fact, anyone still caught in the antique opposition of *praxis* and *poiesis*, of acting and making, will not be inclined to accord the maker of technical products any ethical relevance, because this tradition does not regard it as action; ethics would not then be appropriate for technology. Ethics, namely, is the philosophical theory of morally correct action. Even if the various schools of ethics represent different positions regarding moral correctness, they nonetheless agree that objects having nothing to do with the conditions and expressions of human action are not amenable to ethical consideration.

This misconception, handed down from antiquity, now means to say that production directs itself towards objects outside relations of human action, while only action is oriented towards the human situation. In response to this, I say that this dichotomy, to which Aristotle attached such great value,[14] was not suited even to the antique craft technology, since craft products, too, embody certain social manners that enter into the context of their users' life. For today's industrial mass production, in any case, this splitting into two is altogether untenable, since the produced objects, bearing in mind the differentiated division of labour, are all brought within the sphere of activity of other individuals and, once there, often enough combine with activities that are oriented to the human situation.

True, the views on the significance of technical objects alluded to so far have been explained by philosophy of technology only in the last thirty years.[15] Hitherto, a narrower concept of technology was in fact current in the technological sciences and among the general public, whereby 'Technology' was reduced to artificially made articles, entailing the subliminal notion that technical articles freed themselves, as independent things, from the production process and went on to conduct a sort of life of their 'own'; this is apparently the supposition that Aristotle linked to the products of technology. In reality, however, production with the artificial object creates at the same time a new form of human activity, made up here of human and there of technical components. Anyone, for example,

---

[13] JONAS, this title in LENK and ROPOHL 1993 (1st ed. 1987), pp. 81-91.

[14] Books I and IV of the *Nicomachean Ethics*. This exclusion of object-making possibly merely reflects the aversion of the Greek elite to all forms of manual activity; it would then be about time that the current interpretation of Aristotle took due account of this ideological background.

[15] For representative accounts, see RAPP 1990 and MITCHAM 1994. Despite a very common prejudice, philosophy sometimes makes advances in knowledge!

making cars, does not primarily 'knock out' tin boxes, rather he produces the life form of car-driving; in as much as this life form has become an end in itself for many contemporaries, even Aristotelians must concede that the car industry has produced a new form of action.

The overstated example shows how intimately production and action are related. It therefore strikes me as expedient to jettison the classic idea of action and replace it with a more modern idea of action. Action then means transforming an initial situation into an end situation according to a prioritised maxim.[16] In a borderline situation, this definition embraces the antique idea, namely that a 'doer' transforms his inner state of mind for its own sake but, at the same time, it covers also all other forms of goal-determined activity and allows, in particular, the inter-relation of action and technology. Technology then includes: a) the many utility-oriented, artificial, objective things (artefacts or object systems); b) the many human actions and institutions in which object systems arise; and c) the many human actions in which object systems are used.[17] All actions according to b) and c) fall within the concept of technical action, that is, not only production action but also consumption action. However, since – especially among the conditions of the labour-dividing industrial society – all technical action has its social references, it is always simultaneously social action.[18]

Now, I could be accused of mere definitorial sleight-of-hand to make technology an object for ethics. What I had to do, in fact, is to free the discussion from the Aristotelian division, because this was created for a quite specific ethical conception: it is known that Aristotle represents a school of virtue, and there, of course, great prominence is given to action which is concerned with itself only. Other ethical conceptions, on the other hand, stress the quality of reciprocal relationships between individuals and, consequently, concern themselves first and foremost with social action. When, however, technical action proves also to be social action, then we have the substantial foundation for the ethical relevance of technology.

---

[16] KEMPSKI 1964, p. 297.
[17] VDI 3780, 2; see also my book of 1991, p. 16ff.
[18] More on the subject in my book of 1991.

Ethical Conceptions

At the end of the previous section, I made the point that there are several ethical conceptions which, moreover, are variously classified and described by moral philosophers.[19] Here is not the place to explore these questions, but I should like to give a brief outline of which of these conceptions are represented in the ethics of technology. I would have to qualify that by saying first that the ethics of technology is still too young to sire a coherent, clear picture, for which reason I have resisted, apart from clear exceptions, the urge to give these conceptions the names of their authors.[20] It seems more important, to me, to name possible basic positions and to appreciate, in all brevity, the extent to which they strike me as productive for the ethics of technology.

As far as I can make out, virtue-ethical conceptions were not represented in the ethics of technology until recently. That is not surprising, since technology has primarily to do with the transformation of nature and society, affecting the perfection of the individual Self only in a very relayed form. In different expressions, by contrast, value-ethical conceptions come to the fore, seeking to derive assessment and decision-making criteria for technical action out of general values; these conceptions are sometimes related to responsibility-ethical considerations. Since values claim to have intersubjective validity, they must of course possess a superindividual status. They are partially understood as metaphysical entities, partially also as generalised social constructs. A much-respected canon of values, included in the VDI guideline on technology assessment, states that the values of functionability, safety, economy, welfare, health, environmental quality, development of the personality and quality of society (which can be further divided up into numerous sub-values) can broadly find consensus and can therefore be applied in practice, even when their status of being remains undecided.[21] As further decision criteria in the event of a competition between those values Ch. Hubig recently proposed a kind of basic values that make reference, very generally, to historical continuity

---

[19] See, e.g., PIEPER 1985, pp. 108ff, and NIDA-RÜMELIN 1996, pp. 7-37.

[20] GRUNWALD 1996, who attempted this, was greeted with numerous counter-arguments in the subsequent discussions.

[21] VDI 1991. Incidentally, no basic criticism of the content of the canon of values has been brought to the attention of its authors since its first publication in 1986.

('legacy values') and openness to the future ('option values') of the human situation in life.[22]

Although the strict confrontation of ethics of obligation[23] and ethics of consequence[24] has long been criticised,[25] with good reason, it re-emerges over and over again in the ethics of technology. Against the predominant ethics of consequence, which makes the morally correct dependent upon the compatibility of the consequences with the accepted rules or values, certain authors will argue the case that the consequences of technical action are not sufficiently clearly predictable as regards their nature nor their probability of happening, and they therefore recommend making mechanisation subject to irrefutable obligations. They include, in particular, H. Jonas, who does not interpret the Responsibility Principle in a consequence-ethical manner, but as fundamental social security obligation; if technical action produces the risk of causing considerable damage to the conditions of human life, there is also a categorical obligation to refrain from that action, he states.

From an ecological viewpoint, using comparable arguments, general obligations are being asserted for technical action that must be respected independently of any calculation of consequences. Such unconditional maxims include: error-friendliness, the property of technical systems not to trigger dramatic damage even in the event of human or technically-caused error, limitation of interference, the extent to which and intensity with which naturally existing structures are changed, and revisability, the ability to 'undo' technical projects in the event of clearly wrong development without any unwanted lasting consequences.[26] Deontological conceptions in the ethics of technology suffer from the problems encountered by any ethics of obligation; since they are founded formally, mainly on the basis of the principle of universalisability, they often lack substantial plausibility in the event of actual application.

---

[22] HUBIG 1995, pp. 139ff.

[23] Also described, with the same or similar meaning, as ethics of conviction or, in technical parlance, as deontological ethics.

[24] Also described, with the same or similar meaning, as ethics of responsibility or, in technical parlance, as teleological ethics, consequentialist ethics or, in certain specifications, as utilitarian ethics.

[25] FRANKENA 1963.

[26] See A. VON GLEICH, 'Ökologische Kriterien der Technik- und Stoffbewertung' in VON WESTPHALEN 1997, pp. 499-570.

As I will be discussing the consequentialist conception in some depth in the next section, I shall content myself here with the mention. However, I must also deal with another conception now enjoying interest in the ethics of technology: ethics of discourse.[27] General values and moral rules in a democratic-pluralistic society, it says, are not supposed to have universal validity, because this would mean a paternalistic subordination of the addressees; in addition, they would fail before the complexity of concrete cases of application that only those involved and taking part can fathom. The Morally Correct must therefore be imparted in controlled procedures of discussion, so-called 'discourses', through the final consensus of those taking part in the discussion. However, the discourses that have actually been conducted around particular controversial issues have often not brought forth much in the way of satisfactory results. That seconds the theoretical criticism that the ideal conditions, which are assumed for a successful discourse, cannot be realised in real social situations.

Finally, I must consider a further conceptional distinction that inhabits another dimension. Whereas the conceptions mentioned so far are more or less classified according to the ultima ratio of moral rectitude (virtue, value, duty, consequences, consensus), we come to the subject of the action to which the ethical conception is applied. Large tracts of moral philosophy have an individual ethical stamp. Under the influence of the individualist drift in modern philosophy, individual ethics can conceive only individual persons as agents and thus presupposes that the individual person is capable of behaving morally correctly. Individualist ethical approaches are also prevalent in the ethics of technology.

However, there is also a social ethics (which, incidentally, is usually a denizen of the theological faculties, and rarely of the philosophical faculties of the universities). Social ethics has to do not only with the action of the 'individual person with regard to the social structures', but also with the 'action of the social structures themselves'[28]; it regards organisations and institutions, then, as being moral subjects in their own right. This conception has entered into

---

[27] The authors of the aforementioned research reports, GRUNWALD 1996 and OTT 1996, both sympathise with variants of discourse ethics; op. cit. for more on the subject.
[28] VON NELL-BREUNING 1959, pp. 296f.

consideration in ethics of technology hitherto only in so far as the corporate character of technical action has been acknowledged; I shall return to this point later.

Ethics of Responsibility

Types of Responsibility

The ethics of technology is dominated by consequentialist conceptions and since responsibility figures almost as a key concept in the ethics of technology,[29] I would first like to explain this concept. Like most fundamental concepts, this one, too, has been assigned very divergent meanings,[30] and even where we exclude the unspecific meaning variants, responsibility turns out to be a very complex concept spanning numerous sub-types. 'Unspecific' seems best to describe a use of language that equates responsibility with organizational competence and, in this formal sense, first displays no manner or form of moral content whatsoever. What also appears 'unspecific' is the use of language for which I criticised Jonas, because he sets out from no more than a general feeling of moral obligation, without bringing the respective consequences of action or of deliberate inaction into the equation. The specific consequentialist notion of responsibility was probably formulated first by M. Weber, when he proposed the now-famous distinction between ethics of conviction and ethics of responsibility, according to which 'responsibility' means *'that the individual has to carry the costs of the (foreseeable) consequences of his deeds'*.[31] Closer analysis[32] reveals that responsibility forms a seven-cornered relationship that can be expressed in a '7 Ws' question: (A) WHO bears responsibility in respect of (B) WHAT, (C) for WHAT useful purpose, (D) because of WHAT, (E) concerning WHAT, (F) WHEN and (G) in WHAT WAY? These seven determinative clauses for the concept of responsibility are listed one after the other in the left-hand column of Summary 1; it is not without reason that I used interrogatives, since there may be several expressions for every

[29] Representative, e.g., LENK and MARING 1991.
[30] See esp. H. LENK, 'Über Verantwortungsbegriffe und das Verantwortungsproblem in der Technik' in LENK and ROPOHL 1993, pp. 112-148; also BAYERTZ 1995, esp. pp. 3-71.
[31] WEBER 1919.
[32] For details of this essay, see my book of 1996, pp. 69ff.

characteristic concept. In the interests of clarity, I have entered in the diagram only three particularly characteristic expressions of each characteristic; however, a more detailed discussion threw up several additions that could be quite useful for a specific research objective,[33] but would be to differentiated in an introductory account.

	(1)	(2)	(3)
(A) who takes responsibility	individual	corporation	society
(B) for what	action	product	omission
(C) regarding what	predictable effects	unpredictable effect	long-range/long-term effects
(D) because of what	moral rules	social values	state law
(E) towards what	conscience	judgement of others	jurisdiction
(F) when	before: prospective	at present	afterwards: retrospective
(G) how	in an active manner	virtually	in a passive manner

Table 1: Analysis of the seven determinative clauses of "responsibility"

The salient point of such a concept analysis[34] is now that the characteristic expressions – one from each line – in principle, can be linked together in any possible way at all. This way, many different types of responsibility can be clearly defined. In this way, we obtain the concept of responsibility that is currently prevalent in ethical discussion through combining the expressions of column (1) with each other. If we also consider, as stated earlier, products of actions as ethically relevant, we get this kind of responsibility once we include expression (B2) in the combination. Or we refine the traditional legal concept of responsibility and liability using the combination (A1, B1, C1, D3, E3, F3, G3) if someone is charged with an illegal act and taken to court.

Certainly, not all possible combinations are of equal practical value; for theoretical and factual reasons, some make no sense. A number of other important types of responsibility take shape when the scheme is followed where it goes beyond individualist ethics and considers as subjects of responsibility corporations that is, legal persons such as industrial enterprises or organisations, and the state

---

[33] See the discussion on my essay of 1994.
[34] Introduced by ZWICKY as 'Morphologische Methode', 1966.

and society as a whole, too. A 'conscience' may be assumed as the responsibility authority even for these collective responsibility-bearers, if you understand conscience functionally as an authority of rapid-response reflection that tests possible consequences of action for their tolerability against accepted normative standards; the environmental protection representative in a company, for instance, represents a little bit of corporate conscience.

Finally, I must draw particular attention to characteristic (D) that gives the normative grounds of responsibility. Particularly in the ethics of technology, responsibility is often discussed in a very formalistic manner, without stating more precisely the moral rules or social values on account of which certain consequences of action are to be supported and others are to be avoided. If we pass over the normative grounds of responsibility in silence, we of course sidestep the difficulties that actually arise with the relativisation of moral rules and the pluralisation of values in modern society. However, such flight into non-commitment is obviously not apt to bring the ethics of technology much serious consideration.

It is true that I have somewhat anticipated later considerations with these explanations and I wish first to sum up the predominant basic model of technological ethics of responsibility again. According to this, individuals should check all the conceivable consequences of their technical action beforehand and undertake planned action only when there is no effective risk to 'the safety, health and welfare of the general public'.[35] However, such risks, as we know from our own experience of life and as dozens of case histories show,[36] crop up again and again. So, either countless irresponsible subjects romp about in technology – a premature assumption that had to pay for much regrettable insulting of engineers – or the professional moral demand is more readily made than fulfilled. I tend toward the latter assumption, and in the following sections I will discuss a series of difficulties that confront individual responsibility; these are especially the particularities of technical action, the complications of consequentialist analysis and prognosis, and the normative uncertainties in modern society.

---

[35] *Code of Ethics for Engineers of the American National Society of Professional Engineers*, NSPE Publication No. 1102, January 1990; German reprint in LENK and ROPOHL 1993, pp. 322ff.

[36] FLORES and BAUM; also, my essay of 1997.

Particularities of Technical Action

I would like to explain the particularities of technical action with a constructed example with a real product innovation as its basis. The industry recently placed a vacuum leaf sweeper on the market. A device by which the allotment holder can 'woosh' the dead leaves from his ground in the autumn. A motor drives the blower which, in the same way as a vacuum cleaner, produces a negative pressure that draws the dead leaves into the device and sends them to a chopper unit in which they are cut up and then slung to a collecting bin, which is emptied every so often. It has since been discovered that this product, as useful as it may seem at first glance, has unwelcome side effects. When working, it has a high noise level that maybe harms the user but, in any case, annoys the neighbourhood. Also, the device sucks up not only dead leaves, but also ecologically important itsy-bitsies such as beetles, earthworms and the like, which are destroyed in the chopper. Brushing the surface, the leaf sweeper makes gardening easier but, on closer examination, harms both human health and the natural ecosystem.

I may not know what really happened with the development of this ambivalent product but, with a little background knowledge and awareness of the problem, I believe I can reconstruct probable scenarios that will not be too far removed from reality. There is, first of all, the possibility that the development engineers concerned went about their job in a completely short-sighted and irresponsible manner; seized by the idea of finally being able to mechanise a bothersome garden chore, they did not stop and think about the harmful side effects that their invention would have on human beings and their ecosystem, although these side effects could have been foreseen without undue effort and, in the case of noise, must surely have been blatant when the prototype was being tested. This constellation would appear uppermost in the minds of many representatives of ethics of technology who believe that training reform and professional moral rules can make engineers aware and enlightened to the point where individual wrong-doing of the kind just described can be avoided in the future.

However, this view of the problem ignores the circumstance that the development of a product is more than individual technical action. First, several engineers work together in a team, and the misgivings that one individual may very well have achieve nothing if he is unable to convince the others; technical action is not only individual, but mostly also co-operative action. Second, however, these engineers work for a commercial enterprise that is not primarily concerned with human health or ecological balance, but with economic success. If a development engineer approaches the management with any doubts, he will be told that the new product is needed urgently in order to improve the market position, to stabilise profits and to protect jobs; finally, the consumers can decide by their demand whether they consider the leaf sweeper worth buying. Technical action is usually bound up with corporate action and, since industrial corporations one and all pursue commercial and profit-making goals, also economic action.[37] The technology-ethical question as to whether an engineer can accept responsibility for the development of the ambivalent product is now transformed into the business-ethical question as to whether the company management should order the engineer in their employment to develop such products and make them ready for the market and whether they can answer for placing them on the market.

The engineers and management of the company in question might counter this consideration, pointing out that the negative consequences can be largely avoided through careful use of the product – wearing protective ear muffs, limiting use to normal working hours, careful collection of insects before sweeping – that, in other words, the responsibility does not lie with the producer, but with the consumer. As a matter of fact, technical action is generally intermediary action that, with the production of the product, produces a more or less certain potential for use action. The division of responsibility between producer and user can thus become a problem if the product facilitates but does not oblige harmful use. After all, the producers would in such cases have to give clear warnings in their product information, since noise pollution is determined by the product function and nothing can be done about it.

Finally, technical action is, usually, also collective action: several or many actors decide independently of each other about the production and use of technical systems, and the resultant cumulative conse-

---

[37] On commercial ethics, see HOMANN and BLOME-DRESS 1992; LENK and MARING 1992.

quences of action may clash with the intentions of some or all of the actors. Such paradoxical effects[38] are produced by the distribution of consumer goods once they are used simultaneously in large numbers, if, in this case, dozens of leaf sweepers are droning at the same time and even get on the nerves of the users. Another form of paradoxical effect is noted where, in the uncoordinated collective, the non-use by an individual has hardly any influence on the accumulation of consequences. This is not only the case for the single allotment holder who, whether he turns on his sweeper or not, will not substantially change the noise level already present. By analogy, it likewise concerns the individual producer who, when refraining from the product innovation out of a sense of responsibility, has to take into account the fact that a competitor may put the problem product on the market anyway.

This paradoxical effect, which appreciably curtails the readiness and ability to accept responsibility among individual engineers and producers, follows necessarily from the laws of the market economy and therefore raises the commercial-ethical and social-philosophical question of the best possible socio-economic order. The free competition of the market not only has a 'paradoxical effect', which Adam Smith describes, whereby the general interest is served through the interplay of supply and demand, although all individuals have only their own interests at heart. With certain goods and under certain conditions, regrettably, this can also give rise to general harm, as demonstrated theoretically by fictitious models of the kind of the prisoner's dilemma and empirically by the ecological crisis.

Apart from a few die-hard neoliberals, everyone today knows that such forms of market failure are, in principle, to be compensated only by political and legal measures, which provide the appropriate general set-up, binding for all players in the market. Certainly, if, in the case of the sweeper, we draw the theoretically natural conclusion that every product innovation should undergo public law examination and not be released on the market before proven to be fit for use, the individual state importing its goods under such rules would have considerable competition handicaps in the world market. We would be, in world market terms, faced with the same dilemma as was earlier analysed for the national economy level: the normative ineffectiveness of individual decisions under the conditions of collective competition.

---

[38] BOUDON 1979, see also LENK and MARING 1990, and ensuing discussion.

What began with the problem of responsibility of the individual engineer thus ends, because of the particularities of technical action, with questions of world domestic affairs. Technical action is only to a very limited extent purely individual action; it is rather through its co-operative, corporate, intermediary and collective character, first and foremost also economic and social action. By that, I do not mean to say that the individual is exempted from partial responsibility, to exhaust at least his own modest area of activity, which moreover should be increased through improvements in 'entrepreneurial culture'[39] and industrial law.[40] However, I must stress that we may expect no miracles from an individualist engineer's ethics for the environment- and people-minded technology.

Complications of Consequentialist Analysis and Prognosis

The consequentialist conception in ethics of technology has, if in the first instance we leave out of consideration the qualification of 'use', utilitarian traits. Nevertheless, there has been scarcely any explicit discussion so far regarding what has long been known to utilitarianism: the confrontation of action and rule-utilitarianism.[41] The impression is often that an act-utilitarianism conception is being represented if the individual scientists, engineers and managers are required to draw up a detailed balance sheet of the consequences for each technical project and their result is to form the basis of the decision as to whether or not the project will be implemented. However, discussion has since long shown that act-utilitarianism is impracticable, even in situations of day-to-day action, because individuals are usually not in a position to take into account all conceivable consequences. This objection applies for technical action all the more, since the consequences of technology can occur in the most diverse areas of daily reality in which engineers are in no way expert. The attempt is made to address them with the demand that engineers' training courses should also include qualifications for the areas of effects of technical action, but this is naturally unrealistic if it means special knowledge in all possible disciplines; realistic alone is the demand to equip engineers with more overview knowledge and

---

[39] ZIMMERLI and BRENNECKE 1994.
[40] WENDELING-SCHRÖDER 1994.
[41] On this point, see FRANKENA 1963, pp. 55ff.

judgement competence so that they can judge which specialists in other areas they might have to ask in case of doubt. However, such a procedure already exceeds the individualist conception. The question to what extent the consequences of technology are known at all and can be foreseen requires a careful differentiation. There are determined consequences, already latent in the function of a technical system, which appear unexpected only to those who have missed out a thorough technological systems analysis; these include for example the disadvantages of our leaf sweeper. Then there are the stochastic consequences whose nature is foreseeable, but whose onset can only be assumed with a certain, often very slight, probability. Since a probabilistic ethics, in my opinion, can hardly be substantiated, no individual enjoys the right to risk high damage for others simply because the occurrence of the damage is highly improbable; only the concerned themselves can decide regarding risks, and here the consequentialist view would have to be completed by a discourse-ethical conception.[42]

Cumulative and synergetic consequences also present problems. In these cases, similar or different individual consequences add up, each harmless enough in itself, but such that the resultant chain of consequences exceeds a critical threshold. With cumulative consequences, a developed scientific imagination is often sufficient; synergetic consequences, however, can easily be explained after the event but are seldom clearly visible beforehand. Finally, especially with secondary and tertiary consequences, we also have to reckon with uncertain consequences, which defy prediction by kind or probability. Such technical consequences concern complex ecosystems but are not to be excluded from mass use of technology in socio-economic systems either. In the last-mentioned types of consequence, all act-utilitarian calculus naturally fails.

That is why certain practices can now be recognised in the ethics of technology with a basic regulation-based utilitarian conception. These include, for example, the rule of recyclable design: instead of tracing all the imponderable chains of consequences that may arise from the removal of production residues, we follow the rule of manufacturing products basically in such a way that their components can be used again after their working life and, thus, not encumber the ecosystem. The formal maxims discussed in the

---

[42] See my essay of 1994 and the ensuing discussion.

second section can also be concretised in the sense of such rules which, at the end of the day, are consequential-ethically founded, because they fundamentally tend to avoid consequential complications.

The foundation of generally valid moral rules and values is, as we all know, one of the most difficult tasks of normative ethics, and this task, even now, has not come up with a solution accepted on all sides. This aporia naturally embarrasses the ethics of technology, too. It borders on schizophrenia if we try to oblige the producers and the users of technology to act responsibly, but cannot lay down any binding rules and values against which such responsibility could be measured. Since this calamity is not restricted to the ethics of technology, it is, in my opinion, an urgent task for applied ethics to reconstruct a sort of minimal moral code that can be sure of general acceptance. Personally, I favour for this a negative regulation-based utilitarianism, that would not pursue 'the greatest good for the greatest number', by 'the least suffering for the least number'.[43] The objective of this modest code is not to maximise human happiness; people always have the most varied ideas of that and, in the pluralistic modern society, ideas of happiness are even more differentiated. The objective of the minimal code is rather to protect people from avoidable harm and to minimise human suffering; there is a good deal of agreement on this point, and even the familiar ethical founding fathers seem more bearable when they place the avoidance of suffering in the foreground.

Value-ethically, this can also be expressed in the distinction between minimal values and increase values. Minimal values exist in the absence of human suffering and are unconditionally to be fulfilled; they are incidentally co-extensive with human rights, with the result that morality and right coincide in the normative basis. Increase values include any of the 'good things of life', which are recommended but not obligatory, and should therefore be pursued only to the extent that minimal values are not injured in the heat of the chase. This simple principle of deliberation has far-reaching consequences for technical innovations that offer the One an additional measure of happiness at the same time as heaping some new misery upon the Other and therefore are morally unacceptable. A modern

---

[43] I refer with this conception to GERT 1966; more on this and the following point in my book of 1996, esp. Chaps. 6 and 13.

example is the mobile telephone, which is evidently an enormous benefit for its user but, for the unwilling onlooker, is an unbearable imposition; with this innovation, the minimal code is therefore violated in so far as the producer and the legislator do not take all necessary precautions to prevent the user from inconsiderate misuse (which, as far as I know, has not happened yet!). We could easily make a list of other, less banal conflicts in which unnecessary increases in the affluence for the One are obtained at the cost of immoral and indefensible disadvantages for the Other; it should come as no surprise that such conflicts escalate from time to time and become miniature civil wars.

A well understood ethics of technology could have an altogether pacifying effect here, if it managed to work out the minimal code in question and lead to a broad social consensus.

Prospects of a New Union of Morals and Politics

The ethics of technology is part of Practical Philosophy not only in its technical terminology, but also in a general sense. It aims at shaping the technical world in seeking to subject technical action to moral rules; it intends the ethical steering of technology. Contemporaneously with the ethics of technology, however, another approach began around 1970 that, for its part, envisaged a political steering of technology, known as technology assessment; a genuine normative change in technology can be dated back to that time, a quarter of a century ago.

Technology assessment is an area of scientific political advisory activities and means 'the planned, systematically organised process analysing the state of a technology and its possibilities for development', 'assesses all consequences of this technology and possible alternatives, judges these consequences on the basis of defined objectives and values or calls for other desirable developments', and 'derives and develops action and structuring possibilities therefrom so that well-founded decisions may be facilitated and, where necessary, made and realised by suitable institutions'.[44] Technology assessment does not, therefore, pin its hopes on the responsibility of

---

[44] VDI 3780, 2.

individual engineers, but on the scientifically prepared and supported responsibility of social institutions and, above all, of state policy. Initially, the ethics of technology and technology assessment developed independent of each other,[45] being seen in their relatedness only in recent years.[46] Looking at Summary 1 again, it is easy to see that technology assessment is a variant of responsibility ethics, in which only other responsibility subjects feature. Technology assessment can therefore be seen as a social-ethically socialised operationalisation of consequential ethics, which, for the rest, solves certain problems that the individual-ethical conception has to contend with. Above all, it takes account of the social character of technical action, overcomes the limits of the qualification of the individual through interdisciplinary teamwork in consequential analysis and prognosis and exerts, when taken on by politics, ideally essentially greater effect than the individual engineer in his obligation to the industrial organisation.

I cannot expand on this here, but political technology assessment also has specific weaknesses. These include, in particular, its being situated outside of technical development and unable to have a direct effect on it. I therefore think that the actual players in the process of technical innovation, the industrial corporations, should also realise their social responsibility and occupy themselves with technology assessment in departments of their own. This thought does, however, bind technology ethics to commercial ethics and leads, ultimately, to the social-ethical conception of the concerted steering of technology in which, graded according to the principle of subsidiarity, individuals, corporations and state work together on the environmentally sound and people-friendly structuring of technology.

This programme could also have an integrating effect in another sense; under the impression of the real problems of application, the differences discussed, theoretically, so heatedly between the various ethical conceptions will fade into the background and give way to a pragmatic synthesis in which elements of value, obligation, consequential and discourse ethics and individual and social-ethical perspectives will combine with each other. The normative challenges

---

[45] On the development of technology assessment, see Part 2 of my book of 1996 and the literature cited therein; recently also VON WESTPHALEN 1997.

[46] First in my essay: 'Neue Wege, die Technik zu verantworten' (1987), in LENK and ROPOHL 1993, pp. 149-176; HASTEDT 1991; HUBIG 1995.

could thus bring back again to technological development what separated away in the society and philosophy of the modern age: the unity of morals and politics.

References

BAYERTZ, K. (ed.), *Verantwortung, Prinzip oder Problem*, Darmstadt, Wissenschaftliche Buchgesellschaft, 1995.
BOUDON, R., *Widersprüche sozialen Handelns*, Darmstadt/Neuwied, Luchterhand, 1979.
DESSAUER, F., *Philosophie der Technik*, Bonn, Cohen, 1927.
FLORES, A. and R.J. BAUM (eds.), *Ethical Problems in Engineering*, 2 vols., Troy, NY, Rensselaer Polytechnic Institute, 1979.
FRANKENA, W.K., *Ethics*, Englewood Cliffs, NJ, Prentice Hall, 1963.
GERT, B., *The Moral Rules*, New York, Harper and Row, 1966.
GOLDSTEIN, J., *Die Technik*, Frankfurt a.M., Rütten und Loening, 1912.
GRUNWALD, A., 'Ethik der Technik' in *Ethik und Sozialwissenschaften* 7(1996)2/3, pp. 191-204; Comments and Reply, pp. 205-281.
HASTEDT, H., *Aufklärung und Technik*, Frankfurt a.M., Suhrkamp, 1991.
HÖFFE, O., *Politische Gerechtigkeit*, Frankfurt a.M., Suhrkamp, 1989.
HOMANN, K. and F. BLOME-DREES, *Wirtschafts- und Unternehmensethik*, Göttingen, Vandenhoeck, 1992.
HUBIG, C. *Technik- und Wissenschaftsethik*, 2nd ed., Berlin/Heidelberg/New York, Springer, 1995.
HUBIG, C., A. HUNING and G. ROPOHL (eds.), *Klassiker der Technikphilosophie*, Berlin, Sigma, 2000.
HUNING, A. (ed.), *Ingenieurausbildung und soziale Verantwortung*, Düsseldorf/Pullach, VDI-Verlag, 1974.
JONAS, H., *Das Prinzip Verantwortung*, Frankfurt a.M., Suhrkamp, 1979.
JÜNGER, F.G., *Die Perfektion der Technik*, 1946, 6th ed., Frankfurt a.M., Klostermann, 1980.
KAPP, E., *Grundlinien einer Philosophie der Technik*, Braunschweig, Westermann, 1877, reprint Düsseldorf, Stern, 1978.
LENK, H. and G. ROPOHL (eds.), *Technik und Ethik*, Stuttgart, Reclam, 1987, 2nd ed., 1993.
LENK, H. and M. MARING (eds.), *Technikverantwortung*, Frankfurt/New York, Campus, 1991.
LENK, H. and M. MARING (eds.), *Wirtschaft und Ethik*, Stuttgart, Reclam, 1992.
LENK, H. and M. MARING, 'Verantwortung und soziale Fallen' in *Ethik und Sozialwissenschaften* 1(1990)1, pp. 49-57, Comments and Reply, pp. 57-105.
MARX, K., *Das Kapital*, 1867, vol. 1. K. MARX and F. ENGELS, *Werke*, Berlin, Dietz, 1959 and later, vol. 23.

MARX, K., 'Das Kapital' in K. MARX and F. ENGELS (eds.), *Werke*, vol. 25, Berlin, Dietz, 1964.
MITCHAM, C., *Thinking Through Technology*, Chicago/London, University of Chicago Press, 1994.
MOSER, S. and A. HUNING (eds.), *Werte und Wertordnungen in Technik und Gesellschaft*, Düsseldorf, VDI-Verlag, 1975.
MOSER, S. and A. HUNING (eds.), *Wertpräferenzen in Technik und Gesellschaft*, Düsseldorf, VDI-Verlag, 1976.
NIDA-RÜMELIN, J. (ed.), *Angewandte Ethik*, Stuttgart, Kröner, 1996.
OTT, K., 'Technik und Ethik' in J. NIDA-RÜMELIN (ed.), *Angewandte Ethik*, Stuttgart, Kröner, 1996, pp. 650-717.
PICHT, G., *Mut zur Utopie*, München, Piper, 1969.
PIEPER, A. and U. THURNHERR (eds.), *Angewandte Ethik*. München, Beck, 1998.
PIEPER, A., *Ethik und Moral*, München, Beck, 1985.
RAPP, F. (ed.), *Neue Ethik der Technik?*, Wiesbaden, Deutscher Universitäts-Verlag, 1993.
RAPP, F. (ed.), *Technik und Philosophie, Technik und Kultur*, vol. 1, Düsseldorf, VDI-Verlag, 1990.
ROPOHL, G. *Technologische Aufklärung*, Frankfurt a.M., Suhrkamp, 1991, 2nd ed. 1999.
ROPOHL, G., *Ethik und Technikbewertung*, Frankfurt a.M., Suhrkamp, 1996.
ROPOHL, G., 'Das Risiko im Prinzip Verantwortung' in *Ethik und Sozialwissenschaften* 5(1994)1, pp. 109-120, Comments and Reply, pp. 121-194.
ROPOHL, G., 'Verantwortungskonflikte im technischen Handeln' in J. Hoffmann (ed.), *Irrationale Technikadaptation als Herausforderung an Ethik, Recht und Kultur*, Frankfurt a.M., IKO, 1997, pp. 55-80.
SACHSSE, H., *Technik und Verantwortung*, Freiburg, Rombach, 1972.
VDI-Richtlinie 3780, *Technikbewertung: Begriffe und Grundlagen*, Düsseldorf, VDI-Verlag, 1991. Reprint in F. RAPP (ed.), *Normative Technikbewertung*, Berlin, Sigma 2000, pp. 221-250.
VON KEMPSKI, J., *Brechungen*, Reinbek, Rowohlt, 1964.
VON NELL-BREUNING, O., 'Sozialethik' in W. BRUGGER (ed.), *Philosophisches Wörterbuch*, Freiburg, Herder, 1953, pp. 296f.
VON WESTPHALEN, R. (ed.), *Technikfolgenabschätzung*, 3rd ed., München/Wien, Oldenbourg, 1997.
WEBER, M., 'Politik als Beruf' (1919) in J. WINCKELMANN (ed.), *Soziologie, Universalgeschichtliche Analysen, Politik*, Stuttgart, Kröner, 1973, pp. 167-185.
WENDELING-SCHRÖDER, U., *Autonomie im Arbeitsrecht*, Frankfurt a.M., Klostermann, 1994.
ZIMMERLI, W.C. and V.M. BRENNECKE, (eds.), *Technikverantwortung und Unternehmenskultur*, Stuttgart, Poeschel,1993.
ZWICKY, F., *Entdecken, Erfinden, Forschen im Morphologischen Weltbild*, 1966; München/Zürich, Knaur, 1971.

1

ENGINEERS WITHIN TECHNICAL INSTITUTIONS

INTRODUCTION

Martin Meganck

It is common – and apparently very tempting – to reduce the use of the word 'ethics' to cases where an individual has to decide or judge in situations he is confronted with. In the first place, this has to do with the fact that moral experience grows from the contact of a given situation with the individual conscience. The articulation between the individual as an actor and the acting of groups or communities as would-be persons, is very difficult. The easiest way is to claim that all group action is merely a result of the individual actions of the persons who are member of the group: no group acting without individual actions. The implicit suggestion then is that the moral quality of the group action emerges simply from the moral quality of the constituting individual actions: if everybody does his best, if everybody behaves well, then the group does its best, the group behaves well. The assumption that the sum of micro-rational actions is macro-rational, is held to be as true in the field of ethics as elsewhere.

Unfortunately, things seem to be more complicated than that. It is perfectly possible that a diachronic or synchronic series of micro-rational decisions proves to be macro-irrational. This can be due to limitations of the surrounding systems, to incomplete understanding, to differences of individual goals, to strange (antagonistic or synergetic) interferences, to inherent effects of decision taking structures, and probably to many other reasons. This, together with the psychological effects of group dynamics, makes it often difficult, if not impossible, to impute individual responsibilities for actions that emerge from collectivities. And even for individual lives, a proverb says that 'the way to hell is paved with good intentions'...

If this is true for ethics in general, it is certainly also the case for engineering ethics. Even if the term 'engineering ethics' makes one think of the moral quality of the professional behaviour of people whom we call 'engineers' (see the content of most 'codes of ethics' in engineering), the ethical scope of their work does not only concern their own conscience, values and duties. Most engineers work within institutions and organisations where use is made of their technical and managerial competence. And even for the few engineers who work 'alone', the actual existence and functioning of technological projects in society is not just a neutral fact. Together with the influences from other social, economic and political actors, their decisions change the world in a way which is of moral significance. Hence the attention which is paid in this book to the ethics behind 'the development of technical systems' (part 2) and to 'technology as a societal issue' (part 3).

The awareness and acceptance of this fact by engineers must not be used as a pretext to have their responsibility diluted or absorbed in that of 'structures' or 'the system'. On the contrary, it can be considered as one of their first individual duties to be conscious and accepting of the fact that they play a role in this system, with the inherent responsibility this brings. This responsibility may be described as a 'role responsibility', resulting from the distribution of tasks within an organisation – a responsibility which can often be indicated by the preposition 'as': e.g. 'as a safety engineer', one is responsible for the prevention of industrial accidents. It is also a capability responsibility: by the mere fact of their intellectual competences, their studies and experience, one may expect them to assume certain responsibilities if and when the need or opportunity arise. Finally it is also a causal responsibility: a responsibility that is passed on to one each time one appears in the chain of causes and circumstances which lead to a certain fact to happen. Often this causal responsibility means nothing more or less than that one cannot deny that one 'has got something to do with it'.

This is the reason why the first part of this book deals with the responsibility of 'engineers within technical institutions'. André GRELON describes a history of the development of the engineering profession in Europe till World War I in *'The Emergence and Growth of the Engineering Profession in Europe'*. Knowing one's history is important to understand the development of traditions and values of a community. Even if for most engineers the first 'moral community' for their professional activity is the actual institution or organisation for which they are working (the company, the administration,...), yet the 'engineering world' may put forward its own expectations concerning what an engineer should be and how he should behave.

'Today's Engineer Must Be More than a Technician,' is the message of Hellmuth LANGE and André STÄDLER. They focus on the economic circumstances of engineering employment, and on the influence of recent developments in management concepts. Whereas older engineers could count on stability of employment, on tradition and on hierarchical professional relationships, today's engineer must cope with flexibility, shorter lines of decision, and an erosion of the boundaries between private and professional life. The emphasis now is on 'specialisation-independent, non-technical requirements'.

Five contributions deal with practical experiences and observations of individual engineers. Michel LLORY reflects on *'The Safety of Risk-prone Sociotechnical Systems: Engineers Faced with Ethical Questions'*. Engineers tend to cope with major risks by trying to catch them in structures and formal procedures, often ignoring the practical experience of the actual workers on the floor. Deficient communication and a lack of mutual confidence arise from and give rise to inadequate collaboration. Bernard REBER translates the reflections of a young Swiss computer engineer, who launched a virtual game on the Internet in 1993. It soon appeared that his game had a very strong effect on the private and social lives of the players: *'Virtual Games Inviting Real Ethical Questions'*. In *'From Accusations to Causes: Integrating Controversies and Conflicts into the Innovation Process'*, Madeleine AKRICH describes the discussion between the different parties involved with the failure of a project introducing new technologies in a small village in Costa Rica. Here again, the functioning of the technical systems was hampered by deficient communication. Difficult communication is also at the basis of the problems described in the text of Stanislas DEMBOUR. As a former factory worker, he bears witness of how workers see their own and the engineers' position in actual industrial organisations. He even speaks of *'Workers and Engineers: Two Different Worlds'*. A last example opens the discussion at larger level, on the role and responsibility of scientists in the development of the nuclear bomb. The Pugwash movement holds annual conferences, bringing together scientists from all over the world, who are concerned about the relationship between science and society, particularly in light of the proliferation of nuclear weapons. In 1995 this movement was given the Nobel Peace Prize. Sally WYATT interviewed the former Pugwash president Sir Joseph ROTBLAT. His dream of scientists taking responsibility is also conveyed in his Nobel Peace Price Acceptance Speech.

A more fundamental reflection on the possible attitudes of engineers in ethically relevant situations, is given by Boel BERNER. She reviews some frequent types of reasoning in *'Handling Ethical Dilemmas in Everyday Engineering Work'*, paying attention to *'the possibilities of 'civic courage''*. Christiaan HOGENHUIS and Dick KOELEGA recommend some *'Engineer's Tools for Inclusive Technological Development'*. Whereas a systematic and often ideological criticism of social and political situations may lead to an attitude of hopeless passivity or non-constructive obstruction, they plead for a pragmatic optimism, well in the line of similar currents in business ethics and in politics (e.g. the 'third way'-politics in Holland, Great-Britain, Germany and Belgium). In the case of engineering ethics, they see quality management systems, technology assessment procedures, business ethical committees as just some of the tools and opportunities for engineers to influence the course of actions. Decision taking processes can be organised so that ethical considerations are structurally built in, in order to enhance an 'inclusive technological development'.

One final consideration: ethics is most often spoken about in situations of difficulties, where one may be forced to take a stand as a 'moral hero' against one's own will. Maybe the most important responsibility for engineers within technical institutions is to help create an atmosphere where moral heroes become superfluous: a working environment where it is considered to be normal that the ethical aspects of work are discussed; a working environment where no one has to be afraid that 'anything that you say may be used against you'. Is it utopic to imagine a way of talking about ethics which would render it superfluous?

1.1 DESCRIPTION

1.1.1.

EMERGENCE AND GROWTH OF THE ENGINEERING PROFESSION IN EUROPE IN THE 19TH AND EARLY 20TH CENTURIES

André Grelon

The history of engineers in Europe is like a huge construction site, certainly full of activity, but also widely scattered at many different sites with work progressing at different speeds. No overall approach to the subject has yet appeared, even if comparative studies of certain aspects have been undertaken. Thus, a priori, for no country in Europe does a general history of the profession exist, which takes into account its emergence and the role it played during the vast process of industrialisation, in all its various stages. In this context, any attempt to comment generally on engineers and their history in Europe, is an immense gamble. However, it is one worth taking, albeit cautiously, because it is important to develop a new approach, to have a fresh look at the past and so rediscover the reality of the history of engineers in the context of an as yet unexplored territory: Europe.
This article, then, is an attempt to put forward certain elements relating to a number of historical phenomena which can serve to outline a dynamic of engineers in the 19th and 20th centuries with a specifically European character. In particular, we will look at the rapid development of advanced technical training, the international propagation of teaching models, their progressive insertion in the process of industrial expansion, the development of group identities through the organisation of societies and specific scientific and technical institutions.

For a long time, research in the history of industrialisation aimed to examine macro-economic movements, major technical changes and their impact on industrial development (the coal-steel-steam engine model, the introduction of industrial electricity...), as phenomena in themselves. For its part, social history would oppose two large antagonistic forces: ruling class vs. working class and they would be analysed in terms of their immemorial conflict. But the potential interaction between these two social entities had always been almost completely disregarded. However, there was a gradual shift in perspectives and more attention was devoted to other aspects of the vast process of industrial revolution(s) with engineers emerging as an entity that could shed an interesting light on the issue. On the one hand, they figure not only as the initiators and the champions of numerous technical changes, but also as the careful organisers and managers of these technical systems, guaranteeing their perpetuation and proper implementation. Whilst on the other, they constitute a distinct social figure, a specific professional group, with its own education system, scope of professional activities and career-planning, a group with its own values and vindications.

Many studies have been devoted to engineering in countries industrialised at an early date – most of which notably focus on the institutional aspects: schools, societies, and associations, major State services, technical periodicals, etc., maybe because archives in these fields are more readily available or simply exist. There is ample literature on the issue, although entire aspects of the lives of engineers still need to be explored, e.g. their daily lives at work, their actual standard of living, etc. These studies have led researchers to address the issue from an international point of view, which has opened up new perspectives, bringing to the forefront similarities displayed by engineers across borders, the recognition of the concomitance of the challenges they have to face due to the global character of technical-scientific evolutions. These allow researchers to roughly, but still cautiously, sketch the genesis of the modern 'European engineer'. This is what this article deals with. However, it is not our intention to provide a synthesis on the subject: such an undertaking would take up at least a whole book and would require a thorough acquaintance with a vast body of literature scattered across Europe, which is not always easily accessible and the content of which is very heterogeneous! We shall thus solely evoke a few salient elements of the emergence and the development of the engineering profession which may seem to be common to various European countries – although it should be stressed that we are referring here primarily to continental Europe, given the fact that the British engineering system deserves special treatment. The period under scrutiny starts at the beginning of the 18th century, with the shaping of the profile of the modern engineer as a prophet of the industrial revolution. We shall look at the evolution of engineering until the Great War, the first world cataclysm where technology plays a central role, where war becomes totally industrial. At the end of the conflict, the professional tasks of industrial engineers are firmly established and their social status as wage-earners within a company hierarchy is now taken for granted, both being inextricably interwoven.[1] This is the beginning of a new chapter in the history of engineering. However, a detailed analysis of this new era would fall outside the scope of this article.

As a last preliminary remark, it is worth stressing that paying attention to this evolution implies that we start from the following two assumptions. First, it implies that we consider our present-day engineers as the heirs of those of the past. Only then are we entitled to reconstruct lineages, to act as genealogists, to give a meaning to this evolution or at least to make it understandable with regard to present-day debates. Naturally, one could claim that this undertaking is vindicated because they have the same name. However, we shall see that many people involved in activities that we would commonly associate with engineering work, did not necessarily bear the title of engineers. In other words, this quantitatively very important and very ancient category should not be overlooked. But what about the tasks of engineers? Does it make sense to compare present-day engineering activities with those of people who were called engineers between 1830 and 1914? Asking that question does not mean denying our present-day engineers any historical anteriority, we are merely reflecting on the notion of legacy and the social construction of the engineering profession.

We can carry this reflection one step further if we contend that, beyond state boundaries (which moreover fluctuated over the years), engineers display common traits that justify us in conceiving of them as a transnational professional group. Behind that assertion is the underlying assumption that engineers across Europe had a

---

[1] Even in non-belligerent countries such as Spain, the position of engineers changes from this period onwards. See R. GARRABOU, *Engynyers industrials, modernizació econòmica i burgesia a Catalunya (1850-inicis del segle XX)*, Barcelona, L'Avenç S.A., 1982.

sense of themselves as being of one family. Nevertheless there never was, until recently, i.e. a long time after World War II, any general international organisation of engineers – with the exception of an aborted attempt at an international engineering congress in the first half of the 19th century. Yet we know that historically this mutual recognition was sometimes problematic in practice. Furthermore, this implies that these 'engineers' be identified as such by their employers, which raises the question of the recognition of these professionals in non-initiative taking countries by foreign companies that sent their engineers on a mission to manage institutions or supervise building-sites, under the pretext of the alleged or real lack of skills among native technicians. In the 19th century, for example, Spanish industrial engineers had to struggle to have their abilities recognised by international companies. We can also mention in passing that the problem is still topical today, even though our present-day international labour market of technical experts would obviously require an analysis in different terms.[2]

The 'Modern' Engineer

When can we date the emergence of the modern engineer? Historians generally agree to locate it in the 18th century. However, the engineer is by no means a new figure in the technical world at that time. He has been known in Italy, France and Germany since the Middle Ages as a designer of war machines, as a hydraulic expert capable of supervising the digging of canals and the drying of swamps, as a builder of bridges and roads. He goes across Europe, offering his services to princes.[3] The word engineer itself is first encountered in the 12th century and, as Hélène Vérin points out in her well-documented analysis of the different meanings of the word: 'behind the miscellaneous acceptations of the word is the overriding notion of what truly distinguishes Mankind: its ingenuity, its ability

---

[2] On this subject, see for example the analysis of P. PIERRE, a company recruitment agent: 'Internationalisation de l'entreprise et socialisation professionnelle: étude des stratégies identitaires de cadres de l'industrie pétrolière' in D. GERRITSEN and D. MARTIN (eds.), *Effets et méfaits de la modernisation dans la crise*, Paris, Desclée de Brouwer, 1998, pp. 229-254.

[3] B. GILLE, *Les ingénieurs de la Renaissance*, Paris, Seuil, 1964.

to face new problems, to invent'.[4] In other words we see that, since its inception, the term engineer has coupled the purely technical scope of the expert with intellectual or even moral qualities that define the individual. The engineer may be hired by a prince or a king, he nonetheless remains free in most cases once the job has been done: he is an itinerant expert.

However we witness a gradual shift in the engineer's mode of relationship towards his 'employers'. From the 16th century onwards, the word engineer as it used in official writings and featured in ancient dictionaries comes to denote an army officer in charge of building, defending and investing fortifications. Gradually corps are formed and their number and their tasks increase along with the strengthening of absolute monarchy. This is especially clear in France. As early as 1604, whereas engineers still have not teamed up in a specific unit, Sully, minister of Henry IV and Fortifications Superintendent, decrees a real engineering charter, with specific orders and general rules for all the King's engineers in a Great Regulation, which will be applied until the end of the century.[5]

Since the State is hiring a body of specialists whose activities are extremely expensive (the costs of building a fortress are astronomical), it is only natural that it should ensure that its members have the required expertise. Of course, these engineers- builders very often stem from master-masons circles and they have been trained from early age in the family business, as Anne Blanchard aptly demonstrated. But requirements are getting more exacting and although the noviciate of a young engineer can still only be completed on the kingdom's large building-sites, an aptitude test to access the engineering profession is introduced at the end of the 17th century: they must have a fair knowledge of arithmetics and geometry, they must be able to design section and profile plans of fortifications, they must be acquainted with mechanics and hydraulics... The selection also allows the hiring of only the required number according to the number of available positions in the corps. It does not supersede the practice of recommendation, it complements it.

But we see that this type of organisation prefigures the modern civil servant. The administration functioned at the time with a

---

[4] H. VÉRIN, 'Le mot: ingénieur' in *Culture Technique* 12 (March 1984), p. 22.
[5] A. BLANCHARD, *Les ingénieurs du 'Roy' de Louis XIV à Louis XVI. Étude du corps des fortifications*, Montpellier, 1979.

system of offices, people could buy an office that they could then own and often bequeath to their heirs. The engineer does not own his job, 'he is only a certified expert, appointed to a well-defined position and paid by the king.'[6]

Then what are the changes that occur in the 18th century with regard to the figure of the engineer as a royal civil servant as it was instituted throughout the 17th century? It is the rising awareness across Europe of the necessity of an ad hoc education system. In the various countries, specific institutions will be created in order to train engineers for their professional tasks. This is a very important phenomenon because it signals the passage from a direct and individual transmission of technical knowledge, as was the case in corporate bodies with the strict sequence of stages: from apprentice to mate, from mate to master, (but also with this notion of initiation, i.e. the transmission of a knowledge that is almost in the realm of secrecy, be it knowledge or know-how), to a collective teaching of a democratised knowledge dispensed by teachers who all have their area of expertise – a desacralised teaching that can be transmitted orally but also by means of books, i.e. available to everyone. Of course this formulation is a bit of a simplification and it stands to reason that changes did not occur overnight. However, the fact remains that the question of the diffusion of technical knowledge is in the air. The all-too-famous enterprise of the *Encyclopédie* conducted by Diderot and d'Alembert testifies, among others, to this new interest in the educated circles. Engineering schools thus partakes of this vast movement. The very first are probably the Engineering School of Moscow in 1698 and the School of civilian and military engineering founded in Prague in 1707.[7] Artillery schools also see the light of day in the beginning of the century[8]: even though officers who graduate from this school do not officially bear the title of engineer. The artillery is part of the 'elite weapons' corps and the education provided in schools of artillery must train them

---

[6] Ibid., p. 168.

[7] M. EFMERTOVÁ, 'L'évolution de l'enseignement technique tchèque aux XVIIIe et XIXe siècles' in *Quaderns d'història de l'enginyeria* 3(1999), pp. 51-82.

[8] This is the case in Spain, for instance, with the creation in 1722 of four Academies in the fortresses of Barcelona, Pamplona, Badajoz and Cadix. See H. CAPEL, J.E. SÁNCHEZ and O. MONCADA, *De Palas a Minerva. La formación científica y la estructura institucional de los ingenieros militares en el siglo XVIII*, Barcelona, Serbal/CSIC Publishers, 1988. Especially pp. 147ff.

not only in the art of war and ballistics, but also to know the metals used and the production methods of cannons, as well as to propose and implement innovations in produced pieces (new types of cannons) as well as in their construction process, and to follow up their production in the royal arsenals. This makes them forerunners of industrial engineers. Ken Alder[9] showed that these artillerymen subscribed to a social philosophy born out of the Enlightenment, considering distinction due to merit as the key-criterion for the assessment of achievements. From the middle of the 18th century onwards, technical issues became central to the training: artillery officers learnt descriptive geometry and technical drawing which were the basic tools to found a new artillery. At the same time this training brought about a new labour division and a technical hierarchy based on a knowledge which distinguished these military engineers from their subordinates and confirmed their authority. The idea of a large-scale construction of light, mobile cannons demanded the organisation of production in factories and also redefined the very structure of the Weapon within the armed forces. Road engineers went through a similar evolution, as Antoine Picon pointed out in his brilliant study of their training centre, the first engineering schools to train a *non military* engineering corps in France.[10] These engineers are trained in a body of technical and scientific knowledge which have just entered a long and complex process of interrelationships. On the basis of this body of learned knowledge, they now consider rationality as their value and the rationalisation project as their principle of conduct. The engineer uses a body of knowledge of which he is the only depository, he has a say in the very organisation of his job: '[his] demand for rationality entails an ambition to control the physical process, relying on his knowledge of the physical world and his insight into the production ratio. [His] rationality links professional strategies to scientific knowledge.'[11]

If we consider this professional category from our contemporary vantage point, we must admit that these State engineers, who are

---

[9] K. ALDER, *Engineering the Revolution: Arms and Enlightenment in France, 1763-1815*, Princeton NJ, Princeton University Press, 1997.

[10] A. PICON, *L'invention de l'ingénieur moderne. L'École des Ponts et Chaussées, 1747-1751*, Paris, Presses de l'École Nationale des Ponts et Chaussées, 1992.

[11] Ibid., p. 18

encountered all over Europe,[12] are not the only technical experts around. Those whom we could sometimes dub proto-industrial engineers were not dependent on the administration and were not trained in its schools.[13] There is a group of technicians, builders of mills, or designers of miscellaneous machines that increases in number throughout the 18th century. In France, a certain Vauvanson, designer of tool-machines and automatons, epitomises this new body of unacknowledged civil engineers. They perpetuate the tradition of the Renaissance liberal engineers. But it is in England that proper industrial engineers will emerge: they are at the heart of industrial development and advances in machinery. They will play a central role in the invention of the steam engine, its improvements, its developments for various purposes (pumps to evacuate water from mine wells, loom engines, and its implementation on a mobile support, thus creating the railway). People from abroad will soon visit them and duplicate their creations.[14] They do not have a proper school: they are trained on the shop floor, but some of them get into the habit of gathering in clubs to exchange ideas. Out of these clubs they will found the most ancient and celebrated engineering society in 1818: the Institution of Civil Engineers.

The Engineer and Large-Scale Industry (Mass Production)

At the end of the 18th century France had a system of education centres for civil and military engineers that was without par in Europe:

---

[12] In Spain, after the accession of the Bourbons to the throne, a corps of engineers of the king is created in 1711, it will be in charge of military as well as civilian activities. In Germany, kingdoms and principalities also hire engineering corps...

[13] However the French State hired a corps of 'inspectors of the royal manufactures', but in the old fashion (they buy their office). One of them suggested the creation of a school in order to train these inspectors, leaning on the model provided by the Ecole des ponts et chaussées. He will not be heard. See J. GUILLERME, 'Roland de la Platières et les patiences de l'inspection méthodique' in A. THÉPOT (ed.), *L'ingénieur dans la société française*, Paris, Éditions Ouvrières, 1985.

[14] On a study travel in England in 1788, Spanish engineer Agustin de Bétancourt, trained at the École des Ponts et Chaussées of Paris, stealthily catches a glimpse of a steam engine at Watt and Boulton's and is able to reproduce its plans of memory. He informs the Academy of Sciences of Paris and on his indications the Perrier brothers will construct a 'fire-machine' in 1790-1791 for the Island of Swans Mills in Paris. A. PICON and M. YVON, *L'ingénieur artiste*, Paris, Presses de l'École Nationale des Ponts et Chaussées, 1989, p. 180.

two schools of military engineering, one for the training of shipbuilders, the other for the training of sapper engineers – alongside the training of artillery officers; two non-military schools, the State Road Engineering School and the Mining Engineering School, the latter being notoriously inspired by the Freiburg Mining Academy, founded in 1765. Yet four years after the beginning of the Revolution, little was left of this impressive system. In order to bridge the gap in the training of state technical executives, a commission of scholars proposed a motion and had it adopted by the Convention in 1794 about the founding of a school that would train elite weaponry officers as well as state civil engineers, along with technical experts for industrial development (which is usually overlooked): the *École Centrale des Travaux Publics*. The following year this institution was already re-organised: it would take care of the basic education of future engineers who would later specialise in schools of applied sciences (the *ancien régime* engineering schools which were re-established on this occasion), and the institution then received the name that has survived up to this day: the *École Polytechnique*.

The reason why we have been dwelling on the birth of this new institution is that it paved the way for a whole new generation of education centres where fundamental science and mathematics as the engineer's primary tool loom large. In this perspective, the different technologies recede to the background, they lose their autonomy and become applications of science. To be sure, this is more a general statement than the end-point of a demonstration, but it can be seen as a historical landmark to the extent that it posits science as the foundation of the engineer's rationality, his practical achievements no longer being the sole criterion of gauging his merits. Again, this is merely the beginning of a long and sometimes divided process, which had been announced by the curricula at former engineering schools, but which is now overtly posited as a principle to live by.

The creation of the *École Polytechnique* does not go unnoticed in Europe. Other institutions will borrow the neologism 'polytechnique': *Polytechnische Schulen* will be created in German kingdoms; an imperial polytechnical institute sees the light of day in Vienna in 1815. However, the original model will not be faithfully duplicated, with the exception of a few institutions, like West Point in the United States.[15]

---

[15] B. SEELY, 'European Contributions to American Engineering Education: Blending Old and New' in *Quaderns d'Historia de l'Enginyeria* 3(1999), pp. 25-50.

Nevertheless this influential model compels other institutions to clearly define their conception of engineering education. This is the case in Berlin, with the new direction established by Christian Beuth for the curriculum at the *Gewerbe Institut*, that retains a strong traditional *arts et métiers* orientation;[16] or in Prague where knight Gerstner embarks as early as 1798 on a thorough revision of the organization and curricula at the engineering school, with a view to creating a polytechnical school for the entire Austrian empire,[17] or in Italian States, where *Scuole di Applicazione* are founded upon the model of French applications schools, in Naples in 1811, in Rome in 1817 and in Florence, Modane, Venice, Turin and Milan.[18]

The beginning of the 19th century witnesses the development of mass-production industry in continental Europe. Concomitant with this development is the small-scale emergence of new technical experts whose duty is to facilitate the expansion of companies. England has large numbers of such technicians and at the end of the Napoleonic period, thousands of them are hired in France, Belgium and as far afield as Germany to install and operate steam engines and other devices, and, generally speaking, to organise production. However, it remains essential that every country have its own body of technicians. With this aim in view, educational institutes are founded which set out to provide craftsmen with a basic technical and scientific background. This is the case in Paris with the *Conservatoire des Arts et Métiers* that offers from 1819 onwards evening courses to manufacturers, craftsmen and workers of the capital. Evening courses are also organized in various major towns[19] across

---

[16] P. LUNDGREEN, 'De l'école spéciale à l'université technique, étude sur l'histoire de l'école supérieure technique en Allemagne avant 1870' in *Culture technique*, (March 1984) 12, pp. 305-311. It is after 1870, when *technische Hochschulen* will be founded out of former *écoles polytechniques* that the primacy of science will be asserted, these high schools insisting on stressing their university character

[17] EFMERTOVÁ, 'L'évolution de l'enseignement technique tchèque aux XVIIIe et XIXe siècles'.

[18] F. BUGARINI, 'Ingegneri, Architetti, Geometri, la lunga marcia delle professionni tecniche' in W. TOUSIJN (a cura di), *Le libere professioni in Italia*, Bologna, Il mulino, 1987, pp. 305-335.

[19] The most famous being the well-known 'cours pour les ouvriers de la ville de Metz', where applied geometry, technical drawing and arithmetics are taught. F. HAMELIN, 'L'École d'application de l'artillerie et du génie et les cours industriels de la ville de Metz' in A. GRELON and F. BIRCK (eds.), *Des ingénieurs pour la Lorraine, XIXe-XXe siècles*, Metz, Éditions Serpenoise, 1998. But other towns with a similar school include Lille, Rouen, Bordeaux, Lyon, at the initiative of town councils.

France. In the same spirit, a *Real Conservatorio de Artes* is founded in Madrid by the government in 1824, with branches in several provincial towns.[20] In Britain young workers attend evening courses at mechanics institutes. To be sure, we are not dealing with instruction at engineering level, and these courses are far from being available everywhere, but they are significant enough for us to mention here and they indicate a new awareness of the necessity of training professional workers. At a higher level, education institutions for foremen are created: *polytechnische Schulen* in the first third of the century (Karlsruhe, Dresden, Stuttgart, Hannover, Braunschweig, Darmstadt); in Barcelona, a major industrial city, Chamber of Commerce schools have the same function (school of chemistry, school of mechanics, school of physics); in Sweden, the Chalmers Institute opens its doors in Gothenburg in 1829, in France, we have two, then three *Arts et Métiers* schools, the La Martinière school in Lyon and the foremen school in Douai. Once again, this is not a proper organized system of technical education yet, with a degree at each level, each giving access to a position in a professional hierarchy, but each of these schools plays a pioneering role in the emergence of a national technical education system. Besides, young people who graduate from theses schools usually have an open future, as Charles Day showed in his authoritative study on the French *Arts et Métiers*: if *gadzarts*, graduates from these schools, generally enter companies at a professional worker level, many of them will quickly conquer engineering positions or set up their own flourishing business.[21] Over time, requirements will become more exacting in a number of institutes of this level of education, which will yield civil engineering schools. This is the case for example with the *polytechnische Schulen* which will be turned into *technische Hochschulen*, technical high schools of a university level in the last third of the century. Other schools will follow suit and form this category of technical

---

[20] S. RIERA I TUÈBOLS, 'Industrialisation and Technical Education in Spain, 1850-1914' in R. FOX and A. GUAGNINI (eds.), *Education, Technology and Industrial Performance in Europe, 1850-1939*, Cambridge, Cambridge University Press, Paris, Éditions de la Maison des Sciences de l'Homme, 1993.

[21] C.R. DAY, *Les écoles d'Arts et Métiers, XIXe-XXe siècles*, Paris, Belin, 1991. Gradually the education level of these schools will increase throughout the century and in 1907 the Arts et Métiers will deliver an engineering degree. After World War II, the curriculum will be reinforced until 1975.

intermediary education supplying the European market with technical experts who go by a miscellany of names: *peritos* in Spain, *periti* in Italy, *graduierte Ingenieure* in Germany, *engenheiros tecnicos* in Portugal, *ingénieurs techniciens* in Belgium, etc. – without a proper equivalent in France, which makes it a special case.[22]

Finally, in this first period of expansion in mass production on the European continent emerges a new category of institutes of higher technical education: civil engineering schools. One of the first, if not the first, is probably the Engineering School of Prague. The project formulated by its chairman in 1798 on the basis of the *École polytechnique de Paris* could not be realised: in 1803, he proposes a new motion no longer with a view to training technical civil and military servants for the Austrian-Hungarian Empire, but to create a new body of industrial engineers for Czech countries. Two channels are organised and fully operational as early as 1806, one in chemistry (and especially in dyeing chemistry in order to meet the needs of the textile industry), the other in mechanics and civil engineering. In 1830 the institute already houses 400 students.[23] Another similar example: out of a mechanics school founded in 1798 a royal technical institute is created in Stockholm in 1826. In Denmark, the first engineering school, the *Polyteknisk Laereanstalt* of Copenhagen, sees the light of day in 1829, where the celebrated Danish physicist Hans Christian Örsted read for twenty years. As Henrik Harnow[24] points out, it is very marked in the first period of its existence by the example of celebrated French engineering schools. In Portugal, the State opens a polytechnical academy in Porto in 1837 with a view to training engineers in all the areas of expertise that a country may need: naval construction, roads department, mines, etc.[25] In France, four scientists team up with a businessman to open in Paris in 1829 the

---

[22] This category of technicians will gradually assert itself and in the 20th century it will demand its integration in the engineering community, creating a true 'class conflict' within this socio-professional group. These issues have been studied among others for the Belgian case by R. BRION, 'La querelle des ingénieurs en Belgique' in A. GRELON (ed.), *Les ingénieurs de la crise*, Paris, Éditions de l'École des Hautes Études en Siences Sociales, 1986, pp. 255-270; for Germany by B. LUTZ and G. KAMMERER, *Das Ende des graduierten Ingenieurs?*, Frankfurt/Köln, 1975; and for Portugal by M.D.L. RODRIGUES, *Os Engenheiros en Portugal*, Oeiras, Celta Editori, 1999.

[23] EFMERTOVÁ, 'L'évolution de l'enseignement technique tchèque aux XVIIIe et XIXe siècles'.

[24] H. HARNOW, *Den danske Ingeniors historie 1850-1920*, Arhus, Systime A/S, 1998.

[25] RODRIGUES, *Os Engenheiros en Portugal*.

country's first civil engineering school, a private school, which is more: the *École Centrale des Arts et Manufactures*. Although it was not the first in Europe, it nonetheless epitomises this new category of institutes to the extent that it posits itself on a theoretical level by basing its organisation and curriculum on a new orientation: industrial science, as opposed to the *École Polytechnique*, which is exclusively dedicated to the study of 'pure' science, without considering potential applications. Other institutions will be opened on this model: a mining and metallurgy school in Mons, Belgium, founded by two graduates of the *École Centrale des Arts et Manufactures* in 1836[26] and a little later, in 1853, a special School for industry, public works and civilian constructions in Lausanne, by a group of engineers, two of which at least had been to the *École Centrale*.[27]

We do not claim to provide with these few examples an exhaustive list of the countless institutions which were created across Europe within the space of three decades. Two remarks are in order here. On the one hand we see that this wave of creation affects all of Europe and not just a few so-called 'initiator-countries', even though not every nation is gained by industrialisation in the same way and in the same pace. The awareness of new needs is being raised in every country. We are undoubtedly dealing here with a transnational issue that invites us to rethink history in new terms. At the same time, this awareness raises again the question of the division of Europe into two separate categories.

On the other hand, the creation of institutes dedicated to training engineers directly for the private industry brings about a far-reaching change in the image of the engineer. Those who had been craftsmen trained on the shop floor, who had earned their legitimacy through an experience based on practice, become scholars with a higher education degree, although they can still only become truly accomplished experts through extensive practice in factories. This considerable evolution will never be curbed, far from it, it will only be accentuated. Even though up to this day, some people in all industries may occupy engineering positions without bearing the

---

[26] J.C. BAUDET, *Les ingénieurs belges, de la machine à vapeur à l'an 2000*, Bruxelles, APPS Ed., 1986.

[27] S. PAQUIER, 'Les exemples contrastés de l'École d'ingénieurs de Lausanne et de l'École Polytechnique Fédérale de Zurich (1853-1914)' in A. GRELON, I. GOUZÉVITCH and A. KARVAR (eds.), *La formation des ingénieurs en perspective. Modèles de référence et réseaux de médiation, XIXe-XXe siècles*, Rennes, Presses Universitaires de Rennes, 2001.

title, the profession appears primarily as a learned profession, to be accessed only via higher education institutions. This movement is still timid in the beginning of the century, but it is definitely on the rise. In the last third of the 19th century, with the creation of specialised institutions or channels, it becomes a characteristic of engineers, included in England, where the tradition of the practical technician, trained on the shop floor, under the guidance of a recognised engineer and exchanging experience and analyses within a corporate body, still loomed large.

But their primary concern in Europe in the beginning of the 19th century is to come into the picture in the first place. For one thing, they must vindicate themselves to Official bodies (which sometimes have been around for a long time), that is, technical State Civil servants, such as Military Corps (engineering, artillery, ship building), or Civil (Roads Department, Mining), which enjoy an clear social prestige and a considerable administrative power that the newcomers do not have. Unlike what has often been asserted, the importance of Sate Corps is by no means peculiar to France only, it is equally powerful in many other European countries. This holds true for the Corps of Prussian Mines, which traditionally supervises the entire mining and metallurgy system and which is in charge of refurbishing mines in countries united to the kingdom after 1815.[28] State engineers bear their title as a rank in the hierarchy of Public Utilities: in whose name would the new generation of industrial experts call themselves engineers?[29] On the other hand, the latter must vindicate themselves to time-honoured foremen who have built up an intuitive but extensive knowledge on the shopfloor, people who are steeped in the uses and qualities of materials, who know how to operate machines and which to use, who are able to efficiently manage of a large body of factory workers made up of men, women and children: all this points to the necessity for new experts to come up with totally new types of skills and knowledge. Finally, they must foster their autonomy as a new social body with its own values and its own larger societal project, which threatens to displace outdated ideologies.

---

[28] E. DORN BROSE, *The Politics of Technological Change in Prussia*, Princeton NJ, Princeton University Press, 1993, especially chapter 4.

[29] Let us point out that some of them may leave their administration on a temporary or permanent basis to work as advisers to manufacturers or bankers, or as civil engineers. See e.g. A. THÉPOT, *Les ingénieurs de Mines au XIXe siècle. Histoire d'un corps technique d'État*, Paris, Eska, 1998.

It is no easy job to picture ourselves, from our present-day perspective, what the actual situation of civil engineers would have been like in the first half of the 19th century. First, because we are confronted on a daily basis with the mind-boggling achievements of a triumphant science and industry, mostly wrought by an army of engineers of all kinds in companies and research and development centres. Second because we often fall prey to the stereotypes about the heroic days of 19th century technical progress: steaming factory chimneys, inaugurations of railways with distinguished gentlemen and ladies dressed in crinoline as they are depicted in the numerous engravings and illustrations that celebrate these changes. And this factual information, superimposed on more wistful reminiscences of a past with no precise dating, yields the figure of the engineer as the trailblazer of the industry, a modern times demiurge. Nevertheless, one should bear in mind that engineers in the beginning only make up a tiny minority of the social body. They are but a fistful of men with grand ambitions but comparatively very few achievements so far, and, even though they live in a period of great projects, their social position is far from being secure. In most European countries, industry's contribution to the global economy is still very limited, although it will definitely increase over the next decades. Moreover, investments that engineers request are often so astronomical that sponsors are sometimes led to face ruin. Technological inventiveness is not necessarily synonymous with economic profit.

In order to illustrate this last point, let us consider the vast array of techniques developed by engineers in order to face the challenges posed by the railway. One of their major concerns in this respect was to devise a way to remedy locomotives not being able to pull carriages up steep slopes. In a study published in 1810, Danish engineer Medhurst suggests the following technique: an gigantic piston that would slide along a pipe placed between the tracks in the slope, driven by the vacuum force, and to which the train would be coupled. That process goes by the name of the atmospheric train. A prototype is built in Ireland near Dublin by the engineers Clegg and Samuda and a trial machine is constructed in Wormwood Scrubs near London in 1840. But it is in France that a real large-scale implementation is attempted, between Paris and Saint-Germain, by the Office of engineer-adviser Eugène Flachat, after sending an expert abroad, the passing of a law allotting the necessary funds, creating competition between different procedures, considerable civil engineering works, the arduous manufacturing of kilometres of adequate pipes, the construction of huge 400 h.p. steam engines to create the vacuum in the pipes... It takes two years (1845-1847), astronomical funds and constant revisions of the original plans to install the entire system whereas in the meantime, a new extra powerful steam locomotive is designed and will provide a rational and cost-effective alternative for pulling trains in steep slopes.[30]

This episode is revealing in many respects. First of all, we see that the financial risks involved in technical innovations are very real indeed![31] A great number of engineers get carried away by scientific principles and spawn immediately perfect applications. But the ideal technological invention exists only on paper... Secondly, it becomes clear that ideas, projects and experiments are followed across boundaries, they are launched by one engineer in a given country, echoed in another, and finally realised in a third one. This harks back to the idea of a cross-national 'community' of engineers that benefits from this kind of exchanges. In order to disseminate innovations and apply them locally, engineers across Europe must have a similar educational background that allows them to apprehend basic processes and implement them according to the needs and conditions of the individual countries, but they must also share the same outlook on techno-scientific progress in order to grasp the benefits of a new machine, an emerging technology or a new product and thus to feel compelled to propagate these innovations and advocate them to the broader public. Finally, it is worth mentioning that, while company structures at the time do not always include a permanent team of full-time technical experts, many engineers are working for Engineering Offices, such as the Office of Flachat, who hires a great many young graduates from the new *École Centrale des Arts et Manufactures de Paris*.

---

[30] I borrow the details for this story from A. AUCLAIR's study, *Les ingénieurs et l'équipement de la France. Eugène Flachat (1802-1873)*, Le Creusot-Montceau les Mines, Ecomusée de la communauté urbaine, 1999, pp. 95-99.

[31] We shall note that in the domain of terrestrial transportation, the tradition of utopian projects has been perpetuated to this day. The aborted project conceived by the engineer Bertin in the 1960s of a monorail train system with zero-contact with the tracks resulted in the project of a magnetic monorail train that should have been constructed between Hamburg and Berlin but was abandoned in 1999... On Bertin and his unsuccessful attempt, see C. DIDIER, 'L'aérotrain ou la tragédie de Jean Bertin', in C. DIDIER, A. GIREAU-GENEAUX, B. HÉRIARD DUBREUIL (eds.), *Éthique industrielle. Textes pour un débat*, Brussels, De Boeck, 1998, pp. 323-337.

However, their vital importance is gradually becoming more visible and their number increases. From the middle of the 19th century onwards engineers team up in national societies following the example of their British counterparts who, as we mentioned above, started to organise themselves as early as 1818 with the *Institution of Civil Engineers*: in France, the *Société des ingénieurs civils* is founded in 1848; the *Verein Deutscher Ingenieure* sees the light of day in 1856; in 1861, the *Sociedad de ingenierios industriales* in Madrid and the *Svenska teknologföreningen* in Stockholm are formed; in 1869, the *Associacão dos Engenheiros Portugueses* and so on. Even though these societies are not joined by all those who call themselves engineers – far from it – they nonetheless signal a real institutionalisation of the profession, they are making it more visible. Associations serve as centres of permanent training, as agora as the place where engineers assert their scientific character. Above all, they unofficially – or even at times officially – represent engineers before the authorities and civilian society. One should note that this associative tendency is spreading across Europe, in both so-called central and peripheral countries:[32] the boom in the engineering profession is a wide-ranging phenomenon with a truly European dimension.

The Second Wave of Industrialisation and the Development of New Engineering Disciplines

From the 1860s onwards and particularly in the last third of the 19th century, new technologies are emerging, and engineers will be in charge of organizing their implementations in the various industrial sectors. This affects metallurgy with new processes (Bessemer converter, Martin-Siemens oven, Gilchrist-Thomas process...), chemistry (the Solvay soda process, synthetic colourings), electricity, telephone, internal combustion engine (leading to the creation of the automobile and aeronautical industries). As historian of technological development, Bertrand Gille argues that a new global technical system is emerging, and it will gradually overlap with and displace the former

---

[32] On the contrary, in Belgium, a country that was among the first to be hit by the wave of industrialization, a central engineering association will be put together only in 1885, the *Société Belge des Ingénieurs et des Industriels* (SBII), whereas engineering schools old boys societies were created at an early date (as early as 1847 for graduate engineers of the École de Liège).

technical system:[33] in keeping with this, steam engines will be used to power electricity generators. Rather than an abrupt change, we must speak of a gradual shift in the industry, the economy and generally speaking, in societies at large. Looking back at this period one century later, historians will give it the name of a second industrial revolution, which, in addition to bringing about eventually considerable changes in the internal structure of companies, their methods of production and the manufactured products, will witness an unprecedented boom in the whole industry. It is only then that engineers truly rise into prominence. In the preceding era, technical experts were still limited in number. Even major companies hardly ever hired more than ten of them, more often there were only one or two, and as we mentioned above, a popular stepping-stone for young engineers to enter this particular segment of the labour market was provided by advisory-offices, which would work for industrial companies on a contract basis and serve as intermediaries between designers of machines and their end-users. We can formulate the following hypotheses: there appears to be a growing tendency for expanding companies to set up their own internal teams of experts, and secondly, the increasingly complex industrial processes inevitably call for a variety of specialisations and the constant support of technical experts. These experts, who will increase in number, will soon form a hierarchy and gradually accede to the status of wage-earners. There will be a diversification of tasks and a specialisation of careers.

As a matter of fact, this boom in the industry is accompanied by a huge increase in the number of training courses available. The development of applied sciences in the various disciplines will require adjustments in institutes of higher education: existing institutions will have to modify their curriculum and increase their reception capacities; specialised channels or institutions dedicated to particular industrial branches will be created. A whole new type of technical experts will graduate from these schools and this will yield an increasingly heterogeneous engineering world, which will progressively jeopardise the profession's capacity to speak of one voice. Thus, it is precisely just as engineers are finally managing to assert their vital role in the industrialisation process and are becoming more and more visible within companies that the possibility of

---

[33] B. GILLE, *Histoire des techniques*, Paris, Gallimard, 1978, pp. 772ff.

defining common professional objectives, other than in terms of general principles, is getting undermined. Meanwhile, on a social level, engineers will form a socio-professional group with its own collective vindications; however it will take them a long process of teaming up with other categories to make themselves heard. But this social aspect is not within the scope of this article.

In order to illustrate the importance of the development of engineering academic education in Europe and how it relates to sciences, we shall look in more detail at *the* branch that epitomises these new industrial dynamics: electricity. If electricity had been known since the 18th century as an curious phenomenon observed in physics offices, it is only in the beginning of the 19th century that it became an important object of laboratory studies across Europe with the works of Volta (whose battery was soon duplicated at the *École Polytechnique de Paris*), Ampère, Oersted, Faraday and a few others. Its first industrial applications were in the domain of weak currents, with the installation of telegraphy networks in the middle of the 19th century which necessitated the emergence of new experts in these matters, but most applications were found in the State administration, both civilian and military. However, it is only from 1869 onwards, with the works of Belgian engineer Zénobe Gramme, who designed a marketable dynamo, and with the discovery four years later of the reversibility of this machine (it can *produce* electricity and thus function as a generator and if it *receives* electricity, it becomes an electric engine) that the electrical industry could truly start developing. A decade later, in 1881, the first international electricity exhibition was held in Paris, funded by the French electricity industry and featuring every major company in the world. Around the same time an international scientific congress, the first of its kind, was organised solely to debate the arising question of standardisation that would be crucial to the global diffusion of this new energy mode.

At this point awareness is being raised of the necessity to train proper experts in this field. Until then, the teaching of electricity had been an integral part of physics courses at universities and engineering schools. Electricity periodicals had been born as a result of the rapid growth of the discipline and a few works offered a synthesis of the knowledge on this matter. This literature provided a fair basis for engineers to extend their personal scientific-technical knowledge and to apprehend potential professional applications. But rapid scientific progress in the study of electric current, on a theoretical level as well as in the area of technical implementations – which would come to form a branch in its own right, by the name of electronics, the growing demand from companies for technical executives for the economic exploitation of electricity, its production, distribution, private and industrial applications, call for a swift adjustment of the education system.

The speed of reaction will differ from country to country, according to the existing structures, capacities of adaptation or willingness to create new entities. The first institution of higher education to be entirely devoted to electronics is probably the *Institut électrotechnique Montefiore* of Liège, founded in 1883, i.e. only two years after the first international electricity exhibition, drawing graduate engineers from all over Europe.[34] In the following years, specialised schools will be created or separate departments will be founded in existing institutions, training students as electrical engineers. This is an important phenomenon. In Germany only, an estimated 2,500 electrical engineers were trained in the nine *technische Hochschulen* between 1882 and 1914, most of which would later join the electrical industry. All over Europe countless engineers are hired by companies.[35]

Three remarks are in order here with regard to this new engineering wave. First, they differ from first generation engineers who supervised the entire industrial process and who were required to have general skills. Electricians in the 1890s are the first exponents of a new type of engineer: specialists. A comparison between the curricula in institutions of the first and second category is revealing from this point of view: on the one hand, a series of courses in the various disciplines and their application scopes, and on the other hand, an education dedicated almost exclusively to the many aspects of electricity. To be sure, the education of these groundbreakers in electricity may still appear quite broad if we consider it from the vantage point of our present-day age of hyperspecialisation. The fact remains that an unstoppable process has been set in motion and it will eventually result in the compartmentalisation of the different

---

[34] P. TOMSIN, 'L'Institut électrotechnique Montefiore à l'Université de Liège, des origines à la Seconde Guerre mondiale', in *La naissance de l'ingénieur électricien. Origine et développement des formations électrotechniques*, Paris, AHEF/PUF, 1997, pp. 221-232.

[35] W. KÖNIG, 'The Development of Electrical Engineering Education at the German *technische Hochschulen* before the First World War', in *La naissance de l'ingénieur électricien. Origine et développement des formations électrotechniques*, Paris, AHEF/PUF, 1997, pp. 241-247.

engineering categories. However, one should not conclude that at the end of the 19th century, the emergence of 'specialists' is intent on superseding 'generalists' in companies. The tasks of 'generalists' still remain essential to the production process and new industries hire them as well.

However, it is worth stressing that electricity is probably the first technological realm to be the exclusive province of the engineer. In all the other technical fields where engineers had worked so far, they had been latecomers so to speak: whether for public works, mechanics, metallurgy or the textile industry, there had always been before their advent a body of knowledge inherited from sometimes very ancient traditions, an instituted know-how, or time-honoured practices. In addition to their training at specialised schools they also had had to learn on the shop floor, in the mine or on the building-site, they had to rely on this body of knowledge, to come to the terms and sometimes to contend with it to make way for a new logic based on science. None of that holds for electricity: problems are new, experts must start from scratch and there are no pre-established skills. As soon as the new energy left the physics laboratory to enter the industrial world, engineers appropriated it and made it the true science of the engineer, solving unexpected problems, unpredicted by the physicists. It goes without saying that they still have to rely on the work of researchers, but this productive dialogue once and for all institutes the figure of the scientific engineer.

Finally, the fact that countries across Europe opened specialised education programmes almost simultaneously (in less than two decades) or sent national engineering students to be trained abroad in the new technology testifies to the rapidly universal character of electricity and the central role of engineers.[36] Even though only a few nations participated in the development of this new major innovation, the issue of its broad diffusion engaged engineers on an international level. As a matter of fact, even 'follower-countries' were just as involved as 'initiator-countries'. In Portugal, a poor country in

---

[36] According to W. KÖNIG ('The Development of Electrical Engineering Education at the German *technische Hochschulen* before the First World War'), there were an estimated 2500 foreign engineers trained in electro-technique in Germany before 1914, i.e. as many as native engineers. Shortly after being founded, electro-technique institutes in France would launch a massive recruitment campaign for foreign students, as a matter of fact, the Institut électrotechnique of Toulouse has more than 50% of foreigners in its student population.

Southern Europe, where the dominant ideology seemed to doom the country to semi-autarky and an economy based on agriculture and pastoralism, engineers picked up the issue of electrification and turned it into a real crusade in favour of industrial growth which led to true social development, more generally.[37]

Conclusion

In this article we have attempted to show the relevance of a re-reading of the modern and contemporary history of engineering in the light of a European perspective. If this option makes sense today, it is because the gradual establishment of a European economic and social entity, the creation of a European currency, the increasing number of merging companies, the emergence of European careers for a (still limited, of course) number of professional categories such as technical executives and managers, are all objective factors advocating the fostering of a collective identity for European engineers.

This will be a long and arduous endeavour that must involve an international harmonisation of education programmes, the possibility of compatible careers from one country to the other, or the awareness of a common historical substratum. And yet everything today is still viewed from a national perspective. For instance, although higher education does encourage international exchanges via, among others, European programmes such as the Erasmus programme, the issue of a European harmonisation of engineering degrees is still far from being settled. Besides, a European structure for the engineering labour market is still more of a goal yet to be achieved than a reality. With regard to this problem, the engineering corporate body still being reluctant to recognise those who had engineering functions without bearing the title certainly was a hurdle, at least in those countries where corporate structures like engineering associations are still very cautious on the issue. The division into two separate engineering categories in several States of the Union is also an impeding factor to the European unification of the engineering profession, and, a fortiori, for the emergence of a European identity for engineers.

[3] See M.D.L. RODRIGUES, 'Le génie électrotechnique au Portugal' in *La naissance de l'ingénieur électricien. Origine et développement des formations électrotechniques*, Paris, AHEF/PUF, 1997, pp. 285-318.

Historical science is first and foremost a way of questioning the past according to present challenges. From this point of view, the already incomplete historiography of engineering is still to a very large extent dependent on every country's own vision.

As a matter of fact, it is essential that certain mechanisms should first be articulated within familiar conceptual frameworks in order to envisage transnational comparisons and to establish an approach to the major scientific, technical and economic movements, which brought about far-reaching international changes in the education system, engineering know-how and activities. However, historical research on subjects like the engineering professional group must henceforth be conducted in terms of a European dimension. Similarly, in order to account for the role of these technologists, the historian should bear in mind that the engineer designs above all labour relationships rather than merely machines,[38] and that one should grasp the true meaning of technology: beyond an articulated body of rational knowledge on technological methods, it 'is the negotiated result of cultural, economic, political and social interactions.'[39]

References

ALDER, K., *Engineering the Revolution: Arms and Enlightenment in France, 1763-1815*, Princeton NJ, Princeton University Press, 1997.

AUCLAIR, A., *Les ingénieurs et l'équipement de la France. Eugène Flachat (1802-1873)*, Le Creusot-Montceau les Mines, Écomusée de la communauté urbaine, 1999.

BAUDET, J.C., *Les ingénieurs belges, de la machine à vapeur à l'an 2000*, Brussels, APPS Éd., 1986.

BLANCHARD, A., *Les ingénieurs du 'Roy' de Louis XIV à Louis XVI. Étude du corps des fortifications*, Montpellier, 1979.

BRION, R., 'La querelle des ingénieurs en Belgique' in A. GRELON (ed.), *Les ingénieurs de la crise*, Paris, Éditions de l'École des Hautes Études en Siences Sociales, 1986, pp. 255-270

BUGARINI, F., 'Ingegneri, Architetti, Geometri, la lunga marcia delle professionni tecniche' in W. TOUSIJN (a cura di), *Le libere professioni in Italia*, Bologna, Il mulino, 1987, pp. 305-335.

---

[38] A. GRAS, 'Anthropologie et philosophie de techniques. Le passé d'une illusion' in *Socio-anthropologie* (1998)3, 1st sem., p. 47.

[39] T. SHINN, 'Pillars of French Engineering' in *Social Studies of Science* 29(February 1999)1, p. 135.

CAPEL, H., J.E. SÁNCHEZ and O. MONCADA, *De Palas a Minerva. La formacíon científica y la estructura institucional de los ingenieros militares en el siglo XVIII*, Barcelona, Serbal/CSIC Publishers, 1988.
DAY, C.R., *Les écoles d'arts et métiers, XIXe-XXe siècles*, Paris, Belin, 1991.
DIDIER, C., 'L'aérotrain ou la tragédie de Jean Bertin', in C. DIDIER, A. GIREAU-GENEAUX, B. HÉRIARD DUBREUIL (eds.), *Éthique industrielle. Textes pour un débat*, Brussels, De Boeck, 1998, pp. 323-337.
DORN BROSE, E., *The Politics of Technological Change in Prussia*, Princeton NJ, Princeton University Press, 1993.
EFMERTOVÁ, M., 'L'évolution de l'enseignement technique tchèque aux XVIIIe et XIXe siècles' in *Quaderns d'història de l'enginyeria* 3(1999), pp. 51-82.
GARRABOU, R., *Engynyers industrials, modernizaciò ecònomica i burgesia a Catalunya (1850-inicis del segle XX)*, Barcelona, L'Avenç S.A., 1982.
GILLE, B., *Histoire des techniques*, Paris, Gallimard, 1978.
GILLE, B., *Les ingénieurs de la Renaissance*, Paris, Seuil, 1964.
GRAS, A., 'Anthropologie et philosophie de techniques. Le passé d'une illusion' in *Socio-anthropologie* (1998)3, 1st sem.
GUILLERME, J., 'Roland de la Platières et les patiences de l'inspection méthodique' in A. THÉPOT (ed.), *L'ingénieur dans la société française*, Paris, Éditions ouvrières, 1985.
HAMELIN, F., 'L'École d'application de l'artillerie et du génie et les cours industriels de la ville de Metz' in A. GRELON and F. BIRCK (eds.), *Des ingénieurs pour la Lorraine, XIXe-XXe siècles*, Metz, Éditions Serpenoise, 1998.
HARNOW, H., *Den danske Ingeniors historie 1850-1920*, Arhus, Systime A/S, 1998.
KÖNIG, W., 'The Development of Electrical Engineering Education at the German *technische Hochschulen* before the First World War', in *La naissance de l'ingénieur électricien. Origine et développement des formations électrotechniques*, Paris, AHEF/PUF, 1997, pp. 221-232.
LUNDGREEN, P., 'De l'école spéciale à l'université technique, étude sur l'histoire de l'école supérieure technique en Allemagne avant 1870' in *Culture technique*, (March 1984) 12, pp. 305-311.
LUTZ, B. and G. KAMMERER, *Das Ende des graduierten Ingenieurs?*, Frankfurt/ Köln, 1975.
PAQUIER, S., 'Les exemples contrastés de l'Ecole d'ingénieurs de Lausanne et de l'Ecole polytechnique fédérale de Zurich (1853-1914)' in A. GRELON, I. GOUZÉVITCH and A. KARVAR (eds.), *La formation des ingénieurs en perspective. Modèles de référence et réseaux de médiation, XIXe-XXe siècles*, Rennes, Presses universitaires de Rennes, 2001.
PICON, A. and M. YVON, *L'ingénieur artiste*, Paris, Presses de l'École Nationale des Ponts et Chaussées, 1989.

PICON, A., *L'invention de l'ingénieur moderne. L'École des Ponts et Chaussées, 1747-1751*, Paris, Presses de l'École nationale des ponts et chaussées, 1992.
PIERRE, P., 'Internationalisation de l'entreprise et socialisation professionnelle: étude des stratégies identitaires de cadres de l'industrie pétrolière' in D. GERRTISEN and D. MARTIN (eds.), *Effets et méfaits de la modernisation dans la crise*, Paris, Desclée de Brouwer, 1998, pp. 229-254.
RIERA I TUÈBOLS, S., 'Industrialisation and Technical Education in Spain, 1850-1914' in R. FOX and A. GUAGNINI (eds.), *Education, Technology and Industrial Performance in Europe, 1850-1939*, Cambridge, Cambridge University Press, Paris, Éditions de la Maison des Sciences de l'Homme, 1993.
RODRIGUES, M.D.L., 'Le génie électrotechnique au Portugal' in *La naissance de l'ingénieur électricien. Origine et développement des formations électrotechniques*, Paris, AHEF/PUF, 1997, pp. 285-318.
RODRIGUES, M.D.L., *Os Engenheiros en Portugal*, Oeiras, Celta Editori, 1999.
SEELY, B., 'European Contributions to American Engineering Education: Blending Old and New' in *Quaderns d'Historia de l'Enginyeria* 3(1999), pp. 25-50.
SHINN, T., 'Pillars of French Engineering' in *Social Studies of Science* 29(February 1999)1.
THÉPOT, A., *Les ingénieurs de Mines au XIXe siècle. Histoire d'un corps technique d'État*, Paris, Eska, 1998.
TOMSIN, P., 'L'Institut électrotechnique Montefiore à l'Université de Liège, des origines à la Seconde Guerre mondiale' in *La naissance de l'ingénieur électricien. Origine et développement des formations électrotechniques*, Paris, AHEF/PUF, 1997, pp. 221-232.
VÉRIN, H., 'Le mot: ingénieur' in *Culture Technique* 12 (March 1984).

1.1.2

TODAY'S ENGINEER MUST BE MORE THAN A TECHNICIAN

Corporate Reorganisation Processes, Trends in Modern Engineering and Professional Self-understanding of Technical Experts.

Helmuth Lange and André Städler

Since the beginning of the nineties, market globalisation and the soaring developments in information and communication technology have triggered off radical changes in the economic and management policy of business enterprises. Very often this has resulted in an upgrading of the qualifications demanded and it calls for new forms of collaboration. The work of technical experts too has evolved considerably. New professional requirements such as project work and teamwork, dealing with conflicts, project management, increased contacts with clients and suppliers and business-economic cost accounting make one thing very clear: nowadays an engineer must be more than a technician.

This contribution deals with the present professional modification in the work of engineers and physicists. The first part focuses on the changing job market for this occupational group. How is the employment situation for engineers evolving? The latter part investigates more closely into the new professional requirements. It presents the points of view of authoritative professional federations and reports the findings of new empirical studies of the professional and occupational situation of engineers. It demonstrates that the upgraded assignments clearly question part of the traditional professional role perception of engineers.

Company Reorganisation Processes
and the Professional Self-image of Technical Experts
– Developments in Modern Engineering Work

Following the globalisation of markets and the rapid pace of development in the information and communication technology sector, many commercial businesses have begun, since the early 1990s, to make radical changes in their company procedures. Modern company organisations and the new management concepts associated with them create a need for different forms of collaboration, and have led to a considerable increase in the qualifications that are required of many professional groups.

The work of technical experts has also undergone huge changes in the wake of company reorganisation processes. New professional requirements, such as project work and teamwork, increased contact with customers and suppliers and commercial cost accounting, make it clear that 'an engineer has to be more than a technician these days'. This essay will look at the professional shift which can be seen to be taking place among engineers, under the influence of modernisation processes in companies.

In *Section One* we will first take a brief look at the labour market for this professional group. The focus in *Section Two* will be on new professional challenges. Starting from statements by the key professional associations on this issue, the results of an empirical study of the working and professional situation of engineers will be presented. These show that changes in the organisation of work and the extension of the range of tasks within the context of company reorganisation processes have had a tremendous impact on the traditional understanding of the role of this professional group.

The Labour Market and the Employment Situation of Engineers in Germany

In Germany, engineers are one of the largest professional groups among all employees with higher education degrees (about 16%). From 1987 to 1993 this professional group increased in size from 445,000 to 657,000 employees. The proportion of women rose – first of all due to the national unification of the Federal Republic of Germany and the German Democratic Republic in 1990 – from just over

3% to about 8.7% during the same period.[1] In recent years the situation of engineers in the employment market has been dominated by the cyclical economic fluctuations in those sectors in which they are most often employed. The sectors most affected were mechanical engineering and the electrical sector.

Developments until 1997

From 1985 to 1991 the labour market for engineers in Germany tended to develop more favourably than for other academics, although even in this area there was a constant hard core of unemployment affecting about 20,000 people.[2] From 1991 onwards a clear increase in the unemployment figures was seen even in this professional group.[3] In 1993, the year when joint unemployment statistics for West and East Germany were published for the first time, the number of unemployed engineers doubled. Based on the current figure of 47,327 people unemployed (from technical institutions and universities), unemployment in this professional group then rose to 65,221 in 1997.[4] This crisis in the employment market had a negative effect on student figures in engineering disciplines. Since 1993 the number of students starting courses has been falling. The fall in the traditional disciplines of mechanical engineering and electrical engineering was particularly drastic.

In 1997 the employment market for engineers was characterised by very contrasting developments. Demand has increased sharply, but at the same time unemployment also continued to increase. A

---

[1] The figures for 1987 are based on the results of the micro-census in March 1987. Cf. *Wirtschaft und Statistik* 2/1989. The figures for 1993 are based on the results of the micro-census in April 1993. *Wirtschaft und Statistik* 11/1995. While the proportion of women in engineering professions in the former Federal Republic was always about 3% - 5% (depending on the area of specialisation), in the GDR the proportion of higher education graduates in technical sciences was usually 30% and even in 'hard' engineering subjects such as electrical engineering and mechanical engineering the figures were 15.5% and 21% respectively. Cf. K.-H.MINKS, *Beschäftigungs- und Weiterbildungssituation von Ingenieurinnen in den alten und neuen Ländern*, Hannover: HIS, 1994.

[2] Cf. VDI *survey on the demand for engineers* (VDI-*Studie Ingenieurbedarf*), Appendix 2, p. 11.

[3] In comparison with the previous month in 1991, the number of engineers with higher education degrees registered as unemployed rose by 15.3% in September 1992, and the level of unemployment among those with technical college degrees rose by 17.8%. (Data from: *Annual Report of the* ZAV *1997*, No. 26 dated 1.7.1998, pp. 2356f.)

[4] Data from the *Annual Report of the* ZAV *1997*, No. 26 dated 1.7.1998, pp. 2356f.

[5] *Ibid.*

clear East-West divide can be seen: while engineers in the West were able to benefit from increasing demand, the demand in the East fell back slightly. At the same time the rise in demand did not take place in all areas of specialisation or benefit all age groups. The increase in unemployment seen in this professional group from 1993 onwards mainly relates to older engineers in Eastern Germany, who find it almost impossible to get employment. Last year those aged over 45 years old made up almost two-thirds of all unemployed engineers in Eastern Germany. Measured against the number of all unemployed people educated to higher education standard, the percentage of unemployed engineers reached a sad peak in 1997 as compared with 1985: In that year 21.1% of all unemployed people with a university degree were engineers (1985 = 9%). Among unemployed people educated at technical college, one in two (48.3%) have degrees in engineering subjects (1985 = 32%).[5]

At the end of December 1997, 7,000 unfilled vacancies were reported to the Employment Offices: That was 2,200 more than the number at the same time in the previous year (+ 45%).[6] A survey by VDI-Nachrichten among the 100 German companies with the highest turnover in processing industries and the services sector showed that almost one-third of all companies employed additional engineers in 1997. There were particularly high growth rates in overall staff numbers in the communication media sector (DeTeMobil = +10.7%; Mannesmann Mobilfunk = +18.8%). Siemens, however, is still the largest employer of engineers. With 41,000 engineers, this company employs more engineers than any other German company (Daimler-Benz 26,000[7]; Deutsche Telekom 18,300; RWE AG 11,361). Despite these slightly improved labour market prospects for engineers, the general decline in personnel numbers in most companies continued unabated during the past year. Due to company restructuring and reorganisation measures, up to 10% of all jobs were lost in some companies.[8] Against this background, the increased demand for engineers which became clear in 1997 must be seen as a consequence of the powerful effects of rejuvenation at company level.

[6] Cf. *ibid.*, p. 2304.

[7] Current employee figures for this professional group after the merger in autumn 1998 between Daimler-Benz and the Chrysler group were not yet available.

[8] Survey by VDI-Nachrichten: 'Where do engineers work?' in *VDI-Nachrichten online* dated 28.7.98

[9] Cf. *Annual report of the ZAV 1997*, No. 26 dated 1.7.1998, p. 2305.

The Current Situation since 1998

Engineers with professional experience between the ages of 30 and 45, with considerable additional qualifications (particularly knowledge of computers, PPS** systems and foreign languages, and a personality which is up to companies' demanding standards in the area of 'soft skills' are particularly in demand. Engineers are mainly being sought for production and manufacturing (due to an improved economic climate, particularly in the areas of mechanical and electrical engineering) to manage the optimisation and adaptation of complicated manufacturing processes. Marketing and sales comes in second place. The demand in these areas of work is just as high as in construction. Two thirds of all job offers are in these three areas. The most rapid growth has taken place in customer service. In this area companies were looking for twice as many engineers as last year.[9] The trend described here also continued into the first quarter of 1999. An analysis of vacancies for technical specialists and managers in 35 German publications shows that in comparison with the previous year, 48% (!) more jobs are on offer for people with technical degrees. Looking at the individual sectors, the computer sector comes in first place, mechanical engineering is in second place and the electrical sector comes in third.[10]

Consequences

One traditional characteristic of the labour market and employment situation for engineers in Germany was and is its relatively pronounced dependence on the economic climate in individual employment sectors. This is quite different from the situation in other academic professions. In the past the result was that, on the one hand, the number of unemployed engineers has always followed the cyclical fluctuations in the economic climate, and that on the other hand the medium to long-term demand for engineers could not be forecast in either qualitative or quantitative terms. Although the current labour market data, beginning in 1997, show an unexpectedly positive development for technical specialists and managers, the available data on the engineering profession since the beginning of the 1990s (trends: increase and stabilisation in unemployment at a high level

---

[10] *VDI-Nachrichten online*, media data from 11.5.99 and VDI press release, 'Boom für technische Fach- und Führungskräfte', dated 21.4.99.

for this professional group, more frequent early retirement among older engineers, a dramatic reduction in those starting courses as students) appears as 'a fracture in the professional situation of engineers during the post-war period, which has – subject to economic fluctuations – been stable for decades.[11] What is more, the current favourable situation in the labour market should not deceive anyone into forgetting that the profile of professional requirements for engineers and natural scientists has gradually changed. Thus the leader of the VDI key group 'The Engineer at Work and in Society', Karl-Heinz Simsheuser, observes:

> Those who make efforts even while studying to acquire wide-ranging basic knowledge of scientific and business contexts, learn foreign languages and acquire communication skills in both technical and human domains, will have good opportunities.[12]

Professional Self-image and Structural Changes in Companies – Work Experience of Technical Experts in Reorganisation Processes

The changes in the labour market described above, and the organisational changes at company level have led to further movement in the debate on the education and professional situation of engineers. Even conservative representatives of the engineering profession and professional associations have been talking for some time about the need for a radical change in the traditional professional self-image of this professional group. Hence the German Engineers Association [Verband Deutscher Ingenieure – VDI] writes:

> In order to cope with structural challenges, companies are innovating at an accelerated rate, using production methods that increase quality and reduce costs, and providing better services. To a significant extent they are doing this by reorganising existing management and labour structures in companies and industries. (...) This reorientation of working and decision-making processes, which can

---

[11] See for example W. NEEF, 'Paradigmenwechsel in Beruf und Ausbildung von Ingenieuren' in FRICKE, *Innovation in Technik, Wissenschaft und Gesellschaft, Forum Humane Technikgestaltung*, Vol. 19, Bonn, Werner, 1998, p. 326.

[12] Quoted from VDI press release, 'Boom für technische Fach- und Führungskräfte' dated 21.4.99

be seen in companies everywhere, also requires engineers to change their qualification profile and professional self-image[13]

Image of the Profession, Engineering Training and Professional Self-image

The traditional professional image of engineers in Germany is dominated by specialised understanding, which more or less reduces technology 'to applied physics' (Ropohl 1997). One consequence of the dominant understanding of technology is that the areas of social interaction involved in the creation and use of technical solutions are largely omitted even from training courses. The structure of German engineering courses, with the institutional division between universities and technical colleges, has been oriented mainly towards the needs of large companies with their rigid hierarchies and structures with sharply functional structures based on the division of labour. In the past, through the many years of stability in the working and professional situation it was sufficient, in order to cope successfully with the tasks falling to an engineer, to extend the basic knowledge acquired at college by means of continuing training measures. Due to the growing complexity of engineering tasks in recent years, (both in technical terms but above all also in their social, societal and political dimensions), and against the background of structural changes in companies (in response to the new situation of international competition), the traditional higher education course and the mainly technology-centred profession-specific self-image of this professional group which it promoted is now proving to be increasingly dysfunctional.[14]

---

[13] VDI, *Ingenieurausbildung im Umbruch, Empfehlungen des VDI für eine zukunftsorientierte Ingenieurqualifikation*, Düsseldorf, 1995, p. 2. A large number of activities can also now be seen in Germany which are helping to evaluate German engineering education in the face of the challenges that arise due to increasing internationalisation (Key: attempts to achieve compatibility between qualifications from German degree courses and the Anglo-American Bachelors and Masters courses). On this subject cf. VDI, *Thesen zur Weiterentwicklung der Ingenieurausbildung in Deutschland*, May 1998, and VDI, *Recommendation of the VDI on the accreditation of Bachelor and Master degree courses in engineering*, November 1998.

[14] A summary of the debate on educational reform can be found in: *Ingenieurinnen und Ingenieure für die Zukunft. Aktuelle Entwicklungen von Ingenieurarbeit und Ingenieurausbildung*, Zentraleinrichtung Kooperation der Technischen Universität Berlin, Berlin, Neef/Pelz, 1997.

How can we describe the professional change currently taking place among engineers and the changes in their working situation? What new professional challenges are facing this professional group within the context of company reorganisation processes, and how are they being managed? These questions are at the centre of an empirical study being carried out at the Employment-Environment-Technology Research Centre of the University of Bremen since the end of 1996.[15] The main aim of this project is to research new areas of conflict which are arising through the restructuring – in terms of standards and actual practices – of the professional activity of specialised scientific and technical workers, and to look at different ways of handling this problem, both subjectively and through company policy. In particular this study focuses on the difficulties and opportunities which arise due to, during and as a result of serious practical efforts at reorganisation by companies. It therefore looks at changes that take place during a process of development over a period of time. The initial results show clearly how difficult the way to a new professional self-image in modernised working environments will be.

*Modern Company Organisation,
the Organisation of Work and New Professional Challenges*

Despite the diversity of specific forms taken by the many different concepts of 're-engineering', virtually all of them have the following in common: On the one side, more or less well-defined series of tasks are shifted *outside*, and on the other hand an attempt is made to interconnect the working processes that remain within the company *more closely*. Here the following individual *areas of development in reorganisation* can be identified:

1. During the process of 'slimming down' the company ('lean management') hierarchical stages are abolished and project organisations are implemented on a permanent or temporary basis in parallel with the traditional line organisation.

---

[15] Expert interviews were conducted in five small or medium-sized companies and five large companies in the metal processing industry, the electronics sector and the services sector, involving 52 engineers and physicists, and also including the Works Councils and personnel managers/business managers of the companies. The study was structured as a continuous study. The first interviews were conducted during the fourth quarter of 1997. The second round of interviews with the same people are currently taking place.

2. Individual cost-intensive parts of the business which are not part of the 'core business' are selectively removed ('outsourcing'). At the same time, a reduction in production-based indirect divisions and central services takes place.

3. In order to increase cost transparency within the business, the remaining divisions are restructured as internally independent business units with their own accounts ('profit and cost centre' organisation).

4. Alongside these structural changes, efforts are also made to achieve a far-reaching decentralisation of responsibility and competence, which on the one hand creates the need for a redefinition of internal co-operation relationships between individual business divisions (internal customer-supplier relationships) and at the same time entails more intensive project work and teamwork in order to deal with the required tasks.

5. These changes are almost always accompanied by massive staff cuts.

For technical experts, the result of the changes outlined here is that the required repertoire of qualifications is being radically changed, both in terms of specific *specialised knowledge* and also – which is the actual core of the professional change – with regard to *knowledge and skills independent of their specialisation.*

In terms of *specialised skills,* much more extensive *methodological and system skills* are expected than was previously the case. Particularly important today are knowledge in the areas of market observation, steering product innovation processes, software technology, quality assurance and technology evaluation. A high level of system competence is required in order to contribute effectively towards bringing partial solutions together into a complex finished product while taking relevant interfaces and interactions with other specialised technical areas into account (for example in the integration of the mechanical, electronic and software components of a technical system).[16] One project engineer describes these new challenges as follows:

> In our company there is a shortage of system people, people who can gain an overview of a system (...) break it down into different subsystems and formulate and specify them in such a way that the

[16] Cf. VDE/ZVEI, *Auswirkungen des Strukturwandels in der Elektroindustrie auf die Ingenieurausbildung,* Frankfurt a.M. 1994.

individuals can take them on and deal with them. Such people are in short supply! The old way of working (where a person does five or six different things at once) is no longer possible. (...) So the process of structuring into individual tasks and thinking about the whole system (so that it can be integrated), that is what is missing[17]

The new emphasis on *specialisation-independent, non-technical requirements* is mainly the result of companies' efforts to cut costs and internationalise their business and efforts to achieve greater integration and improved coöperation within the company. In order to cope with the new challenges, engineers now need more knowledge of foreign languages and experience of other countries, much better knowledge of business contexts, extensive organisational knowledge and good social skills.[18] A sales engineer on this subject:

> Yes, even the builder has been forced to open up more. In the past the salesman went out, discussed all the details with the customer and brought the details back. Now it is not the salesperson who worries about those details, it is the builder. (...) The builder drives over to see the customer and discusses it with the detail man there. Communication is not triangular any more, it is direct. That means the builders have to open up more as well. They must not be afraid of travelling to foreign countries. They must not be afraid of meeting people face-to-face and explaining their point of view. That is the big change today.[19]

Overall it is possible to say: 'The importance of purely technical knowledge and skills is moving into the background in many fields of activity, as the range of tasks becomes wider'. As regards the content of the work, specialisation-independent, non-technical qualifications are dominating the picture in the professional changes currently taking place. Teamwork and project work, as well as regular contact with customers, external partner companies and suppliers, require high-level communication skills. What is more, the ability to be flexible and the personal willingness to change increasingly seem to be the basic precondition for handling tasks at work successfully. Flexibility is required both in terms of specialised content (getting to know new issues and tasks, 'lifelong' learning) and also in terms of

---

[17] I 11, Project manager, large company, line 750ff.
[18] Cf. also VDI, *Ingenieurausbildung im Umbruch, Empfehlungen des VDI für eine zukunftsorientierte Ingenieurqualifikation*, Düsseldorf 1995
[19] I 47, Sales engineer, small business, line 758ff.

the willingness to put comparatively tried and tested things (working styles and routines) on trial within a context of restructuring the organisation of work (with a move towards teamwork and project work). The change in professional requirements is therefore mainly characterised by an increase in external relationships and hence by a departure from the technology-centred view of engineers' activities, professional function and position in the company.

New Freedoms vs. New Demands:
An Ambivalent Assessment of Company Modernisation Processes

How do technical experts see the necessary changes, under pressure from processes of change within companies? It is evident that their evaluation of the modernisation processes is ambivalent. Positive aspects are offset by a series of new demands and conflict situations. In the next section the working experiences of the engineers who were interviewed will be illustrated through a few selected accounts.

Less hierarchy, more individual responsibility

> The whole structure as it was before, with all the different positions, from the business manager down to the assistants etc. All that has been flattened out. (...). Direct lines! The business manager has said: You are responsible for your own division and responsible to the business division manager.[20]

Greater transparency, shorter lines of decision

> We are actually quite satisfied with our new management. They have also started to give us more information than used to be the case. Everything was always a secret (...). It was like: My work is so secret that even I don't know what I am doing any more. Employees were not given any information at all ... that has all changed now. Things have become significantly better, yes.[21]

More room for manoeuvre, more freedom to make decisions

> You can (...) also see it within the company. The idea of someone in my position having opportunities, it would not have been like that in the past, because you didn't have a position in the hierarchy. I am just an assistant, but now I get involved with projects, which simply

---

[20] I 23, Project manager, large company, line 143ff.
[21] I 39, Departmental manager, small business, line 730 ff.

was not possible before. This freedom to make my own decisions is great.[22]

Shortage of time and pressure from deadlines

What do I miss? I miss never getting an hour to concentrate on anything any more. I do miss that. You have to make a lot of decisions relatively quickly. It's not too bad here this morning. Normally it's like queuing for the bathroom. You have one deadline chasing another. I do miss that sometimes. Being able to concentrate on something for a while and make a decision on a rather more solid basis. I will, cautiously, say that.[23]

Cost pressure and heavy workload

We were under tremendous pressure to get our costs down in a very short space of time. In the end that had an impact on everything. Above all it (affected) the employees. We did not recruit any new staff for years. (...) The other thing is: We are already struggling to cope with a very high density of work (...) We have these shorter development times. Things move fast. That is not how it used to be. Everything we do here is much more finely tuned.[24]

Erosion of the boundaries between private and professional life

'37.5 hours a week'. That's what it says on paper, yes. (...) I have a PC at home as well, and my briefcase is full of disks. So you take it home with you. Also there are some things that really take time: if the telephone is not ringing like it is now (...), you have to rush something through quickly. For example, take a calculation. You can do that some time during the weekend. In fact it makes no difference whether I do overtime here or whether I do it at home. Except that I can concentrate on my work at home. The whole throughput time for this work is less than half what it would be if you did it here.[25]

Personnel cuts and loss of know-how

They also wanted to reduce the number of staff (...). So they started making a real effort to send people home early. It went right down to (people aged) 55. (...) The resulting problem was that we lost a huge amount of experience. We had to pay for that in the end,

---

[22] I 21, Sales engineer, large company, line 861ff.
[23] I 16, Centre manager, large company, line 301ff.
[24] I 6, Team leader, large company, line 204ff.
[25] I 43, Project engineer, small business, line 634ff.

> because all those things (...) which are there in people's heads somewhere, they are all gone. Not much of it was documented. In those days there just wasn't time for that. Basically you have to start again from scratch now. And with my six years' experience I am actually already one of the oldest people in our company. The average in our field is early to mid-thirties. So we are a very, very young crowd. We even have master engineers who are only about thirty, which does cause a bit of a problem because somehow there isn't that father figure who can guide things, keep the troops calm, you know.[26]

Unstable conditions and confidence to take action

> All I can do is hope that it will all come to an end one day. That we will move into another stage where there is some continuity. We haven't got that at the moment. (...). The company has to calm down at some point. If I keep everything unstable like this, at some point the whole thing will collapse. I must achieve some continuity and calm, at least for a time. (...) I think it is in people's nature that they need to be able to see: a light at the end of the tunnel. If you just keep on going round in circles in the dark, and you never calm down, that's not really how it should be. That is not good for the business. At some point it will flip over, and you have to be careful not to miss that moment when it comes.[27]

The removal of hierarchical levels within companies basically gets a positive reception. Decision-making lines are becoming shorter. The number of checks, which were seen as a burden, is falling. It seems that areas of responsibility and authority are more clearly distributed, due to the delegation of responsibilities.

The demand from companies that people should take more responsibility themselves is well received within the profession. The extension of the range of tasks creates new opportunities to give engineers more room for manoeuvre. Particularly for younger engineers, it gives them an unaccustomed freedom to make decisions. The new organisation of work and project structures make it possible to achieve greater transparency and gain a better overview of the whole product creation process. Most of those we spoke to welcome this as an 'extension of their horizons'.

---

[26] I 23, Commercial engineer, large company, line 69ff.
[27] I 51, Centre manager, large company, line 680ff.

On the other hand, people also referred to new burdens, which one interviewee referred to as the 'dark side' of the working situation of engineers in modern company organisations. As well as relentless time-pressure and cost pressure, they also referred to the generally increased burden of work due to shortages of personnel following restructuring measures.

In most companies there has been a tangible deterioration in the working atmosphere due to what are often quite dramatic staff cuts. Finally, people regret the disappearance of 'private life into professional life' due to greater time pressure and deadlines. The increased intensity of work leads to an erosion of the boundaries between private and professional life. 'Switching off at the end of the day' was causing difficulties for most of the interviewees. In almost all the companies surveyed, the issue of 'overtime and working hours' was a key area of conflict.

There was also criticism of the 'loss of know-how' due to 'outsourcing' individual parts of the business. The introduction of profit centre structures has, in many cases, been accompanied by difficulties in the setting up of 'customer-supplier relationships' within the company. The idea of 'bringing the market into' the company has quite often turned former colleagues into competitors and created new barriers to co-operation between individual centres.

There was criticism – particularly among younger engineers – of the fact that flatter hierarchies make their career or promotion prospects much more unclear. Companies' performance assessment models and bonus systems do not seem adequate to cope with the new organisational structures. Repeated changes in modernisation concepts and managers are also seen as a considerable burden. Frequent restructuring also leads to a great deal of uncertainty about whether or not people will manage to keep their own jobs.

Are Engineers Moving towards a New Professional Self-image?

Two-thirds of the companies in the survey have already undergone one or two 'waves of restructuring'. It can therefore be assumed that the problems outlined here are not simply due to a brief period of 'turbulence' that will be followed by a long period of calm. On the contrary, business managers confirm that constant change will continue to be the normal situation in the future as well. The willingness and ability to be flexible and adapt all the time (in both organisational and individual ways) could therefore become even more important in future.

How do engineers cope with the adjustments that they are already required to make? In comparison with what they have been taught during training, and the type of professional challenges that existed before the reorganisation, a considerable shift is taking place for technical experts in the direction of 'multiple-skilling'. The interviewees themselves did, however, say that they are coping successfully with the changes they are required to make, despite the problems outlined above, and most of them maintained that they had not lost the 'pleasure of working'. Reasons for the high level of job satisfaction encountered here must be sought in two factors:

1. The interviewees are among the 'winners' from the changes, despite the new demands which they referred to. They have every reason to be happy, considering what have in most cases been radical staff cuts in businesses (between 20% and 70%).
2. The willingness to change is seen by most of the engineers surveyed as a central requirement of their everyday professional life. Within this context they mainly see the change as enriching their work (even though it often 'takes some getting used to').

In the current process of reorientation, engineers see themselves first of all as 'independent players', who are mostly given sufficient scope to take action and make decisions, and less as 'victims' of company rationalisation processes. It is true that a purely technically oriented understanding of their own activity only covers a small proportion of the tasks that they have to carry out. On the basis of what we have found here, however, it would not be appropriate to speak of a 'deprofessionalisation' of engineers in the sense of any loss of the 'specialised, identity-creating core of qualified professional work.'[28] It seems that young people do not find the required changes as difficult to cope with as older people, and that all those who have been through more 'modern' courses (e.g. business engineers) find it comparatively easier to cope with the new tasks than those whose image of the profession is defined by a traditional technical course (e.g. mechanical engineering).

---

[28] The words of U. KADRITZKE, in his essay 'Das berufliche Selbstverständnis von Ingenieuren – und die Realität, auf die es trifft' in *Ingenieurinnen und Ingenieure für die Zukunft. Aktuelle Entwicklungen von Ingenieurarbeit und Ingenieurausbildung*, Berlin, Neef/Pelz, 1995, pp. 71-85.

All in all there are many indicators that a far-reaching restructuring of the professional activities of engineers really is taking place. The focus on the product being manufactured, which was mentioned by many of the interviewees, seems to build a kind of bridge that can help them to move from the 'purely object-oriented' focus in the traditional professional self-understanding of engineers and make progress towards the new demands. This may prove to be a helpful resource in coping with the new challenges, as long as it does not reject the economical, social and communicative aspects of the work but – in an engineer's usual sober, results-oriented way – seeks to integrate them in engineers' technical development work. Under these conditions it may also help to create a new self-understanding of the profession which has been adapted in line with the changing task structure. One development engineer in a small company put it like this: *'Overall though, when we are starting up the preliminary series this cost control actually only means maintaining deadline control and cost control together. That is my main task. Of course there is an intellectual link somewhere. But perhaps there may not be, that is the nature of the work. Once you have worked alongside the people to build it (the product), you want it to get to the market so that you can make something out of it. So that is also a very important challenge for me.'*[29]

This rather optimistic perspective should not, however, disguise the fact that individuals' ability and willingness to change is only part of the problem. Whether individual efforts at qualification will result in new co-operation routines, largely depends on factors outside individuals' control. The very rapid change in conditions both within and outside the company is particularly important here. One factor which will play a significant part is the extent to which the personal efforts of staff are rewarded or devalued and disappointed. In many interviews it became clear that one key problem in this connection is a characteristic lack of simultaneity in the processes of change that have to be dealt with: The faster and more radical the changes in conditions (and the shorter the periods of calm and stability in the company), the more difficult it becomes for employees, through their own individual efforts, to adapt and keep up with the external changes.

---

[29] I 34, Development engineer, small business, line 544ff

References

Federal Labour Institute [Bundesanstalt für Arbeit], *Information for Employees and Workers in Engineering Professions*, Volume 25, 1996.
INSTITUTE FOR LABOUR MARKET AND PROFESSIONAL RESEARCH OF THE FEDERAL LABOUR INSTITUTE, MatAB 1.1/1998 (Volume on the theme of engineering subjects).
KADRITZKE, U., 'Das berufliche Selbstverständnis von Ingenieuren – und die Realität auf die es trifft' in *Ingenieurinnen und Ingenieure für die Zukunft. Aktuelle Entwicklungen von Ingenieurarbeit und Ingenieurausbildung*, Berlin, Neef/Pelz, 1995, pp. 71-84.
MINKS, K.-H., *Beschäftigungs- und Weiterbildungssituation von Ingenieurinnen in den alten und neuen Ländern*, Hannover, HIS, 1994.
Ingenieurinnen und Ingenieure für die Zukunft. Aktuelle Entwicklungen von Ingenieurarbeit und Ingenieurausbildung, Berlin: Neef/Pelz, 1997.
NEEF, W., 'Paradigmenwechsel in Beruf und Ausbildung von Ingenieuren' in FRICKE, *Innovation in Technik, Wissenschaft und Gesellschaft, Forum humane Technikgestaltung*, vol. 19, Bonn, Werner, 1998, p. 326.
ROPOHL, G., 'Das neue Paradigma in den Technikwissenschaften' in *Ingenieurinnen und Ingenieure für die Zukunft. Aktuelle Entwicklungen von Ingenieurarbeit und Ingenieurausbildung*, Berlin, Neef/Pelz, 1997, pp. 11-16.
FEDERAL STATISTICS OFFICE (Pub.), *Wirtschaft und Statistik* 2/1989, results of the micro-census in March 1987.
FEDERAL STATISTICS OFFICE (Pub.), *Wirtschaft und Statistik* 11/1995, results of the micro-census in April 1993.
VDE/ZVEI, *Auswirkungen des Strukturwandels in der Elektroindustrie auf die Ingenieurausbildung*, Frankfurt a.M. 1994.
VDI, *Ingenieurausbildung im Umbruch, Empfehlungen des VDI für eine zukunftsorientierte Ingenieurqualifikation*, Düsseldorf, 1995.
VDI, *VDI study on the Demand for Engineers*, 1997.
VDI, *Ideas for the Development of Training for Engineers in Germany*, Düsseldorf, 1998.
VDI, *Recommendation of the VDI on the Accreditation of Bachelor and Master Degree Courses in Engineering*, November 1998.
VDI. VDI press release: 'Boom für technische Fach- und Führungskräfte' dated 21.4.99.
VDI-Nachrichten, 'Where do Engineers Work?' in *VDI-Nachrichten online* dated 28.7.98.
VDI-Nachrichten online (1999), Media information dated 11.5.99.
CENTRE FOR EMPLOYMENT MEDIATION, CENTRE FOR LABOUR MARKET INFORMATION, *Annual Report of the ZAV 1997*, No. 26/1998.

1.2

EXAMPLES

1.2.1

VIRTUAL GAMES
INVITING REAL ETHICAL QUESTIONS

Bernard Reber

In our search for an ethics applied to emergent technologies we shall examine a role-playing game, MUME,¹ which is played in networks on the Internet. It is part of the heterogeneous whole of New Technologies for Information and Communication (NTIC).
One of the engineers who conceived the game started worrying and submitted some questions which became the starting point for the following ethical problemization. We shall see that these new technologies point out the limitations of a classical ethics that treats technical knowledge as a mere instrument. It is our relationship with the world and with the others which is modified, and therefore our concept of mastery.

Games play a prominent role in the boom of computer technologies. They were among the first products to appear on the Internet. One has to keep in mind that in France² games are the most important motivation for buying a PC. Ever more exacting and greedy for memory, these computer games participate in the rapid obsolescence of computer material. Indeed, the capacity of information technology

---

[1] Multi User in the Middle Earth, http://mume.pvv.org
[2] Survey carried out by the SELL (Syndicat des éditeurs de logiciels de loisirs). In this survey, carried out in 1997, 69% of the respondents said they bought a PC to play games at home. Of these, 35% said they bought games for children, 34% for adults.
When asked what the computer is used for at home, 74% answered it was used for playing games. Divided proportionally, this time was: 29% for children's games, 45% for adult games.
Report by Claude HURIET, 'Images de synthèse et monde virtuel: techniques et enjeux de société', OPECST, Sénat, N° 169, Assemblée Nationale, N° 526, December 11, 1997, p. 56.

increases at a dizzying pace, and players are often prepared to buy the latest version or even a more powerful PC to make their game go. The obsolescence is programmed. Interactions have to be ever quicker and images more precise to keep up the illusion of reality. Strangely enough, virtual games have reality as their standard. They are even sold as being 'more real than reality' by means of visual, kinetic and sound effects that create the illusion of 'hyper-realism'.

Fiction is often the source of the most poignant questions concerning our reality. By the novel experiences they create, virtual and computer games very quickly raise crucial ethical and metaphysical questions. We shall see that they cast considerable doubt on our instrumental approach to technical objects.

Our case study will be MUME, a role-playing game mediatized by the Internet, which offers a new concept of role-playing games, be it on the table or large as life. The game has been going on for eight years. The formulation of the problem as it is exposed here wants to be more than a mere theoretical summary.

On the one hand, one of the engineers who conceived MUME has stimulated the ethical and anthropological reflection which follows. The questions we withheld deal with role-playing games and the involvement of the participants via Internet mediatized interactions. These new practices touch upon our very perception of the world.

On the other hand, it considers and analyses a report destined for the French Members of Parliament, and which deals with an ethics of virtuality.

MUME, The Lord of the Rings *on the Internet*.

While many questions concerning the consistency of the virtual world remain unanswered, we shall at present investigate a concrete case of immersion in a virtual world: MUME. The game is based on a client/server model. It allows one to enter into a complex interactive universe, more lively than the ordinary electronic babble called 'chat'.[3]

MUME is a MUD, *Multi-User Dungeon* or *Multi-User Dimension*, viz. a multi-user Internet role-playing game. In these games, each partic-

---

[3] The chat group, originating from the English term for idle, unimportant talk, is a variation on the electronic forums in which the participants dialogue in real time. It concerns a means of synchronic communication in which the whole world sends and receives messages in one hectic discussion among one another.

ipant creates a character which can subsequently grow in an artificial world, inspired by the film *Star Trek* or by the *Knights of the Round Table* for instance. The characters can shift things, move, read documents. They can talk to other characters, travel, collaborate or fight. MUDs have engendered over a thousand more or less sophisticated variants. Through these characters, such simulation games offer new ways of entertaining relationships, of coping with novel situations and of exploring certain aspects of one's personality. It follows that these games are an excellent observation ground for finding out what goes on in cyberspace.

The MUME game we discuss here was developed in the Ecole Polytechnique Federale of Lausanne in Switzerland towards the end of 1993. It is based on the fabulous and imaginary world of *The Lord of the Rings* by J.R.R. Tolkien.[4] It is inspired by this novel and tries to reproduce its cartography as delicately as possible. It wishes to stay close to the spirit of these adventures.

It is related to the role-playing games in that the player 'steers' a character in a virtual world. Yet it offers a closer interaction between the player and the imaginary world of *The Lord of the Rings* since there is no master of ceremonies who interferes. In a classical role-playing game, played on a table, the master of ceremonies distributes the roles, introduces new events and creates the world, the framework within which the characters move. On the basis of the indications the master of ceremonies furnishes, the players imagine a world and propose actions which the master of ceremonies accepts or rejects according to what is possible and what is not. For instance, the master of ceremonies will avoid what would be anachronistic in a classical temporality, or incongruities as to profile and qualities of each of the characters. In the case of a life-size role-playing game, immersion is much more important. Players may take part in such a game in a castle for a whole weekend.

The circumstances are no longer the same with MUME. Indeed, the players do not see their master of ceremonies any more. This already poses certain problems as to the social interactions: the master of ceremonies is invisible, cannot negotiate or interpret by judging the mien of the players. The latter can play for a much longer period. As

---

[4] J.J.R. TOLKIEN, *The Lord of the Rings*, George Allen and Unwin Ltd., 1966. Ever since the publishing of its French translation, the editor proposed to use the novel as a basis for a role-playing game.

a matter of fact, they do not even have to be in the same place for the game to have its course. These elements, which modify the relations with space and time, hence with the body, contribute to a more considerable commitment. MUME is based solely on a textual interface, which *a priori* appears to be a drawback in terms of virtual reality. However, a survey conducted among players demonstrates that they consider the game to be realistic on the social level, and that they do not in the least regard the textual interface as an obstacle to the sensation of virtuality, nor to immersion. The game, moreover, has the advantage of having existed for eight years now, which makes it more than a mere technological curiosity.

At the outset each player has a character at his disposal which will represent him within the world of the game. Practically speaking, when one visits the site[5] for the first time, one first of all has to choose a name for one's character and a password to protect the access. Then one is asked to choose between Human, Dwarf, Elf or Half-elf,[6] and after that a class or profession such as warrior, wizard, cleric or thief.[7] Subsequently, one chooses a profile or a moral alignment.[8] One then finds oneself in a room with instructions one has to read carefully. If one has chosen a human character, one can type *pray fornost* to enter directly into the world of MUME. From then on the players are left to themselves. They can set out for adventure, explore the world of Tolkien, communicate with other players, either to all of them[9] or to a selection[10] of interlocutors. They can then interact with other players by communicating with them in a more or less selective manner, by performing 'social' or civic actions such as smiling or greeting. By joining up with other players, they can take part in their adventures, exchange objects or fight. Should they want to, the players still have the opportunity to modify their appearance by reformulating the description of their character.

---

[5] http://www.mudconnect.com/frameless.html

[6] This list of what the organizers have quite regrettably termed 'race', is not exhaustive. It is surprising and debatable that one does not choose one's sex in this game.

[7] The inventors of the game recommend the beginner to opt for 'human warrior'.

[8] For this choice, the inventors recommend 'undecided' for a first time.

[9] By typing say he talks to all the persons that are with him in the same room.

[10] By typing tell.

Internal Socio-psychological Survey

Today's debate on role-playing games and on video games, i.e. on games similar to MUME, is often centred on the opposition between supporters of two theses: there are those who say that the players do make a clear distinction between reality and fiction (with the exception perhaps of those few 'psychologically fragile' persons); others hold the opposite view and contend that such experiences muddle the borderline between reality and fiction, if indeed they do not incite to bad deeds.[11] Before attacking the question of commitment we shall present here in short the results obtained from a survey conducted among the players of MUME.[12]

This survey was conducted by one of the makers of the game who is at the same time its administrator.[13] Without entering into the details of a complete sociological study, we propose to point out some characteristics encountered in the population of the game.

The majority of the players are of the male sex (96%). 71% of the players are bachelors, 15% are living together, 10% are married and 1% divorced. 44% of the people are students, 35% are employees. The major part of the players come from the United States (39%), 20% from Sweden. The lion's share of the players belong to the group that has been in the game for more than 4 years, and devote an average of 18 hours a week to it.

A poll that is being effectuated at the moment among the players reveals that a majority of them recognizes that MUME has a considerable impact on their private and their social life. Almost all of them realize that the fact of playing this game can have a baleful influence on real life and on the way of establishing social contacts. To the question: 'do you think it would be a good idea to call into being an official organisation to aid players who are suffering from psychological troubles?' a vast majority responded in the affirmative.[14]

---

[11] The debate is not a new one; it is at least as old as the case against Flaubert after the publication of Madame Bovary which, according to its detractors, set the example for adultery and suicide.

[12] http://mume.pvv.org/answers-95.09.html. Remember that this is an internal survey, but that its results have not been contradicted by other European surveys which had a wider scope.

[13] It reproduces the statistics derived from a questionnaire which was returned by 308 of the players, and which gives us an idea of the population 'inhabiting' that universe.

[14] Certain players know each other indeed, and have grown interested in the

These elements allow us to suppose that the impact and the consistency of a virtual world on individuals can be considerable. If at present such manifestations remain confined to a limited group of people, one may well imagine that technical developments and their diffusion among a large public are likely to have more important social consequences. Through the immersion of its players in a virtual reality, MUME allows the transposition of ambitions and dreams one may have in reality. One has for instance the possibility of being God, or of having several different personalities, since one player can embody more than one character.

A Moral Game

This interactive game is an extraordinary example of the revelation of a morality contained in an object. We are well aware of the fact that we are defending a quite uncommon thesis here. And yet it does nothing but take seriously the theses of Heidegger and Simondon[15] on technical tools. Indeed, against those who plead the cause of an exclusively instrumental approach of objects, which is the only approach users take into account for a moral evaluation, we defend the thesis according to which objects do have a morality. An anti-theft device attached to a hotel key, for instance, tells us: 'Do not steal this key.' A safety belt in a Japanese car tells us more clearly than any moral injunction: 'Protect yourself.' Indeed, the vehicle will not start as long as the driver has not fastened his belt. The problem of the weakness of the will has thus been avoided.

game together, but how can one verify whether a player alone before his screen is not beginning to 'waver', and to identify too much with his character?

[15] For instance, at the beginning of one of his essays Simondon writes: 'The present study is inspired by the intention of awaking an awareness of the meaning of technical objects. Culture has become the self-declared system of defence against technics; this defence presents itself as a defence of man in that it supposes technical objects not to contain a human reality. We wanted to show that culture ignores a human reality in the technical reality, and that, in order to play its role to the full, culture has to incorporate the technical beings in the form of knowledge and value judgement.' *Du mode d'existence des objets techniques*, Aubier, p. 9.

This essay, which was published simultaneously with his thesis on individuation, incriminates the scheme borrowed from the instrumental technical approach and applied to human beings by many philosophers in the Greek Platonic or Aristotelian tradition.

It is this whole line of thought which has great influence on the ontology which

As we have said, before beginning MUME the player chooses a moral profile. The system of justice was added after the public opening of the game. It does not reflect any ethical concern, but merely represents a solution to a practical problem. As a matter of fact, its purpose was to protect the new players from the veterans who systematically eliminated them as soon as they entered into the game.[16]

And yet ethics might raise some criticism. Indeed, the moral system of MUME, as manichaean as most games, divides the characters in two categories: the good and the evil. These values are established by means of a binary system: killing good monsters or players lowers the moral positioning into negative values, whereas killing evil monsters makes the position go up again.[17]

Ethical criticism seems easier in the case of games like *Carmageddon I*, which is a kind of car race in which one is attributed bonus points for running over the blind man who crosses the street, or *Doom*, which gives the player the possibility to enter a maternity ward and kill all the babies present there. Instigating acts which the common moral disapproves of in reality seems condemnable, even though the makers quote the cathartic virtue Aristotle attributed to the theatre: it would be a way of deviating our violent urges. The argument seems fallacious.

Fully Responsible or Shirking All Responsibility?

Some players, so it appeared from the socio-psychological study, spend a considerable amount of time in this virtual reality. The impact MUME can have on the life of afficionados ought to induce us to wonder about the psychological, social and moral consequences of such an activity.

There had been no ethical reflection as such prior to the creation of MUME. Such matters transcend those of the rules of the game. They belong to the same domain as the questions about the consistency one can give to virtual reality, since one can attempt to create no

---

Heidegger also criticized, especially in his analysis of modern technics.

[16] One might object that efficiency is not ethical. The debate is open, although a utilitarian ethic might very well be satisfied with this justification.

[17] Of course, murder occurs quite rarely here, as opposed to many other games in which it is the chief activity, but it raises questions: can murder be evaluated positively, e.g. if it is a monster that is killed?

matter which reality by playing on the senses and make it be perceived as real by the user. The analysis of MUME presents a greater problem than that of a mere role-playing game, since it affects our anthropological relation to the world. The differences between the real world and the virtual world, between real life and artificial life, however radical they may be, tend to become vague.

Virtual networks facilitate the development of attitudes based on a feeling of almost absolute impunity due to the anonymity the networks offer. We shall give the example of 'Julie' on the Internet. A woman presents herself to the network of the electronic community of New York as an old, completely paralysed woman who has to type on the keyboard with the help of a stick attached to her forehead. Being of a particularly convivial disposition 'Julie' made many friends on the Net in a few years time, and on a few occasions she was an invaluable psychological help to one of her friends. This woman tried to contact 'Julie' in the real world, and in doing so revealed the secret. 'Julie' was a male psychiatrist who had connected to the network by coincidence, and subsequently had taken on the identity of a woman. Surprised by the kind of exchange he managed to establish from the start with the other women of the network, he had decided to maintain this identity, and moreover he created the image of a weak and bedridden woman to avoid embarrassing encounters. He later defended himself by saying that he wanted to study the female psychology from within: 'I did not know that women talked to each other in this way. There is such a vulnerability, such depth, such complexity (...) Conversation among men on the Internet is much more reserved and superficial, even among intimates. It was fascinating and I wanted to find out more.' 'I felt violated' said one of the women after the discovery of real identity of 'Julie'. This example clearly articulates the problem of the real nature of 'electronic personalities'.

On a more fundamental level, it is the question of our personal, intersubjective responsibility that is raised once again, especially in the case of a virtual game environment. The anecdote of Julie implies the problem of our relation with the body of the others, a relation which until then seemed a necessary condition for the concrete establishing of intellectual, spiritual or affective relations. But the relation to one's own body is a matter at hand as well. It seems that the dichotomy between the carnal body and the mind that inhabits it has to regain particular vigour in our age. The developments within the virtual communities are liable to blur the borderlines between the habitual psychological categories and between the types of relations we entertain with the others.

Living in the virtual world, how will the common sense stand firm before this phenomenon of disincarnation? The classical ethics can hardly explain the gradual anaesthesia of the moral sense which made possible the horrors of genocides or morally reprehensible actions. Distance and derealisation of the 'enemy' make it possible to support one's actions when one's moral conscience would not condone them.[18]

The fascination exerted by the virtual worlds stems from the haze of ambiguity they create between the subjects and the representation they have of themselves. In a way, one may have the impression of 'really' inhabiting the virtual words. The virtual communities as well are real, in a certain way. The events that take place in them play an effective role in the lives of the people who bring them to life. The individuals who invent them create bonds of affection, exchange information, work together. A new repartition of private and public domains ensues. Virtuality becomes a mode of access to reality, and the public domain has failed to prepare itself for this.

And yet our body is neither symbol nor symptom. It is a necessary condition for our experience. This confrontation between utopias and real bodies is one of the knots in the problem of virtuality. The growing gap between presence and representation is accompanied by troubling or overexciting effects, without direct, 'real' compensation. On the other hand, whatever the degree of virtualisation, our body remains real. Eventually, therefore, we have to return to the body that unifies us. To forget it or to deny it would have immediate painful repercussions. This last remark, by the way, holds true for the real world as well.

What is the difference between a real and a virtual place?[19] One of the differences is that a real place gives us a basis, grants us a position. These are conditions for existence and conscience. The position

---

[18] This line of explanation dealing with the anaesthesia of the moral conscience gives ample attention to technology that kills at a distance. Jonathan Clover takes this case to explain the moral bankruptcy of the twentieth century.

[19] We have not aimed at being too precise in our definition of virtuality, and have followed the seven focal elements Michael Heim selected from the divergent circulating definitions: simulation, interaction, artificiality, immersion, telepresence, full-body immersion, networked communications. See *The Metaphysics of Virtual Reality*, Oxford University Press, 1993.

in real space is not merely an attribute of conscience, but a preliminary condition for it. A place in reality is intimately linked with the body. This is not the case with virtual places or spaces.

Granted, virtual spaces can be modelled in such a way as to simulate reality, but they might just as well be modelled arbitrarily, without rhyme or reason. The 'virtual' opposite of a 'real' position is the impression of its abolition; it is the vertigo before the abyss. If the virtual worlds want to impassion us they will have to prove that they can give us the feeling of vertigo, the emotion of the abyss. Virtual vertigoes may well become a new 'opium' for those who crave to escape from this world. They might, however, also be a condition for a keener and more self-assured look on reality.

Players think they hold their distance, but in fact they are more committed than they think, and in two ways: in virtue of the engagement to which they consent, and as to their responsibility.

When Shall We Have an Ethics of Virtuality?

This question lies at the origin of a French report by the OPECST (Office Parlementaire d'Evaluation des Choix Scientifiques et Techniques). The Law commission of the Senate had granted the Office the prerogative of commissioning a study that was initially called 'Technical and Ethical Aspects of Synthetic Images'.[20] The final report was adopted under the modified title 'Synthetic Images and Virtual World: Technics and the Issues of Society'.

In the beginning the approach was a defensive one: does image technology increase the risk of manipulation and contain the threat of damaging one of the fundamental rights of the individual, viz. that of his own image? Eventually, however, the search for an 'ethics for new information and communication technologies' turns out to cover a mere three pages and is followed, moreover, by an interrogation mark. This is inversely proportionate to the space occupied by the ethics in the initial title. Ethics has become a nuisance.

Taking things further, we might develop an ontology of virtual reality, and state that reality resembles the possible, whereas actuality corresponds to the virtual, in keeping with Gilles Deleuze, who is very much inspired by Simondon.

The roots of these definitions reach down to the medieval texts of Duns Scotus who offered an alternative for the ontologies of Aristotle and Plato.

[20] OPECST Report, 'Images de synthèse et monde virtuel: techniques et enjeux de société'.

There is one element that can make us understand why the ethical interest has lost its punch. We find it in the definition the author of the report gives of it in his few lines on 'the ethical approach':[21]

> Ethics is a line of evolutive reflection which aspires neither to universality nor to permanence, and which searches for the most harmonious equilibrium possible between morality and society.(...) Ethical reflection finds its expression in the advice and the recommendations emanating from consultative bodies. This advice, whatever the competence and authority of those who provide it, must not be imposed on whomever, and must not be associated with sanctions.'(*The author defines morality as follows*): 'Since it is essentially spiritual, morality, which is universal and permanent, and of which the source does not lie exclusively in human reason, is not liable to adaptation in function of the evolution of societies.

Of course we are confronted once more with the habitual problem in connection with the distinction between the two terms ethics and morality. And yet ethics is very vague here, exempted somewhat rashly from any attempt at justification, from all imperative universalisation. Thereby it affects the members of ethical bodies, who are easily suspected of irresponsibility.

More surprisingly, morality, which many philosophers classify as a particular concept, here acquires the qualification of universal *and* particular. The text leaves us clueless as to which morality is meant, or which spiritual essence. To continue in the same semantic field as the report: we have the right to ask ourselves whether or not they are virtual. Indeed, apart from mentioning the necessity of reflecting – '*like fifteen years ago on bioethics*' – on a 'computer ethics', the text never gives any further explanation about what is understood by an ethics of the NTIC or by the morality of the NTIC.

The said report achieves the reduction of ethics to mere eyewash in the last two lines of page 178: '*It pertains (to ethics) to elucidate the public opinion. There indeed lies the mission of the OPECST.*' If one does not care for this curtailing of ethics, the resignation of the Office looms, for they ought to have treated the matter more elaborately.

The engineer who developed MUME has done more than just be satisfied with these definitions that are so undemanding on the level of moral philosophy: one who is so close to the 'objects' has put us on the trail of crucial anthropological issues connected with virtual

---

[21] *Ibid*, p. 177.

realities. His moral sensitivity enabled him to function as an excellent alarm system.

These virtual games have proved eminent revealers of the insufficiency of an ethics that approaches objects only as instruments. Modern technologies operate as mediators in our relationship with reality, without our mastering them as we would a traditional tool. However trivial this may seem, one does not analyse a hammer and the Internet in the same way and yet many schemes of ethics still fail to draw the obvious conclusions.

References

DE KERCKHOVE, D. (ed.), *Connected Intelligence: The Arrival of the Web Society*, Toronto, Somerville House Publishing, 1997.
HEIM, M., *The Metaphysics of Virtual Reality*, New York, Oxford University Press, 1993.
LATOUR, B., *La science en action*, La Découverte, 1989.
LÉVY, P., *L'intelligence collective. Pour une anthropologie du cyber espace*, Paris, La Découverte, 1994.
MONOT, P. and M. SIMON, *Habiter le Cybermonde*, Paris, Éd. de l'Atelier, 1998.
SIMONDON, G., *Du mode d'existence des objets techniques*, Paris, Aubier, 1969.

1.2.2

WORKERS AND ENGINEERS

Two Different Worlds?

Stanislas Dembour

Editor's note: One area in which engineers enjoy a sense of their profession and their significance is in their relationships with workers. Whether they like it or not, engineers often become involved in the class conflicts which take place in and through companies.[1] Every engineer is therefore forced to take up a position in relation to the world of workers. Trying to remain neutral is impossible: not to decide is to decide. Even the technologies that engineers produce are not neutral from a social perspective: they condition and organise the lives of workers. It is the reality of class conflicts which forces engineers to define their position and their areas of solidarity with the workers in the businesses in which they work. This is an inevitable ethical choice, which cannot be reduced to any analysis which claims to be purely technical or technocratic.

The brief testimony below has been written by someone who has been employed as a factory worker. It shows how, from the workers' perspective, engineers are far removed from workers, their view of work, their sensitivities and their interests, despite the fact that they come into contact with each other all the time. And despite the fact that many engineers feel more sympathetic towards the world of workers than towards the world of capital. It may not be pleasant for engineers to see themselves embroiled in this

---

[1] The term 'class conflicts' is used because, due to the type of conflict involved, it is not appropriate to analyse it in terms of inter-personal conflicts. On the contrary, this type of conflict can better be understood by taking into account the divergent interests created by the economic system, and in particular by production structures and relationships.

way in a conflict which is not of their choosing. But this is inherent in the situation in which they find themselves. This is no doubt the reason for the ethical question: 'What position can engineers take, either individually or collectively, in the class conflicts which they usually have not wanted, but in which they are nevertheless involved?'

It is difficult to get a testimony from a worker on the way in which workers see engineers.

It would also seem to be just as naïve to say how workers see engineers, as it would be to say how engineers see workers. Every engineer is different from every other, and workers differ in the same way. Sometimes there are sincere human relationships between some engineers and individual workers. However, the world of workers as a whole does make an overall judgement on the world of engineers.

Relationships within a large company are completely different from those in a small or medium-sized business. Are they better or worse? It depends how you look at it! The situation of an engineer in a research division is different again. I will mainly be looking at relationships in large companies.

The workers are suspicious of the way in which the engineers will use their testimony. Won't they turn it against them? Won't it be used to improve the executives' relational techniques, so that they can 'hoodwink' workers better? For many people the experience of quality circles has been a very unpleasant one.

This system creates a lot of suspicion. Perhaps that is the dominant note underlying relationships between workers and engineers.

The workers have no power whatsoever in the company. They have sold their labour and are simply a 'human resource', alongside financial and other resources. Trade union delegates know that even in the works council the management covers up some realities even if they stop short of actually lying. Workers are therefore at the mercy of sudden decisions that will present them with a *fait accompli*. As for the engineers, who are seen as a channel for the transmission of orders and guidelines from the management, they are definitely seen as being on the employers' side.

The division between management and workers is a fact, whether one likes it or not. The world of workers has an awareness of the conflict of interests between the two worlds, and this may be clearer or more diffuse in different individuals. The saying often heard in

Wallonia *'c'est todi les p'tits qu'on spotche'* ('It's always the little ones that get squashed') expresses this idea quite well. A single worker in front of 'the boss' is lost. It is together, as a group, that workers can bring down what the majority of workers see as the arbitrary power of the bosses, as transmitted by the engineer.

The engineers, who usually come from a middle-class background, are often more concerned about form than substance in their relationships. The popular phrase says it all: 'too polite to be honest'. That means that workers will not be deceived for long by the politeness or even kindness shown on the surface by the engineers. It is their actions and general behaviour which will be judged and, these days, it will be above all their attitude towards excessive flexibility, work rates, which are too high, and social conflicts.

Have relations been improved or damaged by recent events (numerous factory closures, mass redundancies etc.)? No doubt some executives are aware of their objective solidarity with the world of workers and their dependence on the boss' power, but most of them, under pressure from events and their managers, are at risk of becoming even harder and more intransigent towards their workers. They are forced to implement new standards in terms of work rates, flexibility etc.

When the engineers walk past workers as they return to their office, they often do not even see them. If they are forced by necessity to go down to the shop floor, they often take the stance of Mr. Know-All. So, discreetly sarcastic but deeply irritated to be counted as worthless, the workers watch Mr. Know-All as they become entangled in difficulties created by their lack of practical experience.[2]

Sometimes, when they are in the middle of their work, exasperated by the orders, advice or comments from the engineers, they will retort: 'Well do it yourself then!'

What do the engineers really know of the actual work, as they sit in their office, passing on orders through the foreman or overseer?

The engineers have completed a course of study which has given

---

[2] Editor's note: This feeling, expressed in a worker's own words, is supported by sociologists when they say that the world of engineers is still largely characterised by a scientific and technocratic model which makes engineers believe that they have a monopoly of knowledge or even rationality. This is intolerable for workers, who know from experience that they also have a valid understanding of their work.

them the best possible preparation for their technical responsibilities, but how can they prepare for the human aspect of their responsibilities in the hierarchy?

A simple choice of words is sometimes enough to turn relations sour. One day the engineer catches a worker doing nothing and says he is idle. This man gets angry, although he would not have minded the engineer saying that he was lazy. For this worker one word was tolerable but the other was not.

Here is another example. Supported by the trade union delegation, maintenance workers refuse to work extra hours during the weekend for an urgent repair on a traversing bridge. After warning them that if they refuse to do it the factory will call in an outside firm, the engineer takes action. The following Monday the engineer tells the workers, who are amazed that the repair has been carried out, that they were wrong to take the advice of the trade union delegation. The conflict has begun.

How many times have I heard workers being threatened when they have expressed criticism of disordered and pointless work: 'If you don't agree, you can always leave; there are hundreds more who would take your place'. This kind of comment creates such strong feelings among workers that it becomes impossible for them to maintain appropriate relationships.

Information travels along the grapevine from one factory to another. Workers communicate about bosses' actions or gestures which have improved or damaged industrial relations. In this way a collective judgement on the world of engineers is gradually formed. Once again, every worker has his or her own personality and judgement, just like every engineer. None of them will identify with all the characteristics outlined above, but workers, through their ongoing experience and in their collective memory, judge engineers as a whole. Is it all about personal relationships? Or is it the overall context of an economic system?

1.2.3

THE SAFETY OF RISK-PRONE SOCIOTECHNICAL SYSTEMS

Engineers Faced with Ethical Questions

Michel Llory

For about two decades, serious accidents that have happened in all industrial and transportation sectors, on the one hand, and field studies of human and social sciences of work, on the other hand, have revealed deep discrepancies between engineers' conception of work and safety and that of field personnel. With these discrepancies come ambiguities and conflicts of an ethical nature. Although it appears to bring progress in safety, the ethical debate between these two categories of personnel is far from having actually begun. Here the outline of this debate is traced, and the conditions of its initiation and satisfactory unfolding are discussed.

The Emergence of the Notion of Major Technological Risk

A number of serious accidents that have happened over about two decades have highlighted the technical and organisational weaknesses of risk-prone industrial systems. They have confirmed a certain number of fears on the part of analysts, observers, and public opinion that technological development is less under control than the big decision-makers or managers of risk-prone systems often claim. Thus the notion of major technological risk has emerged, insofar as the complex malfunctions of industrial systems can result in considerable damage, which are no longer limited to the personnel

---

[1] Original text translated by C. Roy.

or the technical installations of the system where an accident has happened, but can also affect the public and the environment.

Big accidents are most often accompanied by more or less serious crises in the organisations managing the affected technical systems, organisational and even political crises.[2] They have considerable power to destabilise and raise questions within organisations and public opinion, arousing fear vis-à-vis excessively rapid, even anarchic or in any case, badly controlled technological progress. Accidents both upset the status quo and act as telltale markers of underlying problems.

A Sinister Succession of Accidents

Since the end of the 70s, not a single sector of industrial activity and transportation has avoided being touched by serious accidents and organisational crises. The tendency to seek – and to obtain – high technical performances (in terms of power involved, productivity levels, speeds reached, transportation capacity) explains the gravity and the extent of the damage when accidents do happen.

Yet, during the same period, considerable efforts have been made to guarantee greater safety, in spite of increased potential dangers due to the high performances of technical systems today. These efforts have been of a regulatory, technical, and organisational order. Security and safety have grown into activities in their own right, with their experts, specific studies, specialised personnel in the field (security and safety engineers, among others), their conferences and their studies. The overseeing authorities which see to the elaboration and control of industrial activities have multiplied their requirements, demanding of system operators 'safety demonstrations', along with detailed and complex files to prove technical mastery of these systems.

But numerous accidents have regularly come to reawaken worries, to launch anew the same old questions about safety: Are the precautions sufficient? Does the system harbour heretofore-unsuspected flaws? What more can be done? Anglo-Saxons express all this in the concise formula: *'How safe is safe enough?'*

---

[2] T.C PAUCHANT and I.I. MITROFF, *Transforming the Crisis-Prone Organization*, San Francisco, Jossey-Bass Publishers, 1992.

All industrial and transportation sectors have been hit. Three Mile Island and Tchernobyl in the nuclear industry,[3] Seveso,[4] and Bhopal in chemicals, the explosion of the space shuttle *Challenger*[5] and the crash of the DC10 at Ermenonville,[6] as well as the collision of two Boeings 747 on Tenerife airport[7] in the aerospace field, the shipwrecks of the ferryboats *Herald of Free Enterprise*[8] and *Estonia*,[9] as well as the oils slicks provoked by the *Amoco Cadiz* [2] and the *Exxon Valdez*.[10] These are just some from a long list of catastrophes and industrial disasters.

The extent of reactions from the professional body concerned, and more generally the social body, is not necessarily directly proportional to the gravity of the physical, material, immediate consequences. Certain accidents or incidents stand out more than others, be it among other things that they take place at a special historical moment, or that their social, symbolic, and educational import, as *revealing signs* of previously unrecognised, unidentified or silenced problems and malfunctions, is very important.

Thus, how can we explain for instance the intensity, the extent of reactions in the press, in public opinion, and official institutions upon the discovery of the contamination – though relatively slight, and, even according to critics and protesters, not jeopardising the health of workers – of containers serving to transport irradiated fuel from nuclear power plants to the COGEMA reprocessing centre in the

---

[3] M. LLORY, *Accidents industriels: Le coût du silence. Opérateurs privés de parole et cadres introuvables*, Paris, L'Harmattan, 1996.

[4] P. LAGADEC, *États d'urgence. Défaillances techniques et déstabilisation sociale*, Paris, Éditions du Seuil, 1988.

[5] See also the remarkable book by D. VAUGHAN, *The Challenger Launch Decision: Risky Technology, Culture, and Deviance at NASA*, Chicago, The University of Chicago Press, 1996.

[6] P. EDDY, E. POTTER and B. PAGE, *Destination Désastre*, translated from the English, Bernard Grasset, Paris, 1976.

[7] K. E. WEICK, 'The Vulnerable System: An Analysis of the Tenerife Air Disaster' in *Journal of Management* 16(1990)3, pp. 571-593.

[8] DEPARTMENT OF TRANSPORT, *The Roll On/Roll Off Passenger and Freight Ferry Herald of Free Enterprise*, Report of Court n°8074 formal investigation, Her Majesty's Stationery Office, London, U.K., September 1987.

[9] M/S Estonia, September 28, 1994, The final report, The joint accident investigation commission. The government of the Republic of Estonia, Estonia, December 1997.

[10] L. CLARKE, 'The Wreck of the Exxon Valdez' in D. NELKIN (ed.), *Controversy: Politics of Technical Decisions*, 3rd ed., Newbury Park, CA, Sage Publications, 1992, chap. 5, pp. 80-96.

Hague?[11] We would no doubt have to appeal to the revelatory power of the incident, vis-à-vis a certain number of malfunctions – real or supposed – in the organisation of the nuclear industry; to the demonstration of the latter's lack of transparency; to the fact that this type of incident has gone on for many years without being analysed and corrected; to social intolerance vis-à-vis the nuclear industry (its incidents), no doubt greater (or of lesser social acceptability) than for other industrial activities; and finally to the raising of awareness of a problem which, while being minor, affects many nuclear sites, and so happens to be geographically dispersed and not localised. And we do not pretend to put forward an exhaustive interpretation.

As complex, apparently unpredictable, multiform risks are revealed at the level society, the demand for safety tends to increase proportionately. A. Giddens thus thinks that risk is about to become a basic category of sociology, supplanting even the latter's traditional categories.[12]

Long protected by a credit of exacting scientific and technical mastery, the nuclear industry undergoes this demand, all the more so since the two major accidents of TMI and Tchernobyl have undermined the public's confidence. Reports on the situation in Ukraine, following the accident at Tchernobyl, regularly taken to the attention of the public, seem to interfere with a series of recent incidents (including that of the contaminated wagons) and lead to a sometimes-severe critique of the lack of transparency of the nuclear industry.[13]

More generally, from accidents, attention has also shifted to incidents, and then towards the institutional, social, political crises that these critical events trigger. The accident provokes a shock wave, a psychosocial shock.[14] A veil is very often torn upon these occurrences. Malfunctions, vulnerabilities of the sociotechnical system then appear in broad daylight.

---

[11] 'Le Premier ministre réclame un rapport sur les transports contaminés' in *Le Monde*, 8 May 1998, p. 22.

[12] Cited by B.A. TURNER, 'The Future for Risk Research' in *Journal of Contingencies and Crisis Management* 2(September 1994)3, pp. 146-156.

[13] 'M. Jospin rappelle aux industriels du nucléaire la nécessité d'une plus grande transparence. Le gouvernement veut mettre fin à une 'certaine culture du secret'', *Le Monde*, 13 June 1998, p. 17.

[14] M. LLORY, *Accidents industriels: le coût du silence. Opérateurs privés de parole et cadres introuvables*, Paris, L'Harmattan, 1996.

The Human Factor and Human Error

The safety experts and managers of great risk-prone systems have conceived and put into place a rigorous and impressive set of arrangements in order to ensure safety. Each accident is analysed methodically by these officials, almost all of them engineers, to draw up recommendations, and eventually improvements. Accidents are gone over with a fine-tooth comb in their unfolding and underlying causes (what Anglo-Saxons refer to as *'root causes'*). The new arrangements proceeding from this should make sure each accident cannot occur again, and that a certain number of accidents of the same category, of the same type, cannot happen. The same vigilance tends to get applied to the most serious, most significant incidents, which increases by that much again the capacity for prevention: safety is incremental and seems to tend asymptotically towards the absolute ideal of 'zero accident'.

What engineers first learned about big accidents is the existence of the 'human factor'. The reliability of systems they had conceived did not depend only on technical factors, but on 'human reliability'. What accidents seem to reveal is the inadequacy of the behaviour of personnel in the field: human deficiency, human error, looms as large as, and even more than technical deficiency. At one extreme, the system would be technically under control: all that would remain is to control man, this 'unpredictable' factor, to take up a qualifying term often heard used by engineers.

To achieve this, the engineers' chief weapon is a pre-established and codified system of technical norms, of procedures, of safety directives, of work and safety references. These procedures[15] have to be applied rigorously by the personnel on the field, the superintendence operators, or the system pilots, and the maintenance operators. They have been conceived by the experts, the engineers, drawing upon the necessary scientific and technical knowledge – and relying upon them – in the calm of an office environment, thus in particularly favourable conditions of serenity and reflection and far removed from moments of perturbation, system incidents, when stress or the fear of risk threatens the operators.

---

[15] In what follows we will use this term 'procedures' as a generic term designating in fact a complex and voluminous body of texts that are prescriptions.

Each serious incident, and of course each accident, results in complements to the procedures: extra actions to accomplish, new precautions to take for the operators. Safety is based on the formalisation of the work activity; proceduralisation tends to increase, spreading to all sectors of activity in businesses that manage risk-prone systems: from the conduct to have in case of accident, or in case of incident, and in the normal phases of the system's functioning, to the gathering of information and the analysis of incidents, the safety audits performed by the managers to monitor the activity of personnel in the field, to the management of human resources (training, individualised professional paths, etc...). Managers and field personnel thus have at their disposal a substantial stock of procedures, codified methods, tools, in order to ensure safety and work. It seems that not one important moment of everyone's activity would be able to escape this formalisation.

It is the adherence and conformity of operators in the field to the prescriptive system that seem to be infallible guarantees of safety. Managers are thus entrusted with seeing to this conformity, monitoring it, and correcting all behavioural deviations as may be ascertained by them. Changing field personnel's behavioural patterns becomes the major concern of managers, adding itself to the meticulous, almost obsessive observance of prescriptions. Incitements, exhortations, indeed even psychological pressures, are generously dispensed by the managers in order to maintain this 'high level of safety'. They trace the outline of what we have called elsewhere a 'heroism of perfection'.[16] Flawless rigour, vigilance of every instant are the by-words of this conception of safety, largely internalised by agents in the field, who themselves assert 'We do not have the right to make mistakes!'

In parallel to the move towards proceduralisation, there have been developed safety management and safety (or security) culture, which may flexibilise an excessively formalist and rigid conception of safety. Voices have been raised to criticise this position, including in Anglo-Saxon countries.[17] However, too often, safety management limits

---

[16] M. LLORY, Sécurité, prévention des accidents: les cadres entre 'l'assistance à personne en danger' et 'l'héroïsme de la perfection', Performances humaines et techniques, n° HS, September 1998, pp. 10-15.

[17] T. DWYER, 'Industrial Safety Engineering: Challenges of the Future' in *Accident Analysis and Prevention* 23 (1992), pp. 265-273.

itself to global verifications of the observance of the prescribed system, and to pressures on field personnel for it to be respected (safety indicators, audits, visits from superiors); safety culture is defined in most cases in a prescriptive fashion: its approach does not result from an existing and systematic survey of the situation, but from an ideal posture that does not integrate the culture developed by the field personnel, whose characteristics we briefly present below.

What Do We Learn from Accidents and Incidents Anyway?

Flaws exist in what may appear as an impeccable safety apparatus. Flaws through which incidents, and sometimes accidents, creep in. Concerning the causes of accidents, the analyses of the classical paradigm of safety, very largely dominant, and which we have sketched above, themselves present insufficiencies that have been pointed out and explored for some twenty years more and more systematically by a number of researchers in human and social sciences and experts, mainly Anglo-Saxons.

The critical events that accidents and incidents highlight above are all malfunctions in organisations that manage risk-prone systems, and the limits and weaknesses of the prescriptive and managerial safety system: badly conceived field personnel training programmes; too rapid a turn-over of officials, managers crushed by their management workload, who are no longer present enough in the field; weak malfunction signals, and even patent ones, that do not make their way up towards the decision-making echelons of the organisation; organisational policies that introduce too frequent changes and that are also badly understood by personnel in the field, and which destabilise the latter's representations of work; pressures to increase productivity and 'messages' promoting safety pile up over each other in turn without clear articulation of an overall plan; the weakness of exchanges and debates on safety, etc...

The picture of malfunctions gets wider. We are in this case very far from the human error committed by an operator situated at the very end of the organisational chain of safety. The gaze of the researcher and the expert in this new paradigm of safety[18] dwells

---

[18] In all rigour, it may be deemed still premature to oppose a 'new paradigm' to the current, classical paradigm (see reference of the previous footnote).

much more on the managers and the decision-makers, and on the organisational dynamics that develops between them and the field personnel.

Critical events have an incubation period[19] during which the conditions for the serious incident or accident develop insidiously. The historical context and the context of the cultural and institutional environment conjoin to constitute little by little the favourable and dangerous ground upon which the critical event will then develop more rapidly.

According to the Rogers Commission of inquiry on the space shuttle *Challenger* accident, we have to go back thirteen years in order to understand the build-up towards accident. One of the chapters of the Commission's report is entitled: '*An accident rooted in history*'.[20] The experts who analysed the shipwreck of the ferryboat *Herald of Free Enterprise*[21] go back many years previously, and uncover a long list of alarms expressed in the form of letters by the captains of ferryboats to the managers of the Company to which belonged among others the *Herald*. However, not only did the managers remain deaf, but they multiplied productivity demands (increasing the rotation cadence of ferryboats ensuring the traffic back and forth between two cities), sometimes with ambiguous rhetorical formulas such as: '*There seems to be a general tendency of satisfaction if the ship has sailed two or three minutes early. Where a full load is present, then every effort has to be made to sail the ship 15 minutes earlier...*'.[22] This deafness and the inaction of managers who are at the heart of this transportation system have been pinned down by the experts entrusted with the inquiry by the British Court of Justice through the concise formula that characterises the management: '*A vacuum at the centre*'.[23]

---

[19] B.A. TURNER and N.F. PIDGEON, *Man-made Disasters*, 2nd ed., Oxford, Butterworth Heinemann, 1997.

[20] Chapter VI: 'An Accident Rooted in History', Presidential Commission on the Space Shuttle Challenger Accident, Report to the President, Government Printing Office, Washington, D.C., United States, 1986.

[21] DEPARTMENT OF TRANSPORT, *The Roll On/Roll Off Passenger and Freight Ferry Herald of Free Enterprise*.

[22] *Ibid*, p. 11.

[23] 'A Vacuum at the Centre' in Department of Transport, *The Roll On/Roll Off Passenger and Freight Ferry Herald of Free Enterprise*, p. 15.

What Do We Learn from Field Inquiries?

We now have at our disposal the results of studies and inquiries made in the field over a period of about twenty years, aiming at explaining and describing every day work situations, of routine work, outside of periods of great upheavals in the technical system when there are incidents and accidents. These studies have accumulated in the framework of human and social sciences of work (ergonomics, psychology and psychodynamics of work, sociology and anthropology of work) and tend to constitute a coherent body of knowledge on the work of field personnel (operators and direct, ground level supervision). This body constitutes as it were the dual counterpart of the knowledge and teachings drawn from the analysis of incidents and accidents.

Studies are based on field inquiries, not only of a statistical and epidemiological type, but above all of a clinical one, the latter permitting in some cases to prepare and structure the first. These inquiries quite often reveal discrepancies between the work description made by the field personnel of its own work and the description drawn up by officials and managers. This discrepancy may be explained in part in the following fashion:

- Technical processes are never totally under control, including in the unfolding of their malfunctions. In complex technical systems, procedures dealing with the management of incidents and accidents are most often incomplete, or they do not quite correspond to the event to be controlled. They require interpretation on the part of personnel, not systematic and blind application. In other words, the procedures only represent a schematisation, providing a model of what British risk analysis experts call: *'an enormously complex labyrinth of potential sequences and interactions.*[24] To put it in another way: all the details of the work cannot be forecast in advance. The realisation of tasks does not require only the application of procedures, but also tacit knowledge and know-how which have been acquired through experience.
- Work and safety also require a certain number of psychological (keeping one's cool, avoiding getting nervous, controlling fear,

---

[24] THE ROYAL SOCIETY, *Risk Assessment: A Study Group Report*, London, January 1983, p. 31. The emphasis is ours.

coping with stress) and social (conviviality, good team relations climate) conditions. Inquiries show that a certain number of informal or even implicit rules also intervene to guide collective activity, co-ordination and co-operation: rules of 'collective well-being', of solidarity, 'arrangements' between operators and between teams.
- Engineers still keep to the most formalised aspects of safety and work, and remain unaware, in fact, of the existence or the value of know-how, practical abilities and informal rules of human conduct. Thus they overrate the formalised construction of safety, which we can term, by analogy to ergonomists' prescribed work: 'prescribed safety'. This latter is seen as being the only acceptable method, since it can be the object of processes of discussion, of demonstration, analogous to scientific proof procedures. Field safety is in some cases presented as a holdover from the past, like a body of obsolete work practices, since they are founded on *'impalpable'* know-how, on attitudes that may vary from one work team to the next, and are therefore not *'reproducible'*.[25]
- The fundamental problem at hand is that of the recognition by the officials, the engineers, of components and ranges of the personnel's activity that evades an otherwise already excessive instrumentalisation of the analysis of work and human relations. The human dimension is particularly obscured by engineers preoccupied with tools, codified methods, standard behavioural patterns and formal rules.

The Ethical Conflict at the Heart of Discrepancies in Evaluation Between Engineers and Field Personnel

During our field inquiries, led so as to analyse the know-how, particularly that related to safety, known as caution know-how, or undertaken in the aftermath of serious incidents or accidents, we have been struck by the recurrence of complaints and criticisms of field personnel towards its hierarchical superiors, concerning ethical questions.

---

[25] M. LLORY, Sécurité, prévention des accidents: les cadres entre 'l'assistance à personne en danger' et 'l'héroïsme de la perfection', Performances humaines et techniques, n° HS, September 1998, p. 10-15.

First of all, the field personnel's queries concern responsibility, all the more so since the reinforcement of procedures and prescriptions can give officials the illusion that they have transferred responsibility to field personnel. Their mission would simply consist in promulgating norms of behaviour corresponding to the technical behaviour of systems while the responsibility to rigorously apply these norms would devolve upon field personnel.

This conception of the sharing out of responsibilities does not appear to us very coherent, or fair. The mission of managers and experts cannot limit itself to this all too taylorian separation[26] of the responsibilities of each category of personnel. For it is their responsibility to permanently ensure that the optimal conditions of application of the prescriptions are guaranteed, that the evolution of the context does not increase the difficulties of the operators' work, and therefore the risks.

It is in this sense that operators refer to the prescriptions as: *'the umbrella-procedures'* that protect the officials and increase their risks of being penalised. To the fear of having to face a complex incident or accident situation, which would hold traps, bad surprises, and would catch them at fault, is added that of the guilt of not being able to conform themselves *with due rigour* to the procedures. The deep debate between managers and field personnel could among other things clarify what pertains to the moral responsibility of each and what pertains to legal responsibility.

Difficulties of the same type are experienced with regard to the exercise of the right of alert, which is also a duty incumbent on all the personnel; but the abuse of this right is punished as well. Besides, many cases are related in the specialised literature in which the agents who had tried to bring the attention of their superiors, of decision-makers, to the difficulties and malfunctions of the sociotechnical system found themselves without support, or were even severely sanctioned. Some examples are: engineers fired for having expressed worries about safety, as at the American company Lockheed[26]; or like Roger Boisjoly, an engineer at Morton Thiokol who tried in vain to prevent the twenty-fifth launch of the space shuttle *Challenger*; the flight resulted in the explosion of the shuttle and the death of the seven astronauts who were manning the

---

[26] M. LLORY, *Accidents industriels: le coût du silence. Opérateurs privés de parole et cadres introuvables*, Paris, L'Harmattan, 1996.

spacecraft,[27] Boisjoly was quarantined by his colleagues; an engineer of the Millstone nuclear plant, who for three years took steps, without support from his hierarchy, to make known a certain number of deficiencies in maintenance and safety[28]... 'Bad news' is badly received by managers, for they bring organisational processes, decision-making rules, or material and financial means allocated for the management of technical systems into question, and they thwart production targets and calendar forecasts.

One of the major ambiguities resides in the balance between efforts to improve productivity and profitability, on the one hand, and the maintenance of satisfactory levels of safety. Over the last few years, in many industrial sectors, these efforts at improvement have translated into an intensification of work and strong productivity pressures.[29] In spite of the official declarations of the decision-makers, of statements of principle, the corresponding efforts with respect to safety have not been made in order to balance these pressures. Inquiry results show that the daily concerns of managers are not about safety, but rather about productivity and financial management. The interest in safety too often manifests itself only when an incident, or of course an accident, occurs. The attention of middle-level management and high-level management is rarely sustained in this area. Hence, field personnel is quite often 'caught between two stools': follow productive pressures to the detriment of safety, or else enter into with more or less latent conflict with the management and be reproached for a lack of efficiency. In all cases, agents find themselves confronted with a dilemma that is difficult for them to resolve.

---

[27] On the in-depth analysis of the accident, see LLORY, *Accidents industriels: Le coût du silence*. On the reflections of an ethical nature developed in the aftermath of this accident, see: C. WHITBECK, 'The Engineer's Responsibility for Safety: Integrating Ethics Teaching into Courses in Engineering Design', ASME Winter Annual Meeting, Boston, MA, United States, 13-18 December 1987, n° 87 - WA / TS - 2; R.M. BOISJOLY, 'Ethical decisions: Morton Thiokol and the Space Shuttle Challenger Disaster', same references, n° 87 - WA / TS - 4.

[28] L'affaire Millstone, EPRINUC, 29 October 1996, pp. 5-7.

[29] M. and A. LLORY, 'Description gestionnaire et description subjective du travail: des discordances. Le cas d'une usine de montage d'automobiles' in *Revue Internationale de Psychosociologie* 3(1996)5, pp. 33-52.

Ambiguities, Matters of Conscience and Acute Ethical Conflict

These few results of analyses and inquiries in the field allow us to situate the extent of the ethical stakes of risk-prone sociotechnical systems. We have tried to show elsewhere that these stakes are impossible to dissociate from the debate on work conditions, the difficulties and risks of work situations. More precisely, these stakes are rooted in the critical discrepancies noticed not only between prescribed work (such as it is prescribed through the system of procedures) and real work (such as it is actually performed, taking into account a certain number of available means, local conditions and variables in the organisation of work, contextual factors), but also between the representation that officials have of the field personnel's work and the personnel's own perspective. The increase in discrepancies, or even the hardening of the managers' positioning, results in a subjective experience of injustice on the part of field personnel, and the feeling of a growing threat of being compelled to face responsibilities that it cannot fully assume and duties that it can never totally fulfil.

Depending on the work situations and their evolution, relatively permanent ethical ambiguities can lead to matters of conscience or to more acute conflicts. These conflicts can result in situations of tension, experienced individually, but most often collectively. What is one to do when one's conscience dictates one to alert superior hierarchical instances, but they remain deaf to these alerts? Most of these accidents reveal that these isolated persons, or little groups, had detected degradation signals of the sociotechnical system, and had attempted to alert decision-making levels of the organisation: ranging from a letter from an operator, one year prior to the accident at Three Mile Island, drawing attention to difficulties in managing incidents in the control room (ergonomic deficiencies),[30] numerous missives from ferryboat captains to the Board of Directors of their company, which ran among others the *Herald of Free Enterprise* as we have seen above ; to the report by R.V. Ebeling, engineer at Morton Thiokol, worried about the lack of progress of the work group entrusted with the improvement of the O-ring joint system of the boosters of the shuttle *Challenger* five months before the catastrophe, which begins with these words in capitals : 'HELP!'[31]

---

[30] LLORY, *Accidents industriels: le coût du silence*.
[31] Chapter VI: 'An Accident Rooted in History', Presidential Commission on the Space Shuttle Challenger Accident, Report to the President.

An Ethical Debate which Never Took Place

Such discrepancies in the evaluation of work difficulties would necessitate that a permanent debate be initiated between the officials, the engineers, on the one hand, and field personnel, on the other hand. Conversely, we may deem that the absence of this debate has as a consequence the acuteness of discrepancies. But the debate on safety and work generally does not take place. For some years now, this diagnosis of an insufficient presence of managers in the field has been established by many consultants within or outside the businesses, and it has also been pointed out by researchers in human and social sciences. Business leaders and high level managers have in part heard this message. However, the demand for a presence on the field of middle-level and ground level managers was acted upon in the codified, instrumentalised form of 'workplace audits', of 'hierarchical safety visits'. Most of the time, these call upon procedures (questionnaires, forms and reports with a pre-established structure), and follow-up meetings do not allow for engaging and letting unfold a sufficiently free, sustained and constructive dialogue. The episodic and formal presence of managers on shop floors and the workplace in the end only provokes frustration and disappointment among field personnel. The other measures that *could* allow these meetings to take place, such as periodical safety meetings, thematic manifestations, no longer fulfil this role. The field personnel describes them sometimes as 'solemn masses', other times as 'one-way affairs', that is as a place and moment of exclusive, more or less insistent reminding of prescriptions.

To be sure, locally, inquiries highlight the existence of a sustained and fruitful debate between field personnel and middle- and ground level managers. This debate concerns daily work and the difficulties that may be sources of risks. It is the meeting point of two communicative fluxes, descending, *'top-down'* (information on the business's strategies and orientations, global functioning balance-sheets, high-level managers' decisions and demands) and descending, *'bottom-up'* (coming up of doubts, worries concerning the risks, the deficiencies in work organisation). Inevitably, the 'ethical debate' on the ethical aspects and stakes of work, of risks, of potential accidents, finds itself articulated there. It deals with the moral and legal responsibilities of everyone, individual and collective responsibility, the rights and duties of each, the fairness of the share of responsibilities that each agent, and each team assumes. This analysis unquestionably puts the emphasis on the primordial importance of communication between management and field personnel, in the form of exchanges, discussions, 'debates ', which has been noticed by many authors since for instance H. Mintzberg.[32]

Conditions for a Debate: Some Rules

As we have tried to show, the ethical debate cannot be dissociated from the debate on the conception of actual work and daily safety. This debate presupposes the opening of public spaces within risk-prone sociotechnical systems: which presupposes first of all that the material and organisational conditions be fulfilled to guarantee the momentary existence of a sufficiently protected place; a certain privacy of meetings is necessary; the debate cannot be public; the agents will only express their fears, their questionings, their doubts if they feel sufficiently free, and protected, to do it. The debate also presupposes conditions of availability vis-à-vis the workload, the constraints of the tasks (which, as we have seen, only tend to increase). A goodly number of debates do not take place because the personnel, including the officials, being overloaded, are always running around!

However, the success of the debate, from its initiation to its progress and through to its conclusion (always provisional), depends fundamentally on the officials, the engineers, who hold hierarchical power. It is a *comprehensive* posture on their part that appears as the condition par excellence of this success, that is to say seeking the understanding that their collaborators have of the work and its difficulties, as well as of the contradictions of the prescribed work. The officials who maintain such debates get involved in *risky* listening, requiring not only a great openness of mind, but also a willingness to put aside their own presuppositions, or even their stereotypes on the level of control of the sociotechnical system. This form of listening can generate anxiety for officials, as it may lead to a destabilisation, and even a refutation of their conception of work and safety.

---

[32] H. MINTZBERG, *Structure et dynamique des organisations*, Paris, les Éditions d'Organisation, 1982.

The second particularly strong requisite is the recognition of field personnel, of its abilities and its practical knowledge. This recognition presupposes a good understanding of the personnel's contribution, that is to say a just evaluation or appreciation of the work performed. It includes the recognition of the subjective and collective mobilisation of minds as a determining factor of safety. The engineers can thus learn from the field personnel. Officials' knowledge is but one aspect, quite essential to be sure, of the knowledge and more generally the investments necessary to the functioning and daily surveillance of risk-prone systems.

Finally, the debate is built, explicitly and implicitly, on the objectives of a clear redefinition of the responsibilities, rights and duties of each, which presupposes long, open and in-depth discussions, on the actual meaning that these notions will mean in practice on a day-to-day basis. What risks does an agent take when he brings attention to malfunctions? Is the right to make mistakes recognised? Do the analyses of sociotechnical malfunctions include the mistakes or deficiencies of management? Are managers ready to recognise them? What should one do when one discovers that colleagues have seemingly erroneous or deviant practices? What recourses do agents alarmed by signs of deficiency, and unheeded by their immediate superiors, have at their disposal? What risks do agents run when they make use of their right to give the alert in all good faith, but who turn out to be wrong about the supposed seriousness of the fault ascertained? and so on. These are some of the questions which call for clarification, and then for agreement between the managers and the field personnel.

Conclusion: Present Obstacles to the Debate

However, as a general rule, the ethical debate and the debate on work are not initiated. The increasing demands for higher productivity, for profitability, and hence for safety measures, tend instead to *rigidify* the positions of officials: to increase the number and complexity of the system of procedures, to multiply pressures and exhortations addressed to the personnel that this system be respected, in spite of its deficiencies and limitations. In some places though, managers have taken it upon themselves to initiate this debate at local level as far as it is possible; which is why we think that this debate is at once tenable and necessary.

The obstacles seem to us mainly due to the officials' culture, which is positivistic and over-emphasizes technical aspects and factors, to the detriment of human factors and work problems. It is likely that the social evolution of officials is in itself unfavourable: engineers no longer appear to have through their family, as was the case in the past, bonds with working class and/or peasant culture, rooted in concrete and workaday *manual* labour. Besides, the working conditions of officials have evolved in a direction that strongly diminishes their availability, emphasising the management load, multiplying constraints and objectives to obtain with the corresponding demands for formalism which follow. Thus, the officials nowadays have to implement and insure imperatives of quality and quality control, of tight budget control, of human resources management (in the form of individual training and career plans, of formalised periodical interviews, etc.), of middle- and long-range activity forecasts, of necessary means, and of risks. Neither their personal history nor their training (at university or through engineering schools) prepare them to fully assume these very diversified roles, and to face the difficulties of work, which appear to be more extensive and formalised than in the past.

No doubt post-university training is a way of attempting to partially fill these gaps that have to do with the personal history of engineers and the social evolution of this professional community. But the urgency of an integration of philosophical and ethical questions in university curricula and engineering schools seems to us more timely than ever. However, this integration can only be efficient if a new appreciation for *humanities* comes to enrich technical culture. What young engineer nowadays is concerned for instance about the deontological aspects of his trade.[33] The considerable shock provoked by the explosion of *Challenger* has brought back in the limelight in the United States the objective of training engineers in ethics.[34]

Due to the failure to initiate this debate within risk-prone sociotechnical systems, many problems remain in the dark. Ambiguities, concerning the evaluation of risks, between the overseeing staff and the field personnel are rooted in technical and organisational

---

[33] B. HÉRIARD DUBREUIL, *Imaginaire technique et éthique sociale. Essai sur le métier d'ingénieur*, Brussels, De Boeck Université, 1997.

[34] See LLORY, *Accidents industriels: Le coût du silence*.

deficiencies; these can even feed conflicts between these two categories of personnel. The personnel accumulates worries and frustrations, while the officials shut themselves within a management approach remote from the concrete particularities of work and safety, and in silence.[35] An excessively pessimistic vision? A long study of serious industrial accidents,[36] corroborated by numerous authors,[37] has led us to this conclusion. The managers, the engineers do not hear the weak or even the patent signals of organisational malfunction, the 'background noise' of voices of the field personnel attempting to alert them... until the deafening din of a catastrophe.

References

BECKER, G., 'Event Analysis and Regulation: Are We Able to Discover Organisational Factors?' in A. HALE, B. WILPERT and M. FREITAG (eds.), *After the Event: From Accident to Organisational Learning*, Pergamon, Elsevier Science, Oxford, U.K., 1997. Chapter C.3, pp. 197-215.

DIDIER, C., A. GIREAU-GENEAUX and B. HERIARD DUBREUIL (eds.), *Éthique Industrielle. Textes pour un débat*, Brussels, De Boeck Université, 1998.

DWYER, T., 'Industrial Safety Engineering: Challenges of the Future' in *Accident Analysis and Prevention* 23(1992), pp. 265-273.

HÉRIARD DUBREUIL, B., *Imaginaire technique et éthique sociale. Essai sur le métier d'ingénieur*, Brussels, De Boeck Université, 1997.

LLORY, M., *Accidents industriels: le coût du silence. Opérateurs privés de parole et cadres introuvables*, Paris, L'Harmattan, 1996.

LLORY, M., 'Ce que nous apprennent les accidents industriels' in *Revue Générale Nucléaire* (January-February 1998)1, pp. 63-68.

---

[35] LLORY, *Accidents industriels: le coût du silence*; id., 'Ce que nous apprennent les accidents industriels' in *Revue Générale Nucléaire* (January-February 1998)1, pp. 63-68.

[36] LLORY, *Accidents industriels: le coût du silence* ; id., *L'accident de la centrale nucléaire de Three Mile Island*, Paris, L'Harmattan, 1999.

[37] F. PEARCE and S. TOMBS, *Toxic Capitalism: Corporate Crime and the Chemical Industry*, Socio-Legal Studies Series, Ashgate, Dartmouth, Aldershot, 1998; T. DWYER, 'Industrial Safety Engineering: Challenges of the Future' in *Accident Analysis and Prevention* 23(1992), pp. 265-273; G. BECKER, 'Event Analysis and Regulation: Are We Able to Discover Organisational Factors?' in A. HALE, B. WILPERT and M. FREITAG (eds.), *After the Event: From Accident to Organisational Learning*, Oxford, Pergamon, Elsevier Science, 1997, chapter C.3, pp. 197-215; D. VAUGHAN, *The Challenger Launch Decision. Risky Technology, Culture, and Deviance at NASA*.

LLORY, M., *L'accident de la centrale nucléaire de Three Mile Island*, Paris, L'Harmattan, 1999.

TURNER, B.A. and N.F. PIDGEON, *Man-made Disasters*, 2nd edition, Oxford, U.K., Butterworth Heinemann, 1997.

1.2.4

FROM ACCUSATIONS TO CAUSES

*Integrating Controversies and Conflicts
into the Innovation Process*

Madeleine Akrich

Engineers are often confronted by criticism when they are trying to complete an innovation process. This happens especially when some problem arises, a situation which is almost unavoidable. People, things and organisations are accused by some of those involved to be responsible for failures. These attacks, sometimes virulent and apparently without a rational basis in relation to the innovation activities themselves, are hard to handle for engineers who are tempted to discount them. In this paper, drawing upon an example, we will show that criticisms must be considered as attempts to analyse the innovation process in a strategic way, to explain it and to define means of action on it. So they have to be taken seriously, and in fact, their management and confrontation is one of the core tasks of innovators' work.

If one would like to characterise the place of technologies within sociological reflection, one might have to appeal to a wise mixture of omnipresence and evanescence: it rarely happens that a domain of research does not, at some time or other, impose the phrasing of the question of technology, although properly speaking the sociology of techniques has been developed only recently. In a schematic outline one can distinguish between two poles in the existing literature. On the one hand, as a counterpoint to the works of economists on changing techniques, one finds research projects aiming at describing and explaining the processes of innovation; on the other hand, one has to deal with approaches which start off more directly with sociological questions and which inquire into the place and the role of technologies in the social tissue.

The first body of research has committed itself to identifying the determinants of technical choices: in doing so they have shed light on the imbroglio of social, economic, scientific and other factors which qualify for the conception of innovations and which enable us to understand how these innovations spread, how their promoters manage to attract the interest of the actors they need in order to carry out their project. What is at stake here is to relate the form and content of the apparatus with the network of partners – investment bankers, industrials, distributors, researchers... – that will be necessary to enable the project to become reality. Nevertheless these studies are limited: they break off when the final users adopt these technologies. Once the whole socio-technical system is consolidated by the work of the innovator, the final users as it were have no other choice but to come and take the place they have been assigned. In other words, this type of research tends to interpret the final users as the unproblematic extension of the innovators' work.

The second body of research is more specifically interested in the processes involved in the adoption of new technologies, and endeavours to show how the abilities of the actors and the representations they make of themselves and their environment are redefined by the appearance of new technologies. From the point of view of the division, which is operated between explainable variables, and data that need explaining, these works occupy a position that is symmetrical to that of the preceding body. They start out from a universe of rigid technologies, whence they dedicate themselves to observe the interpretative and negociatory work of the actors, who may be described chiefly as users of these technologies in the broadest possible sense.

The first body of research has thus committed itself to identifying the complex social determinations from which the technological choices originate, as opposed to analyses which would postulate a strictly defined technical or economic rationality. The second body, however, has questioned this technical determinism by evaluating the manoeuvring space the actors dispose of once the technologies have been consolidated. These two currents partly complement each other, since the second starts from where the second leaves off. Nevertheless, basing ourselves on a slightly extreme example, we would like to show that the separation that these currents establish between conception and use, does not allow one to find a simple solution to the handling of the numerous conflicts that are caused by today's new technologies.

Some of these conflicts become more wide-ranging and develop into public debates which allow the different actors to expound the way in which they interpret the articulations between technologies and the kind of society they choose. Examples of such debates are the recent discussions on genetically manipulated organisms and on the disposal of nuclear waste. However, there are a multitude of conflicts which remain on a much more modest level and which can simply culminate in a failure of the apparatus or/and the dissatisfaction and even rancour of the different parties involved. How can such situations be unblocked? What role can engineers be expected to play in this process? These conflicts often manifest themselves first of all in the form of complaints that are exchanged between actors, criticisms which allow them to hold each other mutually responsible for the problems. Of course, the parties involved consider their own criticisms to be an objective analysis of the situation, whereas they deny the accusations of the others saying they are unjustified, not to say unjust, and in fact even shocking. Generally speaking the engineers in charge of the project do not escape this trend, and in the end they are dragged along in this spiral of mutual recriminations, which prevents the conflict from being resolved in one way or another.

From the Accusation of Technology
to the Accusation of the Actors

The history of Buena Vista[1], a village in the north of Costa Rica, begins with a vast cooperation programme between the AFME (Agence Française pour la Maitrise de l'Energie) and the Central American countries, coordinated by the OAS (Organisation of American States, an inter-American organisation for cooperation and development): general studies on energy sources are set up, followed by very exhaustive socio-economic studies on particular zones that are destined to undergo experiments in connection with alternative technology. By the time Buena Vista receives a gasogene

---

[1] For more details, see: M. AKRICH, 'Essay of Technosociology: A Gasogene in Costa Rica', in P. LEMONNIER, *Technological Choices. Transformation in Material Cultures since the Neolithic*, London, Routledge, 1993, pp. 289-337.

in order to supply electricity to its inhabitants, the OAS and the Costa Rican go-between know all there is to know about this village: its climate, its geography, the age of its inhabitants, their origin, their income, etc. Going on the results of this inquiry, they have provided an organization that will take charge of the new installation: the ADI (Association de Développement Intégral, Association of Integral Development) of the village is responsible for making invoices and collecting payments. The association pays the operator, a villager who is in charge of operating the station while the ICE (Institut Costaricien de l'Electricité) has to ensure the major maintenance and to effectuate the major repairs. A complex tarification system has been worked out in order to allow all the inhabitants, whose income varies considerably, to have access to the electricity.

Unfortunately, the gasogene breaks down after 6 hours of work. The constructor blames the humidity of the wood, which in its turn is probably due to the humidity of the climate. There is a prompt decision to construct a dryer that should bring the wood to the foreseen degree of humidity. In spite of the construction of a first dryer, then a second, and in spite of visits by several experts, the gasogene still refuses to function correctly. After 14 months it has worked 200 hours in all, which means 20 minutes a day on average, and, on top of that, it has to be specified that only the gasogene properly speaking is concerned here, and not the installation as a whole: the period of time during which the villagers have been provided lighting is considerably shorter. The machine seems to have gone crazy as the months pass by; every day new dysfunctions appear and confronted with this proliferation of symptoms, the succession of experts, technicians and others at the gasogene's sickbed never manage to come to an agreement concerning the causes of this strange illness. Up to the point that, three years later, the engineers of the ICE do not know how to interpret the motor oil analysis any more. In fact, this analysis reveals the oil to be composed of 70% tar, a percentage that is 100 times higher than what is traditionally observed in an excessively polluted engine. One has reached such a level of uncertainty that the technicians prefer to consider the analysis to be valid, be it a priori completely improbable, rather than to suppose that there has been an error in the measurements. We find ourselves in a situation that seems to escape any stabilised form of rationality, in which strange and inexplicable phenomena follow one after the other, without there being a means to stop this series of disturbances. Confronted with this, what are the different actors going to do? No one wants to decide to blame fate for the sake of convenience, all the more because important efforts have been made by the villagers and by the Costa Rican institution which have intervened in the project: they want to find a solution to the problems and in order to do so they first of all have to identify the causes. A French expert who arrived 9 months after the placing of the installation, is the first who tries to fully understand how the gasogene works or why it is not working: up to then the constructor who was afraid that his gasogene would be copied, had refused to furnish its installation plans and to explain its operating principle so the Costa Rican operators did not even have detailed instructions for use of the machine at their disposal. And so the said expert draws up two manuals, an operating and a maintenance manual. Starting from this manual, the OAS creates a comprehensive schedule of 'causes, consequences and corrections of the main functioning problems'. In this way the operator has a guide at his disposal which permits him to diagnose a breakdown and to repair it.

PROBLEM	CAUSE	CONSEQUENCES	CORRECTIVE ACTION
Poor quality gas	Damp wood	Motor breaks down	Dry the wood correctly
Badly trained machine operators	Badly conceived training	Bad operation and maintenance procedures	Train the operators correctly

In fact, if one puts oneself in the position of the operator who is confronted with whichever dysfunction, one reads the schedule in a non-linear way: what one 'sees', be it the result of 'direct measures' – one is not capable of starting the engine, for instance – or of 'instrumentalised measures' – the gas temperature is too high – appears in the schedule under the denominator consequences. From there one can move up towards the problem (which is in fact the cause of the consequence), which is, to some extent, what the machine 'sees' locally. For instance: the engine may stop because the gas is of inferior quality. Next, one moves up towards the cause, that is to say, to what creates the succession of effects that are 'noticed' first by the machine and then by the operator: the gas may be of inferior quality because the wood is wet. Finally, the corrective action simply consists in removing the cause. If we examine the list of causes presented in the schedule, we see that each of them directly refers to maintenance instructions: the maintenance manual is presented as a description of the normal and necessary relations between the machine and the operator, since every violation of this contract by the operator entails a penalty in the form of a dysfunction. The breakdowns appear to be the consequence of a mistake by the operator, who becomes the scapegoat. The representatives of the OAS did not fall into this trap, because they added to the schedule a last line, which at first sight is a bit odd when compared to the preceding lines: one can read under the heading problem that 'insufficiently or badly trained operators' are due to the cause 'badly conceived training'; the consequences of this cause are 'bad operation procedures, an incapability to analyse and repair breakdowns, and bad maintenance procedures (frequency, nature)'. This last series presents itself as an attempt by the OAS to deviate the final impact of the accusation from the operator to the French technicians who were supposed to take care of the operator's training.

Thus, in this first example we see how the actors, because they find themselves in a situation they cannot get under control, are led to look for the causes of the phenomena they observe: the causation or 'accusation' (cause and accusation have the same etymological origin) they formulate in a first move with regard to the climate, then to the operator, and next to the trainers, is to be directly interpreted as an attempt at rationalisation. We also mention how in this operation the technique finds itself 'excused', which means that 'there is no cause against it', by the fact that every break-down, without any exception, is connected with the absence of respect for a functioning or maintenance instruction (which is rather exceptional; take for instance a car breakdown: the driver is seldom accused by his garage man when his starter breaks down...).

The Confrontation of the ICE with the Village:
the Accusation as an Interpretation

This first rationalisation does not succeed in being operational: the training of operators is fully completed but the machine still persists in its dysfunctions. One has to find other reasons for this absurd situation then. Two interpretational systems present themselves, one in Costa Rica where the work done by the machine itself is continued and external culprits are indicated, the other in France where the 'excuse' for the machine will be annihilated. In the one case we focus on the human actors, in the other we dive into technical issues; but even if the angle of assault varies, the target remains the same: the issue is still to continuously relate human or 'social' facts to 'technical' facts in order to regain control over this infernal machine.

In Costa Rica, the ICE and the village enter into the fray: in the hypothesis of a normal functioning of the installation, all the intermediaries (ministries, OAS...) who have allowed these two actors to enter into relations with each other via the French cooperation project, should have disappeared. Because the villagers feel that their future depends on the ICE, they focus on this actor. Their point of view is simple, but supported by numerous facts which are revealed by villagers who interact with the ICE: the ICE is not interested in this kind of technologies and has only accepted to participate in this project because of political pressure from the authorities; the technicians make use of the bureaucratic inertia of the institution to shirk their obligations; they take an enormous amount of time to undertake repairs, they do not take their job seriously, therefore the machine keeps breaking down. The ICE occupies a symmetrical position on the accusation axis: according to the ICE Buena Vista is a village that is profoundly thwarted by conflicts. As soon as the slightest problem arises, the solidarity which is built around the electricity is dissolved; because they do not agree they are unable to arrange for the wood to be cut in advance so that it would have time to dry before being burnt in the gasogene; no one undertakes to check whether the batteries for starting up the engine are sufficiently charged, nor to have them recharged in the neighbouring village should the occasion arise; these negligences entail the bad functioning of the installation and the recurrence of breakdowns. So we have here two groups accusing one another: while expelling the causes of the problems outside the group, each of them reinforces the cohesion of his own group. As long as we only consider these two groups, their strategy seems suicidal: how to actually unblock a situation when neither of them seems to be disposed to accept, at least partially, the other's arguments? In fact, the accusations are never uttered directly in the face of those they are aimed at: it is the 'intermediaries', that is to say first of all the OAS, who receive them. We shall come back to the role of these intermediaries a little bit later, but before this we are going to examine the way in which the controversy is expressed on the French side.

The Confrontation of Experts and Constructors: the Accusation as a Proof

This controversy chiefly opposes two 'actors': the constructor of the gasogene and a group of experts who belong to a research organisation, the CEEMAT.

For the engineer of the CEEMAT the cause is crystal-clear: the choice of the constructor has constituted the fundamental error which has triggered off a series of catastrophic events. What makes his interpretation interesting is that he re-analyses the role of the totality of the actors starting from the configuration of the technical apparatus. We shall only dwell upon one particularly significant element of his analysis here: the problem of humidity. According to him a substantial part of the technical problems which are encountered seem to originate from the fact that the gas used by the engine is dirty: it contains tar, humidity, dust, all kinds of things an engine scarcely tolerates. The gas coming out of a gasogene is always dirty; but in principle the filtering apparatuses which are situated at the outlet of the gasifier allow the gas to be rendered acceptable for the engine: in the case of Buena Vista, the constructor was too sure of himself and of his machine, and provided insufficient filtering. But it becomes worse, since the principal vice is located in the conception of the gasogene: the further the wood goes down in the basin of the gasogene and comes closer to the burner, the higher the temperature rises; because of this, the moisture in the wood evaporates gradually and has a tendency towards climbing up again to the top of the basin in the form of steam: at the top of the basin there is a drain-pipe system which recovers the condensed steam through the effect of a lowering of the temperature. At the bottom of the basin, the gas which is produced by the gasogene is transported to the filtering apparatuses by means of tubes which climb their way up at the outside of the basin. The disposition of these different elements is linked up with a search for efficiency: the gas, which is very hot on its way out, lets the gasogene profit from its heat and in doing so contributes to ameliorate the combustion of the wood. In the Buena Vista installation these tubes climb up to the top of the basin: the effect is dramatic, since the warming up of the top of the basin makes it impossible for the steam to condense, as a result of which it remains trapped in the gasogene. Consequently, because of this humidity, the combustion temperature is lowered, which prevents the tar from being destroyed in the process of combustion. Hence the installation is conspiring to its own dysfunction. How is it possible that the constructor made a mistake which seems grotesque in the eyes of the experts? For the engineer the explanation is simple: the constructor who is used to specialising in gasogenes working on charcoal, suddenly convinces himself that the future lies in gasogenes working on wood. To save time, he copies the models of his competitor and changes some details without being capable of realising the importance of these details, since he has not reflected personally on the functioning mechanisms of his installation. From then on it is clear that the Costa Ricans have nothing to do with the misfortunes of the gasogene which contains within itself its own contradictory principle. There is only one way out of this scheme: to redefine the technical apparatus in order that it should work correctly for its users. The line of argument is astonishingly virtuoso since it allows the continuous ascent of a causality chain that goes from the technical parameters of the installation up to the morality of its maker.

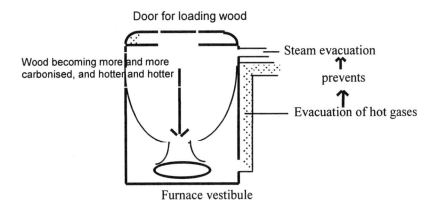

We have no time here to expand on the defence system which is put up by the constructor; let us simply say that it rests on a generalized accusation of incompetence: because the users in the broadest sense do not submit to the directions for use of the machine, some minor malfunctions appear, which could very easily be rectified by the correct application of the violated direction; instead they aggravate the situation by opening the engine and by unsettling it beyond repair, which must necessarily lead one in the direction of an exponential increase of problems.

His line of argument becomes inescapable due to a last precaution: for him, the pollution of the engine, which by the other actors was considered to be the consequence and the proof of the dysfunctions, is normal. So he forbids his critics to prove the dysfunctions by means of the engine, and even attributes the dysfunctions to the opening of the engine.

The engineer would like to engage in a confrontation with the constructor in the field of science; the latter very cleverly disqualifies this tug-of-war by replacing the technical causes by social ones, and interlaces his remarks absolutely logically with a constant menace of going to court. Here as well as before, the 'accusations' appear to be rationalising constructions of the world which have to be tested if they want to become established fact.. These tests may take place in different domains; in this case, we are dealing with science and law, but we might as well quote the entirely different domain of witchcraft. That is one of the reasons why the analysis cannot a priori choose between these interpretations; since the problem at hand in a technological process is precisely to manage to consolidate through successive tests a certain description of the world which would allow one to distribute in a stable way the competences and the responsibilities between things and human beings.

The Intermediaries: From Accusation to Negotiation

There remains another solution for this kind of controversy which is illustrated by the position of the OAS: the latter does not take up a position between the different versions that are proposed; at the crossroads of the different actors – since it is the OAS which negotiates with the French partners, the village and the ICE – it seems to acknowledge an element of truth in the different arguments. More precisely, it uses the different accusations as means of exchange in the negotiating process it tries to start off: 'OK, it says to the villagers, you rightfully complain about the intangibility of the ICE: I shall negotiate a direct access by telephone for you with the director of the ICE, but in exchange for this you have to get organised in order to supply the stock of wood correctly'. What is at stake is no longer to end the controversy by means of proof, but by means of exchange, an exchange which is negotiated time and again between the different accusations, with a view to progressively emptying the conflictual bag. To this purpose, it is also necessary that all parties involved are willing to negotiate, the machine included; because to the OAS it appears in exactly the same way in the accused-accuser standing in the dock. In this constellation, the reciprocal accusations appear to be the basis on which a new understanding can be formed between the various actors and objects in the form of a compromise that can be accepted by all the parties involved. The interpretation of the predicament then becomes the explanation of the means that are needed to turn the tide, to transform it into a success. In this way the situation would be much more dramatic if there were not an incessant volley of accusations but a state of total indifference.

This ordinary as well as exemplary story shows us that accusation and causation are both at the heart of the technological process. How can an engineer who is trapped in these kinds of accusations find the means to resolve the conflicts and make his project go ahead? As we have suggested, he has to take these accusations seriously, in the sense of forms of interpretation of the experience the actors go through. And this instantly returns to them a form of legitimacy. To take them seriously boils down to accepting them as a basis for negotiations between partners. These negotiations aim at shifting the debates progressively, at facilitating their movement from one domain to another – from politics to economics and technical issues, touching legal issues etc. – and at letting them calm down one step at a time.

Because switching fields of expertise constitutes an integral part of the innovation process, the latter cannot reduce itself to a technical project with a social dimension on top of it. Every description, even though it be the most technical one, whether produced by the promoters of the project or by its opponents, is the description of a possible world for this innovation. In other words, it is the description of the relations between the apparatus and its social and physical environment. Each technical choice entails such relations in a more or less rigid way, and proposes one or more possible scenarios for interaction. To take the accusations seriously is to understand that those who are seen by the engineer as opponents to his project do nomore than what he does himself: suggest interpretations of the world that is installed by the project. From accusation to realisation, that is the very dynamism of the technological process in which the engineer who pays attention to the world of others can play a peace-making role.

1.2.5

INTERVIEW WITH
PROFESSOR SIR JOSEPH ROTBLAT

Sally Wyatt

This chapter addresses the question of ethics in science and engineering from the perspective of someone who has been centrally involved in the promotion of socially responsible science and who has been recognised internationally for that part of his work. The second half of the chapter is the 1995 Nobel Peace Prize Acceptance Speech, delivered by Professor Sir Joseph Rotblat, on behalf of himself and the Pugwash Conferences on Science and World Affairs. Pugwash was founded in 1957, a union of scientists concerned about the relationship between science and society, particularly in light of the proliferation of nuclear weapons. Pugwash continues to hold annual conferences, bringing together scientists from all over the world. Even during the height of the Cold War, scientists met to discuss how to use science for the benefit of humankind and how to avoid using it for global destruction.

Sir Joseph Rotblat was born in Warsaw, Poland in 1908. He went to Liverpool, England in 1939 and then participated in the Manhattan Project, the Allied effort to develop nuclear weapons for use in the Second World War. He returned to England after the War and devoted himself to the development of peaceful applications of nuclear physics, especially medical applications. He was one of the founders of Pugwash, its Secretary-General between 1957 and 1973, and its President between 1988 and 1997. He continues to promote the ideals and activities of Pugwash.

---

[1] Editor's note: On 12 January 1999, Sally Wyatt inverviewed Sir Joseph at the London offices of Pugwash. She invited him to reflect on the role of Pugwash, te place of engineers, the development of newer technologies, such as artificial intelligence and biotechnologies and the possibilities for action. This interview has been modified slightly to put it into written form, with the agreement of Joseph Rotblat.

S.W.: What Has Been the Contribution of Pugwash?

J.R.: During the Cold War, our main concern was with the immediate danger of the escalation of the arms race and the possibility of the Cold War turning into a hot war. This was a real danger, with the build-up of the nuclear arsenals and the Soviet Union trying to catch up with the United States and not being able to do so. There was a possibility that some hard-liner would come to power in the Soviet Union who, not being able to keep up with the United States, would try to use the existing arms to try and solve the dispute. It did not occur because a leader emerged in the Soviet Union who realised this danger and acted in the opposite direction. I am referring to Michel Gorbachev. Pugwash played a role in this because Gorbachev was educated in these matters by our Pugwash colleagues in the Soviet Union. Gorbachev told me this himself. It had started even before with Andropov who began to realise this was the only way of dealing with the problem. This was the task of Pugwash during the Cold War years and I believe we were very successful. Maybe I am exaggerating our role, because there were so many other factors. After the end of the Cold War, the immediate danger abated. It was the end of the ideological struggle between capitalism and communism. This divide has not disappeared completely, but it has done so to a very large extent. We were then able to go back to the original objectives of Pugwash. The Russell-Einstein Manifesto states that we should aim to eliminate all war, not just nuclear war. [The Russell-Einstein Manifesto, 1955, was the founding document of Pugwash, signed by eleven leading scientists, including seven Nobel laureates, Joseph Rotblat, Albert Einstein and Bertrand Russell.] Danger will remain as long as nuclear weapons exist in the arsenal. Our short term objective is the elimination of nuclear weapons. In the long term, we have to eliminate all sources of conflict that might lead to military confrontation. At the present time, it is nuclear weapons that pose this global danger to humankind, but our worry is that research in other fields may result in new weapons of mass destruction being developed, perhaps even easier to acquire than nuclear weapons. Therefore the danger to humankind arising out of the progress of science is still with us.

Every scientist must be a philosopher because philosophy has different ways of looking at the world. Bertrand Russell was a mathematician as well as a philosopher. Albert Einstein was a scientist but also a philosopher, a philosopher of life. Russell and Einstein were paragons in this way. It is odd to talk about Russell as a paragon, given some aspects of his life, but he was concerned about issues of science and peace. When most people think about ethics, they think about religion. It is religion which provides a basis for thinking about what is good and what is evil. Scientists do not base their ideas on religion. They have to base them on facts, on things which can be explained. Therefore, scientists have to find different ways of finding reasons for ethics.

The reason for the decrease in activities such as CND [Campaign for Nuclear Disarmament] is because since the end of the Cold War, the danger is not seen as so immediate. It is hard to get people to put it back on the agenda and to make people realise that what is going on is going to lead to nuclear war. The recent examples of India and Pakistan are relevant. If one country feels it needs nuclear weapons for security, then other nations will do the same. Environmental concerns are also important: humankind can disappear either with a bang or with a whimper. Pollution is gradually poisoning us. I am glad this is being recognised as a long term danger.

S.W. What are the Responsibilities of Scientists and Engineers?

J.R.: My main concern is with the dangers of recent, very rapid progress of science and technology in this century; particularly in relation to nuclear weapons, which for the first time created the possibility of the whole of the human race being obliterated in a single act. I am concerned specifically with this issue, but of course ethical issues should be of concern to all scientists and engineers, and to every citizen. My main focus has been on scientists but engineers are much nearer to the direct application of science and therefore maybe even more important.

The work of scientists may lead to complete catastrophe. We had weapons of mass destruction before [Hiroshima and Nagasaki] but not on a scale as to endanger the whole of the human species. Genocide can be carried out with very simple things, as Hitler did during the Holocaust, with simple chemicals. Science can lead to the sort of work in which the whole of humanity is in danger. This has created a new situation. The most important feature of this century was that for the first time the whole of the human race was in danger. This should not be allowed to happen. Every citizen should be concerned by this, not only scientists. Scientists have a particular responsibility because it is their work which leads to these dangers. There is a possibility that some outcome of research in genetic engineering may lead to weapons of mass destruction. For example, one could perhaps decide to put a virus into a gene that is characteristic of a certain race.

Scientists have a special responsibility. Science is in the forefront of our intellectual exercise, of progress in our knowledge and therefore anything dangerous may result from science. Many things are beneficial; but being at this frontier of knowledge could also lead to very, very harmful outcomes. This is why science has a special responsibility.

Traditionally, engineers applied the discoveries of scientists. Now, the difference between pure and applied science, the difference between science and engineering is practically disappearing. In the past, it was almost a generation before something that appeared in the literature percolated into application. It is almost simultaneous now, which is why what I am saying applies equally to scientists and engineers.

S.W.: What Does the Future Hold?

J.R.: Conflicts arise out of ignorance. Some politicians feed off ignorance and misinformation and create an atmosphere so that people will want to go to war. With greater communication, almost everyone has access to everything that is going on in the world. It is much more difficult to manipulate people than it was in the past. There is now so much interdependence and people are more aware of each other. This will create an atmosphere more conducive to peace.

Developments in communication technology should result in a feeling of belonging to the human race, a new feeling of loyalty to humanity in general. All of us have many loyalties. Each of us has been brought up to depend our family. Gradually with the development of civilisation, we increased the circle of loyalties – a whole clan, a village. Gradually, this circle has been increasing, but it finishes now at the nation. We need to add to this circle, a much larger circle – humankind. We should realise humankind is in danger – this needs to be part of education in schools and universities.

S.W.: What Can Scientists and Engineers Do?

J.R.: We need to move away from a culture of violence toward a culture of peace. The millennium is a symbol. People use New Year's Day to make resolutions. They do not always keep them, but it is a good thing to make them. The millennium is such a symbol, a milestone in the lives of most people. We should use this occasion to try and bring in this culture of peace. We are still following the culture of security: if you want peace, prepare for war. It should be that if you want peace, prepare for peace. This is what I feel should be in education, and it should begin with teaching in elementary school.

Take the example of medicine, of the Hippocratic Oath. This has been in existence for 2500 years. It arose from the responsibility of the doctor. The life of the patient is literally in the hands of the doctor, the surgeon, or the physician. This person has a special responsibility. The oath states that the primary concern of the doctor is care of the patient. Now, the work of a scientist may affect everybody, the future, the destiny of humankind. The whole community is now the patient of the scientist. The time has come for scientists to try and formulate some ethical principles. Although the Hippocratic Oath is an ancient formula which contains some silly things about the gods and the cosmos, it remains an important symbol for a doctor to take this oath.

There are many cases in which scientists can see that what they are going to do can only be harmful; for example, people choosing a career in military research in order to develop more weapons of mass destruction. Unfortunately, many thousands of young people are being drawn into these jobs, at laboratories such as Los Alamos and Livermore. If you had such an oath, young people might consider their options. If we could convince scientists to take such an oath and not work in such establishments, these weapons would disappear because nobody would be working on them. Of course, this is utopian. There are many reasons why young people take such jobs, even if they are socially conscious. When they look for a job, the first thing is to try to find a means of living; such jobs offer better salaries than other places. Nevertheless, for some people the nature of the work is a consideration.

S.W.: What Can One Do More Specifically?

J.R.: I spoke about the Hippocratic Oath. This should apply primarily to people who are starting their careers, as with doctors. Some action is being taken. Some universities in the United States ask their students to take the following oath during degree ceremonies:

> I promise to work for a better world, where science and technology are used in socially responsible ways. I will not use my education for any purpose intended to harm human beings or the environment. Throughout my career, I will consider the ethical implications of my work before I take action. While the demands placed upon me may be great, I sign this declaration because I recognise that individual responsibility is the first step on the path to peace. (Student Pugwash USA, quoted by Rotblat, 1997, p. 247.)

S.W.: What Can Established Scientists Do?

J.R.: In medicine, if we wanted to start a research project which involved patients, the project had to be submitted to an ethics committee. The committee examined it from the point of view of what sort of harm might be done. A research project could not begin without the approval of this committee. We should apply similar procedures to other research work. In the first instance, the field which at the present time probably could lead to problems is genetic engineering. What I would like to see is that we have ethics committees set up, where we have people who are well known, respectable people in their profession, with some knowledge. A research project should be submitted to such a committee. They should look at what might be the outcome of this work. There is so much to do, we can choose. We can say, 'this project may lead to danger'. This has to be done on an international basis. It is no use having certain guidelines in one country which can be violated in another country. Take the example of an American scientist who said he would clone humans. Immediately President Clinton said this will not be allowed in the United States and the scientist said, 'I'll go to Mexico'. Of course, the Mexicans did not want to be involved. But another country might accept it. Therefore, it has to be on an international basis. I do not underestimate how difficult it is going to be. It will be very difficult.

References

ROTBLAT, J., *Scientists in the Quest for Peace, A History of the Pugwash Conferences*, Cambridge, MA, MIT Press, 1972.
ROTBLAT, J., 'Valedictory Address to the 47th annual Pugwash Conference' in *Pugwash Newsletter* 34(November 1997)4, pp. 245-250.

NOBEL PEACE PRIZE ACCEPTANCE SPEECH

Joseph Rotblat
1995 Nobel Peace Prize Laureate

Oslo, December 10, 1995

Remember Your Humanity

At this momentous event in my life – the acceptance of the Nobel Peace Prize, I want to speak as a scientist, but also as a human being. From my earliest days I had a passion for science. But science, the exercise of the supreme power of the human intellect, was always linked in my mind with benefit to people. I saw science as being in harmony with humanity. I did not imagine that the second half of my life would be spent on efforts to avert a mortal danger to humanity created by science.

The practical release of nuclear energy was the outcome of many years of experimental and theoretical research. It had great potential for the common good. But the first the general public learned about this discovery was the news of the destruction of Hiroshima by the atom bomb. A splendid achievement of science and technology had turned malign. Science became identified with death and destruction.

It is painful to me to admit that this depiction of science was deserved. The decision to use the atom bomb on Japanese cities, and the consequent build up of enormous nuclear arsenals, was made by governments, on the basis of political and military perceptions. But scientists on both sides of the iron curtain played a very significant role in maintaining the momentum of the nuclear arms race throughout the four decades of the Cold War.

The role of scientists in the nuclear arms race was expressed bluntly by Lord Zuckerman, for many years Chief Scientific Adviser to the British Government:

> When it comes to nuclear weapons... it is the man or the woman in the laboratory who at the start proposes that for this or that arcane reason it would be useful to improve an old or to devise a new nuclear warhead. It is them, the technicians, not the commander in the field, who are at the heart of the arms race.

Long before the terrifying potential of the arms race was recognised, there was a widespread instinctive abhorrence of nuclear weapons, and a strong desire to get rid of them. Indeed, the very first resolution of the General Assembly of the United Nations – adopted unanimously – called for the elimination of nuclear weapons. But the world was then polarised by the bitter ideological struggle between East and West. There was no chance to meet this call. The chief task was to stop the arms race before it brought utter disaster. However, after the collapse of communism and the disintegration of the Soviet Union, any rationale for having nuclear weapons disappeared. The quest for their total elimination could be resumed, but the nuclear powers still cling tenaciously to their weapons.

Let me remind you that nuclear disarmament is not just an ardent desire of the people, as expressed in many resolutions of the United Nations. It is a legal commitment by the five official nuclear states, entered into when they signed the Non-Proliferation Treaty. Only a few months ago when the indefinite extension of the Treaty was agreed, the nuclear powers committed themselves again to complete nuclear disarmament. This is still their declared goal. But the declarations are not matched by their policies, and this divergence seems to be intrinsic.

Since the end of the Cold War the two main nuclear powers have begun to make big reductions in their nuclear arsenals. Each of them is dismantling about 2000 nuclear warheads a year. If this program continued, all nuclear warheads could be dismantled in little over ten years from now. We have the technical means to create a nuclear-weapon-free world in about a decade. Alas, the present program does not provide for this. When the START-2 treaty has been implemented and remember it has not yet been ratified – we will be left with some 15,000 nuclear warheads, active and in reserve. Fifteen thousand weapons with an average yield of 20 Hiroshima bombs.

Unless there is a change in the basic philosophy, we will not see a reduction of nuclear arsenals to zero for a very long time, if ever. The present basic philosophy is nuclear deterrence. This was stated clearly in the US Nuclear Posture Review which concluded: 'Post-

Cold War environment requires nuclear deterrence', and this is echoed by other nuclear states. Nuclear weapons are kept as a hedge against some unspecified dangers.

This policy is simply an inertial continuation from the Cold War era. The Cold War is over but Cold War thinking survives. Then, we were told that a world war was prevented by the existence of nuclear weapons. Now, we are told that nuclear weapons prevent all kinds of war. These are armaments that purport to prove a negative. I am reminded of a story told in my boyhood, at the time when radio communication began.

> Two wise men were arguing about the ancient civilisation in their respective countries. One said: 'My country has a long history of technological development; we have carried out deep excavations and found a wire, which shows that already in the old days we had the telegraph.' The other man retorted: 'We too made excavations; we dug much deeper than you and found ... nothing, which proves that already in those days we had wireless communication!'

There is no direct evidence that nuclear weapons prevented a world war. Conversely, it is known that they nearly caused one. The most terrifying moment in my life was October 1962, during the Cuban Missile Crisis. I did not know all the facts – we have learned only recently how close we were to war – but I knew enough to make me tremble.

The lives of millions of people were about to end abruptly; millions of others were to suffer a lingering death; much of our civilisation was to be destroyed. It all hung on the decision of one man, Nikita Krushchev: would he or would he not yield to the US ultimatum? This is the reality of nuclear weapons: they may trigger a world war; a war which, unlike previous ones, destroys all of civilisation.

As for the assertion that nuclear weapons prevent wars, how many more wars are needed to refute this argument? Tens of millions have died in the many wars that have taken place since 1945. In a number of them nuclear states were directly involved. In two they were actually defeated. Having nuclear weapons was of no use to them.

To sum up, there is no evidence that a world without nuclear weapons would be a dangerous world. On the contrary, it would be a safer world, as I will show later.

We are told that the possession of nuclear weapons – in some cases even the testing of these weapons is essential for national security. But this argument can be made by other countries as well. If the militarily most powerful – and least threatened – states need nuclear weapons for their security, how can one deny such security to countries that are truly insecure? The present nuclear policy is a recipe for proliferation. It is a policy for disaster.

To prevent this disaster for the sake of humanity – we must get rid of all nuclear weapons.

Achieving this goal will take time, but it will never happen unless we make a start. Some essential steps towards it can be taken now. Several studies, and a number of public statements by senior military and political personalities, testify that, except for disputes between the present nuclear states – all military conflicts, as well as threats to peace, can be dealt with using conventional weapons. This means that the only function of nuclear weapons, while they exist, is to deter a nuclear attack. All nuclear weapon states should now recognise that this is so, and declare – in Treaty form – that they will never be the first to use nuclear weapons. This would open the way to the gradual, mutual reduction of nuclear arsenals, down to zero. It would also open the way for a Nuclear Weapons Convention. This would be universal; it would prohibit all possession of nuclear weapons.

We will need to work out the necessary verification system to safeguard the Convention. A Pugwash study produced suggestions on these matters. The mechanism for negotiating such a Convention already exist. Entering into negotiations does not commit the parties. There is no reason why they should not begin now. If not now, when?

So I ask the nuclear powers to abandon the out-of date thinking of the Cold War period and take a fresh look. Above all, I appeal to them to bear in mind the long-term threat that nuclear weapons pose to humankind and to begin action towards their elimination. Remember your duty to humanity.

My second appeal is to my fellow scientists. I described earlier the disgraceful role played by a few scientists, caricatured as 'Dr. Strangeloves' in fuelling the arms race. They did great damage to the image of science.

On the other side there are the scientists, in Pugwash and other bodies, who devote much of their time and ingenuity to averting the dangers created by advances in science and technology. However, they embrace only a small part of the scientific community. I want to address the scientific community as a whole.

You are doing fundamental work pushing forward the frontiers of knowledge, but often you do it without giving much thought to the impact of your work on society. Precepts such as 'science is neutral' or 'science has nothing to do with politics,' still prevail. They are remnants of the ivory tower mentality, although the ivory tower was finally demolished by the Hiroshima bomb.

Here, for instance, is a question: Should any scientist work on the development of weapons of mass destruction? A clear 'no' was the answer recently given by Hans Bethe. Professor Bethe, a Nobel Laureate, is the most senior of the surviving members of the Manhattan Project. On the occasion of the 50th Anniversary of Hiroshima, he issued a statement that I will quote in full.

> As the Director of the Theoretical Division of Los Alamos, I participated at the most senior level in the World War II Manhattan Project that produced the first atomic weapons.
>
> Now, at age 88, I am one of the few remaining such senior persons alive. Looking back at the half century since that time, I feel the most intense relief that these weapons have not been used since World War II, mixed with the horror that tens of thousands of such weapons have been built since that time – one hundred times more than any of us at Los Alamos could ever had imagined.
>
> Today we are rightly in an era of disarmament and dismantlement of nuclear weapons. But in some countries nuclear weapons development still continues. Whether and when the various nations of the World can agree to stop this is uncertain. But individual scientists can still influence this process by withholding their skills.
>
> Accordingly, I call on all scientists in all countries to cease and desist from work creating, developing, improving and manufacturing further nuclear weapons – and, for that matter, other weapons of potential mass destruction such as chemical and biological weapons.

If all scientists heeded this call there would be no more new nuclear warheads; no French scientists at Moruroa; no new chemical and biological poisons. The arms race would be over.

But there are other areas of scientific research that may directly or indirectly lead to harm to society. This calls for constant vigilance. The purpose of some government or industrial research is sometimes concealed, and misleading information is presented to the public. It should be the duty of scientists to expose such malfeasance. 'Whistle-blowing' should become part of the scientist's ethos. This may bring reprisals; a price to be paid for one's convictions. The price may be very heavy, as illustrated by the disproportionately severe punishment of Mordechai Vanunu. I believe he has suffered enough.

The time has come to formulate guidelines for the ethical conduct of scientists, perhaps in the form of a voluntary Hippocratic Oath. This would be particularly valuable for young scientists when they embark on a scientific career. The US Student Pugwash Group has taken up this idea – and that is very heartening.

At a time when science plays such a powerful role in the life of society, when the destiny of the whole of mankind may hinge on the result of scientific research, it is incumbent on all scientists to be fully conscious of that role, and conduct themselves accordingly. I appeal to my fellow scientists to remember their responsibility to humanity.

My third appeal is to my fellow citizens in all countries: Help us to establish lasting peace in the world.

I have to bring to your notice a terrifying reality: with the development of nuclear weapons, man has acquired, for the first time in history, the technical means to destroy the whole of civilisation in a single act. Indeed, the whole human species is endangered, by nuclear weapons or by other means of wholesale destruction which further advances in science are likely to produce.

I have argued that we must eliminate nuclear weapons. While this would remove the immediate threat, it will not provide permanent security. Nuclear weapons cannot be disinvented. The knowledge of how to make them cannot be erased. Even in a nuclear-weapon-free world, should any of the great powers become involved in a military confrontation, they would be tempted to rebuild their nuclear arsenals. That would still be a better situation than the one we have now, because the rebuilding would take a considerable time, and in that time the dispute might be settled. A nuclear-weapon-free world would be safer than the present one. But the danger of the ultimate catastrophe would still be there.

The only way to prevent it is to abolish war altogether. War must cease to be an admissible social institution. We must learn to resolve our disputes by means other than military confrontation.

This need was recognised forty years ago when we said in the Russell-Einstein Manifesto:

> Here then is the problem which we present to you, stark and dreadful, and inescapable: shall we put an end to the human race, or shall mankind renounce war?

The abolition of war is also the commitment of the nuclear weapon states; Article VI of the NPT calls for a treaty on general and complete disarmament under strict and effective international control.

Any international treaty entails some surrender of national sovereignty, and is generally unpopular. As we said in the Russell-Einstein Manifesto:

> The abolition of war will demand distasteful limitations of national sovereignty.

Whatever system of governance is eventually adopted, it is important that it carries the people with it. We need to convey the message that safeguarding our common property, humankind, will require developing in each of us a new loyalty: a loyalty to mankind. It calls for the nurturing of a feeling of belonging to the human race. We have to become world citizens.

Notwithstanding the fragmentation that has occurred since the end of the Cold War, and the many wars for recognition of national or ethnic identities, I believe that the prospects for the acceptance of this new loyalty are now better than at the time of the Russell-Einstein Manifesto. This is so largely because of the enormous progress made by science and technology during these 40 years. The fantastic advances in communication and transportation have shrunk our globe. All nations of the world have become close neighbours. Modern information techniques enable us to learn instantly about every event in every part of the globe. We can talk to each other via the various networks. This facility will improve enormously with time, because the achievements so far have only scratched the surface, Technology is driving us together. In many ways we are becoming like one family.

In advocating the new loyalty to mankind, I am not suggesting that we give up national loyalties. Each of us has loyalties to several groups – from the smallest, the family, to the largest, at present, the nation. Many of these groups provide protection for their members. With the global threats resulting from science and technology, the whole of humankind now needs protection. We have to extend our loyalty to the whole of the human race.

What we are advocating in Pugwash, a war-free world, will be seen by many as a Utopian dream. It is not Utopian. There already exist in the world large regions, for example, the European Union, within which war is inconceivable. What is needed is to extend these to cover the world's major powers.

In any case, we have no choice. The alternative is unacceptable. Let me quote the last sentence of the Russell-Einstein Manifesto:

> We appeal, as human beings, to human beings: Remember your humanity and forget the rest. If you can do so, the way lies open for a new paradise; if you cannot, there lies before you the risk of universal death.

The quest for a war-free world has a basic purpose: survival. But if in the process we learn how to achieve it by love rather than by fear, by kindness rather than by compulsion; if in the process we learn to combine the essential with the enjoyable, the expedient with the benevolent, the practical with the beautiful, this will be an extra incentive to embark on this great task.

Above all, remember your humanity.

1.3

REFLECTION

1.3.1

HANDLING ETHICAL DILEMMAS IN EVERYDAY ENGINEERING WORK

Boel Berner

The article discusses ethical dilemmas arising in everyday engineering work. It takes as its point of departure ethical notions of precaution and not causing harm and discusses what possibilities there are for individual engineers to act in accordance with these principles in situations of stress and contradictory demands. Hirschman's notions of Exit, Voice and Loyalty are discussed as concrete alternatives facing engineers. The article especially focuses on arguments for and against a Loyalty option; i.e. one which does not involve taking an ethically concerned stance. The discussion is then broadened to include issues of collective responsibility among engineers and public measures to help individual engineers show 'civil courage' in ethically difficult situations. Finally, organisational changes are discussed which may support individual engineers and create ethically responsible environments for engineering work.

Introduction: Loyalty and Responsibility in a Historical Perspective

> The engineer stands – because of his scientific training – on the same social level as the physician and the lawyer. His occupation is basically a free occupation, or what the English call a 'profession'. This is, without question, a valuable privilege and one that is to be safeguarded and honoured.

These words were expressed by a Swedish engineer in 1928 during a debate on engineering ethics within the Swedish Association of Graduate Engineers, an organisation for all engineers with higher education degrees. A Code of Honour was adopted the following year. The code and the discussion reflected a self-assuredness on the part of a professional group that, according to the last paragraph in the code, clearly saw itself as being 'in the service of society, the fatherland, and humanity.'[1]

The Age of Technology had arrived. Having specialised technical-scientific training, as do engineers, was a privilege in the rising industrial society – a ticket into the industrial and national elite. Engineers formed a relatively small profession, just as they did elsewhere in Europe and the United States. They comprised only one or two percent of the gainfully employed men.[2] The university-educated engineers, in particular, saw their work as a 'free' occupation, also when they, as was mostly the case, were employed within private enterprise or public administration. Expertise brought authority and a career. Engineers were the boss's right hand – or they were the bosses themselves, as company owner or manager.

This was the situation when Swedish engineers first discussed a common ethical stance toward their technical work. The quotation above indicates that this strong position was a privilege that had to be safeguarded. Upstarts and swindlers must not be allowed to sully the prestige the nation's engineers sought and, to a great extent, had achieved. Consequently, the code of honour stressed the importance of honesty in business, impartiality, and collegial behaviour. *Loyalty* was a key word – towards colleagues, employers and employees – and the honour of the profession was to be held high.

It was during this period, in the early decades of the 20th century, that the status and self-perception of the engineering profession were established in many European countries. The image of a profession 'of the future' was reinforced over the years, most notably during the period following World War II. Engineering jobs and education expanded. Few people questioned the positive force of technology. Plastics, household appliances, automobiles, and nuclear power plants were expected to bring prosperity for all. The 1950s and 60s brought a debate about the arms race and about nuclear tests, but few engineers raised their voice. In general, the ethics of technology was not discussed among Swedish engineers.

---

[1] *Code of Honour of the Swedish Association of Graduate Engineers*, 1929, Article 1.
[2] G. AHLSTRÖM, *Engineers and Industrial Growth*, London, Croom Helm, 1982.

Today, toward the end of the century, the situation is different. Three important changes tend to make the position of the engineer more problematic than before. I base my discussion on Sweden, but the situation seems to be similar in other European countries.[3]

Diversity and New Demands

First, we should note that there are many more, and different kinds of engineers than before. The 1990s have seen another great expansion of engineering education. Today, about 7% of Sweden's gainfully employed persons are engineers with some kind of middle or higher level of training. To be a mechanical or electrical engineer is the single most important occupation for Swedish men. In recent years, women have also entered the profession in unprecedented numbers. They now comprise about 16% of university trained engineers and about one fourth of the students at technical universities.[4]

Thus, we can no longer view even the university-educated engineers as a homogeneous group with a common culture or ethos to support them. Engineers come from varied social and gendered backgrounds. They work in many different technical areas, with research, design, production, consultancy, sales, administration etc.

Secondly, the social context of engineers' work has changed. Most university trained engineers today work in large, bureaucratic organisations within industry, consulting firms, and public agencies. Their relationship to workers and to clients is often indirect. New forms of 'flat', project-based organisations may increasingly replace the traditional hierarchical ones. But the pressure to perform and compete, and to conform is still very strong. Many engineers consider themselves as mere 'cogs in the wheels', while others have leading and responsible positions. All in all, however, engineering may no longer be the 'free' and 'privileged' profession it was once deemed to be.

---

[3] See B. BERNER, 'Professional or Wage Worker? Engineers and Economic Transformation in Sweden,' and other articles in P. MEIKSINS and C. SMITH (eds.), *Engineering Labour: Technical Workers in Comparative Perspective*, London, Verso, 1996.

[4] B. BERNER, 'Explaining Exclusion: Women and Swedish Engineering Education from the 1890s to the 1920s' in *History and Technology* 14(1997), pp. 7-29; B. BERNER and U. MELLSTRÖM, 'Looking for Mister Engineer. Understanding Masculinity and Technology at two Fin de Siècles' in Boel BERNER (ed.), *Gendered Practices*, Stockholm, Almqvist and Wiksell International, 1997, pp 39-68.

Thirdly, technology is today not seen as an unquestioned good. Nor is it any more considered the exclusive domain of engineers alone. New kinds of technology related risks and conflicts have appeared in society. The protests against what engineers have created have been numerous in recent decades - from the protests against chemical hazards and 'The Silent Spring' in the early 1960's, over workers' strikes during the 1960s and 70s against inhuman work conditions, to the movements against nuclear power and the technological arms race from the 1970s onwards. Currently, it is computer technology and biotechnology that arouse public fear and concern.

In the past, many engineers considered themselves to have done their part for 'society, the fatherland, and humanity' by doing their job carefully and loyally, and by developing the technology virtually everyone desired. Today, after Hiroshima, Seveso, and Chernobyl, the very rationality of what an engineer does has been called into question. In the past, also, it was enough for the engineer to be loyal to colleagues, clients, and employers. Today such loyalty is often seen as a way of shirking wider responsibilities for one's work. Engineers today also have to take other 'stakeholders' demands into consideration. They have to think of the possibly wasteful or harmful effects of their technology upon society at large, and on the environment in times to come. And in their everyday work, they have to consider an increasing number of government regulations on acceptable emissions into the air and water, on noise levels, on food additives, on safety measures at factories, and so on. Engineering decisions are definitely no longer about technical details alone.

What does this new situation mean for the individual engineer? What is his or her responsibility in a complex and potentially dangerous socio-technical world? How should he or she act, and whose interests does he or she represent?

Bases for an Engineering Ethic

Let me start my discussion of these issues with an article by Kenneth D. Alpern, entitled 'Engineers as Moral Heroes' from the book *Beyond Whistleblowing: Defining Engineers' Responsibilities*. Alpern assumes that it is immoral to hurt others. This principle, which is part of all ethical theory, can be restated as follows:

Other things being equal, one should exercise due care to avoid contributing to significantly harming others.[5]

'Due care' means, among other things, that one must be aware of the injury one's actions can cause, take reasonable steps to avoid such injury, and be ready and willing to make reasonable sacrifices to minimise imminent danger.

'Injury' and 'harm' are vague concepts. I take them here to mean not only bodily and psychological harm to individuals, but also undue disruption of communities, environmental damage, and disobedience of those rules and laws intended to protect individuals, communities and the environment from harm.

The principle is formulated more in terms of *contributing to* rather than *causing* injury, since the responsibility is to avoid having *any part* in the occurrence of injury: to avoid playing a directly causative role, and to avoid creating conditions from which injury could reasonably be expected to arise. The care that must be exercised is a function both of the magnitude of the injury that can arise and of how *central* one's own role is to the occurrence of the injury.

This principle is mainly negative. One's responsibility with regard to actively promoting *positive* conditions, such as equity, justice, or environmental sustainability, is not explicitly stated. I believe such considerations important. The *imperative of responsibility* for the well being of future generations, formulated by the philosopher Hans Jonas, will thus be the second pillar for the engineering ethics advocated here.[6]

It is, of course, often difficult to determine in advance the effects, both positive and negative, of a new technology or engineering decision. Another difficulty is that others may use one's knowledge or products in ways one does not agree with. We are here touching upon an issue which will be important on the pages to follow.

Engineers are proud of their central role in creating new technologies and constructing the infrastructure and industries society

---

[5] K.D. ALPERN, 'Engineers as Moral Heroes' in V. WEIL (ed.), *Beyond Whistleblowing: Defining Engineers' Responsibilities*, Chicago, Illinois Institute of Technology, 1982, p. 41.

[6] H. JONAS, *The Imperative of Responsibility: In Search of an Ethic for the Technological Age*, Chicago, Chicago University Press, 1984.

needs. On the other hand, many dispute their centrality when it comes to failures and harmful technology. 'Why place the responsibility on us, insignificant cogs in the machinery, when technical shortcomings and unethical decisions are actually the fault of *others*?', they say. Why should *we* be the target of criticism - and why must *we* be the moral heroes and blow the whistle over dangerous technology?

The Importance of Civil Courage

In fact is there really a special responsibility for the engineer? And must he or she be a moral hero? Let us take the second question first. Nobody expects the engineer to make unreasonable sacrifices and express a sort of saintly heroism. What seems reasonable to demand, however, is for him or her to show strength of mind, or *civil courage* in everyday life. Joseph Weizenbaum has phrased it well in his book *Computer Power and Human Reason*:

> It is a widely held but a grievously mistaken belief that civil courage finds exercise only in the context of world-shaking events. To the contrary, its most arduous exercise is often in those small contexts in which the challenge is to overcome the fears induced by petty concerns over career, over our relationships to those who appear to have power over us, over whatever may disturb the tranquillity of our mundane existence.[7]

The idea is that it is sufficient to meet *ordinary* moral requirements, even under what may be *unusual or abnormal situations* of stress, power play and contradictory demands.

I will in this article discuss the alternatives facing an engineer placed in extraordinary situations where civil courage may be demanded of him or her. I am well aware that the dilemmas faced by engineers are often encountered also by other, non-technical personnel. Technology is created, produced, sold and maintained in socially complex processes with many social groups involved – through what we may call 'heterogeneous engineering'.[8] I will

---

[7] J. WEIZENBAUM, *Computer Power and Human Reason: From Judgment to Calculation*, San Francisco, Freeman, 1976, p. 276.

[8] See e.g. J. LAW, 'Technology and Heterogeneous Engineering: The Case of Portugese Expansion' in W.E. BIJKER, T.P. HUGHES and T. PINCH (eds.), *The Social Con-*

nevertheless argue that engineers have a special responsibility for creating 'good' technology and avoiding harming others. Through their training and experience they usually have more information, knowledge, and control over the shaping of technology than most ordinary people, or their non-technical colleagues. They have, it seems to me, a special responsibility to evaluate the risks and problems of technology, and act with insight and care. If there are alternatives and if their actions have an impact on what will happen, then they are morally responsible for the results of those actions.

At the same time, I am well aware, that the very exercise of their profession may place engineers in out of the ordinary situations that give rise to dilemmas between differing requirements. Such situations arise, firstly, from the fact that engineers often deal with as yet unproven methods or technical solutions that can produce unforeseen effects on humans and the environment. Secondly, they arise from the conditions of engineering work. Engineers often work in large projects where not taking a certain decision may mean many millions of kronors lost to employers or shareholders. They belong to organisational milieus where there can be strong pressures from superiors and colleagues *not* to question the rationality of a possibly dubious decision.

Engineers, like other individuals, will have to think through what they individually believe is right or wrong. What should engineers do when their personal interests or loyalty to their employer come into conflict with other 'stakeholders'' demands or with generally accepted ideas of right and morality? How can they have the 'courage of their convictions' if powerful interests demand otherwise?

I shall take up some arguments and alternatives possible for the individual engineer faced with an ethical dilemma; some of them imply *not* living up to the moral standards which we may expect. This discussion will, hopefully, help the concerned engineers sort out their attitudes and possibilities. I will argue that individual responsibility is important. But I will also argue that the ethically aware engineer must be supported in his or her decisions by organisational arrangements at the workplace and in society at large. At

struction of Technological Systems, Cambridge, MA, MIT Press, 1987, pp. 111-134; B. BERNER, *Perpetuum Mobile? Teknikens utmaningar och historiens gång* [*Perpetuum Mobile? Technology's Challenges and the Course of History*], Lund, Arkiv, 1999.

the end of the article I will therefore discuss the social support and solutions necessary to prevent dilemmas from arising, and which may help individual engineers make morally responsible choices in their everyday work.

The Alternatives

I will in the following use a typology developed by the American political scientist Alfred Hirschman, who discussed people's behaviour when faced with a difficult choice. For example, in the choice between various products on the market there are basically three ways in which a person can behave: by EXIT, VOICE, or LOYALTY.[9]

Similarly, it is conceivable that an engineer who is put into a situation that requires him or her to show the courage of his or her convictions would react in one of these three ways.

First of all, the uncertain engineer could resign and look elsewhere for work that will not place them in that type of moral dilemma (EXIT). This choice need not represent a significant sacrifice for an engineer with unique skills in an expanding job market. The company, on the other hand, suffers a genuine loss of a useful engineer. For another engineer who may have put down roots in a place where there are few other jobs, who is a bit older or who specialises in an area that is not so attractive on the job market, the EXIT alternative may mean a tragic and forced escape. It is a form of silent protest that mainly affects the individual employee themself.

A well-publicised case in Sweden during the 1980s involved graduate engineer Ingvar Bratt, who worked at the arms manufacturer Bofors for many years before he decided to leave for ideological reasons.[10] He no longer wished to manufacture and sell arms to poor countries. He viewed protesting against the company's objectives within the framework of his job as meaningless and psychologically highly demanding. The ethos of the entire company, and of the community where the company was located, was based on pride in Bofors' success and technical capabilities.

---

[9] A. HIRSCHMAN, *Exit, Voice and Loyalty: Responses to Declines in Firms, Organisations and States*, Cambridge, MA, Harvard University Press, 1970.

[10] I. BRATT, *Mot rädslan* [*Against Fear*], Stockholm, Carlssons, 1988.

Secondly, the morally responsible engineer can refuse to carry out unethical tasks, demand changes in company policy or practices, or – if this fails – publicly 'expose' unethical or illegal conduct (VOICE). This is what in the US is called 'whistleblowing'.

Let us return to Ingvar Bratt. He also discovered while working at Bofors that the company was involved in arms deals that were in conflict with Swedish law. After leaving the company and after a period of anguish – and with the help of the respected peace group Swedish Peace and Arbitration Association – he decided to sound the alarm. He demonstrated the courage of his convictions while facing a difficult decision. The result was an extensive social debate and a trial against the company. Ingvar Bratt was seen both as a hero and as a social pariah in his hometown, where many were loyal to Bofors.

The third alternative is for the engineer to avoid exit and voice. They will put the interests of the company, their colleagues or the organisation first, even though this could mean injury to others or violate the laws of the land. They thus may order others to commit unethical actions or will commit them themselves. They will keep quiet about problems, or will not bother to investigate suspected irregularities. They do so because they do not want or do not dare to protest, because they see no other alternative, or because they believe a certain action to be reasonable, even though it could conceivably cause injury to others or be illegal. They may also think that loyalty in some cases is the best way to prevent a larger harm, or may change company or agency policy in the long run.

I do not know how common such LOYALTY is among engineers facing a moral dilemma. But many arguments have been developed to justify such a position. I will now spend some time discussing three of them and present several counterarguments.[11]

Arguments Against the Courage of One's Convictions

1. *'It is not my job'*

Or as the American songwriter Tom Lehrer put it:

[11] The discussion is influenced by ALPERN, art. cit. and D.F. THOMPSON, 'Moral Responsibility of Public Officials: The Problem of Many Hands' in *American Political Science Review* 74(1980), pp. 905-916.

> Once the rockets go up, who cares where they come down.
> That's not my department, says Wernher von Braun...

What reasons may be given for not accepting responsibility for a certain product or decision?

a) *Ignorance.* Engineers sometimes claim that they do not really know what is happening in the other departments of the company, how the results of their work will be used, in what larger context it will be included, or what the cumulative effects will be.

Modern technology is frequently highly complex. Many engineers work with detailed assignments or with individual components, and are often separated from one another geographically and departmentally. It is difficult for the individual engineer to obtain the big picture. They have to concentrate on their own task and will interpret the probability of various risks from this perspective. But they may not see the other parts or the context in which their component will be used – or they do not care. That is 'someone else's department.'

Is this a reasonable attitude? Can engineers reasonably claim ignorance of the technology they are working with?

Certain technologies are the objects of public debate, scientific controversy, or legal investigation. It is reasonable to expect that a responsible engineer will be familiar with the state of the art, the discussions, and the arguments of the various sides. They will reflect over whether they want their knowledge to be used for possibly controversial or ethically uncertain purposes.

A responsible engineer also, it seems to me, does not limit their interest to the little part or problem that they work with every day. They try to take a broader point of view. This is necessary, not least because technology today often consists of linked, complementary systems. Problems, symptoms, or courses of events in one part will interact with those in another; communication with others and attempts to get an overview are required of those who use and design such technology. The conventional 'tunnel vision' of one's task is no longer adequate.

The events in Chernobyl in April of 1986 is an, albeit extreme, case in point. It shows the catastrophic results of having too narrow a definition of what one has to know in order to perform one's task. The engineers running an experiment involving certain reactor processes were ignorant of the effects of the experiment on other parts of the system – with the now well-known catastrophic results.

b) *Lack of influence.* The 'loyal' engineer sometimes claims that they cannot influence events, even if they are aware of the problems. It is *someone else* who makes the decisions. The responsibility should be placed on that person or level in the hierarchy.

It may be argued against this position that many engineers actually are in decision-making positions. To a very high degree, compared to other occupations, they have what the statisticians call 'leading' or 'independent' tasks. They have a large measure of discretion, even though they do not always make the final decisions. Due to their frequently exclusive knowledge they, as advisors within the company or as outside consultants, have significant possibilities of influencing what the alternatives facing the decision-maker will be.

On the other hand, of course, many engineers work within narrow bounds that are specified for them. The hierarchical structure at many workplaces prevents any independent decisions concerning the type of technology they are to develop and what is to be done with the results. Engineers can try to influence their tasks and complain to their superiors about problems, but it is these superiors who decide what is to be done with their criticism.

Many engineers assume that the individuals who make the decisions at a reputable company will normally do so in a morally responsible manner. They will correct mistakes and appreciate having mistakes pointed out to them. It can come as a shock to the protesting individual when this does not happen but perhaps, to the contrary, they are punished or harassed by their superiors because they made their views known.

In situations when the individual experiences or has reason to suspect their criticism will not be appreciated – but when the faults they have discovered are serious – their loyalty is put to the test. A second type of arguments will come to the fore, against demonstrating the courage of one's convictions.

2. *'If I am not loyal, I will lose my job'*

There are cases when a superior threatens the engineer with dismissal, transfer, an end to their career, or worse working conditions, if they are not loyal. Similarly, a consultant may find that if they fail to give the client what they want, i.e. a statement in their interest – even when it goes against the public interest – then they may not be hired again. Consequently, they might feel obliged to censure themselves, omit important issues in their report, overestimate the advantages of the project (a road, tunnel, bridge, chemical plant, etc.) and underestimate its cost or environmental impact. All this to promote their career or avoid losing assignments in the future.[12]

Other situations may force the engineer him- or herself to initiate an immoral action. He or she falsifies test reports in a direction favourable to the company, ignores unfavourable information about the company's products, disregards information concerning a dangerous working environment, orders others to circumvent the law and the regulations. This may be done for 'good' reasons, to help the company stay in business, or avoid laying off employees. The engineer may have seen others commit such irregularities or come to understand that they are expected to occur routinely. He or she is merely following the implicit rules. And not doing so may jeopardise his or her job.

Normally, most people agree that the general interest should take precedence before the personal interest. A certain action is immoral, even when the engineer feels forced to do it, by order or implicit understandings. In some such cases, the engineer is a *victim*. They must either abandon their convictions or make a personal sacrifice due to *someone else's* immorality.

It may be argued that situations of this kind are within the rules of the game, so to speak. The engineer should have known when they sought this position or contract that such moral dilemmas could arise. It should have been part of their consideration of the advantages and disadvantages of a certain job, and if possible they should have stated their position clearly at that stage. On the other hand, it often happens that the problems and pressures do not appear until after a certain period of time. They need not even be unpleasant, but simply a friendly reminder from colleagues or superiors that 'we have always done things this way.' Thus, it is difficult to protest or withdraw.

A third argument against VOICE or EXIT, and for LOYALTY, is then likely to appear.

---

[12] S. BEDER, 'Engineers, Ethics and Etiquette' in *New Scientist* 25(September 1993), pp. 36-41.

3. 'If I do not do it, then someone else will'

This argument can be rephrased: 'Why should *I* sacrifice myself when the damage is going to be done anyway?'

The answer, of course, is that one's moral self-respect demands it. We cannot accept just anything that violates our convictions, consideration of other stakeholders' needs, or respect for the law and the regulations.

It could, however, be argued that a certain amount of loyalty in conflict with one's own views may sometimes be reasonable. 'Someone else', i.e. less scrupulous or competent colleagues, may actually make the situation worse. They might be less aware of the risks or illegalities involved, or less inclined to do something about them.

Much of the validity of this argument depends on how the 'loyal' engineer then acts upon his convictions. We may talk of a 'critical loyalty'. This means that they should accept dubious assignments under a certain protest, rather than with their cheerful or indifferent consent. They should use their imagination and try to 'get around' and alleviate the effects of an incorrect decision. If bureaucratic routines or cultural patterns 'force' employees to act in a morally dubious manner, they should draw attention to the defects, even if they can not correct them themselves. It may also be possible for them to initiate an internal debate at the workplace and try to prevent similar situations from arising in the future. In other words, the responsible engineer has a *continuing* obligation to consider and question the uses to which their contribution is put.

If the damage can be minimised, such 'reluctant consent' may be the best moral alternative for a loyal, but nevertheless responsible engineer. We shall now turn to the organisational and social support needed for such a stance, and for those engineers that choose the EXIT or VOICE alternatives.

Collective Support

EXIT is often an individual and rather unnoticed action. It may have no significant effect on the organisation that is acting unethically (unless the person leaving is a key figure). It mainly affects the individual.

Collective Exit

Forms of collective EXIT are possible, however. A labour organisation can call on its members to conduct a work stoppage for a colleague who has been harassed. Mass resignations are a possible, although highly unusual, form of collective struggle. The professional organisation for engineers – or parts thereof – can recommend that its members refuse to carry out certain specific tasks. One example of this is the appeal made by a large number of American scientists and engineers during the 1980s in which they announced their decision not to accept research money related to the strategic defence initiative 'Star Wars'. With their exit, they demonstrated their repudiation of a costly and dangerous project.[13]

Voice and Collective Support

Thus far we have discussed the taking of an ethical position as a *duty*. Engineers should consider the impact of their work and decisions. But making an ethically informed choice should also be considered a *right*. Employees should be able to complain or refuse to do things without being harassed or fired.

The social support a 'whistleblower' needs is often quite substantial. As the experience of Ingvar Bratt bears witness, the individual acting alone is never particularly strong. In principle, the most important psychological support should be the one from one's fellow workers, the local union, and one's own professional group. In practice, however, this support is not always forthcoming. There is a considerable risk that the opposite will occur: the individual is frozen out or harassed. Professional honour and esprit de corps demand that the worried engineer should *not* sound the alarm. They should keep their opinions and their criticism to themselves, within the group, or inside the firm. No outsiders are to become involved. This becomes a dilemma when the problem is *not* solved 'within the family', when irregularities are ignored and criticism silenced.

Without the collective support of a knowledgeable and dedicated peace association, the knowledge Ingvar Bratt had concerning the company's illegal activities probably never would have gained a public voice. Another precondition in his case was the legal protection given in Sweden for people providing sensitive information to the public. An

---

[13] *Nature* 321(22 May 1986), p. 369.

agency or company that is criticised should not be allowed to force those involved to reveal the name of someone who has reported irregularities to their professional organisation or the mass media.

In recent years, several professional engineering organisations have taken on the role as 'spokespersons' for concerned engineers. New codes of ethics have been elaborated, e.g. within the Swedish Association of Graduate Engineers who, unlike the one of 1929 mentioned above, take other stakeholders' legitimate demands into consideration. The engineer's loyalty is no longer primarily to their own group or employer. They also have to consider e.g. short and long-term effects on the environment. In themselves, such codes accomplish nothing. But they do provide guidelines and a cultural legitimacy for those who wish to demonstrate thoughtfulness and the courage of their convictions; they may also help create a debate on controversial issues at the local or national level.

A code becomes more effective if there are also collective mechanisms for dealing with the conflicts that many engineers experience – between loyalty to the employer or client, on the one hand, and to the public, on the other, between commercial concerns and protection of the environment, etc. One example is provided by the American professional organisation IEEE. It has set up rewards for courageous engineers and established funds to defend ethical engineers in the courts. It has also set up procedures for investigating cases, where employers have acted improperly towards engineers, and published ratings of employers on the basis of whether they encourage ethical behaviour, or not.[14]

Making the Moral Hero Superfluous

Finally, I would like to turn this discussion around. How dependent should society be on the civil courage of engineers? Or to phrase it differently: Do we really want important moral decisions, ethically difficult compromises and complex risk assessments about technology to be the domain of individuals – who may be unaware and unprepared – within closed corporate and public environments?[15]

---

[14] BEDER, art. cit., p. 41.
[15] The problem of how societies can control trust relationships which are not embedded in structures of personal relations is discussed in S.P. SHAPIRO, 'The Social Control of Impersonal Trust' in *American Journal of Sociology* 93(1987)3, pp. 623-658.

Engineers are not infallible. In situations involving numerous possible interpretations and considerable stress they, like other people, act in a contradictory and sometimes rash manner. The character of much of today's technology does not make things easier for them. Complex technical systems place great demands on their knowledge, alertness, and sometimes integrity.

A great deal of the everyday stress on many engineers could possibly disappear if technology that is too complex, difficult to handle, and potentially catastrophic was simply never developed or put to use.[16] These are political issues which should be the object of public decision-making and debate. I will not discuss them here but refer the reader to two useful texts.[17]

Another 'political' issue is the character of the organisations involved. Also morally alert engineers very often participate in unethical decisions because of the social and economic context in which they work. It is very difficult to withstand economic and social pressures and take a morally upright stand. Can the structure of work be changed to make the moral hero superfluous? Let me mention three desirable changes at the workplace. To be sure, they do not directly affect the profit motives or political interests that ultimately lie behind much unethical behaviour. But they would set new standards, protect critical persons and assist engineers in their day-to-day decisions.

Creating a New Culture and New Practices

Today, the media and outside stakeholders have far less respect for companies and experts than they had just a few decades ago. They know that controversy exists and they are eager to expose unethical behaviour. Consequently, companies and agencies must create a good reputation both externally and internally. It does not look good if they get a reputation for not listening to well-founded criticism, so that their employees must publicly call attention to harmful products or irresponsible actions – with or without the help of outside 'muckrakers.'

---

[16] See the arguments in C. PERROW, *Normal Accidents: Living with High Risk Technologies*, New York, Basic Books, 1984.

[17] J.G. MORONE and E.J. WOODHOUSE, *Averting Catastrophe: Strategies for Regulating Risky Technologies*, Berkeley, University of California Press, 1987; A. IRWIN, *Citizen Science: A Study of People, Expertise and Sustainable Development*, London, Routledge, 1995.

What smart organisations do in this situation is to send a clear signal that they *want* criticism of established structures and procedures. Thus, some companies have created mechanisms and guidelines for how product quality and a good working environment are to be assured. Internationally imposed quality requirements may also assist the concerned engineer in her work.

Other companies try to establish what is sometimes called a 'safety culture' at work. Safety is given top priority when it comes to structures and products. Special engineering personnel are assigned to work continuously with safety issues. Engineers with other tasks are also involved in a system of safety discussions and recurrent controls, and receive extensive support via internal and external safety regulations, manuals, and incentives for continuing training. Safety considerations thus permeate the *entire* organisation and are not dependent on individual alertness and morality alone.[18]

Allowing Different Voices to Be Heard

Cultural change, however, is not enough. Organisational change is also needed.

An organisation that accepts its ethical responsibility is characterised by *'requisite variety'*.[19] Today, technical decisions are made by a very limited type of people. Engineering workplaces are dominated by white, middle-class men with the same narrow technical background. Only one type of experience is voiced – although there are many other interests, opinions, and skills in society at large. The organisation is simply not qualified to understand their legitimate demands.

It seems unreasonable to demand of an individual engineer that they represent all these demands. Consequently, the organisation as a whole must be able to match the variety it encounters in its environments with a sufficiently sophisticated internal variety. This means, that decision-making groups should consist of people with different professional backgrounds, experiences, genders, ages, abilities, and personalities. In a socially mixed environment, there are more information pathways, broader knowledge, and more thoroughly discussed alternatives. In this way, the decisions can be more

---

[18] A. FLORES, 'Designing for Safety: Organisational Influences on Engineers' in V. WEIL, (ed.), *Beyond Whistleblowing*, op. cit., pp. 153-167.

[19] B.W. HUSTED, 'Reliability and the Design of Ethical Organisations: A Rational Systems Approach' in *Journal of Business Ethics* 12(1993), p. 763.

in line with what is seen as ethically correct outside the company or agency.

'Requisite variety' may also prevent the tunnel vision that sometimes arises in work groups that are too homogeneous. The risk is that people with the same background will look at a technical problem with the same conceptual lenses. Anomalies and contradictory information may thus not be noted by anyone. As Diane Vaughan showed in her analysis of the *Challenger* accident, this consensus behind an ultimately faulty paradigm was a contributing cause of the disaster.[20]

Today, many dynamic enterprises involved with front-line technologies organise work along *ad hoc* project lines, and delegate power to working groups. Many researchers predict that such 'adhocracies' will have an advantage over old-fashioned bureaucracies in the highly competitive, uncertain, and knowledge-based environments of today and tomorrow. Different ways of thinking, different values and opinions are there considered a resource rather than a threat to established ways.[21] Hopefully, this kind of open and learning environment will also include ethical concerns and engineers will take the chance to learn about, discuss and promote visions of an ethically responsible technology.

Creating Organisational Checks and Balances

Finally, there is a danger in the present management interest in 'streamlining' and 'slimming organisations'. All 'unnecessary' personnel is to be taken away. This means stress and vulnerability for those remaining – and a potentially long-term loss of reflexivity within the organisation as a whole. To create an ethically responsible organisation, more people may be needed, not less.

Research on so-called 'High Reliability Organisations' has shown the importance of *redundancy* in creating safety in socio-technical systems.[22] Back-ups, overlapping or duplicate units increase safety

[20] D. VAUGHAN, *The Challenger Launch Decision. Risky Technology, Culture and Deviance at* NASA, Chicago and London, Chicago University Press, 1996.

[21] See B. BERNER, 'Women Engineers and the Transformation of the Engineering Profession in Sweden Today' in Shirley GORENSTEIN (ed.) *Research in Science and Technology Studies: Gender and Work* (Knowledge and Society Series, 12) Greenwich (CN), J/A, 2000.

[22] K. ROBERTS, 'Some Characteristics of One Type of High Reliability Organisation' in *Organisation Science* 1(1990), pp. 160-176.

and quality of work. Such redundant units act as a system of checks and balances to assure reflexivity and avoid unethical or unsafe decisions.

One example of redundancy is making units both in the staff and the line part of the organisation responsible for dealing with important issues such as safety or the work environment. They will do it in different ways and serve as important complements to each other. They provide several paths along which important information and legitimate concerns can be communicated to decision-makers. The risk will be smaller that critique and information will stop halfway up.

Another example is to appoint a special manager or unit for sensitive or important problems. There are in many companies today formal systems in which, for example, a senior executive outside the regular chain of command is available on a 'hot line' to deal with employee grievances and alarms on a confidential basis.[23] In other companies, a complaint department or special board is set up to actively support engineers and others who react to unethical activities and to investigate their complaints. Such advisory committees should be independent and involve people who do not have material stakes in the organisation.

Needless to say, engineers would also have a *duty* to turn to such a person or department in problematic situations. In this way, misguided loyalty could be avoided – while at the same time protecting those who want, but are unable, to raise their voice.[24]

Such an 'extra' unit or person need not only represent the interests of protesting engineers. Their task could also be to voice the concerns of workers affected by technical change. All too often, what engineers design or decide is unacceptable with regard to production workers' work environment. Skills are destroyed, workers suffer physical or psychological injury. Those affected must be given real power to influence their working environment – even if this infringes on the engineers' sphere of responsibility and right to make decisions. Such concerns must be given a voice within the company, whether through workers' representatives, or special units responsible for creating a good work environment – and thus perhaps duplicating or going against what many engineers do.

---

[23] S.C. FLORMAN, *The Introspective Engineer*, New York, St. Martin's Press, 1996, pp. 159f.

[24] HUSTED, art.cit., pp. 763ff.

Finally, 'redundant' units may be responsible for the concerns and demands of *outside* stakeholders and make them heard within the company: those concerning consumer safety, environmental problems, etc. The point of this, and other forms of redundancy is for difficult decisions or dilemmas to be dealt with by several units, and also not to 'fall between two stools'. When more interests are given consideration and a legitimate voice, there can be a more genuine process of compromise and adjustment. Knowledge will not be the privilege of the few and important information will reach more people concerned. The company and society at large will not be dependent only upon whether the individual engineer thinks or acts in a responsible way.

To sum up, cultural and organisational arrangements, along the lines suggested here, can help make the moral hero superfluous. They can act as checks on the individual engineer, give him support and a broader perspective. They do not absolve him from responsibility and care. But ideally, they should be able to help him or her act courageously in the many problematic situations that the work of the engineer will continue to hold.

References

AHLSTRÖM, G., *Engineers and Industrial Growth*, London, Croom Helm, 1982.

ALPERN, K.D., 'Engineers as Moral Heroes' in V. WEIL (ed.), *Beyond Whistleblowing: Defining Engineers' Responsibilities,* Chicago, Illinois Institute of Technology, 1982.

BEDER, S., 'Engineers, Ethics and Etiquette' in *New Scientist* 25(September 1993), pp. 36-41.

BERNER, B. and U. MELLSTRÖM, 'Looking for Mister Engineer: Understanding Masculinity and Technology at two Fin de Siècles' in B. BERNER (ed.), *Gendered Practices*, Stockholm, Almqvist & Wiksell International, 1997.

BERNER, B., 'Explaining Exclusion: Women and Swedish Engineering Education from the 1890s to the 1920s' in *History and Technology* 14(1997), pp. 7-29.

BERNER, B., 'Professional or Wage Worker? Engineers and Economic Transformation in Sweden,' and other articles in P. MEIKSINS and C. SMITH (eds.), *Engineering Labour: Technical Workers in Comparative Perspective*, London, Verso, 1996.

BERNER, B., 'Women Engineers and the Transformation of the Engineering Profession in Sweden Today' in *Knowledge and Society* (2000), forthcoming.
BERNER, B., *Perpetuum Mobile? Teknikens utmaningar och historiens gång* [*Perpetuum Mobile? Technology's Challenges and the Course of History*], Lund, Arkiv, 1999.
BRATT, I., *Mot rädslan* [*Against Fear*], Stockholm, Carlssons, 1988.
Code of Honour of the Swedish Association of Graduate Engineers, 1929, Article 1.
FLORES, A., 'Designing for Safety: Organisational Influences on Engineers,' in V. WEIL, (ed.), *Beyond Whistleblowing*, Chicago, Illinois Institute of Technology, 1983, pp. 153-167.
FLORMAN, S.C., *The Introspective Engineer*, New York, St. Martin's Press, 1996.
HIRSCHMAN, A., *Exit, Voice and Loyalty: Responses to Declines in Firms, Organisations and States*, Cambridge, MA, Harvard University Press, 1970.
HUSTED, B.W., 'Reliability and the Design of Ethical Organisations: A Rational Systems Approach' in *Journal of Business Ethics* 12(1993).
IRWIN, A., *Citizen Science: A Study of People, Expertise and Sustainable Development*, London, Routledge, 1995.
JONAS, H., *The Imperative of Responsibility: In Search of an Ethic for the Technological Age*, Chicago, Chicago University Press, 1984.
LAW, J., 'Technology and Heterogeneous Engineering: The Case of Portugese Expansion' in W.E. BIJKER, T.P. HUGHES, and T. PINCH (eds.), *The Social Construction of Technological Systems*, Cambridge, MA, MIT Press, 1987.
MORONE, J.G. and E.J. WOODHOUSE, *Averting Catastrophe: Strategies for Regulating Risky Technologies*, Berkeley, University of California Press, 1987.
PERROW, C., *Normal Accidents: Living with High Risk Technologies*, New York, Basic Books, 1984.
ROBERTS, K., 'Some Characteristics of One Type of High Reliability Organisation' in *Organisation Science* 1(1990), pp. 160-176.
SHAPIRO, S.P., 'The Social Control of Impersonal Trust' in *American Journal of Sociology* 93(1987) 3, pp. 623-658.
THOMPSON, D.F., 'Moral Responsibility of Public Officials: The Problem of Many Hands' in *American Political Science Review* 74(1980), pp. 905-916.
VAUGHAN, D., *The Challenger Launch Decision: Risky Technology, Culture and Deviance at NASA*, Chicago and London, Chicago University Press, 1996.
WEIZENBAUM, J., *Computer Power and Human Reason: from Judgment to Calculation*, San Francisco, Freeman, 1976.

1.3.2

ENGINEERS' TOOLS FOR INCLUSIVE TECHNOLOGICAL DEVELOPMENT

Christiaan T. Hogenhuis and Dick G.A. Koelega

This article defends the thesis that engineers already have quite a number of possibilities for integrating moral considerations with respect to technological development in their everyday work, both within industrial companies and research institutes as within consultancy firms and governmental agencies. First it discusses in a rather general way the questions of how technological development 'works', in what sense its direction can be changed and what is needed for that. Next it sketches the many and different roles engineers play in that process, and finally it presents a broad range of 'tools for inclusive technology' that engineers already have at their disposal in their working environment and which only wait to be used practically.

Introduction

Around technological development many ethical questions arise that need special attention. Questions like: How does one deal with the risks of long term social, economic and environmental effects of the technological development one is involved in? How can one come to a just evaluation of the fact that it may contribute to a gradual change in our view of life, ranging from seeing it as a sacred gift towards considering it as just a commodity that may be manipulated? How can one protect the interests of the poor and deprived around the world and those of future generations? How can one pay respect to the inherent worth of non-human nature?

In this article we will focus on the question of how engineers can actually deal with such ethical issues in their daily work, how they can handle ethical issues there and how they can actually work towards practical 'solutions'. However, technological development is not merely brought about by engineers. Many others in society contribute to it as well: scientists, investors, lawmakers, military, merchants, consumers, politicians, members of protest groups, et cetera. Technology, in other words, is always the collective work of many people in many different roles. Nevertheless engineers do play an important role in developing technology, be it within industrial companies, research institutes, consultancy firms or government agencies. There they are confronted with ethical issues, like the ones mentioned above. Whereas engineers often think that integrating ethics in their work is as a burden, we promote the idea that it can be a joyful and very rewarding process when one considers participating in technological development more as an 'art for life'.[1]

Autonomous Technology?

In the 19th century technology was widely seen as merely a powerful instrument for creating a better life and welcomed as something to be very proud of.[2] However, during the 20th century in which technology developed at an increasing pace, many people have become less positive and confident about technology. Nowadays they see technology as an autonomous force to which they have to submit themselves. They stress the fact that, although it offers them many useful devices, it also brings about many problems and useless gadgets. Instead of technology being the servant of mankind, they feel they have become the servants of technology. This view is called the deterministic view of technology[3] and was most fervently defended by Jacques Ellul.[4]

---

[1] C.T. HOGENHUIS and D.G.A. KOELEGA (eds.), *Technologie als levenskunst. Visies op instrumenten voor inclusieve technologie-ontwikkeling*, Kampen 1996.

[2] J.H. VAN DER POT, *Die Bewehrtung des technischen Fortschritts. Eine systhematische Übersicht der Theorien*, 2 Bände, Assen 1985, 805.

[3] L. WINNER, *Autonomous Technology*, Cambridge, MA, MIT Press, 1977, p. 53.

[4] J. ELLUL, *La technique ou l'enjeu du siècle*, Paris, Colin, 1954.

During the last two decades, however, the understanding of the way technologies develop has grown considerably. Through detailed historic, sociological as well as anthropological studies the 'black box' that technological development formerly was, has been opened.[5] These studies show that technological development is not autono-mous but a social construction within a specific social, cultural, political and natural context. At the same time these studies changed the concept of technology. Technology is not seen anymore as simply an individual tool or machine or a specific way (or technique) of manipulating materials and people. The new 'definition' of technology is that it is an interconnected body of (scientific) knowledge, skills, instruments (hardware and software), products, organisational methods and structures in research, production, marketing and use, and of the values that are expressed in all these parts.[6] These studies show that almost everyone is involved in processes of developing, implementing and revising technology, be it in their role as a scientist, engineer or businessman or as consumer, politician, voter, trade unionist, member of an environmental group et cetera. Technological development therefore is a collective activity and the borders between technology and its context have become permeable. Or as Ron Westrum[7] puts it: 'Technology and society intertwine'.

This multifaceted, connected and embedded or network character gives technology a certain inertia that resists change. However, all the interests and wishes of the people and organisations that play a part, give technology a certain dynamics as well. Some want it to be further developed or renewed, others want to develop new domains of application for it; still others want it to be regulated in a better way, etc. Once we accept the broader meaning of the concept of technology, all of these changes can be called development of technology or technological development. All of the changes together cause technology to develop in a certain way, with a certain direction and speed. With reference to another concept of physics one could say

---

[5] T.P. HUGHES, *Networks of Power. Electrification in Western Society*, 1880-1930, Baltimore, John Hopkins University Press, 1983; W.E. BIJKER, T.P. HUGHES and T.J. PINCH (eds.), *The Social Construction of Technological Systems: New Directions in the Sociology and History of technology*, Cambridge, MA, MIT Press, 1987.

[6] C.T. HOGENHUIS and D.G.A. KOELEGA (eds.), *Technologie als levenskunst. Visies op instrumenten voor inclusieve technologie-ontwikkeling*, Kampen, Kok, 1996, p. 27.

[7] R. WESTRUM, *Technology and Society: The Shaping of People and Things*, Belmont, Wadsworth, 1991, p. 4.

that technological development gets a momentum of its own, as Thomas Hughes stated.[8]

So it is the interaction of choices of many people in the past and in present times with each other and with existing social and natural constraints that gives a specific technology its shape and inertia and that give technological development its momentum, thereby creating an appearance of autonomy. In distinction of the deterministic view one could call this an 'interactionistic' view on technology.[9] The crucial difference between the deterministic view and the interactionistic view is that in the former technology has its own dynamics whatever the choices of human beings, whereas in the latter this is largely a result of their interacting choices. Whereas in the first case it is a matter of complete indifference which choices are being made by humans – which implies technological development cannot be influenced at all – in the second case human choices still can have a crucial impact on the direction and speed of technological development.

Unwanted Technology

But when technological development is the result of our choices, why then does it so often take a track no one really wants? For instance because it threatens the quality of work or the existence of living species, destroys the natural environment or affects the economic prospects of specific groups of people.

The first part of the answer is obvious: technological development is a collective activity. There is no specific willing person or 'general will' in charge. Due to the large number of contributors and the large (social, geographical en temporal) distances between them these contributions are not co-ordinated very well and often counteract each other or push in different directions. Moreover many of us contributors have different opinions as to which direction technological development should take and therefore consensus can seldom be reached. Therefore, like in all democratic decision making processes, it is unavoidable that sometimes decisions are taken that none of us really wants.

---

[8] T.P. HUGHES, *Networks of Power. Electrification in Western Society*, 1880-1930, Baltimore, John Hopkins University Press, 1983, p. 140.

[9] We prefer 'interactionist' instead of 'constructivist', because the former word shows in a clearer way the stress this approach puts on the fact technology is the outcome of the mutual interaction between human actors and between humans and nature.

There are, however, some other factors that cause 'unwanted technology' as well. In the first place we often make choices under conditions of uncertainty. Both individually and collectively, we do not always have all the information that is needed for a good decision. Moreover, the amount of information we do have or still need in order make good choices, is often enormous due to the complexity and large-scale character of many technological developments. And finally most of the time we are not aware of the fact that even the simple decisions we take in daily life are connected to a huge and mostly invisible network of technologies. When we buy a bunch of roses, for example, we do not see or think of the enormous technological network that enabled us to buy these roses. A network consisting of: the trucks that brought the flowers to the shop, the road infrastructure, the coolers, the electronic auction systems, the greenhouses with their systems for air conditioning and water purification and distribution, the fertilisers and herbicides, the computers and software programmes used for financing and insuring the greenhouses, the research laboratories which bred this new species of rose, et cetera.

In the second place, we often make choices which we thought were right at the moment we made them and according to the information available at that time, but which turn out to be undesirable afterwards, because the circumstances changed in a way we could not have foreseen. Social processes, future human wishes and natural processes are to a large extent unpredictable, however comprehensive our knowledge is.

And finally our freedom of choice is often limited, due to social constraints. We live and make our choices within a specific social environment, consisting of laws, rules and regulations, cultural norms and values, organisational patterns, habits, et cetera. Sociologists call these 'institutions'.[10] Institutions facilitate our life and behaviour, but they also promote a certain type of life or behaviour and therefore limit our freedom of action as well. One important example of a social institution is the social role. Roles are connected to social positions, like the position of a mother in a family, a teacher in a school, a nurse in a hospital, a chief executive in a firm or an engineer in a department for research and development. Roles

---

[10] T. PARSONS, *The System of Modern Society*, Englewood Cliffs, NJ, Prentice Hall, 1971; P.L. BERGER and B. BERGER, *Sociology*, New York, Basic Books, 1972.

express social expectations about how one should behave in certain social positions. These expectations are the result of the lessons that people in the past have learned and that are conveyed to us by our parents, teachers and peers and through the media, festivities, rituals et cetera. Because they have old historic roots and have precipitated in so many parts of society and in so many hearts and minds, they have a great impact on humans and cannot be changed easily. Therefore social roles determine our behaviour and choices to a large degree.

However, social roles do not determine our behaviour completely. They express general, not detailed expectations. As a result they leave some room for personal interpretation. Just like two actors who do not have to play a certain role in the same way. One can play a role in a purely traditional or conventional way, but one can also choose to try an alternative, new, more personal and challenging interpretation. Thus social roles always leave some freedom of choice. However, we do not always use this freedom either because we do not realise that it exists, or because we do not know alternative ways of action or because we do not have the courage or creativity to use them.

The factors mentioned above all contribute to the fact that it is often very difficult to guide technological developments in a 'wanted' or 'desired' direction. However, the foregoing discussion also shows that still a lot can be done to (re)direct technological development. Its direction can be changed if we find ways of making visible the specific technological systems that lay hidden behind our daily routines and products and finding ways of co-operation between those involved in specific technological development processes. In other words: if we find ways of co-ordinating the actions of as many people as possible, as well as possible and ways of sharing information and experience about as many alternatives and consequences as possible.

At first glance this may look quite unrealistic, but when one looks more closely one will see there already exist many starting points for such a strategy, even in the daily working environment of engineers, as we will show.

The important question then, is where we can find such starting points for co-operation and clarification. However, before we address that question, in particular in relation to the position and role of engineers, we will first explore what is needed for co-operation, and especially for the most problematic part of it: common decision making.

Inclusive Technological Development

Now that we know that, in principal, technological development can be influenced to some extent, the next question is what kind of technology we want, in other words: what do we think is 'appropriate technology'. That, of course, is a difficult question, because, as we have already noted, opinions on that subject differ. Most of us do agree on general values like justice, equality, peace, liberty and sustainability, but disagreement starts when it comes to choosing more specific values and goals of technological development. Are we also able to reach agreement on such a level?

Under ideal conditions, agreement could be reached through a fully democratic debate and decision making process. In such a process all people involved in or (potentially) affected by a technological development (in the following we will summarise this as 'all relevant people') would be able to participate and argue with good reasons instead of exerting power or using irrational arguments. Such a *'herrschaftsfreie Diskussion'*[11] may lead to consensus, or at least to a compromise or decision to which all participants concede or have 'no regret'. This can be called discursive and direct democracy, because the emphasis is on rational reasoning and participation of all relevant people.

Clearly, in daily social life it is not possible to use this ideal model of reaching an agreement. In the first place on many specific issues related to modern technology it is not always possible to reach a consensus or even a reasonable compromise. There one has to use different procedures of decision-making, like letting every participant issue a vote with equal weight and letting the majority rule. This can be called numerical democracy. In the second place not all the relevant people or groups are actually able to participate themselves in a decision making process. The great number of relevant people and the many different social positions and opinions they have, obstruct direct communication. Furthermore, future generations and people in many underdeveloped countries can not be

[11] J. HABERMAS, *Der Philosophische Diskurs der Moderne*, Frankfurt a.M., Suhrkamp, 1988.

present. And finally, as we have already seen, it is often not clear which people are affected due to the fact that we cannot oversee the vast network which surrounds every specific technology and all the consequences of all the choices which are being made in the process of technological development. The best thing we can do is to involve as many people as we can and see to it that these people take into account as many of the needs, interests, wishes, norms, values and ideals of all relevant people as possible, recognising that their wishes, values and ideals can reflect the needs and interests of still other people and of nature (for the sake of convenience we will refer to this broad range of needs, interests, values, wishes, norms and ideals as 'broader interests' in the following). This we call the requirement of inclusivity. It is a requirement that applies to everyone who one way or another is involved in or affected by a technological development, be it as an engineer or a consumer, etc.[12]

Of course no one can be neutral in the way he or she takes the interests of others into account and in practice power and irrationalities always play a role. Besides, not everyone is willing to be completely frank about his or her interests, to consider the interests of other people fairly and to reach a reasonable compromise. In fact some kind of neutral principles would be needed to enable us to evaluate all interests and arguments in an objective and just way as best as possible. For years ethicists have been trying to find such principles. However, nowadays most of them agree that universal principles on which all moral problems can be solved in an objective way do not exist and that the way of decision making closest to the ideal of objectivity is one of reflective equilibrium. This consists of a continuing process of going between the relevant specific facts and opinions of as many people as possible on the one hand and more general and socially shared moral principles on the other hand.[13] Moral justification in other words is seen as a way of 'improvising'[14];

---

[12] The requirement of inclusivity resembles the principle of equal representation in a parliamentary democracy. A difference is that the latter demands that the representatives are being chosen by all people and thus represent all people. In what we argue for, however, this is not possible: representatives cannot be chosen by every one who is involved in or affected by every stage of development of a specific technology.

[13] J. RAWLS, *A Theory of Justice*, New York, Harvard University Press, 1971, pp. 48-51.

[14] B.R. BARBER, 'Liberal Democracy and the Costs of Consent' in N.L. ROSENBLUM (ed.), *Liberalism and the Moral Life*, Cambridge, MA, Harvard University Press, 1989, pp. 54-68.

but that does not mean it is not rational and well structured and that moral principles, like the principle of justice, of equality, of not harming and of doing good do not play a part in it.[15] In showing how my actions and choices are related to such principles I do not give completely objective and decisive reasons for them, but still I give as good as possible reasons and that may convince other people or will show why our opinions differ in the end. Moral justification therefore should be part of our common decision-making about technological development (as about other issues).

There is another (often forgotten or undervalued) aspect of ethics which, in our view, plays (or should play) an important part in engineering ethics: ethics as the art of doing good and of thinking about what it is to live responsibly. Ethics is not only to reason well on the basis of accepted norms and values, but also to reflect critically on these norms and values and about what we consider crucial in our vision of the good life or living well. This may be called moral philosophy or reflection on the good life. Such reflection and the worldview that results from it form the background for our moral justifications. Although generally they operate at a larger distance to the practical decisions we make, they can become of great practical importance, especially when new technological developments, as often happens, challenge our current norms and values. That is why we think that in decision-making processes around new technologies from time to time there should also be room for (individual and collective) reflection about our fundamental views about the good life.

Moral justification and moral philosophy together can be described as moral deliberation. So we can state that moral deliberation should be part of a good decision making process, acknowledging that in most cases emphasis will be on moral justification.

As the foregoing discussion shows, moral deliberation often does not lead to uncontroversial and unanimous arguments for specific decisions. Therefore one often has to resort to numerical democracy. The requirement of moral deliberation is not a replacement of the requirement of democracy, but it places numerical democracy in the right perspective and shows its limited, though often indispensable value. Together moral deliberation, inclusivity and numerical

---

[15] W.K. FRANKENA, *Ethics*, Englewood Cliffs, NJ, Prentice Hall, 1973.

democracy approach the ideal of 'herrschaftsfreie' or direct discursive democracy as best as possible. Technological development that meets these requirements results in what we call Inclusive Technology.[16] Inclusive, because it takes into account the broader interests of as many relevant people as possible as well as of non-humans.

In many cases technological development is not inclusive yet. In the first place technological development is not inclusive in the sense that many people who are involved in or may be affected by technological development have no real influence in the decision-making process, nor are their interests taken into account seriously. Of course in their role as consumers they can express their wishes by buying (or not buying) certain products and marketers do their best to map those wishes. However, they usually only focus on wishes that can be translated into economic terms, whereas not all of the broader interests mentioned earlier can be translated in such terms. Next as voters people choose specific politicians. However, most of the time their choice is not guided by the opinions politicians have about specific aspects or goals of technological development, but mainly by opinions about more general socio-economic issues (employment, taxes, inflation, etc). Moreover, politicians also have limited influence: they can often only set the margins for technological developments.

In the second place technological development is not inclusive yet because the decisions about its direction are still mainly taken by a small group of investors and managers within companies.

In the third place in most companies and research institutes where decisions are being taken about technological development time and room for moral deliberation is almost absent.

In order to make technological development more inclusive and to give moral deliberation a more structural – or 'normal' – place in it, we need 'instruments' to achieve that goal. Instruments, for example for letting more relevant people and groups participate in the decision making process, for taking into account more interests and aspects of a technology and for making more room for moral deliberation 'on the job'.

---

[16] C.T. HOGENHUIS and D.G.A. KOELEGA (eds.), *Technologie als levenskunst. Visies op instrumenten voor inclusieve technologie-ontwikkeling*, Kampen, 1996, Kok, p. 248.

The Position of the Engineer

What we have said so far, about technological development as collective work of many people and about the need for moral deliberation in this process, was necessary to clear the ground for the main question we want to address in this article: what can engineers, working within technological organisations, actually do themselves to contribute to a more inclusive technological development and what kind of instruments do they (already) have at their disposal for that?

Why do we concentrate on engineers within technological organisations? The main reason is that most engineers work in this kind of context. They do technological research, develop technological products or give technological advice. And although according to our interactionistic view on technological development practically everybody is involved in technological development in several ways, engineers still have a special and important role in it. They are the only persons that have the specialised expertise for determining and evaluating the necessary technical specifications and for assessing their technical consequences. This special role implies a special responsibility, in terms of the task and obligation to see to it that the technical specifications and consequences are 'inclusive' to all interests at stake, in the sense we discussed in the preceding paragraph.[17]

Of course the first responsibility of engineers is to elaborate and analyse the technical specifications, indicating the choices that are being made in the process of formulating the specifications and technical alternatives for these, and communicating all this information to their superiors, the decision makers. At first sight it seems that it is not their responsibility to indicate and evaluate the social impacts or moral implications of the technical specifications and alternatives that are chosen. However, if one takes a better look, one sees that those tasks really do come under the engineers' responsibility as well. Engineers very often have a leading position in their organisations; many of them are managers or directors. Even those engineers

---

[17] We like to stress that responsibility is not meant here in terms of guilt and accountability. The idea that engineers are to blame for technological developments is simplistic, because according to the interactionistic view, there are many other contributors to technological development, as well as there are many natural and social constraints which have an influence.

who are not in such positions do, when they do their work, make choices that may affect people and nature. Often these choices are implicit or habitual and therefore cannot all be traced and evaluated easily by decision-makers higher in the organisational hierarchy.[18]

In other words, engineers do take moral decisions almost all the time, although they are not aware of it most of the time. Therefore they do have a moral responsibility of their own. The question, however, is: do they have practical instruments for taking up this responsibility? Do they have at their disposal ways for introducing the interests of as many relevant people outside the organisation as possible or for raising awareness of environmental consequences and for integrating moral deliberation in the decision making process inside the organisation? Also, do they have instruments for getting a clear picture of the complex network of the specific technology they work on, for sharing information on it with others inside and outside the organisation and for co-ordinating actions meant to redirect the technological development? After studying the working environment of engineers, our conclusion is that they have quite a large number of such instruments already at their disposition.

Before we sketch these instruments, however, we want to mention some other reasons why engineers have a responsibility to participate in decision-making about technological development in a more 'inclusive' way. In the first place in this way information about interests of relevant people outside the organisation can be gathered and put to use in a better way. Normal instruments for gathering information of relevant people outside the organisation, like marketing surveys or government regulations, are narrow-sighted, because they only 'see' economic or political wishes of large groups and not the broader interests and worries of specific groups or persons. Furthermore these instruments mostly function in a top down approach: they inform and address the top management of organisations. Unfortunately the top cannot evaluate the technical relevancy of all that information and does not really know what the technical alternatives are, whereas the engineers at the 'bottom' can, but do not get this information. Therefore engineers (as well as other employees within organisations) should have direct contact with 'relevant' groups outside their organisation and be allowed to participate directly in the decision-making process inside their organisation.

---

[18] L.L. BUCCIARELLI, *Designing Engineers*, Cambridge, MA, MIR Press, 1994.

In the second place engineers themselves are a 'window' to society. Apart from being engineer within some organisation, they live in other social contexts as well. For instance, as a parent, a church member, a member of environmental groups, a car owner, a member of a scientific organisation or a professional society, a house owner in a specific neighbourhood. Through these social roles they have gained knowledge about opinions, worries and critiques that exist in society outside their company or research institute. This knowledge can be made fruitful inside the organisation by letting engineers participate in the decision-making process, possibly complemented by the participation of representatives of public groups. An important prerequisite for this, however, is that the engineers' knowledge of opinions and interests that exist in society is made explicit, so that the way in which this knowledge is influenced by their habitual professional interests and reference frame can be clearly seen. This underscores the importance of moral deliberation 'on the job'.

Modern management theories about self-regulating teams have recognised the value of an approach as sketched above some time ago. On the other hand these theories made clear that new ways of co-ordinating the actions of different self-regulating teams were to be introduced, instead of the old hierarchy. This contributed to the popularity of procedures like Quality Care, certification and auditing.

Engineers' Tools for Inclusive Technological Development

In the previous sections we have advocated that engineers themselves can make specific and important contributions to a more Inclusive Technological Development, by:
- Denoting the technical specifications of the artefacts they are developing, their possible alternatives and the technical choices (mostly implicit) that have been made during the process of technological development;
- Helping to keep in mind the social network in which these artefacts will function eventually, to assess their (actual and potential) social, environmental and moral consequences;
- Participating in the organisation's and/or public processes of moral deliberation, without overlooking or forgetting their own values, norms and beliefs;

- Stimulating a working attitude and atmosphere – for themselves and in their organisation – that takes into account these consequences and the broader interests of as many relevant people as possible and of nature, that welcomes the engineers' and public participation in the decision-making process regarding technological development on different levels in the organisation – and this from the early stages on – and that secures that the 'inclusive' decisions taken will be carried through effectively.

Of course, individual engineers cannot do all these things on their own. That would be unrealistic and it would also be unfair to ask it from them. However, as we said earlier, they already have quite a number of what we have called 'instruments for responsibility' or 'tools for inclusive technological development' at their disposal, even in the organisations in which they work. The challenge that lies ahead is to become aware of the presence of these tools and to start to make use of them in a more structural and habitual way.

The first 'tools' we would like to point to are all linked to what we called moral deliberation. Moral deliberation is not a simple procedure that one can follow, although there are some rules for good moral deliberation, nor is it something one can do individually. Rather it is a process one has to pursue together with other people. What we plead for is that moral deliberation should be institutionalised inside organisations. That is, that it would become a regular practice there and an integral part of the management process. Such an Institutionalised Moral Deliberation[19] as it has been called, would be an excellent occasion for engineers to participate in the process of making the technologies they are developing more inclusive.

In many companies and research organisations some starting points for such an Institutionalised Moral Deliberation do already exist. Recently, for example, what is called Integrated Design[20] has become popular. It is a design approach in which from the early stages on technical specifications and economic choices are being made explicit, as well as potential social and environmental consequences and interests of relevant 'stakeholders' being taken into account. Ecodesign to

---

[19] A. BAART c.s, *Werkschrift moreel beraad in kerken. Een nadere begripsbepaling*, Driebergen, MCKS, 1990.

[20] R.J.H.G. VAN HEUR and A.H. MARINISSEN, 'Een goed ontwerp is het halve werk. De verantwoordelijkheid van een industrieel ontwerper', in HOGENHUIS and KOELEGA, 1996, pp. 80-98.

date is probably the best known example of this approach. In Ecodesign environmental aspects are taken account of in the designing process from the start on, instead of at the end of the line. This approach or 'tool' can be supported by Technology Assessment and related tools like Life Cycle Analysis, Environmental Impact Assessment and Risk Analysis. The difference with Integral Design is that such studies are mostly conducted by specialised institutions and experts outside the company. However, especially what has been called Constructive Technology Assessment[21] comes very close to Integrated Design, although it focuses at a more general level.

However, what is still missing in the way these tools are being used until now is explicit and systematic attention for the moral aspects of the decisions that being are taken. Reflecting on such aspects is exactly the central element in Institutionalised Moral Deliberation. This is not to say that we blame engineers for not paying attention to the moral aspects. Making an analysis and evaluation of such aspects asks for special experience and expertise, which engineers do not get in their training. To some extent though they can gain the needed experience and expertise when they engage in moral deliberation processes. In addition, they can also be supported by the advice of ethicists and ethical committees. In many countries for instance Medical Ethical Committees do already exist, in hospitals and medical research institutes. Ethical committees also exist in pharmaceutical and biotechnological companies that use animals for experiments. It would be a good development to implement ethical committees in companies and research institutions that work in other branches or technology as well. One could also think of inviting the Business Ethical Committees for help on this subject, which already function within many companies.[22]

A next step could be that representatives from relevant public groups would be invited to participate in such committees. Organisations should not forget that these could be a very good instrument for getting to know the broader interests that exist in various parts in society. Another instrument for that purpose could be to engage in

---

[21] A. RIP, J. MISA and J. SCHOT, *Managing Technology in Society. The Approach of Constructive Technology Assessment*, London, Pinter Publishers, 1995.

[22] J.M.G. VORSTENBOSCH, 'Ethische commissies: instrumenten van verantwoordelijkheid, verantwoorde instrumenten?' in HOGENHUIS and KOELEGA, 1996, pp. 142-163.

dialogue with public movements and special interest groups in society. Several companies, like Shell and Monsanto, have already done so, publicly or in a more private setting. Of course it must be seen to it that such dialogue follows the rules for 'Herrschafstsfreiheit'. Apart from that it is valuable that special interest groups exert direct pressure on companies in order to draw their attention to broader interests. On the other hand, dialogue offers special interest groups the opportunity to influence in a direct and effective way how the broader interests will be taken into account in the companies. In this way dialogue and direct action can benefit from each other, as has been shown successfully by Friends of the Earth in the Netherlands. In some countries, by the way, public debate between industry and society is organised by governmental organisations and churches.[23]

Next moral deliberation within organisations can also be supported by moral codes like Professional and Business Codes of Ethics or Professional Moral Guidelines.[24] In these codes, which again already exist, professional groups and companies have written down their conclusions drawn in moral deliberation processes they carried out before. Of course, one has to acknowledge that these conclusions are always temporary, because circumstances change, new problems and understandings arise and therefore moral standpoints can always change. In order to be of any help in moral deliberation these codes should not be simple aspirational codes, containing only some general principles and ideals, nor should they be extensive regulating codes, containing detailed prescriptions for professional behaviour that will be sanctioned if not followed, but they should be advisory or educational codes, in which arguments are put forward to show how general principles can be related to actual moral problems in professional practices.[25]

---

[23] For governmental activities, see, for instance, Van Eindhoven 1998 (The Netherlands) and Agersnap 1992 (Denmark). For activities by church organisations, see the work of the Institute Church and Society – formerly MCKS (Multidisciplinary Centre for Church and Society) in Driebergen, the Netherlands; the Society, Religion and Technology project of the Church of Scotland in Edinburgh and the work of FEST (Forsungsstätte der Evangelischen Studiengemeinschaft) in Heidelberg, Germany.

[24] M.S. FRANKEL, 'Professional Codes: Why, How and With What Impact?' in *Journal of Business Ethics* 8(1989), pp. 109-115; M. DAVIS, 'Thinking Like an Engineer: The Place of a Code of Ethics in the Practice of a Profession' in *Philosophy and Public Affairs* (1991), pp. 150-167; S. UNGER, *Controlling Technology: Ethics and the Responsible Engineer*, New York, Wiley, 1994.

[25] C.T. HOGENHUIS, *Beroepscodes en morele verantwoordelijkheid in technische en natuurwetenschappelijke beroepen*, Driebergen/Zoetermeer, MCKS, 1993; C.T. HOGENHUIS, 'Een

Available to engineers are also already a number of well-known tools for participation in decision-making procedures. First of all, there are the regular Team Meeting and the Employees Council. More recently, 'town meetings' and 'taskforce groups' have become common, related to what is called Bottom Up Management.[26] In town meetings in principle all employees are gathered for some time in order to discuss specific company issues that exceed the domain and competencies of single teams or divisions. In taskforce groups, some employees from different divisions and company levels are put together in order to work out a solution for a specific problem. These procedures, meetings, councils etc. are also occasions for moral deliberation. Because of the fact that moral deliberation can only prosper when there is mutual trust between the participants, which takes some time to be established, the regular team meetings and employees council in particular are valuable tools for Institutionalised Moral Deliberation.

Finally, the last important tool engineers can use to contribute to a more inclusive technological development is the instrument which is known as Quality Care or Quality Management. This tool is especially useful for making sure that decisions taken in the process of Institutionalised Moral Deliberation are actually executed in the right way. A development in recent years which we can only welcome is that Quality Care is becoming more integrated with procedures to guarantee Health, Safety and Environmental Care. This is called Total Quality Management.[27]

Inclusive Technological Development in Business Reality

Some of the 'tools' we described above are relatively new and still in juvenile stage. But, as we pointed out, several of them are in use already for quite some time and are in high esteem as well. A main problem we see is that they often are not used in an integrated way. We think they can become much more effective if they were con-

---

nieuwe opening. Morele codes in de beroepspraktijk', in HOGENHUIS and KOELEGA, 1996, pp. 175-196.

[26] A.J.M. ROOBEEK and M.N.F. DE BRUIJNE, *Strategisch management van onderop*, Amsterdam, 1993.

[27] O.A.M. FISSCHER, M.C. POT and A.H.J. NIJHOF, 'Kwaliteitszorg: op weg naar een volwassen organisatie', in HOGENHUIS and KOELEGA, 1996, pp. 197-216.

nected and adapted to one another. Just as a carpenter cannot do a good job when he only uses a hammer, working at a more inclusive technological development cannot be done by only using instruments like Technology Assessment or Ethical Codes, as is often thought.

Obviously engineers need the support and co-operation of their superiors if they want to introduce and use these tools successfully. Even when their superiors sympathise with the inclusive approach sketched here, they will have to take into consideration the financial consequences of it. They will come up with questions like: 'Will the use of these instruments not lead to time consuming bureaucratic procedures, which slow down the decision making process and raise costs? Will it not harm the innovativeness and competitiveness of our company? Will the inclusive technological products not be too expensive and therefore less attractive to consumers? Will the intensifying of contacts with critical groups and organisations in society not enlarge the risk that precious strategic information becomes known outside the company? Will these instruments be effective at all? Why should we introduce those instruments in our company when our competitors don't do it?' These questions are, of course, realistic and cannot be neglected. Unfortunately we cannot give extensive answers to them here. We can only give some brief comments.

Firstly, these instruments can pay for themselves because they help companies to adapt their technological products in better ways to the interests and values in society. Such a policy can reduce marketing costs, juridical claims and opposition from society and therefore also the financial risks of a technological development. Using these instruments can lead to some costs, but these costs can often be reduced in the long run if the instruments are implemented in an integrated way that removes overlapping procedures between various instruments and enlarge their effectiveness. Integration of Quality Care procedures with Health, Safety and Environmental Care procedures, for example, can decrease the total amount of paperwork and planning sessions considerably.

Furthermore it is not necessary that all detailed choices and decisions be subjected to all the costly and time consuming procedures and deliberations explicitly and separately. By having Institutionalised Moral Deliberation and by the Business Ethical Committee guidelines can be formulated that can be applied in the context of different technological developments. Eventually these guidelines can become a part of the company's Code of Ethics. Another opportunity to cut costs is to make use of the fact that most Technology Assessment studies have such a broad scope that their results often can be used for several technological developments. Moreover, these studies can also be organised by branch organisations, thereby reducing costs for individual companies.

Secondly, a dialogue with people or groups from outside the company does not necessarily lead to leaks of strategic information to competitors or society at large. To prevent it, a stepwise approach can be used in which the dialogue with the public first takes place in an outer circle in which only the general characteristics of a specific technological development are being discussed. Next the results of that dialogue could then be transferred to an inner circle of people working within the company itself. It hardly needs saying that in the way a company uses these results internally as well as in the way it participates in the dialogue with society a company should operate in an honest and sincere way. Doing otherwise in a modern society of well-informed and critical citizens will undermine its public credibility and will cause strong opposition. Finally we would even like to suggest that a small number of relevant persons from outside the company could be invited to participate in the internal dialogue and decision-making process, for instance in the Business Ethical Committee. They could be placed under the obligations and rules of secrecy that are already common in many other places in society.

Thirdly, while it is true that individual companies cannot themselves change the main direction of a technological development by using the tools sketched above, these tools can nevertheless enable them to make first steps by making their specific products and production processes more inclusive. That will not necessarily make their products and production more expensive, as we have argued. But even if it did, that does not need to be a problem. If a company starts to take into account a broad range of interests and values in society, social interest groups will often be prepared to make good publicity for such a company. Examples already exist. In Germany Greenpeace supported both the introduction of a green refrigerator and of a superefficient car a few years ago. Another possibility is to make direct arrangements with consumers. In the Netherlands consumers subscribe to a weekly portion of organic vegetables, which guarantees the producers of these products secure sales. Subscriptions to portions of sustainably produced energy or wood also already exist. Annual Social and Environmental Impact Statements, Certification and Ethical and Social Audits[28] can strengthen the social image of the company, which makes the company more attractive to specific consumer groups. Even the government will sometimes be prepared to support the development and introduction of specific products. A Swedish company for instance was able to introduce an energy saving but more expensive refrigerator with support of the government in the form of a design competition with an award, including a guaranteed order of the designed products.[29]

Clearly co-operation within branches of industries and making covenants with government or with non-governmental organisations will enlarge the possibilities for inclusive technological development. Special Quality Labels can support this as well. Examples here are the introduction of the catalytic converter and the use of unleaded petrol in motor cars, cleaner leather production by the Kenyan Leather Development Centre, the Fair Trade Organisation and the Forest Stewardship Council label for non rainforest wood, which is supported by the environmental organisation Friends of the Earth International.

The influence of such initiatives on the development of technology in many cases will be small unless many companies and governments co-operate world-wide. In order to promote this goal several fora exist already: the International Labour Organisation (ILO), the World Trading Organisation (WTO), the Food and Agriculture Organisation (FAO), the Organisation for Economic Cooperation and Development (OECD), the World Health Organisation (WHO), the International Monetary Fund (IMF), the Worldbank, et cetera. Slowly but gradually these organisations are showing more interest in social and environmental criteria for trading and technological development. As a consequence, we see a growing number of international regulations and treaties that illustrate this: the Montreal Protocol against the production of CFC's, the Kyoto treaty on reduction of CO_2 emission, the International Tropical Timber Agreement, the ILO agreement against child labour, the WHO agreement on promotion of milkpowder for baby nurture, et cetera.

[28] M. KAPTEIN, *Ethics Management: Auditing and Developing the Ethical Content of Organisations*, Dordrecht, Kluwer, 1998.
[29] S. SCHMIDHEINY, *Changing Course: A Global Business Perspective on Development and the Environment*, Cambridge, MA, MIT Press, 1992.

Furthermore there is a movement in the direction of 'caring capitalism', illustrated for instance by the Responsible Care Program in the chemical industry, the Fair Trade Organisation, the Social Venture Network Europe, the Body Shop and the popularity of Business Ethics. Central to this movement is the idea that companies do not have a license to operate automatically, but have to gain and nurture it.

In summary, there is no fundamental reason why companies should not engage in efforts towards a more Inclusive Technological Development and support initiatives from their workers in that direction. Many companies already do so.

Conclusion

Engineers, when trying to influence the direction of technological developments, face many problems such as the multifaceted, interconnected and embedded network character of technology and the inertia and momentum of technological development. Nevertheless they already have several tools at their disposal for making technological development more inclusive, that is: more adapted to the needs, interests, wishes, norms, values and ideals of all people concerned. These tools also can be used in the context of the organisations in which they work, such as industrial companies, research institutes, consultancy firms or governmental agencies. Moreover it can be argued that Inclusive Technological Development and the use of the instruments mentioned, contrary to being harmful to the organisation, often is lucrative, both morally and economically. Management of technological organisations therefore has many opportunities for initiating Inclusive Technological Development as well. Several companies throughout the world already have shown this to be true.

Furthermore the interactionistic view on technology teaches that lots of other individual persons, groups and organisations in society are involved in technological development and consequently also have a responsibility of their own: governments and their agencies, non governmental organisations, consumers, voters. Their responsibility, among others, is to support initiatives from engineers and management of technological organisations towards Inclusive Technology as well.

References

AGERSNAP, T., 'Consensus Conferences for Technology Assessment', paper read at the conference *Technology and Democracy*, Copenhagen 1992.
BAART c.s, A., *Werkschrift moreel beraad in kerken. Een nadere begripsbepaling*, MCKS, Driebergen, 1990.
BARBER, B.R., 'Liberal Democracy and the Costs of Consent' in N.L. ROSENBLUM (ed.), *Liberalism and the Moral Life*, Cambridge, MA, Harvard University Press, 1989, pp. 54-68.
BERGER, P.L. and B. BERGER, *Sociology*, New York, 1972.
BIJKER, W.E., T.P. HUGHES and T.J. PINCH (eds.), *The Social Construction of Technological Systems: New Directions in the Sociology and History of Technology*, Cambridge, MA, MIT Press, 1987.
BUCCIARELLI, L.L., *Designing Engineers*, Cambridge, MA, MIT Press, 1994.
DAVIS, M., 'Thinking Like an Engineer: The Place of a Code of Ethics in the Practice of a Profession' in *Philosophy and Public Affairs* (1991), pp. 150-167.
ELLUL, J., *La technique ou l'enjeu du siècle*, Paris, Colin, 1954.
FISSCHER, O.A.M., M.C. POT and A.H.J. NIJHOF, 'Kwaliteitszorg: op weg naar een volwassen organisatie', in HOGENHUIS and KOELEGA, 1996, pp. 197-216.
FRANKEL, M.S., 'Professional Codes: Why, How and With What Impact?' in *Journal of Business Ethics* 8(1989), pp. 109-115.
FRANKENA, W.K., *Ethics*, Englewood Cliffs, NJ, Prentice Hall, 1973.
HABERMAS, J., *Der Philosophische Diskurs der Moderne*, Frankfurt am Main, Suhrkamp, 1988.
HOGENHUIS, C.T. and D.G.A. KOELEGA (eds.), *Technologie als levenskunst. Visies op instrumenten voor inclusieve technologie-ontwikkeling*, Kampen, Kok, 1996.
HOGENHUIS, C.T., 'Een nieuwe opening. Morele codes in de beroepspraktijk', in HOGENHUIS and KOELEGA, 1996, pp. 175-196.
HOGENHUIS, C.T., *Beroepscodes en morele verantwoordelijkheid in technische en natuurwetenschappelijke beroepen*, Driebergen/Zoetermeer, MCKS, 1993.
HUGHES, T.P., *Networks of Power. Electrification in Western Society, 1880-1930*, Baltimore, John Hopkins University Press, 1983.
KAPTEIN, M., *Ethics Management. Auditing and Developing the Ethical Content of Organisations*, Dordrecht, Kluwer, 1998.
LENK, H. and G. ROPOHL (eds.), *Technik und Ethik*, Stuttgart, Reclam, 1987.
MÜNCH, R., *Dialektik der Kommunikationsgesellschaft*, Frankfurt am Main, Suhrkamp, 1991.

PARSONS, T., *The System of Modern Society*, Englewood Cliff, NJ, Prentice Hall, 1971.
RAWLS, J., *A Theory of Justice*, New York, Harvard University Press, 1971.
RIP, A., J. MISA and J. SCHOT, *Managing Technology in Society: The Approach of Constructive Technology Assessment*, London, Pinter Publishers, 1995.
ROOBEEK, A.J.M. and M.N.F. DE BRUIJNE, *Strategisch management van onderop*, Amsterdam, 1993.
SCHMIDHEINY, S. *Changing Course: A Global Business Perspective on Development and the Environment*, Cambridge, MA, MIT Press, 1992.
UNGER, S. *Controlling Technology. Ethics and the Responsible Engineer*, New York, Wiley, 1994.
VAN DER POT, J.H., *Die Bewehrtung des technischen Fortschritts. Eine systhematische Übersicht der Theorien*, 2 Bände, Assen, 1985.
VAN EINDHOVEN c.s., J., *Ethics: To Coerce or Counsel? Annual Report 1997*, Rathenau Institute, The Hague, 1998.
VAN HEUR, R.J.H.G. and A.H. MARINISSEN, 'Een goed ontwerp is het halve werk. De verantwoordelijkheid van een industrieel ontwerper', in HOGENHUIS and KOELEGA, 1996, pp. 80-98.
VORSTENBOSCH, J.M.G., 'Ethische commissies: instrumenten van verantwoordelijkheid, verantwoorde instrumenten?', in HOGENHUIS and KOELEGA, 1996, pp. 142-163.
WESTRUM, R., *Technology and Society: The Shaping of People and Things*, Belmont, 1991.
WINNER, L., *Autonomous Technology*, Cambridge, MA, Wadsworth, MIT Press, 1977.

2

THE DEVELOPMENT OF TECHNICAL SYSTEMS

INTRODUCTION

Sally Wyatt

Developments in information, transport and power technologies mean that many technical systems are becoming increasingly complex and large scale. These pose new questions, some of which concern engineers. Other questions apply to society more broadly. These questions are relevant both to specific technological institutions and to national political and regulatory bodies; the main sites where risk assessments are conducted and where societal problems are reformulated as technical problems. It is from this vantage point that the chapters in this Part explore questions such as: can individual corporations be held responsible for issues arising from the development of technological systems? At what level should codes of ethics apply – to individual engineers, to particular projects or technologies, to decision-making procedures or to whole corporations? Does the greater scale impose new burdens on the state, or does the increased transnationalisation of technical systems limit the scope for action even of national bodies?

There are no simple answers, and the authors do not always agree with one another. Some focus on how to make engineers more cognizant of social and ethical issues; others focus on how to increase the participation and representation of non-engineers in technical decision-making. The extent to which different authors see the social and technical as mutually constituted or as separate spheres also varies. Whether ethical reflection is considered as integral to engineering practice or as part of societal debate remains open. Nonetheless, all of the authors promote the importance of greater reflexivity on the part of engineers regarding the social and ethical dimensions of their practices.

Johan Schot argues that the social problems of technology can be addressed through broadening the design process. He proposes a methodology for this, *'constructive technology assessment'* (CTA), which aims at anticipation, reflexivity and social learning. CTA is not only a method but also a critique of existing, modernist management practices in which technology is separated from social effects. As a result, regulatory bodies have to be put in place, not to guide science and technology from their beginnings but to ameliorate any subsequent negative effects. The objective of CTA is not to reduce the dominant role of technology in society, but rather to change the mode of its control so that a broader range of social groups will be involved at all stages, creating more possibilities for experimentation with alternatives.

André Talmant and Pierre Calame focus on the responsibilities of engineers who work for the State, drawing upon their own experiences of working for a number of different ministries in France. They recognise the potentially contradictory responsibilities faced by State engineers: maintenance of technical standards, promotion of the greatest good for the greatest number of citizens and the encouragement of social dialogue. They point to the dangers of becoming disconnected from society, of feeling superior to other citizens and of searching for technical solutions to social problems. Talmant and Calame discuss the difficulties of promoting dialogue in an asymmetrical situation, namely when engineers have greater authority and more formal expertise than citizens.

After these introductory chapters, four examples are presented. Pierre Rossel examines recent debates in Switzerland regarding the development of *Swissmetro*, a type of high speed, magnetic train. High-speed trains similar to those developed in France (TGV) and Germany (ICE) are not suitable for the Swiss landscape. Rossel argues that *Swissmetro* is a classic example of an engineering project, in that it aims to resolve a socio-economic problem with a technical solution. However, it has proved difficult to move *Swissmetro* out of the design offices and into the ground. *Swissmetro* might have become a vehicle for cultural change because it fits more closely with an emerging environmental objective to reduce air and car travel.

The decision-making processes surrounding the *Superphenix* nuclear reactor in France, a controversy which has dominated French energy debates for over thirty years, is the object of analysis in the chapter by Bertrand Hériard Dubreuil. With the benefit of hindsight, he argues that the Superphenix was a waste of effort and money. Then Hériard Dubreuil focuses on the role of ethics in the decision-making *process*. He concludes with some general lessons to be learned from the Superphenix experience. The first is that it is necessary to follow the appropriate decision-making procedures established within a democratic framework. Second, measures have to be found in order to avoid aggregation mistakes which can arise when actual systems are significantly scaled-up from designs.

In the following two chapters, the focus switches to information and communication technologies (ICTs). Simon Rogerson argues that they are a particularly important site for ethical consideration because of their potential to transform all areas of human activity. He examines different systems development methodologies in light of their commitment to eight ethical principles he identifies: honour, honesty, bias, professional adequacy, due care, fairness, consideration of social cost, effective and efficient action. He concentrates on the definition phase of information systems projects, and argues that the failure to consider social and ethical issues can be considered not only unprofessional but also socially damaging. He acknowledges that more attention needs to be paid to the question of conflict resolution.

Jacques Berleur continues the focus on ICTs, through an analysis of the attempt by IFIP (International Federation of Information Processing) to develop a formal, universal code of ethics. After much discussion during the late 1980s and early 1990s, it was decided *not* to adopt an international code. The explanation often put forward for this refusal is the difficulty, if not impossibility, of developing mechanisms to enforce it. Berleur suggests that many of the problems experienced were the result of different understandings of the concept of 'ethics' between those parts of the world influenced by analytic philosophy and those influenced by the more hermeneutic tradition of continental Europe. He argues against the development of a universal code and for more localised discussion of ethical issues, relevant to the situations of particular times and places.

Part 2 concludes with reflections by Göran Möller and José Luis Fernández Fernández. Möller reflects on the nature of risk, and the calculations individuals and societies make in deciding which risks are acceptable. Decisions about transport, power generation and information systems (discussed in previous chapters) all involve risk assessments. Möller argues that there are no absolute rules to be followed when making risk assessments. One important variable for individuals is the degree of voluntarism, and the extent to which risks can be confined to oneself and not result in threats to the well being of others. In addition, the level of resources available and overall risk level in any particular situation are important in determining the extent of socially acceptable risk.

Fernández also discusses the riskiness of engineering activities, and attempts to answer the question of who should bear the moral responsibility: individual engineers or corporations? He begins by criticising three common ways of understanding the ethical issues raised by engineering: he himself distinguishes between technical problems with technical solutions and ethical problems requiring different sorts of solutions, but recognises that there is often a normative dimension to the way in which technical problems are framed. His resolution of this self-imposed (though widely shared) dichotomy is to examine the ethical dimensions of technical problems. He concludes that both individual engineers and corporations can and should be held accountable for their actions.

2.1 DESCRIPTION

2.1.1.

CONSTRUCTIVE TECHNOLOGY ASSESSMENT AS REFLEXIVE TECHNOLOGY POLITICS[1]

Johan Schot

The core of the constructive technology assessment perspective is that social problems surrounding technology can and must be addressed through the broadening of the design process.
It is argued that not every effort to broaden the design process (or the dialogue between social and technological actors) can be labelled CTA. *CTA aims to shape the design process in order to encourage anticipation, reflexivity and social learning. Its normativity rests on these features. Second,* CTA *practices may be viewed as new alternative forms of management, which replace the problematic modernist manner of managing technology.* CTA *is thus a critique of current 'management practices'.* CTA *breaks the modernist management assumption that technology and its social effects can be separated.*

The core of the CTA perspective is the idea that the social problems surrounding technology can and must be addressed through the broadening of the design process. Here broadening implies the involvement of social actors. Social actors are those who experience the effects of evolving technologies but are not actively involved in the development of the technologies. They can be consumers, citizens, employees, companies, social groups, etc. Design should not be viewed merely as the first phase of the innovation process, but

---

[1] An extended version of this text is published in A. JAMISON (ed.) *Public Participation and Sustainable Development* PESTO *Papers* II, Aalborg University Press, 1998. I would like to thank Andy Jamison, Arie Rip, Richard Rogers, Christiaan Hogenhuis and two anonymous reviewers for their comments on earlier drafts.

also as the phase of implementation where redesign often occurs. An example: the development and application of modern biotechnology is controversial. Especially the genetic modification of animals and the related applications in the food industry raise social questions. Questions are posed by environmental and consumer groups and relate to public health, environmental effects and ethics. CTA would involve steering the development and introduction of genetically modified products (or indeed their rejection) by organising activities whereby the social questions are further articulated and coupled to the development process. The activities can take the form of dialogue workshops, consensus conferences (public debates), scenario workshops, or citizen reports. These are methods which can be used to organise structured discussions between social actors and designers (or technological actors).

The activities depart from traditional technology assessment, which is aimed at charting the effects of the given technological options, and not at influencing and broadening the design process. CTA shares a number of features with more recent forms of technology assessment, especially interactive technology assessment. However, its emphasis on influencing design and technical change through changing the nature of the technology development process is characteristic for CTA and is much less emphasised by interactive technology assessment.[2] Regular technology policies are not aimed at the integration of societal aspects into technical change, although some organisations and authors have called for such integration. They have argued that technology policies aim much at promoting those technologies that promise positive societal effects or externalities as economists would call it.[3]

CTA philosophy and practice are on the move. This article is an attempt to take stock. Now that the first studies have been completed, it is useful to evaluate the state of CTA. Such an evaluation is the starting point for further development. In this article I will evaluate CTA from a long-term perspective. This has two advantages. First, not every effort to broaden the design process (or the dialogue between social and technological actors) can be labelled CTA. CTA

---

[2] J. GRIN, H. VAN DE GRAAF and R. HOPPE, *Technology Assessment through Interaction: A Guide*, Working document 57 of the Rathenau Institute, The Hague, 1997.

[3] L. SOETE, '(Constructive) Technology Assessment: An Economic Perspective' in A. RIP, Th.J. MISA and J. SCHOT, *Managing Technology in Society: The Approach of Constructive Technology Assessment*, London, Pinter Publishers, 1995, pp. 37-52.

aims to shape the design process in order to encourage anticipation, reflexivity and social learning. Its normativity rests on these features. Second, CTA practices may be viewed as new alternative forms of management, which replace the problematic modernist manner of managing technology. CTA is thus a critique of current 'management practices'.[4]

The CTA Perspective

In my view the efforts of CTA (and the activities which could be called CTA) break conclusively with the modernist management of technology. The core of modernist management lies in the separation of technology and its social effects. The separation came into being in the early modern period. The lack of what I call negotiating space between the actors involved in the design process and spokespersons for actors who are directly affected by the technology is a feature of the modernisation process as it has manifested itself till now. In the modern regime of technology management, two tracks are apparent: promotion and regulation. On the one hand there have emerged separate sites – called laboratories – where designers are given plenty of room to tinker with new technologies without having to think about the effects of their introduction. They are not even allowed to think about the effects, for creativity may suffer. After they have been tried and tested, the black boxes are sent off into the world to bring about welfare and progress. Just plug it in; playing with the technology is even considered dangerous, and thus not valued. On the other hand there has emerged a regulatory arena to mitigate the appearance of negative effects. Regulation does not concern itself with steering the scientific and technical developments, but rather with setting limits to their application.

Since the 1970s more and more problems and limitations have become associated with this dual-track approach. Particularly in the decades after the Second World War, people were promised that science and technology would solve their problems. However, their salutary promise did not pay out the expected dividends. More and

---

[4] The word 'management' refers not only to management in terms of 'strategies to achieve a desired result' but also to the form of management, or the manner in which we cope with technology.

more problems cropped up and so-called negative side-effects of existing technologies were not be solved through ex post-regulation. They only worsened. Environmental problems are good examples of this. In the past twenty years we have witnessed an explosion of new governmental regulations as well as a great increase in knowledge of environmental problems and solutions. Environmental advisory agencies have flourished. Yet the environmental problems appear only to have worsened. Chimney filters and catalytic converters appear unsatisfactory. It has become clear that environmental problems must be addressed through a drastic reduction of energy and resource use. Another form of production and consumption is required; sustainable development is the term used. An alternative form of production and consumption implies not only making environmentally-friendly technologies, but also an alternative form of making technology – according to CTA, as I will argue.

Because science and technology have not matched up to their promises, the responsible institutions have increasingly lost their credibility. Experts are no longer automatically taken at their word. Some new technologies, such as biotechnology, are being met with new, robust forms of resistance. Nuclear energy already has been effectively frustrated. Paradoxically, science and technology are being called upon to solve the problems. A quick implementation of new technology is now also a political priority. Convincing people to accept new technology is high on the agenda of a government wishing to stimulate technology.

In order to overcome the problem of acceptance, firms are increasingly anticipating the prospective social effects, especially when developing 'sensitive' technologies. This is, for example, visible in the environmental field, firms are mapping all kind of effects through tools such as life-cycle analysis. They also enter into discussions with social groups at an early stage. Social groups also sometimes seek contact with the firms, and some governments and agencies attempt to support this kind of cooperation. Researchers and policy-makers attempting to develop the CTA perspective are following this trend and are trying to shape the process.[5] The development

---

[5] For an example of CTA in practice, see P. VERGRAGT and M. VAN DER WEL, 'Backcasting: An Example of Sustainable Washing' in N.J. ROOME, *Sustainable Strategies for Industry: The Future of Corporate Practice*, Washington, D.C., Island Press, 1998, pp. 171-184.

of the perspective is to be understood as an attempt to articulate what is going on and which steps could be taken to improve the integration of technology and society. The diagnosis is that such an integration cannot be achieved by undertaking research into effects, for example by doing technology assessment. Rather, the character of the design process is in need of change. It must be broadened to include social aspects and actors. Ultimately such a broadening could lead to a change in the current pattern of technology management (the dual-track approach). New institutions should emerge which will become platforms for the constructive integration of technology and society. It is constructive not in the sense of conflict avoidance, but in the sense that all of the affected are in a position to take responsibility for the construction of technology and its effects. In the existing dual-track regime, no one really takes responsibility for the effects. To technology developers society must see to the effects; they do not follow from the technology. Societal actors do not feel responsible for the effects either; they subsequently call for protection from the government. By institutionalising CTA practices proponents and opponents both will become responsible for technology and its effects.

The Features of CTA

The view that design processes must be broadened is not based on the presumption that social effects play no role in the design process. On the contrary, they are present in the form of (sometimes implicit) assumptions about the world in which the product will function. Thus, when technologies are designed, assumptions are made about users, regulations, available infrastructures, and responsibilities between various actors involved. In technology studies the notion of scripts is used to refer to this set of assumptions. The effect of broadening (and thus of the application of CTA) is that the designers' scripts[6] are articulated and laid bare as early as possible to the users, governments and other parties who have their own scripts, and who will feel the effects of the technology. From the point of view of CTA,

---

[6] For the notion of script see M. AKRICH, 'The Description of Technical Objects' in W.E. BIJKER and J. LAW, *Shaping Technology/Building Society: Studies in Sociotechnical Change*, Cambridge, MIT Press, 1992, pp. 205-224.

itis important to make room for such an early and more regular confrontation and exchange of all the scripts. Thus CTA processes acquire their three beneficial features: (1) anticipation, (2) reflexivity, and (3) social learning.

The Importance of Anticipation

Whenever users, social groups and citizens take part in the design processes, they are more likely to bring in social aspects at an early stage than are designers. Designers rarely anticipate social effects; they even have a hard enough time anticipating market conditions in a timely fashion. They do not seem to seek the relevant market information, and even when they do they do not seem to be in the position to put it to good use. They react to market signals and social effects only when they occur, which leads to ad hoc problem-solving. In the field of management studies this lack of sensitivity toward user needs has been identified as a barrier for successful innovation.[7]

When involving users, social actors, citizens etc. in the design processes in order to anticipate effects, it is important not to structure the process too much in advance. The existing method of user research is to ask the users to react to pre-set product ideas. Users are not invited or given any space in which to come up with their own ideas. Car users, for example, are invited at the so-called 'clinics' to try out the latest model for a couple of hours and fill in a pre-structured questionnaire. They are not put in a position to define the problems themselves and experiment with various modes of mobility. Consequently consumers seem to ask only for more comfort, speed and acceleration capacity. A different procedure was sought during the traffic discussions in the city Groningen in 1995 in The Netherlands. Conversations and discussions were organized between citizens. In four so called working ateliers four visions were developed to achieve an accessible city in future. Significantly innovative and feasible traffic systems were suggested.[8] From the perspective of CTA, it should be noted however, that traffic system designers were not present at the discussions. In order to encourage anticipative behavior

---

[7] See for example J. FLECK, 'Learning by Trying: The Implementation of Configurational Technology' in *Research Policy* 23(1994), pp. 637-652.

[8] *Draagvlak voor verkeersbeleid, Het Groningse openplanproces als voorbeeld*, rapport van Centrum voor Ruimtelijke Ordening en Wegenbouw, Venlo, 1998.

in users and designers, it is important to organize social experiments where participants are stimulated to consider possible synergies between design, market conditions and social effects.

Despite the emphasis on anticipation, there is no presumption here that all social effects can be predicted. On the contrary, it must be assumed that technological development is non-linear and unpredictable. During development all kinds of unexpected sideroads and branchings emerge. Path dependencies will appear; certain solutions chosen for local reasons will continue to drive technological development. However, this unpredictability also does not mean that anticipation as such is not possible or is senseless. There are methods now under development which attempt to take into account the non-linear, capricious character of technological development and build upon the notion of path dependencies.[9]

The given unpredictability of technological development has two implications. First, anticipation must be organized into a regular activity, also during the phase of implementation. That is when new unforeseen effects emerge by way of new interactions and applications. Owing to the importance of anticipating social effects as early as possible firms can be advised to organize a trajectory to develop scenarios for coping with social effects, alongside product development trajectories. Second, the technology development process should be structured flexibly so that choices can be deferred or altered. If flexibility and alternation are built into the standard design process, the effect is that the 'things' themselves take on the form of an experiment. This way they become more open to input from social groups. Herbold has shown how the design of a disposal site could be designed to allow for changes to reduce risk in later stages. This gave designers and social groups the time to negotiate a definition of a safe site.[10] Another example is the development of modular design strategies, for example by firms as Oce van der Grinten and Xerox. Photocopiers were designed to allow parts (or modules) to be replaced. This enables consumers to replace parts relatively easily, if they become dissatisfied with the environmental performance in the future.

---

[9] See for example, M. WEBER, R. HOOGMA, B. LANE and J. SCHOT, *Experimenting with Sustainable Transport Innovation: A Workbook for Strategic Niche Management*, Sevilla/Enschede, University of Twente, 1999.

[10] R. HERBOLD, 'Technologies as Societal Experiments: The Construction and Implementation of a High-tech Waste Disposal Site' in A. RIP, Th.J. MISA and J. SCHOT, *Managing Technology in Society*, London, Pinter Publishers, 1995.

Reflexivity

Broadening the design process results in being able to notice earlier and more clearly that social effects are coupled to specific technical options and that designers design not only technological but social effects. Scripts can no longer remain hidden. The effects that emerge are dependent not only on the designers' scripts but also often on the outcomes of complex interactions between designers, users, third parties and the context in which these actors operate. CTA activities aim to stimulate actors to take account of the presence of scripts and realise that technological developments and social effects are co-produced. Actors thereby become reflexive. They must integrate things and their effects into their thoughts and actions.[11] Consensus may be reached, but controversies could very well occur. CTA could bring about controversy for hidden scripts are exposed and placed next to one another. That needn't be such a great problem in societies where controversies are a routine and normal part of the process of technology development. Analyses of controversies made by Wynne, for example, have shown that often attempts are made to suppress reflexivity. Attempts are made to separate technical facts from assumptions about the social reality in which the technologies function.[12] Controversies subsequently take the unproductive course of the dual-track regime, either emphasising promotion or regulation of new technologies.

Social Learning Processes

Technological development can be described as a process where new couplings are forged between technical standards, production structures, market conditions and cultural notions. Thus the development of the electric car can be viewed in terms of the development of new

---

[11] This is a specific way of defining reflexivity, referring to the awareness of coproduction of technology and society. For a discussion see the introduction of B. WYNNE and S. LASH, in U. BECK, *Risk Society: Towards a New Modernity*, London, Sage Publications, 1992.

[12] B. WYNNE, 'Technology Assessment and Reflexive Social Learning: Observations from the Risk Field' in A. RIP, Th.J. MISA and J. SCHOT, *Managing Technology in Society*, London, Pinter Publishers, 1995.

batteries and drive chains; consumers change their mobility patterns in favor of a smaller radius of activity; and society comes to appreciate quiet transport. (Slow electric cars are silent.) In the process actors *learn* to specify technical standards, the market conditions, the sufficiency of the current institutional structures, etc.

In current design processes learning processes are organized in a linear manner. First there is an attempt to optimise the technology of a process or a product. Second the production requirements are better specified, third the market demand is evaluated, and finally the social effects are taken into account, as they affect the firm. Of course no design process is as linear as this. There is an attempt to work with feedback mechanisms. Feedback also occurs unexpectedly, as do problems during application, forcing redesign. However, such adjustments do not greatly change the basic linear character of the process. Societal learning with respect to the social effects always relates to learning about a technical configuration which is more or less known. This has serious consequences. Not only is it difficult to make large changes (environmental problems, for example, are solved by installing a filter in a chimney or a catalytic converter in the exhaust system) but the learning process also is determined by this basic structure.

Learning could happen on two levels. The first level relates to developing a better ability to specify and define one's own design. Second-order learning means learning about one's own assumptions and scripts, learning that one is creating new couplings and demands. CTA relates to both forms of learning. It is important to embed technological development in social learning processes as early as possible so that users, designers and third parties have the opportunity to scrutinize their own presumptions and come to new specifications. In practice design processes become then more symmetrical from the beginning. As much attention is paid to technical as market and social issues. Design processes become open (so actors are ready to partake) and space is made for experimentation, for trying out various couplings and problem definitions.

CTA Revisited

CTA activities are not directed in the first instance at such substantive goals as the reduction of environmental pollution, the creation of more privacy, etc. Thus for instance the development of wind energy or a security system to guard against bank fraud cannot be automatically labeled CTA. The purpose of CTA is to shape technological development processes in such a way that social aspects are symmetrically considered. When design processes assume the character of CTA fewer undesired and more desired effects will result. Such a claim is based on two arguments. (1) By incorporating anticipation, reflexivity and social learning, technology development becomes more transparent and more compliant to the wishes of various social actors. They will address the social effects that are relevant to them. (2) In a society where CTA processes have become the norm, technology developers and those likely to be affected by the technology will be in the position to negotiate about the technology. An ability to formulate sociotechnical critique and contribute to design will become widespread. Resistance to specific social aspects will not be viewed as technophobia, but as a signal to take up and as an opportunity to optimize the design (or achieve a better fit in society).

The effect of CTA will not be to bring technology under control so that it plays a less dominant role in society. What changes is the form of control and how technology development is played out. CTA concerns changing change; designs would be designed differently. The goal is to anticipate earlier and more frequently, to set up design processes to stimulate reflexivity and learning, and thus to create greater space for experimentation. Where possible technologies should be made opener and more flexible so users can easily have control over them. Technological development will also become more complex. More coordination and new competencies will be required. In some cases the processes will slow. New institutions will emerge to encourage negotiation between developers, users and third parties. Should design processes acquire the character of CTA, technologists will not suddenly see their work disappear or have it constantly evaluated by all sorts of commissions. Almost all of the incremental design changes will not require negotiation. In the program of requirements allowance will have been made routinely for social aspects (including flexibility). However, the variety of technological designs probably will increase, for more groups will be addressed in their capacity as knowledge producer and technology developer.

The three quality criteria for CTA processes make apparent that broadening the design process is not an end in itself and that 'broader' does not necessarily mean 'better'. Broader is better only in those design processes where space has been created for anticipation, reflexivity and learning. That provides some guarantee that processes should result in better technology, which is to say technology with more positive and fewer negative effects. With the aid of the three features also existing CTA activities can be subjected to a litmus test. They can be evaluated and suggestions can be given for improvement.

References

DEUTEN, J.J., A. RIP and J. JELSMA, 'Societal Embedment and Product Creation Management' in *Technology Analysis and Strategic Management* 9(1997)2, pp. 219-236.

GRIN, J. and H. DE GRAAF, 'Technology Assessment as Learning' in *Science, Technology and Human Values* 21(1996), pp. 72-99.

REMMEN, A., 'Constructive Technology Assessment' in T. CRONBERG et al. (eds.), *Danish Experiment: Social Constructions of Technology*, Copenhagen, New Social Science Monographs, 1991, pp. 185-200.

RIP, A., Th.J. MISA and J. SCHOT, *Managing Technology in Society: The Approach of Constructive Technology Assessment*, London, Pinter Publishers, 1995.

SCHOT, J. and A. RIP, 'The Past and Future of Constructive Technology Assessment' in *Technological Forecasting and Social Change* 54(1998), pp. 251-268.

SCHWARZ and M. THOMPSON, *Divided We Stand: Redefining Politics, Technology and Social Choice*, New York, Harvester Wheatsheaf, 1990.

2.1.2

THE STATE AS CONSTRUCTOR OF THE LIVING ENVIRONMENT

Its Abilities and Limitations[1]

André Talmant and Pierre Calame

Those engineers who work as Civil Servants bear, in addition to their scientific knowledge, the authority of the institution for which they work. This institution – the State- enjoys the power to exert pressure on society; but at the same time, thanks to its ability to summon meetings, it provides the opportunity for decision makers to enrich their knowledge and improve their action. Relying on their own experience, the authors of this article want to stress the possibility – and therefore the obligation – for the engineers in charge of technical projects to envisage the manifold consequences of each of their deeds. One should not minimise the difficulties that may arise from ethical dilemmas nor the difficulties there are for them to acknowledge each of the social actors as a real partner. The examples that follow highlight the absolute necessity for such engineers to establish dialogues and to fully understand reality as their basic rules of ethics. Such 'ethics' relies on the necessity of acknowledging complexity and of being prepared to take it into account. For the engineers, this means assuming responsibility for the risk of working outside of their own spheres of ability. It also implies the necessity of fundamental changes in State procedures.

---

[1] Original text translated by Françoise ASTIER

Introduction

The State is a very complex entity so we definitely have to delineate the limits of our discussion. What we here want to reflect upon is how to think through the relations between State and civil society anew. We believe that, as former civil servants, we are in a position to reflect on the threefold logic which underpinned our activities but which also sometimes pervaded us without our being aware of it.

For we were in the service of a powerful institution, 'keeper' of the 'live-together-as-a-country-rule'. The projects it designed and which shaped the environment of our fellow citizens were all aimed at reaching that ideal of unity: when we enforced the laws and regulations, it was obvious the standards we imposed always cropped up from a discourse that purposed to bring about 'the greater good (for the greatest number)'. We also belonged to the Body of the State Engineers, which means we were part of a network that was the guarantor of technical quality, scientific exactness and objectivity. And lastly we were part of a society over which we had authority. We also had the conviction that the State has a duty to help the various actors of society voice their wishes and see them fulfilled, much more than it has to exert an authority obliging them to submit to its command. We conceded that, all too often, there was a risk that we might feel disconnected from society and even superior to it.

Our reflection, steered by the desire we share to 'do well', has led us to take up our responsibility and eventually propose a renewed vision of the State, therefore a new ethics, a new pattern of behaviour, for its engineers.

Accordingly, we will first describe the State, its engineers and their power (part 2); we will in the next part show why the servants of this powerful institution run the risk of feeling disconnected from the society they yet wish to serve (part 3); we will then be able to draft (part 4) the elements of a new ethics – indispensable to deal with the complexity of real life situations (part 5).

It seems relevant however to first set forth some specifications concerning the administrative and political culture of our country, France, so our testimony can be set against a background that makes it both precise and humble. We feel confident the reader will then be able to pick out those features among those we describe which can be recognised as universal.

The State and Its Engineers

The Role of the State in France

The centralism of the French State is a well-known fact. The fact that since the Revolution – in the late 18th century – the prime concern of governments has been the unity of the country, accounts for the prominent place given to the prefectoral corps in the French administration. Appointed by the Government, the Prefects – the administrative arms in the approximately one hundred departments in France – are the representatives of the Government in all and every part of the territory. They equally have the responsibility for securing unity and coherence between the various administrative agencies. The result is a subtle mix of the technical and the political in the running of the actions of the State.

It is worth emphasizing that previous to the 1983 decentralisation law, the prefectorial corps presided over the local authorities, even the largest. For example, the Prefect was the executive arm of the Department council – yet elected on a universal suffrage basis. Fifteen years of decentralisation have not erased the weight of the State on local authorities.

Another major feature of the French state is that the territory is administered on three different levels, i.e. national level, department level – there are about 100 departments as we mentioned before – and a further, lower level about which there is no need to go into detail here. The need felt thirty years ago, however, to group departments on a regional basis – France is now divided into twenty two regions – was an attempt to decentralise but the traditionally permanent presence of the State is still there.

It is this tradition to centralise which explains why the rules and standards which are issued at Government level are enforced exactly in the same way in every single point of the territory. The hierarchical pyramid, both powerful and efficient, provides the channels necessary to circulate them.

It is the all-pervasive rooting of the State everywhere on French territory which enables the civil servants to apply the rules and standards made at national level. These rules and these standards are for the civil servant both a strength and a limitation, even a shackle, because while they give him or her real means, they also put him or her under the obligation to use those means. This is known in France as *'obligations de moyens'* i.e. 'obligations of means'.

The decentralisation law was a major change. It freed the local authorities from the weight of the State and may therefore appear as an epitome of the principle of subsidiarity.[2] But if the law makes it clear how authority is to be shared between the various levels it does not erase the principle of 'obligation of means'. What is more, the lawmaker has striven to make sure no authority should be under supervision of another: the principle of local autonomy remains absolute notwithstanding the fact that France alone has more local authorities than the rest of Europe. One should stress however that real efforts are being made to bring sense and efficiency at city level.

Such is the context, both complex and still in the making, in which we can now address the place and the role of the State Engineers.

The State Engineers and the Authority
Their Technological Background Gives Them

The Ministry or Ministries we worked for were in charge of public works such as infrastructures, town planning, housing and transport. Supervision of these works at the various territorial levels was and still is in a large part reserved for State engineers. There are of course administrative executives but their authority is traditionally outweighed by that of the engineers. At the outset and especially until the 2nd World War, engineers received an excellent training in legal and administrative matters to the extent that some became real specialists in these fields.

This status the engineer enjoys clearly shows that the structures of the State as we know them sprang from the mechanistic conception of the Enlightenment period, in the spirit of the 18th century French philosophers. The idea that prevailed was that society is a huge machine whose machinery should be dealt with according to the best rational approach available. And who then but the engineers could be better prepared to guarantee an objective approach to all the 'engineering' society needs to work smoothly. Napoleon always summoning scientists around him is a symbolic image of such a self-evident truth!

The celebration in 1997 of the 250th anniversary of one of the oldest State schools for engineers, l'École Nationale des Ponts et Chaussées, i.e. the National School for Roads and Bridges, is telling

---

[2] Principle of subsidiarity: see B. HÉRIARD DUBREUIL, *Imaginaire technique et éthique sociale*, Brussels, De Boeck Université, 1997, p. 65, note 45.

too. The main brochure that was issued on that occasion displays the portraits of a hundred Engineers over the two and a half centuries. One can only naturally feel proud to be one of a body which boasts so many engineers who were also scientists. They have passed onto us a tradition of extreme precision which is the hallmark of the scientific spirit.

The State engineers feel both secure with the knowledge they have acquired and also endowed with authority due to their pertaining to an institution, the State, which towers over society. The last 'portrait' in the brochure is that of Robert Hirsch, an engineer we knew at the time when he was the Head of the École des Ponts et Chaussées. He was a man who was renowned and liked for his great qualities, be they professional or human. He had been entrusted by Paul Delouvrier – and through him it could be said, by General De Gaulle – with the task of designing the new city of Cergy Pontoise. Comparing this appointment to others he had had before, he once wrote: 'There I was, appointed director of a new city, that is to say of something that was still to be invented and which furthermore had no legal existence.' No words could have better expressed, however humble Bernard Hirsch may have been known to be, his awareness of his ability to create and of his almost unrestricted power.

Now, how to exert their authority is at the core of the ethics of the engineers – above all when they are State engineers. Whatever the many virtues of the system, the State through its civil servants exerts an authority the potential 'violence' of which must be acknowledged and considered with great attention and awareness.

Endowed with the ability to constrain that is handed over to them by their employer, the State, it is hard for the civil servants to discard that feeling of prestige of being the ones who inform decision-making. And what is more, in virtue of their scientific and professional status, the engineers' decisions cannot be disputed.

Let us therefore consider how State engineers, aware of the power they represent, but also reaching a state of awareness where they realise that their patterns of thought and behaviour are somehow conditioned, will realise that major ethical questions face them in their job as decision-makers.

State Engineers Confronting Reality as It Is

Take for example state engineers posted in a territory. They are in charge of the construction and maintenance of public works which are of paramount importance to the elected councillors and citizens, as well as the local economy. They are also in charge of enforcing numerous rules, and assessing projects being carried out by all types of social actors. In other words, they have been given the formidable power to say 'no'. The engineers then suddenly realise that they are persons of distinction and of political importance!

They have been trained to resolve difficult technical problems. They have a real desire to do their utmost. At this point, allow us to recall a personal experience: after several months of work, we had just completed the design of a metal bridge which had proved a bit difficult, we then asked for an appointment with the specialist of metal bridges in our ministry. We submitted the project to him and he thought it was quite satisfactory. However he then added that the metal industry had started designing new types of steel which if wisely used in various parts of our bridge would prove efficient in terms of technical achievement and cost. Believe it or not, we did review our project! On that occasion we experienced two things: first, the advantage in that precise example of working in a network of engineers; and second the mechanism of judgement by one's own peers.

What is specific to this example though is that the 'object' to be made, a bridge, is isolated from any social context: it seems the engineer is alone with his project and does not need anything but good calculations to conceive a good design, the best possible design. As a matter of fact, this is seldom the case.

Let us take the example of something simple: planning roads in a housing estate. What is concrete obviously is the reality of the people who are going to use the roads. It is the route to the school, it is the use which will be made by the children of the public space, it is the fact that each metre will be devoted to something different, that the needs will vary.

For many engineers however, such considerations concerning everyday life, are discounted as being abstractions, simply because they do not fit into the engineers' usual work pattern or fit together with the elements they feel able to grasp. What to them appears as 'concrete' is applying the given standards of how wide a causeway should be, applying the rules stating how to lay out the roads in the Jacques Brel housing estate where Mrs. Dupont and Mrs. Durand will live, exactly in the same way as in the Yves Montand housing estate at the other end of France and for perhaps completely different people.

Thus reversing the notions of what is concrete and what is abstract, as shown in this example, is ample proof of a resistance to taking into account the complexity there is in human matters. In this case, the engineer whose good will could absolutely not be questioned, listened to the users; but though he met them as fully responsible partners, they were in his mind no more than still other equations. Yet together with them he could have established the list of their requirements, useful for elaborating his own technical work.

Why taking that step proves so difficult probably comes from the sentiment the engineers have that they benefit from a knowledge which those they speak with cannot share, and also the sentiment that they represent an institution which is totally impartial and should therefore not yield to new particular cases.

We can now envisage what relationships there are between engineers and society and what it means in terms of the relationships between the State – i.e. central government structures – and that society. Here we will find the key to a new ethics.

The State, Its Engineers and Society

Isn't Impartiality but an Illusion?
The Case of the Relations with the Economic Sphere

When widening of the canals was started in the North of France, studies as to the optimal size of the locks were carried out together with the users – nothing more natural: what nonsense it would have been to implement infrastructures devoted to transport without knowing precisely the size of the boats that would be using them!

One can compare a barge entering a lock to a box having to fit into another box only slightly bigger – the problem is to compute how much play there should be. Now, as a matter of fact, the bigger the play the easier it is for the boats to enter the lock. What is at stake for the users is to gain time thanks to the widest possible entrance; what is at stake for the constructor is not to overdimension the work and therefore the cost. After many a discussion, an agreement was reached between government agencies and the representatives of the big transport companies: such agreement is proof enough that hydraulics alone could not set the optimum standard!!

Now, what would follow was always the same. Once the work was completed, or even while it was in progress, the transporters would design units, bigger than those which had been the reference when striking the agreement. When the locks were put into service they would complain that the next ones should definitely be made bigger to enable them to gain time and therefore commercial efficiency. The new locks would be made bigger – the cost would obviously be higher too – and the same cycle would start again.

Exactly the same description could apply to why the road systems get bigger and bigger and consequently more and more dangerous.

This example is interesting in many respects. First of all because while insisting that it should 'keep its independence' and therefore remain 'neutral' between competing interests, the process we have just described clearly shows that the State actually served the interests of the transport companies – which quite legitimately wanted a new concept of river transport. This however meant a widening gap between profitability for the industrial freight companies and profitability for the self employed ship owners who are the other group who live from transportation on the river. Can you still claim then, that you are impartial and equitable?

But the 'trap' of the transport companies had another cause, namely that there had not been a real 'dialogue': while the big companies had manoeuvred to gain benefits for their commercial strategy, the state agents had only considered them as one more variable to be taken into account in their computing and had failed to consider the economic issues at stake. What was wrong then, was not an objective alliance of the civil servants with any private interests, but their inability to look beyond their technical horizon. It is an absolute necessity that engineers should be reminded over and over again to always try to scrutinize beyond the apparent limits of the problems they have to solve. But, is such an invitation to fully determine the various actors congruent with the unconscious desire to remain aloof from the complexity of life?

Acting on Society while Remaining an Outsider:
A Temptation

To better grasp the mentality of the engineers and the way they deal with ethical challenges, one must bear in mind that the technically complex objects are in fact simple things because they can be mastered while on the other hand, the negotiating process is something complex as it requires bringing together a diversity of 'cultures', interests and expectations. Building a superb curved bridge of prestressed concrete will require countless calculations and a sophisticated knowledge of the performance of materials as to the limit of their elasticity. Needless to say it requires technical expertise, but unlike the 19th century engineer who was rather alone to conceive a project this expertise is now public knowledge and is collected in the bosom of major scientific organisations. On the contrary, to initiate and make viable a dialogue on a project, to deal with the workers, to plan how to transform a neighbourhood, engineers are alone in front of the others and can no longer claim the dominant position their scientific knowledge once gave them. It is little wonder therefore that they should continuously try to turn second-type situations into first-type situations. While we were still working, there had been a project which proposed replacing road workers with automatic systems to survey the state of the roads. It is so much easier to analyse the results of a video than to handle teams of workers! This process meant cutting jobs but still worse it meant suppressing the visible presence of the State in the management of its domain.

A much bigger change is now in process with the spreading use of 'geo-referred information systems': they consist in utilising the transmission of remote information, particularly recording of data by satellite, to collect all the information necessary to carry out a technical object without having to be physically on the spot. It is The Dream of the State Engineer come true: acting on and for society without having to come into contact with it. Some of those who favour such methods say it flatly when referring to third world cities: what is at stake is collecting technical information about neighbourhoods where it is dangerous to set foot. Technical achievement of the computerised management of the data apart, the side effects of such an approach can of course be seen right away; for as a matter of fact, the reason why the neighbourhoods are dangerous is just this widening gap between human groups considered as outcasts and the representatives of state authority. When using technical devices instead of people the gap is made still wider and the engineers are thereby contributing to the worsening of the problem.

How many an engineer however will sincerely be sad at what they will resent as a moral criticism of their action, when they, on the contrary, wish to display the purity of their intentions and the sincerity of their commitment toward society!

We wrote in an other book[3] that the main difference between war and peace is that warfare requires summoning sophisticated technical means to reach simple aims, while the peace process only requires simple technical means, but to reach aims that are highly complex. Likewise, if they fail to impose upon themselves an ethical reflection on their practise, the state engineers will always be – though without realising it – siding with the logic of war. The question is: when progress is so demanding, is there still room for a debate on and in society? This is precisely the question we put to ourselves when working on the widening of the canals.

The Quest for Progress and/or for a Debate on Society?

Waterways play an important part in quite a number of countries, and after the Second World War a real effort was made to modernise them in many places. This led to talks at European level long before the construction of Europe proper was started and the ministers in charge of transport then promoted technical unification in the European countries. This however did have economic consequences.

In the 1950s – slightly later than our Northwest European neighbours – we embarked on a modernisation era. The size of the transport units was increasing quickly; the navigation companies were developing and giving water transportation an industrial dimension; the salaried worker was replacing the self-employed ship owner. The running of companies was experiencing big changes. The State had committed itself to building completely new infrastructures for the waterways. The State engineers could not but be enthusiastic at the idea of participating in the progress. They mustered all their know-how.

[3] P. CALAME, *Mission Possible*, Paris, Desclée de Brouwer, Collection Culture de paix, 1993, p. 167.

Soon however, the small-sized canals came to be considered as an obsolete past; only the big-sized ones became worthy of interest. Some engineers suggested the situation should be considered as a whole: it was obvious the network would not jump from past into future in the blink of an eye. Common sense showed that modernisation should be made, but not without reflecting on the optimum result to be obtained from the whole network – not just from what was then considered as modern. Should innovation and the potentialities offered by the new ideas justify overlooking the social use of the small ship owners? Was this fair? Could suddenly stopping the maintenance of State property such as the small-sized canals – still indispensable for many years – be justified from a general economic point of view, on the grounds that all available finance should be concentrated on the new network? Should a gradual phasing out time not even be considered?

In this debate two points proved decisive. The first displayed the natural passion engineers have to complete works pioneering new techniques. These always require sophisticated scientific abilities, stir up scientific debate at high-level international meetings and promote the scientific and technical culture of a country.

The second was once again that tradition of 'neutrality': the State constructs large infrastructures, but in no way wants to take into account the commercial dimension which it considers the responsibility of the transport companies. It is out of the question that the State should be liable even if breakdowns or misfunctions impede transportation and therefore break the terms of the contract-delivery time for example. Under such conditions, it is easy to ignore the fact that lack of upkeep of the canals – not dredging them for example – leads to diminishing their depth and, therefore reducing the weight that can be carried in each barge which in turn entails a lack of profitability on the small sized canals. Thus, by refusing to take into account the change in society that went together with technical progress, the State in fact opted to help industrial companies at the expense of the self-employed population.

We of course agreed on the necessity of modernising the waterways. But we also believed that the State should have taken into account all the questions that sprang up due to such necessary changes. We believed the transitions should have been made more gradually, and the weaker, for whom change was to prove difficult, helped. Failing to do so, could one claim that there was any respect for the people affected by such rapid upheavals?

A New Way for the State to Be 'Present' in Society:
An Active Subsidiarity

Such examples show how strong is our belief that the State must be the backbone of society and such conviction is the reason why we think it is urgent to reconsider how the State in today's world, can still be 'present' in society. For as a matter of fact, the discourse on rule of law and on human rights must be visible in the way we handle the changes in our society. Arrant determination to tackle the complexity of the world so as to be in the service of society is the stepping-stone to lending sense to the 'social dialogue' which only the State can guarantee on a permanent basis. It is the conditions of such dialogue which we now want to survey.

Allow us first of all to assume that many a question cannot be answered at one single level. Implementing a housing policy for example requires dealing with three different levels, for in fact the housing policy was left out of the 1983 decentralisation law, thus meaning it was such an important matter that it should remain on a national basis and be dealt with at 'national level'. Most families, who decide to move however, move within a region, that is to say, within an area wider than the 'city level'. And lastly, housing means many questions at 'district level', such a level being where citizenship is being learnt and exercised and where the most vivid and concrete questions are being put, questions relating to schools, public space or neighbourhood services. Managing a housing policy therefore requires that the various responsibilities at the various levels be interrelated

This calls for a debate in which the various partners are acknowledged as equals. On this basis, the partners will be able to agree on a common analysis of the problems, and on common values; they will define which objectives each of them should strive to reach, agree on the necessity to make regular assessments and repeatedly remind one another of the aim to be reached. This is what we have in other documents called 'active subsidiarity' and which really means giving priority within the governance field to interrelations between the various levels rather than to management at one and only level. This calls for dealing with the questions that arise on the border line between two levels of responsibility, and obliges one to focus on 'the obligation of results' rather than on 'the obligation of means'.[4]

'Active subsidiarity' therefore appears both as a philosophy and a practice of governance able to meet the demands of our time, i.e. the need to reconcile unity and diversity. It requires the strengthening of democratic life through enabling the citizens to have real access to information and to decision-making. It requires from the Civil Servants a new ethical standpoint – we will, as we have previously in this article, refer to it as 'dialogue' – some features of which we now want to highlight so as to show what this means in practice.

The Ethics of Dialogue

The Principle of Asymmetry: To Empower the Partners
i.e. To Help the Partners Shape Their Own Tools

The State engineers have three formidable means for creating conditions of complete inequality when dealing with their partners: They have a whole network of technical information at their disposal, their authority bears the stamp of the French government and it is they who decide the matter and terms of the dialogue. The ethics of dialogue should induce them to better balance the conditions of a dialogue, and therefore consider it a priority to empower their partners. It is what in English is called 'empowerment'. Let us take as an example a situation we experienced on many occasions, that of the dialogue with the local councillors on apparently technical matters like city streets. The State engineers will rely on the specialised technical Centre provided by their administration and on the regular train-in-service classes they have attended. The elected councillor will probably not be able to call forth – to contribute to the debate – anything more than few memories from his/her holiday travels! Our approach had been to organise trips for groups of local executives so

---

[4] What we call 'obligation of results' – as opposed to 'obligation of means' (cf. p. 253) – is a situation in which it is the aims to be reached which are paramount in the action. It is the actor's responsibility to choose, within a general framework, what means will enable him/her to implement his/her aims and s/he therefore represents, in the specifications which bind two levels of action, the commitment of a higher level in the charter which defines the rights and duties of the two levels of responsibility. It is his/her hierarchy's duty to evaluate with him/her if the aims are really obtained or not.

they would be able to build up their own references and thus be better 'armed', therefore more demanding, in their relations with us.

When – a long time back now – we prepared a Programme of Modernisation and Equipment in the Northern France Valenciennes region, it was very easy for us to compel our partners to discuss on the basis of lines of credit. As a matter of fact our partners thought it was when speaking of schools, housing or new infrastructures that they were in a better position to get State funding. The State while rigidly granting its help for standardised technical objects imposed its logic – often without being aware of it – on the local authorities. Our duty, ethically speaking, was then on the contrary to 'deconstruct' the administrative categories, to start at grassroots level from the real needs of the region and understand as an outcome of the dialogue, and not from ready made categories, what technical objects these needs required.

The Principle of Diversity: Reconcile Unity and Diversity

A human group is not a clock on which you can work from the outside provided you know its mechanism. And yet, there is a strong temptation for the engineers to apply to society exactly the same patterns they have learnt to apply to technical objects. A good example for that matter is the strange perception there is of how information is passed around inside the technical agencies of the State. Their principle generally speaking is as follows: you experiment, you create a model, and you spread the model everywhere. Such principles are valid as long as the objects are technical and the aim of experiment is to discover the universal laws of matter. In most cases however, whether dealing with water systems, transportation, town planning, protection of the environment or housing, what is at stake is not to produce technical objects, but to invent, together with the other forces of society, answers that will be adapted to each specific context. In other words, what can be standardised is not the 'solution', but 'the questions to be solved'; what can be standardised is not a 'product' but a 'collective process' through which a project can be built. The experience acquired from other situations can be extremely valuable if it helps construct a project with its own specificity. It is conversely highly damaging if it leads to imposing standardised solutions ill adapted to a given context and ill – appropriated by a society which has not contributed to its designing.

If we manage this, we will indeed contribute to fulfilling a need which is one of the conditions essential to the *living together* process. For in all life as a society, it is an absolute necessity to reconcile individual interests and expectations with the constraints of life as a group. While experiencing the dialectics between unity and diversity, we will reconcile the necessity of taking into account the interdependence that unites us with respect for the great variety of people and places that enriches us. We deeply believe that herein lies one of the principles of the art of governing and of government practise adapted to the coming century.

The Principle of Equity: The Quest for Justice

State engineers, whose work is devoted to public service, often have a feeling that they have no personal responsibility concerning equity. They seem to take it for granted that it is being taken care of once and for all by the principle stating 'equality of the citizens before the law'. Yet in many respects, this equality is mere hypocrisy because, as the saying goes, 'all the French are equal but some are more equal than others', and also because state engineers in fact enjoy much flexibility in the way they interpret the laws, which is fortunate, for no law or regulation, however precise it be can take into account the immense variety of situations.

Granting planning permissions became for us one of the privileged domains in which we promoted the principle of equity. The more detailed town planning regulations are, the less adapted and the more stifling they are – sometimes with a result exactly the reverse of the goal aimed at. In practise, we enjoy ample room to manoeuvre but the problem is how to use it to the best advantage. We had noticed how deeply attached all French citizens were to this principle. Being refused planning permission is never readily accepted but it can become really intolerable if you are under the impression that someone in the same situation has had his/her project accepted. In order to meet this desire, we had set up a method of collective jurisprudence by keeping a local record of the answers which had been given in our department in such and such situation. We thus had the possibility to refer every difficult case to our local jurisprudence in order to make sure our decisions were consistent.

The Principle of Facing Reality: The Love of the Truth

Most engineers quite sincerely believe that the superb object they dream of building, to which they dream of seeing their name attached, is useful to society. They are always tempted to interpret the reluctance of society to 'follow' them as a sign of mediocrity or of illegitimate resistance to 'progress'. Now, the engineers have and are almost the only ones to have the elements either to convince or to reinforce reluctance. It is they, for example, who assess the cost of a project and the benefits to be expected. They can then often be tempted to 'help fate' by minimising the costs or by sticking to the most favourable hypothesis when assessing the benefits. Strangely enough they do not feel they are acting against the ethics of truth, they believe rather – and it is their way of avoiding the problem – that they somehow have a duty to make society advance, even against its will. Abiding by the ethics of truth though, it must be said, can reach the dimension of real dilemmas. At the time when the actual project was launched, Eurotunnel executives were well aware that the cost of the project had been underestimated and the benefits expected overestimated. They knew they were lying, not to derive personal profit but because they believed that the symbol of the tunnel was so important for the building of Europe that it was well worth a lie! When you consider how what followed has for long discredited the role of private investment in major infrastructure projects one can ponder on the consequences of their choice.

Even when the cases are not so extreme, there is a duty to provide information – the true information – to those who are concerned. The love of the truth for engineers means accepting that reality should be paramount and must be faced as it is the only way to secure clarity in the dialogue and therefore participate to the making of a 'culture of complexity'.

Conclusion: Towards a 'Culture of Complexity'

The points we have tackled show that State engineers are necessarily obliged to be in contact with society and that this is when and where they are confronted with complexity. Because of the authority they have, both administrative and technical, they could be tempted to ignore it. Having to establish a partnership with other groups means they have to take into account interests other than those of government agencies, a logic different from that of technique, ways of thinking different from those of the State. And yet, it is precisely in confronting their technical objects and their responsibilities to members of the other groups which makes their action relevant. They have to admit the uniqueness of each situation and of each territory when all they have been taught before induces them to apply – for reason's sake – the universal laws of physics, of chemistry, of the resistance of materials and the universal principles of government action. As we are about to conclude allow us to once again stress the necessity there is to promote a new culture, the 'culture of complexity'. And to repeat that to see it emerge it is paramount that we should change our viewpoint, that we should instil a new insight of the State and of the role of its agents. And to reach that aim, we want to suggest acceptance of three acknowledgements.

The first acknowledgement is to never forget that the civil servant is also a citizen, a 'political being'. The State is composed of men and women whose personal commitment is also one of the conditions of the changes many call for. A short scene will exemplify both that wish and the institutional barriers as they are now. We had gathered for a meeting to discuss new equipment projects, together with the local authorities. While waiting for the meeting to start, everybody was debating on what should be done. Our expectations were high when hearing so many brilliant ideas on what would be relevant for the State to do. Unfortunately no sooner were we sitting around the table and the meeting had *officially* started than we heard nothing more than the usual commonplace proposals rooted in the logics of the various departments each belonged to! To discern the *citizen* in each civil servant demands that civil servants should always make sure they remain aloof from the institution, as it is the only way for them to be aware of ethical dilemmas set forth by the situations in which their actions are being carried out.

Second acknowledgement: to accept complexity in order to see the 'culture of complexity' prevailing is that civil servants should *admit that they have to learn*. It is amazing the number of engineers we met who would hardly be aware of their *duty to learn,* or who would dwarf it to simply making out the forces at stake in the small circle of the local gentry! Having become aware that learning is an absolute necessity, they will become acquainted with the methods, the know-how and above all the behaviour that will enable them to put their abilities and their authority at the actual service of society when it is all too often society which tends to be obliged to yield to government decisions.

The third acknowledgement is that the State has a duty to innervate society, to exercise a power of social adjustment. We deliberately chose to base this article on the *questions* that arise when reflecting on the work of State engineers. This led us to ponder on the relationships between State and society. The power of the State to summon is huge; but the State is *within* society and not above it; which means the ability of the State to innervate society implies that its agents, the civil servants, be welcome to put questions which do not relate strictly to their administrative ability. They will thus be more confident to suggest relationships that have never been experienced and to create authentic partnerships for debates in which all do feel on an equal footing.

If we owe allegiance to this institution whose prime value is to serve each and everybody, we will definitely come across another 'obligation': watch out that every agent acts in a free and responsible way. Such *task* is a paramount duty for all the civil servants as they *are* the State.

References

BARBER, B., *Démocratie forte*, Paris, Desclée de Brouwer, Collection Gouvernances Démocratiques, 1997.

CALAME, P., *Un territoire pour l'homme*, La Tour d'Aigues, Éditions de l'Aube, Collection Territoires et Société, 1994.

CALAME, P., *Mission possible*, Paris, Desclée de Brouwer, Collection Culture de paix, 1995.

CALAME, P. and A. TALMANT, *L'État au cœur*, Paris, Desclée de Brouwer, Collection Gouvernances Démocratiques, 1997.

GAUDIN, T., *L'aménagement du territoire vu de 2100*, La Tour d'Aigues, Éditions de l'Aube, Collection Territoires et Société, 1994.

MILLON-DELSOL, C., *L'Etat subsidiaire*, Paris, PUF, 1992.

VIVERET, P., *Démocratie, passions et frontières*, Paris, Éditions Charles Léopold Mayer, 1995.

VELTZ, P., *Des territoires pour apprendre à innover*, La Tour d'Aigues, Éditions de l'Aube, Collection Territoires et Société, 1994.

2.2

EXAMPLES

2.2.1

SWISSMETRO

Engineers' Fancy or Major Project?

Pierre Rossel

Swissmetro is a transport project using the technique of magnetic suspension in a depressurised tunnel. Combining, as it does, in an innovative way technological methods which have been mastered to a greater or lesser extent, this project would definitely solve certain transport problems, in Switzerland and elsewhere. At the same time it would also raise other problems, which is why there is considerable controversy over its emergence. At this stage, when a response is expected soon from the Swiss government concerning the granting of a license, two main difficulties appear more and more often on the agenda. First, because of the project's radical nature, its conceivers have to develop the system in the light of technological methods which are still developing and some of which will only exist in 5 or 6 years' time. Secondly, the project engineers must make a smooth transition from a modernist logic whose main goal was simply to travel faster, to one where environmental issues and social acceptability must be taken into account. In this latter paradigm, it is particularly important to find out by trial and error the modalities of a system more environmentally friendly than the TGV or the plane. Such a system needs to cover a range of distances and speeds, as integrated as possible within the existing rail system, both for the decision-makers and the potential users of such a system for the next fifty years.

Switzerland and High-speed Railways

During the 1970s and 1980s, Switzerland, like all its neighbouring countries, was considering ways in which it could build high-speed railways. Studies that took place during this period (on what were called New Rail Links) revealed that Switzerland is not a favourable environment for this type of project. Switzerland therefore abandoned this project and began to look at other railway developments, particularly an Alpine crossing that could integrate piggy-back motorail services (Alptransit) and a medium-speed 'network' based concept which may only be partly implemented (Rail 2000). These decisions were confirmed, the people voted in favour and the initial work, at least on Rail 2000, is making progress. The case of Alptransit was complicated by difficult European negotiations, involving many parameters, both intrinsic and extrinsic. However, in this country which rejected the concept of TGV in the 1990s (although it is allowed to run at reduced speeds there), the 20 billions to be allocated to these projects, although disturbing, were accepted by public voting in autumn 1999.

The Origin of the Swissmetro Project

For more than twenty-five years a group of engineers from the École Polytechnique Fédérale de Lausanne (EPFL), encouraged by a former manager from the Swiss Railways, Rodolphe Nieth (who was, in particular, responsible for building the rail link to Geneva-Cointrain airport), has been dauntlessly engaged in research underpinning progress with the scientific and technical aspects of a revolutionary transport project: Swissmetro. This is an underground means of transport, magnetically levitated and guided, which can travel at 400 km/h or more, in a depressurised atmosphere.

At the beginning, the hope that was attached to Swissmetro, when it was still only a project for a project, was based on the following factors:
- routing a TGV in Switzerland raises problems, and in any case it is increasingly inconceivable to build on the surface (particularly due to the very numerous objections which inevitably arise);
- in economic terms, there is a demand for more mobility at medium and high speeds; this is shown by the extrapolation of current figures into future years; all that is needed is to find a way of meeting this demand;
- it is technically possible and the time savings are impressive

(both significant in quantitative terms and also in qualitative terms, with regard to the transportation of people);[1]
- financial viability was then explored (during the preliminary study), on the basis of an argument developed by Professor Perret of the EPFL, working from the concept of net present value and the possible profitability scenarios corresponding to these.

Of course all these assertions have been the subject of many debates. For a systematic study of the controversies relating to this technology, cf. the studies by Bovy,[2] and Rossel, Bosset, Glassey and Mantilleri.[3]

The Basic Concept

The Swissmetro concept consists of four basic technologies:

- an *entirely underground infrastructure*, mainly in the form of two bored tunnels, with a small diameter, about 5m, and stations connected to the existing public transport networks.
- a *partial vacuum* maintained in the tunnels to make savings on the energy needed to propel pressurised vehicles, according to the same principle as that of an airframe.
- a system of *propulsion using linear electric motors* which are built into the tunnels themselves.
- a system of *magnetic suspension and steering* allowing travel at speeds of the order of 400 to 500 km/h between the urban centres linked by the underground network.

It will be noted that none of these technologies is intrinsically revolutionary, and indeed the basic principles underlying some of them date back several decades. Bringing them together into a coherent, functional whole, on the other hand, is certainly a new challenge.

---

[1] The second and third points correspond to a very pushy approach to the introduction of a technology, and form the basis for a significant belief in technological progress based on the idea that whatever offers superior, achievable performance will ultimately be successful.

[2] P. BOVY, 'Le Swissmetro en neuf questions', Lausanne, Polyrama EPFL, 1994, N° 96.

[3] P. ROSSEL, F. BOSSET, O. GLASSEY and R. MANTILLERI, 'Les enjeux des transports à grande vitesse: Des méthodes pour l'évaluation des innovations technologiques, L'exemple de Swissmetro', Bern, FNRS/PNR41, Rapport F3, 1999.

The route, the cross-section of the tunnel, the capacity of the vehicles, the frequency (currently fixed at an interval of about 6 minutes, for inter-station journeys of 12 minutes), and whether or not the engines are on the vehicle itself, are still some of the parameters on which some change may take place.

Swissmetro, or the Evaluation of a Technology
and a Specific Project

In late November 1997 a significant event took place from the point of view of the Swissmetro project: a request for a license for a pilot route from Lausanne-Geneva was submitted to the Federal Transport Office (OFT), for examination by the Federal Council, the body which is authorised to make a decision on this question.

For the first fifteen years of the project, and particularly throughout the entire preliminary study phase, i.e. until 1992, the Swissmetro project was led by a group of engineers teaching at the École Polytechnique Fédérale de Lausanne, with the active collaboration of the idea's creator, Rodolphe Nieth. When the project entered the main research stage, i.e. from about 1994, although the initial group continued to play a significant innovative role, in order for the concept to take a specific shape in the form of an effective, functional mode of transport, the private company Swissmetro SA to some extent took over a number of operations, particularly with regard to the license application, a stage which was considered indispensable to the development of the planned innovation. The two processes of scientific research and technical and commercial promotion took place in parallel, and one might say, so far at least, with a relative degree of synergy. We must also point out that in order to reach the current level of know-how, 15 institutes from the Écoles Polytechniques Fédérales in Lausanne and Zurich and 80 private companies have so far taken part in the project, which gives an initial impression of its size.

In the Summer of 1999, the Federal Council made its decision, basically answering: 'your project is interesting and should be pursued, but as it is, your financial plan should be revised'. According to the different points of view, this was considered either as a 'yes, but...' or a 'no, but...'. In any case, Swissmetro promoters took it as being very encouraging. As for ourselves, we have good reasons to think that as new Swissmetro-based proposals will be submitted, a fracture could emerge in the management of the project and we are going to see why, and above all why this may be helpful.

Competing Projects, both inside and outside Switzerland

It is clear that the large amounts of investment, totalling at the end between 20 to 40 billions of Swiss francs, required for the major transport projects which are currently being developed, constitute a serious obstacle to the implementation of Swissmetro. These projects essentially comprise: Rail 2000, although this does not yet exist in its ideal 'networked' form, but in a truncated and relatively centralised version, and above all the NLFAS (New Alpine Rail Links, or Alptransit, when referring to the concept), consist of boring two tunnels under the Alps for passenger and motorail transport, both locally and from Northern to Southern Europe. In addition to these projects there are various plans to link Switzerland with the French Eastern dorsal route (TGV Nord-Sud) or with the German ICE.[4] Competition or complementarity? This would actually seem to depend on the level of interest shown by the various players involved in these projects in relation to one or other of these approaches. It is undeniable that network complementarity or even technical collaboration could take place. The know-how available in the area of how trains behave in tunnels at high speeds could therefore be reinforced by experimental studies under the auspices of Swissmetro which could benefit the other projects. There is, however, also a question of political will involved.

Swissmetro also has foreign competitors, mainly the German Transrapid, which has been scheduled to run on a pilot route from Berlin to Hamburg, but is now being severely questioned, and the two Japanese projects, one of them intended more for low and medium speeds, and the other which has been developed in stages by the Japanese Railways to provide an eventual successor for the Shinkansen (TGV-type technology, which is, like its European equivalent, beginning to reach a ceiling).

[4] For the sake of completeness we should also mention the major investments which are taking place or are about to take place in the area of RER [regional express network] in the country's main urban centres, which will no doubt impose a significant burden on the possibility for infrastructure financing from public funds.

The German Transrapid and the Japanese JR Maglev are more advanced projects than Swissmetro, but they are not without some problems, due to some of the choices which have been made. The Transrapid is actually a surface vehicle, which will give rise to numerous objections. Any underground section of its route will significantly increase the cost. Furthermore, although its suspension is magnetic, the Transrapid is still a kind of articulated train, with bogies, and is therefore relatively heavy. The Japanese Maglev is a system which, like an aircraft, takes off and lands by raising and lowering its wheels, and which uses superconductivity for its propulsion and suspension. There will, in all probability, be problems to be resolved when leaving tunnels, and also where there are crossings. The experiences gathered by Swissmetro, using the specific approach of this concept (entirely in the tunnel, partial vacuum, in principle a semi-articulated vehicle, acting as a long self-deforming cigar), are raising a fair amount of interest among its competitors.[5] The question of knowing whether these magnetically levitated projects are actually competitors or potential partners is, however, completely unresolved.

Opposition

We should review the added value that comes out of the license application: a saving of about twenty minutes on the Lausanne-Geneva route (since these two towns are situated about 12 minutes apart in theory), reducing the burden on motorways and railways which are currently saturated, for a price of 2.8 billion francs gross. The pilot project, in order to ensure its viability, stipulates 1) complete transfer of the existing CFF Intercity trains in favour of Swissmetro and 2) financing of the infrastructure by the public authorities. These two points could raise problems. The first could raise problems because the CFF is in the process of being deregulated and the public authorities will soon no longer be able to tell this company what to do. As a result of this, the CFF, which has always been opposed to Swissmetro, will above all have to evaluate whether the

---

[5] Since recently the Swissmetro concept has even aroused interest at GEC-Alsthom, the builder of TGV, which has taken up a significant stake in Swissmetro SA.

proposal is commercially attractive, which will not necessarily be the case.⁶ The second could do so because public money is becoming scarce and difficult to mobilise without a very convincing reason. This brings us to the existence of various opponents to the project. If we recall the some 50-60 controversies that have arisen in relation to the Swissmetro project, there are four themes that constantly reoccur.

- *There are still quite a few technical difficulties, particularly in the area of safety.*
- *Commitments have already been made for other facilities.*
- *It is an unnecessarily fast and expensive means of transport.*
- *The project causes a centralising effect.*

To these it should be added that at present Swissmetro would be of very little benefit to German-speaking Switzerland.

In order to counterbalance these objections and seek to develop a more integrative process (no longer based exclusively on the arguments of speed and progress), it will be important to initiate an innovative process along other lines: problems associated with the centralisation of economic activities, interconnection within the existing transport network, environmental effects, etc. From this perspective it would seem that a more systematic examination is required of the effect of Swissmetro on mobility and traffic,⁷ also including benchmarking of the planned system against others, including the external effects. Whether in the area of mobility, the environment, territorial or economic effects, this question is actually still very open and this makes it possible to develop the last and also the most essential of these points.

---

⁶ There should be no illusions about the fact that, in the license application, the infrastructure cost is to be assumed by the public authorities: although indirectly, these costs will largely be recouped through the operating prices and will therefore be in competition with the other commercial choices that the future deregulated company CFF may wish to make.

⁷ Like the one initiated by SCHULER and KAUFMANN, 1996, and discussed in detail in the study by ROSSEL, BOSSET, GLASSEY and MANTILLERI, 1999.

This is a three-year study, actually carried out within the framework of a national research programme by the Fonds National Suisse de la Recherche Scientifique, on the theme of transport and the environment (PNR41).

A Probable Development or a Desirable Revolution?

So far we have presented the main aspects of the Swissmetro project. At this stage, particularly in the area of management of the project, a conflict of paradigms is about to erupt, even amongst the promoters of the project (here we are of course to some extent acting as an oracle, entirely at our own risk). In fact it is our opinion that the response from the Federal Council to the application for a license submitted by Swissmetro SA will speed up the process of choosing a clearly defined philosophy of innovation, thus amplifying the differences that exist, although these have been managed internally until now.

As the response from the Federal Council was rather negative, it will be more necessary than ever to distinguish between the appreciation of the project and the necessarily specific way in which it has been expressed in the application for a license, and on the other hand, to examine the future of this technology in a more general sense.

The problem that we want to raise here is therefore as follows:

Swissmetro is a real engineers' project, not only because it aims to make the best possible use of this company's knowledge and R&D know-how, but because it claims to be able to solve a socio-economic problem by providing a technical solution. But there is more. Swissmetro is a project situated at a point where several different developments are converging. First of all we should mention the historical chain formed by the 'reverse salient' of rail-wheel friction, i.e. levitation, first aerodynamic (on an air cushion, particularly in the Bertin system) and then magnetic. The idea of the 'reverse salient', borrowed from Hughes, McKenzie and a few others, is used, in the area of the innovation dynamic, to refer to this point of technological resistance which is stimulating the imagination and inventiveness of the various players. Throughout this century there has been no shortage of projects aimed at resolving not only the question of levitation but also those of propulsion, steering and finally and most importantly, air resistance. The Nieth project (in the mid-1970s) is in fact contemporary with the American Planetran, and more or less similar. We have checked this with him and it seems that what we have here is a case of simultaneous formulation, rather like the telephone and the competition between Bell and Edison. In general, this speaks volumes about the ideas which are in the air at a given time. Furthermore, at the time, although certainly in a less innovative form, the German and even the Japanese projects had already reached a stage of concrete development. Magnetic levitation, as a new form of railway, therefore appeared to be the method *par excellence* for going beyond the limits that could already be perceived of 'classic' high-speed trains (in particular the French TGV and the German ICE cannot currently go beyond 400 km/hour without significant distortion of the rails and above all of the ballast, not to mention the noise which is caused and the overall loss of energy efficiency above 300 km/h). Above all, Swissmetro relies on a paradigm of modernity which requires that the speeds of all technologies put together will always increase, that it is not possible to stop Progress, and that its manifestations, particularly in the area of speed, constitute a benefit for society.

Without wishing to demonise the question of speed, as for example Virilio has tended to do, it should be pointed out that so far Swissmetro, like Transrapid and JR Maglev, has above all sought to implement, as rapidly and cheaply as possible, in a speed range which is considered to be innovative and high-performance (400 km/hour), a means of transport that meets the criteria of the four constituent technologies. During the course of this process, two problems have arisen.

The first one is ultimately quite banal, whenever a radical innovation is involved (we should remember that there are no precedents which can be relied upon, and this is one of the difficulties encountered by the Federal Council and the Federal Transport Office when it comes to identifying to which existing system to relate the project which has been submitted). Transrapid, which has been under development for twenty-five years, has already carried more than 200,000 passengers and has covered 10 million kilometres, is a typical expression of this dilemma. In order to work quickly and be the first, this system was indeed developed essentially using the technologies available at the time. Today, however, twenty-five years later, Transrapid is working and seems to be ready for industrial development and commercial operation. But some of the technologies built for this project are already outdated (particularly in the area of analog electronics) and they can only be developed through new R&D efforts, while others are perhaps fixed forever (the choice of long stators, for example, making it necessary to use high frequencies for energy transmission, resulting in significant losses); on the other hand Transrapid is also difficult to sell from the environmental perspective. We should remember first of all that Germany has preferred a method of energy generation based on fossil fuels, hence giving rise to significant levels of atmospheric pollution, and also generating greenhouse gases. Transrapid is therefore hardly favourable from the energy perspective; it also consumes large amounts of land and countryside area, and generates large amounts of noise (which forces it to slow down considerably on the edge of built-up areas), thus expressing a high level of obsolescence even before it comes into commercial service. The benefits of being first (authorisation has been granted to build the Berlin-Hamburg route, before having been stopped as too costly by the new German Government and main industrial operators) seem to be largely counterbalanced by some significant disadvantages. The dilemma facing Transrapid, for these two reasons, is exactly the same as the one now facing the promoters of Swissmetro.

With regard to the first of these two aspects, it is certainly not a trivial thing to build a radically new technological system, which will take about twenty years at least, using technologies some of which do not yet exist or at least have not been mastered. On the other hand, if only known technologies are used and Swissmetro is built like a traditional civil engineering project, this could penalise the innovative effort involved in the project. At any time this could result in it only ultimately using resources which have already been brought under control, thus distorting the original aim and concept, for an overall gain which may even be doubtful in relation to traditional means of transport. The problem of Transrapid, in particular, is that at best it only offers an increase in speed of the order of 50-100 km/h over 'traditional' high-speed trains. In fact there are two completely different philosophies of research and development involved, whose links to the world of finance are virtually incompatible.

To a large extent, this basic choice related to the desire to build a real route from the outset (Lausanne-Geneva), with the risks of competition and opposition that this presupposes, rather than a shorter trial route, for example after the fashion of the Japanese JR Maglev. The relationship with the main players, opponents and the general public as a whole is quite different, as is the time-scale involved in each. In fact the modernist paradigm, when it is expressed as a primary value (to go faster, as soon as possible, at the lowest cost, in order to replace and supplant existing systems which are seen as outdated), interferes with the dominant regime. In Switzerland this paradigm is 'CFF' (the Federal Railways which, even in deregulated mode, has aimed to build a certain type of network and technology, namely essentially medium-speed commuter trains), and in Europe, put simply, it is 'TGV/ICE'. These regimes will define the investments, the state of knowledge, the lobbies, the expectations in terms of services to the public, and also in terms of return on investment, for many years to come.

The second aspect may possibly outstrip the first one, as long as it is identified, handled and formulated correctly. This is the environmental paradigm. Mobility is indeed developing, but it is mainly short-range mobility. In order for a high-speed means of transport to be successful, it must not only deliver improved performance in terms of speed or cost/distance ratio, but above all offer decisive long-term benefits, as for example in terms of energy, in comparison with other means of transport. In the case of Swissmetro, due to some of its fundamental technological aspects,[8] the system could perhaps be developed as a concept more favourable than air travel and even than the TGV, for a distance range of between 200 and 1200 km and speeds of 300 to 800 km/hour, and indirectly (due to the possible transfers coming from liberation from the traditional 'railway passband'[9]), than the car. The project still needs to be organised consistently. When depreciated over one hundred years (which is normal for a tunnel), the infrastructure would represent between 15% and 35% of the energy of the system, the rest being involved in operation and maintenance.

The weight of the vehicles will therefore be particularly significant (this is where maglev systems like Swissmetro should be able to press an advantage), as well as the very useful research that has

---

[8] P. ROSSEL, F. BOSSET, O. GLASSEY and R. MANTILLERI, 'Les enjeux des transports à grande vitesse: Des méthodes pour l'évaluation des innovations technologiques, L'exemple de Swissmetro', Berne, FNRS/PNR 41, Rapport F3, 1999.

[9] In this metaphor we are referring to the possible maximum traffic capacity along a given route, putting all the systems together. We should recall that at present intercity express-type trains, goods trains and regional trains, which are slower than intercity trains are often at the bottom of the list of priorities when the route becomes saturated, thus limiting the attractiveness of the local transport supply because of the low frequencies on offer and hence the low level of transfers from the car to public transport.

already been done on the vacuum, safety, passenger flows and various impacts which have already been analysed. The question, from this perspective, is no longer how to build Swissmetro as fast as possible, but: under what conditions and after what research and checking processes could such a concept be the embodiment of a dynamic alternative[10] to the systems which exist or are soon to arrive (consider future low-consumption aircraft), for a specific speed and distance range, and for a network operating mode and therefore for specific services (a great deal of research remains to be done in this last area)? Within the specific context of Switzerland, with its tradition of direct democracy (with the possibility of referenda at every level of political life) this perspective based on a closer correlation between the demand for mobility and environmental demands, also has the merit of allowing the necessary stages and iterations required to gradually increase the interest among the population (we should also remember that Switzerland has, to some extent, forgotten how to take risks, particularly with regard to major projects).

Hence it is clear that this change in paradigm, which not only consists of 'discovering' the value of the environmental issue, as all the maglev projects are currently doing, but of reversing causalities and the order of priorities, may be the pivotal perspective around which a concept of radical innovation can form. This development which is currently taking place in Swissmetro thus raises the challenge of transport in broad, open and evolving terms, which will be useful even for the competing projects, whether these are in the area of maglev or traditional technology. Probably the most difficult aspect of translating this preoccupation into practice lies in the change of perspective linked to the way in which work is done by the researchers and engineers involved in the project, both those acting on behalf of Swissmetro SA and those working at a more fundamental stage of research, particularly at the Écoles Polytechniques Fédérales in Lausanne and Zurich.

---

[10] It remains to be demonstrated, of course, that the Swissmetro system is able to evolve and absorb growing demand, as well as satisfying the demands for linkage with other transport systems, which are also liable to develop, even of course after being built.

Swissmetro: A Cultural Vehicle

As we reach the end of this brief tour around the Swissmetro project and the challenges facing it, it is worthwhile to insist that Swissmetro not only appears as a means of transport, but also a as a vehicle for cultural change. In fact it forces us to rethink certain habits:

1. In the area of 'transport', where it is necessary to move from analysis on a project by project basis or based essentially on making past choices profitable, towards a systematic, strategic vision which is as global as it is local;
2. In Switzerland in particular, where there is new know-how to be acquired in the areas of risk management and the setting up of major projects, or even the linkage between national choices and European options;
3. In the socio-economic domain, where there is a need to be able to internalise the costs of mobility and infrastructure, which have hitherto been external, in the actual concept of a specific mode of transport.
4. On the ethical level, all the mapping of stakes and controversies, all the existing knowledge and comparative work on high-speed transportation and in particular on the Swissmetro project should be made explicit as to allow many more people than just engineers to participate in the decision-making process, with in mind not only issues of costs, perceived risks and mobility needs, but also sustainable choices and global understanding of how these choices will affect mobility and society in the future.

It is clear that the world of engineers is not necessarily ready to take notice of the importance of these parameters, and also that integrating them will require a series of inter-disciplinary, multi-player interactive processes which go far beyond the know-how generally available within the profession. At the price of a small, prestigious project, however Swissmetro may become more of a reality, or at least a democratic learning process of how to deal with engineering proposals regarding such basic aspects of society as high-speed mobility systems.

In any case, due to the requirements and pressures which have already been brought to bear on this project and the evaluation of the impact/opportunity report which it represented, the ways of thinking developed through its evaluation should, in parallel (the famous Bloor requirement for symmetry), become a standard for the evaluation of all transport projects of a certain size.

References

BASSAND, M. and P. ROSSEL, 'Swissmetro pour repenser notre vision de la Suisse', Lausanne, Polyrama EPFL, 1994, N° 96.
BONNAFOUS, A., 'Transport et environnement: comment valoriser et maîtriser les effets externes?' in *Économie et statistiques*, Paris, 1992, N° 258-259.
BOVY, P., 'Le Swissmetro en neuf questions', Lausanne, Polyrama EPFL, 1994, N° 96.
CEDRE, *Le défi régional de la grande vitesse*, Paris, Syros-alternative, 1992.
CEMT (European Conference of Transport Ministers), *Tendances du transport européen et besoins en infrastructures*, Paris, CEMT, 1995.
CESRW, *Revue du Conseil Économique et Social de la Région Wallonne*, Liège, 1996 (for an overall analysis matrix for transport projects).
JUFER, M., *Swissmetro – synthèse de l'étude préliminaire*, Report DFTCE-SET 201, Lausanne, LEME/EPFL, 1993.
JUFER, M. and F.L. PERRET, 'Swissmetro, une chance de renouveau pour l'industrie suisse' in *La Vie économique*, Genève, Feb. 1994.
LEPORI, B., *L'avenir des transports en Europe: résultats choisis de la recherche européenne*, Bern, Swiss Science Council, TA 5/1995.
MAGLEV 98, *Proceedings of the 16th International Conference on Magnetically Levitated Systems*, Yamanashi, Japan, 10-14 April 1998.
RAOUL, J.C. 'How High-Speed Trains Make Tracks' in *Scientific American* (October 1997).
ROSSEL, P., 'Dimensions sociologiques de Swissmetro' in M. JUFER et al., *Rapport pour l'étude préliminaire*, Lausanne, IREC-EPFL, 1992.
ROSSEL, P., F. BOSSET, O. GLASSEY and R. MANTILLERI, 'Les enjeux des transports à grande vitesse: Des méthodes pour l'évaluation des innovations technologiques, L'exemple de Swissmetro', Bern, FNRS/PNR 41, Rapport F3, 1999.
SCHULER, M. and V. KAUFMANN, *'Pendularité à longue distance: la vitesse comme facteur structurant l'urbain'*, DISP 126, Zurich, ORL, ETH, 1966.
STIX, G., 'Maglev Racing to Oblivion?' in *Scientific American* (Oct. 1997).
TANGL, P., *A Short History of Expectations on Magnetically Levitated Trains*, Master's Thesis, ESST, University of East London, 1996.

2.2.2

THE SUPERPHÉNIX CONTROVERSY

An Ethical Perspective

Bertrand Hériard Dubreuil[1]

Spectacular developments in technology pose new questions for an old discipline like moral philosophy: the scale of technological decisions has changed, both in space – they affect more people – and in time – they often have irreversible consequences. A good example of that may be the Superphénix nuclear reactor whose controversy has lasted thirty years and reached the highest levels of the French state. We will suggest that ethics can make a contribution on two levels: it can allow for discussion about an overall judgement on the sense of such a venture; it can also shed some critical light on the decision procedures which led to the construction of the installation. We will arrive to the conclusion that the importance of professional ethics is not diluted by these new dimensions of technical decision, but radicalised. In other words, the larger decisions become in space and time, the more particular prudence must be respected and the more conditions for real debate must be assured.

Introduction

The spectacular development of technologies raises new questions for an old discipline like moral philosophy – witness the numerous publications in medical ethics. To be sure, the physician's power over the patient is an old question, since the Hippocratic oath already regulated it, but we see that the new possibilities opened up by advances in medical technology raise other prob-

---

[1] Original text translated by C. ROY.

lems. Engineers, too, have been interested in ethics for a long time. This is evidenced in the technical norms that the profession was able to promote or in the deontology codes through which it has regulated itself. However, these means seem very small in face of the scope and the newness of the problems raised by modern technology. What can ethics bring to engineers responsible for applying brand new technologies?

This chapter would like to show what an ethical perspective could bring to a long and controversial story: that of Superphénix.

- The controversy has lasted for thirty years since the nuclear installation at Creys-Malville was conceived in the 1960s, built in the 1970s, started up in the 1980s, recommissioned for a scientific project in the 1990s, and will no doubt be dismantled in the next decade.
- The controversy is a deeply political one, for the Superphénix was conceived by the *Commissariat à l'Énergie Atomique* (CEA –Atomic Energy Commission) in order to have a French technology following the abandonment of the graphite-gas reactor systems; construction was finalised in line with the Messmer plan which had financed the construction of five or six reactors a year from 1974 to 1981; placing Superphénix on line gave rise to several legal challenges which were settled by the *Conseil d'État* (State Council); recommissioning was decided by Pierre Beregovoy on the recommendation of the minister of research; and the dismantling was decided by Lionel Jospin against the advice of Jacques Chirac.

There are few technical devices that can boast of such treatment by the French and even international media. There are few technical devices whose name is spelled out in political platforms. There are few technical devices whose fate depends on the outcome of so impassioned a debate. What can a philosophical discipline say about such a complex and politically charged matter?

I think that ethics can make a contribution on two levels:

- it can enable us to ground a global judgement on the meaning of such a venture;
- it can shed some critical light on the decision-making process which led to the construction of the plant.

To defend these two theses, I will use the concepts of narrative ethics as they have been introduced by Paul Ricoeur[2] and used to shed light on technological problems by Peter Kemp.[3] According to these two authors, we are steeped in stories allowing decision-makers to express the meaning of their action and other agents to reinterpret this meaning in their own temporality. The conflict of interpretation thus concerns the global vision in so far as it expresses the meaning of the venture, as well as the capacity of agents to introduce this vision into the long and complex process of technical decisions.

Can an Overall Judgement Be Made about the Sense of Such a Venture?

If the plant is dismantled without having functioned, then there will be something nonsensical about the whole human venture that carried it. I know that the plant is currently being dismantled, but I say 'if', since the controversy is far from over and the 'bird may yet be reborn from its ashes'. In making the hypothesis that the current renovation of its smaller sister Phénix will have no industrial posterity, I am placing myself at the end of the story of fast breeder reactors and I am making an overall judgement on the entire venture. How can such a judgement be grounded?

This intuitive judgement rests on the following question: what will have been the use of mobilising hundreds of scientists, thousands of technicians and more than 60 billion francs[4] of public funds to produce less than 3 billion kWh while taking a considerable safety gamble? Let us note that this intuition is formulated in terms of utility: what will have been the use? For the moment I am talking from the standpoint of the plant's promoters and I am questioning the validity of a means with respect to the presumed end of social utility.

What in fact was the aim of the Superphénix, what did its promoters want to accomplish, and what arguments did they put forward to convince the national community to share that aim?

---

[2] P. RICOEUR, *Soi-même comme un autre*, Paris, Seuil, 1990.

[3] P. KEMP, *L'Irremplaçable: une éthique de la technologie*, Paris, Cerf, 1991.

[4] Cours des comptes, 'Les comptes et la gestion de la NERSA: La Centrale Européenne à Neutrons Rapides (Superphénix)', 1996, op. cit. in R. BELL, *Les péchés capitaux de la haute technologie*, Paris, Seuil, 1998, p. 71.

Figure 1: Physical principle of electronuclear reactor systems

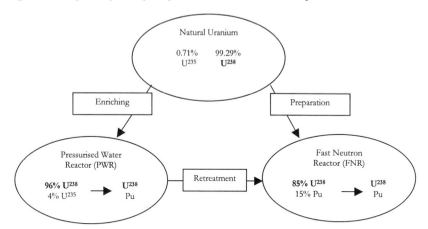

The first scientists to look for a civil application of nuclear energy wanted to achieve the fission of U^{238}, an element making up more than 99% of uranium ore. They were outdone, however, by those who were content with fissioning U^{235} – the fissile isotope and therefore the least present in natural uranium – following an ore enriching procedure. In France, the abandonment of the graphite-gas reactors allowed research efforts to be concentrated on U^{238} and a medium power prototype was rapidly developed: the Phénix. On the strength of this success, the CEA convinced the major European producers of electricity to launch a new reactor system for electrical production. The industry would exploit a market niche where the Americans were lagging behind, would assure greater European independence with regard to uranium ores, and would offer a competitive cost price. In this way, scientific know-how opened up a technical possibility, which was quickly translated into political and economic arguments.

The aims of this program were constructed in the context of the 1970s. Over the course of the twenty years it took to build and to start the installation running, the context changed considerably. Political concerns shifted from energy self-sufficiency to questions of nuclear non-proliferation. The energy crisis prompted the discovery and exploitation of other fossil fuels, notably natural gas. Public opinion showed itself a lot more demanding and other nuclear technologies arrived to compete with breeding. We cannot blame yesterday's decision-makers for not having responded to today's demands. But they can be faulted for the plant's failure to reach the objectives they had negotiated with the rest of society at the time it was built.

We are entering here into several controversies. Let us leave aside that of knowing with whom lies the legal liability of the project and the cost overruns of the dismantling. Let us also avoid getting mired in properly historical debates about just what were the actual objectives at the time of construction. Let us work from the hypothesis that a technology at least has technical production objectives, the political objective organising its community of operation and its relation with the community of users, the objective of profitability – be it only to earn a living for those who operate it – and the symbolic objective of being accepted in its social environment. Let us now see if these four types of objectives have been reached or not. We have just dissected the controversy into four types of questions[5]:

- What was the level of availability expected, what was the one actually reached, and to whom should the difference be ascribed? The question is seemingly a technical one, but the answer is too global to be left solely to the technicians, since stoppages can be imputed to technical incidents as well as to administrative and legal problems they have generated.
- What was the future of the fast breeder reactors expected in 1974, and what can we say about it today? The answer was debated by energy policy specialists, but it was a delicate one since the future of the system is as much a political as an industrial question.
- What is the price of a kilowatt/hour expected from fast breeder reactors and what is the one actually delivered by Superphénix? Here again the question is disputed by energy economists, and it is a difficult one as the economic landscape is changing so much.
- Did this reactor system have a chance of getting accepted by public opinion? On what scale: local, national, international? What constraints did its construction impose on the natural and social environment? On which conditions were or will these constraints be acceptable by the general public at the moment of its construction, of its operation or of its dismantling?

---

[5] For a theoretical justification of this division, see B. HÉRIARD DUBREUIL, *Imaginaire technique et éthique sociale, essai sur le métier d'ingénieur*, Brussels, De Boeck Université, 1997, pp. 41-48.

We cannot answer all these questions without the mediation of disciplines where the different experts may ground their judgement and without collective debates where their recommendations may be compared. This debate would allow what Gadamer calls an 'objective' hermeneutics to be carried out – objective in the sense that we now have hindsight at our disposal, 'hermeneutics' though, since it would be difficult to compare interpretations from various disciplinary fields. When at the end of a megaproject the agents hold each other mutually accountable, they are comparing global judgements without being able however to abstract themselves from the social groups of which they are a part.

In what way can a discipline like ethics contribute to this debate? It is one discipline among many, older to be sure, yet little used to studying the arcane details of technical decision-making. Nevertheless, it can bring its contribution to the debate by reminding people of the demand for universality running through its entire tradition. This conflict of interpretations could not be concluded without giving everyone who was involved – at whatever degree, decision-maker or decision-bearer, immediate or remote, today or tomorrow – the right to speak. This demand thus has an actually utopian character. More specifically, how can we anticipate the judgement of future generations? What will they retain from the whole venture? In what way will they have benefited from it? What is the fallout for which they have to bear the consequences in terms of sites to protect, of waste to monitor, of cost overruns to pay?

Yet it is on the basis of this demand for universality that the judgement expressed in an intuitive fashion at the beginning of this part can be founded. This demand for universality is first of all a demand for scientific standards. The ethicist turns into an epistemologist for a moment in order to privilege the judgement that is best rooted in each discipline, the most discussed between the disciplines, and the one that has convinced the most people. This is why, among all the theses defended at the University – whose role is precisely to ensure the quality of the debate – we have chosen that of Dominique Finon, not only because it earned a most prestigious academic degree, but because it is the most multidisciplinary. His State thesis[6] compares

---

[6] D. FINON, *Les États face à la grande technologie dans le domaine civil: le cas des programmes surgénérateurs*, Thèse de doctorat d'État ès Sciences Économiques, Université de Grenoble II, 1988.

all the fast breeder reactor programmes in the world. It relates the technical choices that have been made, the ones that have been validated by experience and those which did not work; it analyses the political strategy chosen by these promoters and shows why the other countries gradually abandoned it; it shows the 'impossible economic competitiveness of the fast breeder reactor'; finally, it shows how narrow is the community that has defended it and the hard time it had getting accepted in France and abroad.

This last argument is the most convincing one from the ethical perspective. For if the promoters have not convinced their contemporaries, how will they be able to convince future generations? If the promoters were not able to get people to share in their aim, how will they be able to give an account of the meaning of their venture?

Is not this appeal to meaning a polite way to take the side of the victors in the controversy? Certainly, but it is not without reason that practically all the countries engaged in this technological race have finally abandoned it. The meaning evoked here is the global signification of the venture, the one that can be made out of a complete narrative once it is finished, the one from which we can draw lessons. It is also the common sense that any narrative seeks to convey, the consensus sought by any controversy. Finally, it is the direction, the aim that was pursued by the entrepreneurs in building the plant. The quest for meaning is thus a point in common between the agents, the protesters and those who study the controversy. All are using narratives to state their aim, question it and find a consensus. The end of the narrative is thus the final judgement, which the agents have anticipated and which will qualify their morality. The full success of the venture would have made of them heroes, while a catastrophic failure would have made them criminals. The in-between makes them humans who have erred on certain points and were right about others.

Thus, the global judgement that I have stated *a posteriori* is the exact opposite of the practical judgement that was made at the time of the building of the plant. But if professional ethics has a role, it is not only to criticise past decisions from the outside, but above all to help the agents to better take present decisions. What did the Superphénix decision-makers miss? Why were they not able to anticipate the elements that are held against them today? Where did the initial aim of social utility get lost? This is what we have to look at in detail by analysing the decision-making process.

Can the Decision-Making Process Be Criticised?

In analysing the decision-making process, we come up against three methodological problems:

1. The decisions are part of a history in which it is difficult to separate the initial decisions from the subsequent ones. It is difficult because Superphénix is a complicated object, which only functions when its parts (machines and people) carry out their tasks effectively. It is also difficult because, as with any prototype, every single decision is the object of internal controversies, re-examinations and modifications. It is only from an external viewpoint that one can speak of one single decision, since the plant, in the end, was built. In the name of this unity, I am going to arbitrarily isolate the period from 1974 to 1976, after which construction was more or less assured, particularly after the first public enquiry.

2. It is difficult to know who took the decision of launching the project, so numerous were the decision-makers and so complex their interactions. Not that it matters so much incidentally, as the point is not to pin blame on this one or that one so much as to stress the moral responsibilities. For this reason, I will be content to describe the decision-making process as if the institutions (CEA, EDF,[7] NERSA,[8] SCSIN[9]), were 'moral persons', not in the legal sense, but in the moral sense, which supposes that they have a free will and an awareness of what they want (cf. Fernandez, 2.3.2).

3. Finally, it is difficult to isolate the four dimensions – technical, political, economic and cultural – referred to in the global judgement. In fact, this separation is arbitrary and depends mainly on the way in which knowledge was constituted into disciplines. I will analyse the process according to these four dimensions because the global judgement took them as presupposed.

I will thus give four independent accounts in order to underline in turn the political, the economic, the cultural dimensions, finishing with the most controversial, the technical dimension. The thesis that

---

[7] Electricité de France.

[8] The EDF created in 1974 with the main German and Italian producers the joint-stock company 'Centrale Nucléaire Européenne à Neutrons Rapides' (NERSA) to build and operate an industrial scale fast breeder reactor.

[9] Service Central de la Sûreté des Installations Nucléaires (SCSIN). It became Direction de Sureté des Installation Nucléaires (DSIN).

this second part wants to defend consists in saying that if the plant is a whole, none of these dimensions can be neglected by these promoters in the decision-making process.

Figure 2:
Synchronological analysis of the decisions between 1974 and 1976

DIMENSION	POLITICAL	ECONOMIC	CULTURAL	TECHNICAL
CONTEXT 1974-1976	Energy self-sufficiency	Crisis perceived as context-bound	Productivist values	Technological race
By delegation from... ↕ Technical institutions responsible	Citizens ↕ Organisers EDF/NOVATOME	Stock-holders & Consumers ↕ Makers EDF / NERSA	General public ↕ Experts SCSIN	Users ... ↕ Designers CEA
Arguments 1974-1976	U^{238} fission	Competitiveness?	Safety	Industrial prototype

Political Dimension

Let us begin with the political dimension, for it seems to be the decisive one. Just after the Yom Kippur War, the EDF directors feared an explosion of uranium prices that would seriously hamper the principle of energy self-sufficiency outlined by the Messmer plan. The technicians proposed a simple solution: the fission of U^{238}, which makes up more than 99% of natural uranium.

Dominique Finon maintains that the fear of an explosion of prices was more a construct of the nuclear lobby than something actually anticipated by the politicians.[10] At the very least, this interpretation demonstrates the difficulty of separating technical aims from their strategic overtones. In any case, the analysis was superficial since the uranium reserves turned out to be more abundant than anyone had predicted, and the producing countries lacked the means to establish a cartel.

---

[10] D. FINON, *L'échec des surgénérateurs: autopsie d'un grand programme*, Grenoble, Presses Universitaires de Grenoble, 1988, pp. 25-30.

Could these elements, which appear so evident today, really have been anticipated at the moment the decision was made? Two arguments point toward an affirmative answer. An argument from principle: it is the role of the institution in charge of industrialising the process, in this case the EDF, to ground its strategic analysis, or at least to criticise the one that is submitted to it; a more circumstantial argument: such an analysis can already be found in the Ford-Mitre report which was decisive in halting the American programme in April of 1977.

It was a different geopolitical event, however, which determined the American decision. Following the explosion of the Indian bomb in May 1974, the Arms Control and Disarmament Agency became concerned about the risks of nuclear proliferation linked with plutonium, one of the components which is both produced and consumed in fast neutron reactors. Several independent reports commissioned by the State Department and Congress strengthened opposition, which became conclusive with the election of Jimmy Carter. Can the French be accused of having neglected a problem which, since the collapse of the Soviet block, has become ever more significant? Let us not forget that France only signed the 1968 nuclear non-proliferation treaty in 1992.

Yet one could reproach the French for having been too hasty in the decision-making process. One detail is particularly revealing in this respect: the decision to build the Superphénix was made without even going through the proper channel of a commission in charge of directing electrical production of nuclear origin (PEON). The industrial context was indeed a race in which the first to develop the technology would be able to impose its own norms on the market. But France found itself at the head of a pack which eventually dwindled to nothing.

The absence of public debate in France leads me to conclude that there was not only a strategic error, but also a real carelessness. For the fact that the decision was made through improper channels shows that the supporters had the deliberate intention of doing things quickly. Prudence, however, is the cardinal political virtue, and the haste of the French turned out to be unjustified. Might not this mistake have been averted by respecting the commonly accepted norms for public decisions of such importance?

In other words: we can congratulate the French technicians for having been the first to industrialise a new possibility for the production of electricity; we can admire their political sense in defending this project at a time when international circumstances posed a grave threat to France's energy self-sufficiency. But by short-circuiting public debate, they foreclosed the possibility of allowing their decision to be influenced by elements which they may *a posteriori* be faulted for ignoring. They proved themselves to be excellent politicians in defending their project, but not so good at taking into account superior interests. They and the government were, in Aristotelian terms, imprudent.

Economic Dimension

We can see, still a posteriori, that the most neglected dimension of the decision-making process was the economic one: we have seen that the projected sales of the fast breeder reactor were overestimated, this was the case with the entire market in electronuclear equipment; we know too that the cost of investment in the plant went out of control, but in proportions comparable to the first French pressurised water reactors; finally we also know that the estimated cost price per kWh is still contested today. Let us examine this calculation:

The plant's principal promoter, Georges Vendryes, speaks little of this calculation.[11] By contrast, Dominique Finon[12] believes that most of the prospective calculations were made on the assumption that plutonium would be available at a reduced cost. Now this assumption called for two other ones:

- the necessity to reprocess irradiated fuel from classical plants, while the global industry today is overwhelmingly oriented toward burying waste directly, a much more economic choice even if it is much disputed;
- the main civilian use for plutonium would be in fast breeder reactors, a decision which delayed many other developments, such as the development of mixed fuels (MOX).

So the competitiveness of this reactor system is not so much an error in calculation as an error in the reference, a contradiction which did

---

[11] G. VENDRYES, *Superphénix pourquoi?*, Paris, Nucléon, 1997. It is true that, in his chapter 7 entitled: 'the cost of Superphénix', he reasons in terms of marginal cost to defend the plant's new scientific mission.

[12] D. FINON, *L'échec des surgénérateurs*, pp. 215-18.

not fail to appear to some economists at the CEA.[13] But what is even stranger is that the company in charge of industrialisation, the EDF, adopted these calculations without discussion.

For errors in calculation are indeed frequent with new technology, which is why people protect themselves. The fact that it was not until 1979 that the EDF carried out its own expertise, and not until 1984 that its president dared to raise doubts publicly about the industrial future of the reactor system, is therefore problematic. It is true that by virtue of an agreement made in 1971, the State committed itself to covering the extra costs over and above the cost of a classical reactor of the same power. But neither this umbrella, nor the necessary solidarity with the CEA can excuse the EDF for having taken their responsibility so lightly. Who else could have defended the interests of future consumers of electricity if not the institution which has a monopoly on its distribution in France?

Cultural Dimension

Social acceptability is the most difficult to gauge: witness the evolution of public opinion about nuclear power over the last 30 years. Its promoters have navigated between two extremes: that of ignoring it as too complex and that of managing it like an advertising image is managed. In doing so, they testified that it was a dimension which had to be addressed if the reactor and all its social constraints were to find a place in the French landscape.

In 1974, we were still in the trail of post-war boom period and public opinion could not have foreseen the danger which was lurking behind the first oil crisis. Electrical power was seen as one of the engines of development and few were concerned about the way in which it was produced. The ecological movement was linked with the extreme left wing, and its proposals were considered exaggerated. The ideological debate surrounding the reactor became radical, marginalising all other criticism.[14]

There was yet a smallest common denominator on which no one wanted to compromise: safety. Experience shows that this required very varied, very concrete compromises, from the steel grades of the

---

[13] 'An economist from the CEA told me, with regret, that the reactors cost estimates had been generated on the basis of what still had to be established, namely the reactor's competitiveness.' FINON, *L'échec des surgénérateurs*, p. 302.

[14] A. TOURAINE, Z. HEGEDUS, F. DUBET, and M. WIEVIORKA, *La prophétie antinucléaire*, Paris, Seuil, 1980.

sodium containers to the protection of the installation against terrorist acts. But at the time when the French authorities were to authorise construction, the concept of safety was still rather abstract, since:

- the physical models of fast neutron fission were still under discussion;
- the experience gained was either out of proportion (Rapsodie) or else over too short a period of time (Phenix, PFR, BN 350);
- very little problematic data were available, given the small number and geographical dispersion of reactors of this capacity.[15]

One can imagine the difficulties involved in transposing this small body of knowledge to an industrial scale. This was indeed the wager of a prototype. To proceed with construction implied extending the validity of the models, and taking into account the experimental results and the problematic data coming from other countries. We can imagine the difficulty of such an endeavour since such knowledge was scattered and the technological race did nothing to promote its diffusion.[16]

A posteriori, the controversy about safety was never concluded. The proponents minimised the safety problems, even if some recognised that some of them could have been avoided.[17] The opponents maximised the same incidents, stating that 'the probability of new breakdown was significant.' The controversy gained the DSIN, where

---

[15] With the notable exception of the partial core fusion that put an end in October 1966 to the commercial experiment of Fermi, a fast reactor of 250MW built near Detroit. Cf. J. FULLER, *We Almost Lost Detroit*, New York, Reader's Digest Press, 1975.

[16] In the face of such uncertainties and taking into account the tragic consequences of an accident, the applicability of the principle of precaution could be discussed here. But it would be unfair to ask of those who have decided to build the plant to have anticipated a principle that was only officially proclaimed at the Rio summit in 1992 and which remains very ambiguous. Cf. O. GODARD (ed.), *Le principe de précaution dans la conduite des affaires humaines*, Paris, Éditions de la Maison des Sciences de l'homme, 1997.

[17] 'During its first years of being operational, Superphénix ran into two problems in connection with the nuclear plant or its annexes which required lengthy corrective action... The first concerned a component called the drum . . . the choice of steel, limited to certain external parts of the installation proved unfortunate. Our German colleagues had encountered the same kind of problem on certain SNR 300 components. It is regrettable that their experience was not put to better use on this occasion.' G. VENDRYES, *Superphénix pourquoi?*, Paris, Nucléon, 1997, pp. 81-82.

two directors successively criticised the conception.[18] The most disturbing aspect concerned the pressure the cover can contain in case of an over rapid acceleration of the nuclear reactor. An incident of this type occurred four times on Phénix between 1989 and 1990, and had to be resolved before a nuclear accident could really take place.

Technical Dimension Proper

While the public discourse promoting acceptance of the plant was often limited to guaranteeing safety, the rationalisation of the technological race brought about a certain number of ambitious technical gambles:

- the direct jump to an industrial scale;
- the decision to simplify the total design in order to limit investment costs;
- the decision to start construction without even waiting for the experience gained from running the Phenix.

These gambles presupposed an unshakeable faith in the industrial future of the reactor system, and confidence in the ability to overcome any difficulties that might arise. These beliefs have to do with what one might call a technological passion, with all the ambiguity that the philosophical tradition attaches to that word. It proceeds from the vital force which sustains the technicians and links them together, but which makes them particularly vulnerable to the excesses which the Greek tradition has been fighting since the myth of Icarus.

Now the technical gambles mentioned above are indeed the sign of a certain excessiveness, which brings some technologists to the conclusion that Superphénix was 'bad' engineering.[19] This judgement seems to me peremptory. What is more striking is that technical passion made the promoters blind to the arguments of the spokesmen of other disciplines which they needed in order to organise the technical system, to assure its economic viability, to negotiate its social acceptability or at least not remain stuck in a blind alley. But, as we have seen, this dialogue had been considerably simplified, permitting the technicians to impose their logic at the cost of the other dimensions.

---

[18] R. BELL, *Les péchés capitaux de la haute technologie*, Paris, Seuil, 1998, pp. 80-86.
[19] J. NEYRINCK, *Les cendres de Superphénix*, Paris, Desclée de Brouwer, 1997.

The presentation of the Superphénix as the 'cornerstone of the French electronuclear system' was quite emblematic of this logic. According to this view, which was supported until the end of the 1970s,[20] the fast breeder reactor line considerably augmented the overall efficiency of the system by extracting more electricity from natural uranium and managing the waste from classical reactors. However, its decline in the 1980s, and the necessity for finding a new commission for the Superphénix in the 1990s, reveal the existential nature of the argument – to defend the functioning of the plant and the survival of the group who had promoted it.

Of course, the defence of a group and their working tools is a noble cause – if, that is, one does not hide the group's interests behind seemingly technical constraints. Therein lies the danger: while it is sometimes difficult to see the boundary between technical constraints and social constraints, confusing them with each other leads to a mixing up of ends and means and to getting caught up in one's own technical paradigm. Now technical paradigms are many, always intersecting with other dimensions that permit them to exist socially. It is within these intersections, within these fault lines, that degrees of freedom can be found which allow a social group to aim at what is right and lead an ethical existence.

Conclusion

My initial question was the following: what can moral philosophy contribute to the analysis of a decision-making process as complex and as long as that of the construction of the Superphénix?

To answer it, I have first of all tried to apply an objective hermeneutics to uncover the meaning of such a venture and to make a global judgement. This judgement uses hindsight in order to articulate those dimensions without which the Superphénix could not exist socially. It is a moral judgement in the sense that it qualifies the actions that led to the construction of the plant.

But no one made a global decision. It was the product of particular decisions taken at each step along the way. The real question for professional ethics is then the following: how to anticipate the hori-

---

[20] It is curious that we can still find the FNR pictured at the centre of the fuel cycle in a 1993 work: R. CARLE, *L'électricité nucléaire*, coll. Que sais-je?, Paris, PUF, 1993, p. 40.

zon of meaning which an *a posteriori* judgement describes each time a decision is made? How does one aim at a global judgement when confronted with the particularity of action?

To go forward with this question, I have taken up the analysis again one dimension after the other. Since the global judgement concludes that the venture seemed nonsensical, I looked to see where the direction got lost, and discovered a certain number of shortcuts where particular decisions did not exhibit the prudence demanded by the importance of the global decision. I have thus denounced:

- a political irregularity, due to the absence of debate;
- an economic irregularity, due to the confusion of roles;
- a cultural irregularity, due to the reduction of social acceptability to safety;
- a technical irregularity, due to pushing the 'state of the art' too far.

These irregularities demonstrate the incapacity of particular decision-makers to envision in each concrete decision the coherence of the whole. This is of course a major problem in modern societies, where labour has never been so fragmented. But the fact we can detect these difficulties is vice's homage to virtue, revealing the ultimate aim of the professional ethics transgressed: to allow the translation of the global scope of a problem, or at least to verify the capacity for taking it into account at the time of each decision.

It is startling to see how these intersecting ethics – ethics of public decision-making, business ethics, ethics of the scientific world, ethics of technical decision-making – all defend in their own area the preconditions for deliberation, which – as Aristotle aptly remarked – is the locus *par excellence* for the exercise of prudence. By contrast, in doing without public debate by closing off the decision process, by simplifying political, economic and cultural analyses, the proponents of Superphénix withdrew into their technology. Is this not one of the main causes of their failure?

Thus, the importance of professional ethics, as so with many safeguards, is not diminished by these new dimensions of technical decision-making, but radicalised. In other words, the longer in range and the more complex decisions become in space and time, the more necessary it becomes to respect specific measures of prudence and to ensure that the conditions for real debate are maintained. Now there is a fine challenge for tomorrow's engineers.

References

BELL, R., *Les péchés capitaux de la haute technologie*, Paris, Seuil, 1998.
FINON, D., *L'échec des surgénérateurs: autopsie d'un grand programme*, Grenoble, Presses Universitaires de Grenoble, 1988.
GODARD, O. (ed.), *Le principe de précaution dans la conduite des affaires humaines*, Paris, Editions de la Maison des Sciences de l'Homme, 1997.
HÉRIARD DUBREUIL, B., *Imaginaire technique et éthique sociale, essai sur le métier d'ingénieur*, Bruxelles, De Boeck Université, 1997.
KEMP, P., *L'Irremplaçable: une éthique de la technologie*, Paris, Cerf, 1991.
NEYRINCK, J., *Les cendres de Superphénix*, Paris, Desclée de Brouwer, 1997.
TOURAINE, A., Z. HEGEDUS, F. DUBET and M. WIEVIORKA, *La prophétie antinucléaire*, Paris, Seuil, 1980.
VENDRYES, G., *Superphénix pourquoi?*, Paris, Nucléon, 1997.

Table 1: Chronology of the fast breeder reactor program.

	Civil / International Nuclear Energy (Accident)	Military	Nuclear in France — Technical	Nuclear in France — Incident	Nuclear in France — Judicial	Fast Neutron Reactor in France — Technical	Fast Neutron Reactor in France — Incident	Fast Neutron Reactor in France — Judicial
1945	Hiroshima (U^{235})/Nagasaki (Pu)							
1946					Creation of the CEA			
1947					Nationalisation of EDF			
1949		Soviet Bomb						
1951	EBR I, Idaho (USA)		Start up of Marcoule					
1952		British Bomb						
1953								
1954	Policy "Atoms for Peace"							
1955	British Programme							
1956	Creation of the AIEA			Chinon / Creation of the PEON		Study on liquid metals		
1957	Creation of Euratom							
1958	2nd Geneva conference					Studies on fast neutrons		
1959								
1960		British Bomb	Start of la Hague			Creation of Cadarache		
1961								
1962						Start of Rapsodie		
1963	Commercial take-off of PWR-BWR Treaty on Nuclear Tests Fermi (USA)			Beginning of the graphite-gas system	Decree 63-1228			
1964						Retreatment workshop AT1		
1965								
1966	*Fermi*			*Conflict CEA-EDF*		Phénix studies (Marcoule)		
1967								
1968	Treaty of non-proliferation		Abandonment of the graphite-gas system			Rapsodie divergence (24MW)		
1969								

1970			Start of the PWR programme	European Cooperation
1971	BN 350 (USSR)			Divergence of Phénix (250MW)
1972	First oil crisis			
1973				
1974	PFR (GB)	Indian Bomb	Messmer Plan	Creation of NERSA (Creys)
1975				**First public inquiry**
1976	PWR-BWR commercial crisis		Creation of the Cogéma	Creation of Novatome
				Autorisation of Superphénix
1977	Carter: Suspension of Clinch River		Anti-nuclear demonstrations	Civil engineering
1978	KNK II (RFA)			
1979	Second oil crisis			
	Three Mile Island			
1980	*Almeria*			Secure containers
1981			Mitterrand	
1982				Steam-generator
1983			Slow-down of the PWR prog.	End of the fill up sodium
1984			EFR programme	Divergence of Superphénix (1250MW)
1985			MELOX programme	
1986	*Chernobyl*			*Leak from the drum*
1987				
1988			Mitterrand	*Pollution of the primary sodium*
1989	Fall of Berlin Wall			
1990	Gulf war			
1991			Law 91-318 on nuclear waste	PTC
1992				Curien report
1994				**Second public inquiry (en 1993)**
1994				Decision of restart
				Leak of argon / of steam
1995	Monju (Japan)		Chirac	4 month restart
	Monju			12 month restart
1996				Ten year shut-down
1997			(Jospin)	**Definitive shut-down**
1998	Pakistan Bomb			Dismantling process

2.2.3

A PRACTICAL PERSPECTIVE OF INFORMATION ETHICS

Simon Rogerson

Information is the new lifeblood of society and its organisations. Our dependence on information grows daily with the advance of information and communication technology and its global application. The integrity of information relies upon the development and operation of computer based information systems. Those who undertake the planning, development and operation of these information systems have obligations to assure information integrity and contribute to the public good.

This chapter addresses these issues from three practical perspectives with the field of information ethics. Firstly, the manner in which strategy is formulated and the dilemmas that might arise are considered. This is followed by a discussion as to how an ethical dimension might be added to the project management of computing applications. Finally, the methods used for information systems development are reviewed from an ethical standpoint. These issues are important to information engineers and software engineers who are now playing a key role in the running of modern organisations.

Introduction

Information is the new lifeblood of society and its organisations. Our dependence on information grows daily with the advance of information and communication technology and its global application. The integrity of information relies upon the development and operation of computer based information systems. Those who undertake the planning, development and operation of these information systems have obligations to assure information integrity and contribute to the public good. This is the field of information ethics, a brief discussion of which is included after this introduction.

It is interesting to note that much of the work in this area has been concentrated in the philosophy and sociology disciplines with restricted input from the information systems and computer science disciplines. This may be the reason why Walsham[1] found that '...*there is little published work which directly relates these [specific information systems related ethical] issues to more general ethical theory...*'. The work has tended to be conceptual and more of a commentary on computer phenomena rather than an attempt to develop strategies to identify and address societal and ethical issues associated with information systems and the underpinning information and communication technology.[2]

Given the overall focus of this book it has been decided to adopt an organisational view of information ethics in this chapter rather than consider specific ethical issues such as privacy, access and property which are dealt with extensively in other publications in this field.

This chapter addresses these issues from three practical perspectives within the field of information ethics. In the section *Information Strategy*, the manner in which strategy is formulated and the dilemmas that might arise are considered. This is followed by a discussion in the section *Project Management* as to how an ethical dimension might be added to the project management of software development. Finally in the section *Information Systems Development*, the methods used for information systems development are reviewed from an ethical standpoint. These issues are important to information engineers and software engineers who are now playing a key role in the running of modern organisations.

Information Ethics

According to Van Luijk,[3] ethics comprises both practice and reflection. Practice is the conscious appeal to norms and values to which individuals are obliged to conform, whilst reflection on practice is the elaboration of norms and values that colour daily activity. Norms are

---

[1] G. WALSHAM, 'Ethical Theory, Codes of Ethics and IS Practice' in *Information Systems Journal* 6(1996), pp. 69-81.

[2] S. ROGERSON, and T.W. BYNUM, *Information Ethics: The Second Generation*, UK Academy for Information Systems Conference, UK, 1996.

[3] H. VAN LUIJK, 'Business Ethics: The Field and its Importance' in B. HARVEY, (ed.), *Business Ethics: A European Approach*, New York, Prentice Hall, 1994.

collective expectations regarding a certain type of behaviour whilst values are collective representations of what constitutes a good society. The existence of a plan and a controlling mechanism is the accepted norm in project management which itself is an accepted value in software development. For the purpose of this chapter it is sufficient to consider only ethics practice because it is concerned with professional action rather than conceptual reflection. Conceptual reflection might manifest itself in, for example, codes of conduct which are concerned with establishing what are the generalised ways of working that are acceptable to a wider community. In other words, the chapter is concerned with how to use and when to apply norms and values rather than establishing what these norms and values are.

Bynum[4] maintains that information ethics is the most important field of applied philosophy since 'the world is transforming exponentially ... under the influence of the most powerful and the most flexible technology ever devised'. The main thrust of information ethics must be to integrate computing and human values in such a way that computing advances and protects human values rather than harms them.[5] There are however complex problems to overcome if such a goal is to be realised. It was James Moor[6] who asserted that computers are 'logically malleable' in the sense that 'they can be shaped and moulded to do any activity that can be characterised in terms of inputs, outputs, and connecting logical operations'. This might lead to policy vacuums caused by possibilities that did not exist before computers. In these situations there are often no useful analogies to draw upon for help and as Maner[7] explains we are therefore forced, 'to formulate new moral principles, develop new policies, and find new ways to think about the issues presented to us.' In such situations Gotterbarn[8] suggests that professionals

---

[4] T.W. BYNUM, The Development of Computer Ethics as a Philosophical Field of Study' in *Australian Journal of Professional and Applied Ethics* 1 (july 1999) 1, pp. 1-29.

[5] T.W. BYNUM, 'Computer Ethics in the Computer Science Curriculum' in T.W. BYNUM, W. MANER and J.L. FODOR (eds.), *Teaching Computer Ethics*, New Haven, CT, Research Center on Computing & Society, Southern Connecticut State University, 1992.

[6] J.H. MOOR, 'What is Computer Ethics?' in *Metaphilosophy* 16(1985)4, pp. 266-279.

[7] W. MANER, 'Unique Ethical Problems in Information Technology' in *Science and Engineering Ethics* 2(1996)2, pp. 137-155.

[8] D. GOTTERBARN, 'The Use and Abuse of Computer Ethics' in T.W. BYNUM, W. MANER and J.L. FODOR (eds.), *Teaching Computer Ethics*, New Haven, CT, Research Center on Computing and Society, Southern Connecticut State University, 1992, pp. 73-83.

must be aware of their professional responsibilities, have available methods for resolving non-technical ethics questions and develop proactive skills to reduce the likelihood of ethical problems occurring. Given the global nature of computing it is important to base professional responses on Moor's suggestion[9] of common core values and a respect for others. Those involved in computer-based information provision are one group significantly affected by such ethically volatile situations.

An ethical framework for computer professionals has been developed comprising a list of eight ethical principles that professionals need to be aware of, and a method by which these ethical principles can be applied to a particular application domain.[10] The eight ethical principles are:

- *honour* – is the action considered beyond reproach?
- *honesty* – will the action violate any explicit or implicit agreement or trust?
- *bias* – are there any external considerations that may bias the action to be taken?
- *professional adequacy* – is the action within the limits of capability?
- *due care* – is the action to be exposed to the best possible quality assurance standards?
- *fairness* – are all stakeholders' views considered with regard to the action?
- *consideration of social cost* – is the appropriate accountability and responsibility accepted with respect to this action?
- *effective and efficient action* – is the action suitable, given the objectives set, and is it to be completed using the least expenditure of resources?

Looking at some of the principles in more detail, the principle of honour can be considered as the 'umbrella' principle, to which all other principles contribute. It is concerned with the verification that all actions are beyond reproach. The principle of bias focuses on due

---

[9] J.H. MOOR, 'Reason, Relativity, and Responsibility in Computer Ethics' in *Computers and Society* 28(March 1998)1, pp. 14-21.

[10] S. ROGERSON, 'Software Project Management Ethics' in C. MYERS, T. HALL, D. PITT (eds.), *The Responsible Software Engineer*, London, Springer-Verlag, 1996, Chapter 11, pp. 100-106.

consideration of any factors that may influence the action to be taken. The principle of due care is concerned with putting into place the measures by which any undesirable effects can be prevented, which may require additional attention beyond that agreed formally within a contract. The principle of social cost recognises that those involved in an action should take responsibility and be held accountable for the action. Such principles are enshrined in the codes of conduct of professional bodies such as the British Computer Society, the Australian Computer Society, the ACM and the influential Software Engineering Code of Ethics and Professional Practice.[11]

These ethical principles are not mutually exclusive. They were developed to establish a checklist of ethical aspects to be applied whenever an action associated with computer systems takes place. The term action is used to represent any process or task undertaken, which normally includes a human element as the performer of the task or as the beneficiary, or as both performer and beneficiary. Within the information and communication technology domain, actions may be associated with the development, management or use of information and communication technology.

These principles can be used to analyse, inform, and colour practice within the whole of computing. They are useful in identifying areas of high ethical sensitivity known as the ethical hotspots.[12] There are many ethical issues surrounding the decisions and actions within the information systems field and there are limitations on professional time to respond to such issues. It is important to prioritise these issues on the basis of impact on the public good and so focus effort on the ethical hotspots.

Information Strategy

The alignment of corporate strategy and information systems strategy continues to be of critical concern to organisations. This is one of the key directions identified as being fundamental to corporate well being into the next century. Such alignment can be achieved through

---

[11] D. GOTTERBARN, K. MILLER, S. ROGERSON, 'Software Engineering Code of Ethics' in *Communications of the ACM* 40(November 1997)11, pp. 110-118.

[12] S. ROGERSON, 'Software Project Management Ethics' in C. MYERS, T. HALL, D. PITT, (eds.), *The Responsible Software Engineer*, Springer-Verlag, 1996, Chapter 11, pp. 100-106.

Strategic Information Systems Planning (SISP) which is the means of identifying applications systems that support and enhance organisational strategy and provides for the effective implementation of these systems.[13] For SISP to be effective both the SISP process and the products identified as a result of SISP need to be accepted by the organisation and its wider community. Acceptance is multi-faceted and requires the examination of technical, social, legal, and economic aspects. More recently acceptance has been coloured by greater attention being paid to ethical issues associated with information and communication technology.[14] This is particularly so given the advent of technologies with the potential to re-shape dramatically many aspects of our current lifestyles.[15]

Therefore, it is important to identify those key areas of SISP activity that warrant particular attention to ethical issues.

A good approach to SISP, within a particular organisation setting, depends on several factors, including:

- the environment of the organisation and the external pressures constraining IS/IT application;[16]
- the relative maturity of the organisation with respect to IS/IT use;[17]

---

[13] C.S. FIDLER and S. ROGERSON, *Strategic Management Support Systems*, London, Pitman Publishing, 1996.

[14] S.R. HAWK, 'The Effects of Computerised Performance Monitoring: An Ethical Perspective' in *Journal of Business Ethics* 13(1994), pp. 949-957; T. FORESTER, 'Megatrends or Megamistakes? What ever Happened to the Information Society?' in *The Information Society* 18(1992), pp. 133-146; K.A. FORCHT, D.S. THOMAS, 'Information Compilation and Disbursement: Moral, Legal and Ethical Considerations' in *Information Management and Computer Security* 12(1994)2, pp. 23-28; D. LANGFORD, J. WUSTEMAN, 'The Increasing Importance of Ethics in Computer Science' in *Business Ethics – A European Review* 3(1994)4, pp. 219-222; E. OZ, 'Ethical Standards for Computer Professionals: A Comparative Analysis of Four Major Codes' in *Journal of Business Ethics* 12(1993)9, pp. 709-728.

[15] S. ROGERSON and T.W. BYNUM, 'Cyberspace the Ethical Frontier', Multimedia pp iv, *The Times Higher Education Supplement*, No 1179, June 9, 1995; R. WIDDIFIELD and V. GROVER, 'Internet and the Implications of the Information Superhighway for Business' in *Journal of Systems Management* (May/June 1995), pp. 16-21, 65.

[16] D.J. FLYNN and P.A. HEPBURN, 'Strategic Planning for Information Systems – A Case Study of a Metropolitan Council' in *European Journal of Information Systems* 3(1994)3, pp. 207-217.

[17] J.M. BURN, 'Information Systems Strategies and the Management of Organisational Change – A Strategic Alignment Model' in *Journal of Information Technology* 8(1993), pp. 205-216.

- the size and stability of the organisation;[18]
- whether a SISP methodology and its supportive techniques are already determined.

Despite the variability of approach between organisations,[19] a list of generic activities typically encountered within SISP can be identified:

1. defining the scope of the SISP exercise
2. understanding and interpreting the business requirements
3. defining the organisational information needs and the underpinning Systems Architecture
4. formulating the information, information systems and information technology strategies
5. presenting the final output to the client
6. reviewing the SISP experience

The first activity, defining the scope of the SISP exercise, is concerned with obtaining the authorisation to proceed, establishing the SISP team participants, setting out an agenda and creating a timetable for activities, and allocating responsibilities. Essentially, this is the planning phase of SISP. Understanding and interpreting the business requirements involves establishing the corporate strategy and corporate Critical Success Factors, reviewing the current state of IS/IT within the organisation, and identifying potential IS/IT applications by taking both an internal and an external perspective to the organisation. Defining the organisational information needs and the underpinning systems architecture involves the mapping of systems to needs. This may provide some initial insight into the priority of subsequent systems developments or enhancements. From this, the strategies concerning the provision of information, the applications to support the provision of information and the underlying technologies to support the applications can be derived. These strategies are presented to the client. Finally, the experiences of the overall SISP exercise are reflected upon. This may lead to proposals for changes to be incorporated into future SISP activity.

---

[18] F. BERGERON and L. RAYMOND, Planning of Information Systems to Gain a Competitive Edge' in *Journal of Small Business Management* 30(1992)1, pp. 21-26.

[19] M.J. EARL, 'Experiences in Strategic Information Systems Planning' in *MIS Quarterly* (March 1993).

Although the activities appear to follow a natural sequence, in practice many iterations to previous activities may occur, due to changes in requirements or in the operational environment. Furthermore, several activities may occur in parallel. SISP activity may also lead to changes in strategy, which leads subsequently to changes in the defined scope for the SISP exercise.

Undertaking an analysis using the eight ethical principles suggests that there are two ethical hotspots in SISP, namely, the first and fifth activities.

Activity 1 – Defining the scope of the SISP exercise

As previously mentioned, this activity is concerned with the planning of subsequent SISP activities. Setting out the scope of the exercise, deciding on the team membership and their respective tasks, and obtaining authorisation to proceed based on the plan, involves many decisions regarding the requirements and involvement of others. Ethical issues are therefore highly significant within this activity. For example, it is vitally important at this initial stage to be above board so that a good working relationship is established with the client. The principle of honesty addresses this point. It is also important to ensure that decisions concerning the team membership, for example, are not biased by previous stereotypical situations, but are for the good of the organisation in hand and are as fair as possible for all. The principles of bias and fairness are significant here. Any shortfalls in the team's experience in performing SISP should be made explicit at the outset of the project, and measures taken to make good this shortfall. This relates to the principle of adequacy. Appropriate quality control procedures should be considered as part of the planning process to ensure due care is given to the quality of each ensuing activity. Thus the principal of due care is invoked. Attention to the social cost principle is also warranted, to ensure that the plan incorporates issues of broader significance and that the accountability of the project team is appropriately defined. Actions on the part of those team members involved in the planning of the SISP exercise should always be considered beyond reproach, which directly relates to the principle of honour.

Activity 5 – Presenting the final output to the client

This activity involves significant client liaison, and many of the ethical principles are dominant within this activity. The output must be presented efficiently and effectively so that deliberations are focused on the principal aspects of importance to the organisation and so that costly management time is not wasted. This relates to the principle of action. The presentation of results should paint an accurate and fair picture of the situation. This relates to the principles of bias, fairness and honesty. All stakeholders' views should be taken into account within the plan, although diplomacy concerning stakeholders' views is required at all times. Overall, the SISP team's approach to this activity must be professional and beyond reproach. This relates directly to the principle of honour.

SISP provides general guidance on the activities to be undertaken in order to achieve strategically aligned information systems and the ICT infrastructure. Any ethical consideration tends to be implicit rather than explicit which has a tendency to devalue the importance of the ethical dimension. Therefore, a stronger more explicit emphasis on ethical issues is needed. The mapping of ethical principles to SISP activities provides a general framework for explicitly including an ethical perspective within the SISP process.

Project Management

It appears universally accepted that the most effective way to develop information systems software is through the use of a project-based organisational structure which encourages individuals to participate in teams with the goal of achieving some common objective. Much has been written about the management of software development projects and no doubt much will be written in the future.

In his book, *How to Run Successful Projects*, in the British Computer Society Practitioner Series, O'Connell[20] provides details of the Structured Project Management (SPM) approach. He explains that SPM is a practical methodology that, as DeMarco[21] states, is a *'basic approach one takes to getting a job done'*. SPM has been chosen for discussion as it is practical rather than conceptual, generalist rather than specific and provides practitioners with realistic guidance in undertaking the vastly complex activity of project management. SPM comprises the following ten steps:

---

[20] F. O'CONNELL, *How to Run Successful Projects*, Englewood Cliffs, NJ, Prentice Hall, 1994.
[21] T. DEMARCO and T. LISTER, *Peopleware*, New York, Dorset House Publishing, 1987.

1. visualise what the goal is,
2. make a list of the jobs that need to be done,
3. ensure there is one leader,
4. assign people to jobs,
5. manage expectations, allow a margin of error and have a fallback position,
6. use an appropriate leadership style,
7. know what is going on,
8. tell people what is going on,
9. repeat Step 1 through 8 until Step 10 is achievable,
10. realise the project goal,

The first five steps are concerned with planning and the remaining five deal with implementing the plan and achieving the goal. O'Connell states that most projects succeed or fail because of decisions made during the planning stage thereby justifying the fact that half of the effort expended in the SPM approach is on preparation. It is this planning element of project management which lays down the foundations on which the project ethos is built. Here the scope of consideration is established, albeit implicitly or explicitly, which in turn locates the horizon beyond which issues are deemed not to influence the project or be influenced by the project. How the project is conducted will depend heavily upon the perceived goal. The visualisation of this goal takes place in Step 1. The first two points in the visualisation checklist given by O'Connell are:

- What will the goal of the project mean to all the people involved in the project when the project completes?
- What are the things the project will actually produce? Where will these things go? What will happen to them? Who will use them? How will they be affected by them?

These are important because through answering these questions an acceptable project ethos and scope of consideration should be achieved. The problem is that in practice these fundamental questions are often overlooked. It is more likely that a narrower perspective is adopted with only the obvious issues in close proximity to the project being considered. The holistic view promoted by the two checklist points requires greater vision, analysis and reflection. The project manager is under pressure to deliver and so the tendency is to reduce the horizon and establish an artificial boundary around the project. Steps 2 to 5 are concerned with adding detail and refinements thus arriving at a workable and acceptable plan. Steps 6 to 8 are concerned with implementing the plan, monitoring performance and keeping those associated with the project informed of progress. Step 9 defines the control feedback loops which ensure that the plan remains focused, current and realistic. Finally, Step 10 is the delivery of the project output to the client and an opportunity to reflect upon what has and has not been achieved.

Once again undertaking an analysis suggests that the ethical hotspots are the first and eighth step and that probably the most important ethical issues lie within the first step. This first step, visualising what the goal is, as previously mentioned establishes the project ethos and consequently there are several ethical issues that need to be borne in mind. This is the start of the project and it is vitally important to be above board at the onset so that a good working relationship is established with client. The principles of honour and honesty address this point. Bias in decision making and the undertaking of actions is a major concern throughout the project including Step 1. It is important to take a balanced viewpoint based on economic, technological and sociological information. The view often portrayed is skewed towards technology and economics which can have disastrous results leading to major system failure or rejection, as was the case, for example, at the London Ambulance Service regarding LAS-CAD its computer aided despatch system and at the London Stock Exchange regarding TAURUS its computerised trading system.[22] This leads to three further dominant principles of due care, fairness and social cost. Computer systems have an impact directly and indirectly on many people and it is important to include all parties in decisions that affect the way in which the project is conducted. Software development projects involve many stakeholders and each is worthy of fair treatment. The principles of due care and social cost will ensure that a longer term and broader perspective is adopted.

Stakeholders

Establishing the right scope of consideration is essential in defining acceptable project goals. The scope of consideration is influenced by

[22] R. RAHANU, J. DAVIES and S. ROGERSON, 'Ethical Analysis of Software Failure Cases' in *Failure & Lessons Learned in Information Technology Management* 3(1990).

the identification and involvement of stakeholders. In traditional software project management the stated needs of the customer are the primary items of concern in defining the project objectives. There has been some recognition that in defining how software will address those needs the customer is also presented with a predefined set of constraints which limit the customer's freedom of expression.[23] There is a mutual incompatibility between some customer needs, for example, the amount of code required to make a system easy to use makes a system difficult to modify. The balancing of these items is an ethical dimension in the development of a software product. Such considerations have been limited in scope to the customer. Investigating 16 organisational IS-related projects led[24] to conclude that regarding evaluation of IT investment, '...*the perception of what needed to be considered was disappointingly narrow, whether it concerned the possible scope and level of use of the system, [or] the range of people who could or should have been involved...*'. They discovered, with the exception of vendors, all stakeholders involved in evaluation were internal to the organisations. The reason for this restricted involvement is that these are the only stakeholders originally identified in the traditional project goals. However, consideration of stakeholders should not be limited to those who are financing the project or are politically influential but broaden it to be consistent with models of ethical analysis. Stakeholders must include individuals or groups who may be directly or indirectly affected by the project and thus have a stake in the development activities. Those stakeholders who are negatively affected are particularly important regarding ethical sensitivity because they are often the ones overlooked.

The Software Development Impact Statement

One way of addressing the need to modify project goals in a formal way is to use a modification of a social impact statement. A social impact statement is modelled on an environmental impact statement which is required before major construction is undertaken. The environmental impact statement is supposed to specify the potential negative impacts on the environment of the proposed construction and specify what actions will be taken to minimise those

---

[23] J. MCCARTHY, *Dynamics of Software Development*, Redmont, WA, Microsoft Press, 1996.

[24] B. FARBEY, F. LAND and D. TARGETT, *How to Assess Your IT Investment*, Oxford, Butterworth Heinemann, 1993.

impacts. Proposed social impact statements have been described for identifying the impact of information systems on direct and indirect system users.[25] The limitations of these proposals are addressed by the Software Development Impact Statement (SODIS) which is intended to reflect the software development process as well as the more general ethical obligations to various stakeholders.[26]

There are two types of SODIS. The first is a Generic SODIS which has as its primary function the identification of stakeholders and related ethical issues. In the light of the identified issues a preliminary project management plan is developed. A second more detailed SODIS is then employed. This is the Specific SODIS. There will be a number of Specific SODIS within a particular methodology. Each SODIS is tied to a particular development methodology and to each step in that methodology. Even though each Specific SODIS is tied to a development methodology, they all include the means of revisiting and re-evaluating ethical issues in the light of the unfolding development process. This organic nature of the SODIS is very different for the environmental impact statement model.

Just as producing software of high quality should be second nature to the software engineer so should producing software that is ethically sensitive. Indeed there is clearly an overlap in these two requirements. The project management process for software development must accommodate an ethical perspective. The major criticism of current practice is that any ethical consideration tends to be implicit rather than explicit which has a tendency to devalue the importance of the ethical dimension. By using ethical principles, identifying of ethical hotspots and using SODIS it is possible to ensure that the key ethical issues are properly addressed as an integral part of the software development process. Quite simply, project management should be guided by a sense of justice, a sense of equal distributions of benefits and burdens and a sense of equal opportunity.

---

[25] B. SHNEIDERMAN and A. ROSE, *Social Impact Statements: Engaging Public Participation in Information Technology Design*, Technical Report of the Human Computer Interaction Laboratory, September, 1995, pp 1-13.

[26] S. ROGERSON and D. GOTTERBARN, 'The Ethics of Software Project Management' in G. COLLSTE (ed.), *Ethics and Information Technology*, Delhi, India, New Academic Publishers, 1998, pp. 137-154.

Information Systems Development

There are numerous methodological approaches to information systems development. Few deal adequately with the ethical dimensions of the development process. Avison[27] criticises the development methodologies, such as Structured Systems Analysis and Design Method (SSADM), Merise and Yourdon, that are adopted by most organisations because they tend to stress formal and technical aspects. He argues that, 'The emphasis ... must move away from technical systems which have behavioural and social problems to social systems which rely to an increasing extent on information technology.' He suggests that the human, social and organisational aspects are often overlooked. The consideration of this broader perspective only seems to occur in the event of systems failure or underperformance. This issue is addressed by Wood-Harper et al.[28] who identify a number of dilemmas which a systems analyst might face when undertaking a systems development activity using a methodological approach. These dilemmas are summarised as:

- Whose ethical perspective will dominate the study of the situation and development of IS?
- Will ethical viewpoints be included in the study?
- What methodology should be used for the study?
- What approach should the analyst use if there is an obvious conflict of interests?

Wood-Harper et al.[29] have further articulated these concerns by suggesting four principles that can be used to identify issues related to ethically sensitised information systems development. These are:

- Ethical reasoning should be conducted throughout the life of an information system, including inception, testing, distribution, modification and termination.

---

[27] D.E. AVISON, *'What is IS?'*, An inaugural lecture delivered at the University of Southampton, 3 Nov 1994, 1995.

[28] A.T. WOOD-HARPER, S. CORDER, J.R.G. WOOD and H. WATSON, 'How We Profess: The Ethical Systems Analyst' in *Communications of the ACM* 39(March 1996)3, pp. 69-77.

[29] A.T. WOOD-HARPER, S. CORDER, B. BYRNE, 'Ethically Situated Information Systems Development' in C.R. SIMPSON, (ed.), *AICE99 Conference Proceedings*, Australian Institute of Computer Ethics, 1999.

- Every information systems should improve the ethical actions of its users.
- The benefits of an information system should be distributed to all people who have an ethical need for its use.
- The design of every information system should include the design of its ethical use, the design of its ethical distribution, the design of its ethical risk and the methods of justifying ethical criteria.

It is important to recognise that there are a few methodological approaches, notably ETHICS from Mumford,[30] Soft Systems Methodology from Checkland[31] and Multiview from Avison and Wood-Harper[32] that attempt to include consideration of ethical and societal issues. In evaluating ETHICS, Jayaratna[33] suggests that, '[it] offers many design guidelines useful for the understanding and the design of human-centred systems, but ...does not offer models or ways for performing ... the steps. Nor does it offer any models for handling interpersonal and political conflicts.' He concludes that, 'ETHICS is committed to a particular ethical stance [and] does not offer any means of discussing or resolving many of the ethical dilemmas ... in systems development.' This appears to be a recurrent criticism of such methodologies. Whilst it is laudable that ethical sensitivity is raised as an issue worthy of investigation, the manner in which investigation is undertaken and, ultimately, an ethically defensible position derived is vague. Methodologies need to be enhanced to address these criticisms.

SSADM is now used as an illustrative method to further discussion of the ethics of information systems development because it is felt that if ethical enrichment can be achieved in a 'hard' systems approach then it is likely to be achievable in most approaches. SSADM is a set of procedural, technical and documentation standards for information systems development.[34] SSADM comprises five core modules; feasibility study, requirements analysis, requirements specifica-

---

[30] E. MUMFORD, *Designing Participatively*, Manchester Business School, 1983.
[31] P.B. CHECKLAND, *Systems Thinking, Systems Practice*, New York, Wiley, 1981.
[32] D.E. AVISON, A.T. WOOD-HARPER, *Multiview: An Exploration in Information Systems Development*, Henley-on-Thames, Alfred Waller, 1990.
[33] N. JAYARATNA, *Understanding and Evaluating Methodologies*, Maidenhead, McGraw-Hill, 1994.
[34] S. SKIDMORE, R. FARMER and G. MILLS, *SSADM Version 4 Models and Methods*, 2nd ed., Manchester, Blackwell, NCC, 1994.

tion, logical systems specification and physical design module. SSADM adopts a product oriented approach where each element, be it a module, stage or step produces predictable outputs from given inputs.

Winter, Brown and Checkland[35] explain that computer-based information systems are systems that serve purposeful human actions. They argue that there is a very heavy emphasis on the serving information system in the systems development life cycle with, in the main, only limited and implicit account taken of the purposeful human action or the so-called served system. This is particularly the case in the 'hard' systems thinking approaches of structured methods that are highly technical and rational. In examining SSADM, Winter, Brown and Checkland found no explicit organisational activity model and that it appeared to accept without question the image, activities and performance aspects of the organisation in question. Furthermore, they were critical of SSADM's concept of information requirements as concerned with that required for the system to function rather than the information needed by the people undertaking organisational activities. The approach adopted by Winter, Brown and Checkland is organisationally oriented. It does not appear to go beyond the environs of the organisation where one finds a greater range of system stakeholders. This lack of stakeholder involvement is a further shortcoming of SSADM and it is unclear whether distant stakeholders will be identified let alone involved. The implications of such restricted stakeholder involvement on achieving a socially and ethically sensitive approach are obvious.[36]

The ethical enhancement of SSADM is a considerable task that is beyond the scope of this chapter. The discussion is confined to making some suggestions as to how that task might be achieved. It is clear that a combination of teleology and deontology should be used. This is because teleological approaches focus on outcomes whilst deontological approaches focus on actions. White[37] explains that together, *'they reveal a wide array of internal and external factors of*

---

[35] M.C. WINTER, D.H. BROWN and P.B. CHECKLAND, 'A Role for Soft Systems Methodology in Information Systems Development' in *European Journal of Information Systems* 4(1995)3, pp. 130-142.

[36] S. ROGERSON, J. WECKERT and C. SIMPSON, 'An Ethical Review of Information Systems Development: The Australian Computer Society's Code of Ethics and SSADM' in *Information Technology and People*, forthcoming, ISSN 0959 3845, 1999.

[37] T.I. WHITE, *Business Ethics: A Philosophical Reader*, New York, Macmillan, 1993.

human actions that have moral consequences. Although these two outlooks conflict in theory, they complement one another in practice. ... each acts as a check on the limitations of the other.'

With this in mind it is important to address both the process and product of SSADM. The Product Breakdown Structure provides the impetus to address issues in a certain way. The completeness of these products needs to be considered given their powerful influence in the development process. Consideration must also be given to how the systems developers should think when undertaking the various development tasks. The SSADM culture promotes technological and economic thinking but not ethical thinking explicitly. The later enhanced version 4+ appears to be moving to a more balanced view with more emphasis on business orientation and the inclusion of user culture in the systems development template.[38]

This albeit minor change in emphasis towards people offers more scope to include ethical sensitivity.

It is now recognised that quality should permeate the whole of the information systems development process and not simply be considered at discrete points within the process. This is reflected in SSADM by the inclusion of a quality products set comprising a number of files that demonstrate quality has been built into the system.[39] Product descriptions are part of this set. There are product descriptions for all the products specified in SSADM. The details include quality criteria against which the product can be checked. Product descriptions are used to monitor progress and success of the project.

Including quality in each description ensures it permeates the whole process and promotes a quality culture throughout the development team. Similarly ethical and societal consideration should permeate the whole process. It follows that each product description should include ethical and social criteria to promote this awareness and consideration. This might be systematically addressed by using clauses from a code of conduct to form the ethical criteria for products within each of the core modules. A suitable code to use would be the Software Engineering Code of Ethics and Professional Practice which has its clauses in eight groups relating

---

[38] J. HALL, *How is SSADM4.2 Different from SSADMV4?: SSADM4+ for Academics*, CCTA and International SSADM User Group Workshop, Leeds Metropolitan University, 1995.

[39] CCTA, *SSADM Version 4 Reference Manual: Volume 4 Dictionary*, Oxford, Blackwell, NCC, 1990.

to the public, the client and employer, the software product, professional judgement, management, the profession, colleagues, and oneself.[40]

Conclusion

The provision of information is a complex activity involving many decisions, many people and much time. This chapter has highlighted the major elements of this activity and considered the level of ethical sensitivity of each element. It appears that current practice is deficient in addressing the broader social implication of information provision through the use of information and communication technology. Suggestions have been made in using information ethics to address such shortcomings through the modification and enhancement of current processes and procedures. Failure by information engineers and software engineers to consider the broader social and ethical issues is at best unprofessional and at worst societally disastrous.

References

AVISON, D.E., 'What is IS?', An inaugural lecture delivered at the University of Southampton, 3 Nov 1994, 1995.

AVISON, D.E. and A.T. WOOD-HARPER, *Multiview: An Exploration in Information Systems Development*, Henley-on-Thames, Alfred Waller, 1990.

BERGERON, F. and L. RAYMOND, 'Planning of Information Systems to Gain a Competitive Edge' in *Journal of Small Business Management* 30(1992)1, pp. 21-26.

BURN, J.M., 'Information Systems Strategies and the Management of Organisational Change – A Strategic Alignment Model' in *Journal of Information Technology* 8(1993), pp. 205-216.

BYNUM, T.W., 'The Development of Computer Ethics as a Philosophical Field of Study' in *Australian Journal of Professional and Applied Ethics* 1(July 1999)1, pp. 1-29.

BYNUM, T.W., 'Computer Ethics in the Computer Science Curriculum' in T.W. BYNUM, W. MANER, and J.L. FODOR (eds.), *Teaching Computer*

---

[40] D. GOTTERBARN, K. MILLER and S. ROGERSON, 'Software Engineering Code of Ethics' in *Communications of the ACM* 40(November 1997)11, pp. 110-118.

Ethics, New Haven, CT, Research Center on Computing & Society, 1992.
CCTA, SSADM *Version 4 Reference Manual: Volume 4 Dictionary*, Oxford, Blackwell, NCC, 1990.
CHECKLAND, P.B., *Systems Thinking, Systems Practice*, New York, Wiley, 1981.
DEMARCO, T. and T. LISTER, *Peopleware*, New York, Dorset House Publishing, 1987.
EARL, M.J., 'Experiences in Strategic Information Systems Planning' in MIS *Quarterly*, March, 1993.
FARBEY, B., F. LAND and D. TARGETT, *How To Assess Your IT Investment*, Oxford, Butterworth Heinemann, 1993.
FIDLER, C.S. and S ROGERSON, *Strategic Management Support Systems*, London, Pitman Publishing, 1996.
FLYNN, D. J. and P.A. HEPBURN, 'Strategic Planning for Information Systems – A Case Study of a Metropolitan Council' in *European Journal of Information Systems* 3(1994)3, pp. 207-217.
FORCHT, K.A. and D.S. THOMAS, Information Compilation and Disbursement: Moral, Legal and Ethical Considerations' in *Information Management and Computer Security* 12(1994)2, pp. 23-28.
FORESTER, T., 'Megatrends or Megamistakes? What ever Happened to the Information Society?' in *The Information Society* 18(1992), pp. 133-146.
GOTTERBARN, D., 'The Use and Abuse of Computer Ethics' in T.W. BYNUM, W. MANER, and J.L. FODOR, (eds.), *Teaching Computer Ethics*, New Haven, CT, Research Center on Computing and Society, Southern Connecticut State University, 1992, pp. 73-83.
GOTTERBARN, D., K. MILLER and S. ROGERSON, 'Software Engineering Code of Ethics' in *Communications of the* ACM 40(November 1997)11, pp. 110-118.
HALL, J., *How is* SSADM*4.2 Different from* SSADMV*4?:* SSADM*4+ for Academics*, CCTA & International SSADM User Group Workshop, Leeds Metropolitan University, 1995.
HAWK, S.R., 'The Effects of Computerised Performance Monitoring: An Ethical Perspective' in *Journal of Business Ethics* 13(1994), pp. 949-957.
JAYARATNA, N., *Understanding and Evaluating Methodologies*, Maidenhead, McGraw-Hill, 1994.
LANGFORD, D. and J. WUSTEMAN, 'The Increasing Importance of Ethics in Computer Science' in *Business Ethics – A European Review* 3(1994)4, pp. 219-222.
MANER, W., 'Unique Ethical Problems in Information Technology' in *Science and Engineering Ethics* 2(1996)2, pp. 137-155.

McCarthy, J., *Dynamics of Software Development*, Redmont, WA, Microsoft Press, 1996.

Moor, J.H., 'What is Computer Ethics?' in *Metaphilosophy* 16(1985)4, pp. 266-279.

Moor, J.H, 'Reason, Relativity, and Responsibility in Computer Ethics' in *Computers and Society* 28(March 1998)1, pp 14-21.

Mumford, E., *Designing Participatively*, Manchester Business School, 1983.

O'Connell, F., *How to Run Successful Projects*, Englewood Cliffs, Prentice Hall, 1994.

Oz, E., 'Ethical Standards for Computer Professionals: A Comparative Analysis of Four Major Codes' in *Journal of Business Ethics* 12(1993)9, pp. 709-728.

Rahanu, R., J. Davies, and S. Rogerson, 'Ethical Analysis of Software Failure Cases' in *Failure & Lessons Learned in Information Technology Management* 3(1990).

Rogerson, S., 'Software Project Management Ethics' in C. Myers, T. Hall and D. Pitt (eds.), *The Responsible Software Engineer*, London, Springer-Verlag, 1996, Chapter 11, pp. 100-106.

Rogerson, S. and T.W. Bynum, *Information Ethics: The Second Generation*, UK Academy for Information Systems conference, UK, 1996.

Rogerson, S. and T.W. Bynum, 'Cyberspace the Ethical Frontier, Multimedia pp iv' in *The Times Higher Education Supplement*, No. 1179, June 9, 1995.

Rogerson, S. and D. Gotterbarn, 'The Ethics of Software Project Management' in G. Collste (ed.), *Ethics and Information Technology*, Delhi, India, New Academic Publishers, 1998, pp. 137-154.

Rogerson, S., J. Weckert and C. Simpson, 'An Ethical Review of Information Systems Development: The Australian Computer Society's Code of Ethics and SSADM' in *Information Technology and People*, 1999.

Shneiderman, B. and A. Rose, *Social Impact Statements: Engaging Public Participation in Information Technology Design*, Technical Report of the Human Computer Interaction Laboratory, September, 1995, pp. 1-13.

Skidmore, S., R. Farmer and G. Mills, *SSADM Version 4 Models and Methods*, 2nd ed., Manchester, Blackwell, NCC, 1994.

Van Luijk, H., 'Business Ethics: The Field and its Importance' in B. Harvey (ed.), *Business Ethics: A European Approach*, New York, Prentice Hall, 1994.

Walsham, G., 'Ethical Theory, Codes of Ethics and IS Practice' in *Information Systems Journal* 6(1996), pp. 69-81.

White, T.I., *Business Ethics: A Philosophical Reader*, New York, Macmillan, 1993.

WIDDIFIELD, R. and V. GROVER, 'Internet and the Implications of the Information Superhighway for Business' in *Journal of Systems Management* (May/June 1995), pp. 16-21, 65.

WINTER, M.C., D.H. BROWN and P.B. CHECKLAND, 'A Role for Soft Systems Methodology in Information Systems Development' in *European Journal of Information Systems* 4(1995)3, pp. 130-142.

WOOD-HARPER, A.T., S. CORDER, J.R.G. WOOD and H. WATSON, 'How We Profess: The Ethical Systems Analyst' in *Communications of the ACM* 39(March 1996)3, pp. 69-77.

WOOD-HARPER, A.T., S. CORDER and B. BYRNE, 'Ethically Situated Information Systems Development' in C.R. SIMPSON (ed.), *AICE99 Conference Proceedings*, Australian Institute of Computer Ethics, 1999.

2.2.4

CODES OF ETHICS: CONDUCT FOR COMPUTER SOCIETIES

The Experience of IFIP

Jacques Berleur
and Marie d'Udekem-Gevers

Remembering the debates started in 1988 within IFIP (International Federation for Information Processing) about a suggested international 'Code of Ethics' (or/and of Conduct?), lessons may be derived in terms of ways of building up a code, as well as in terms of claims of respect for cultural, social and legal environments. Further steps such as the recommendations of IFIP 1994 General Assembly may also enlighten us as to how the members of an international Ethics Network need to act and support each other in creating 'spaces for discussion' where the ethical debate is permanently promoted and supported.[1]

IFIP 1988-1992 Debates

The first official consideration within IFIP, at the level of its General Assembly (GA), of a proposal for developing an international Code of Ethics dates back to 1988, when the New Delhi General Assembly

---

[1] This paper extends one which has been published under the title: 'IFIP Framework for Ethics', in *Science and Engineering Ethics* (1996)2, pp. 155-165, A special Issue on Global Information Ethics. The author is indebted to Opragen Publications, and particularly to Merilyn Spier, Publishing Manager, for having accepted its partial reproduction. The same thanks and acknowledgments go to Simon Rogerson and Terrell W. Bynum, Directors of the Centre for Computing and Social Responsibility, de Montfort University, Leicester, UK, Guest Editors of that Special Issue which includes a selection of ETHICOMP'95 papers.

decided to investigate the desirability of creating such a Code.² Previous work had been done in the late seventies within Technical Committee 9 (TC9, Computers and Society) and its Working Group 9.2 (WG9.2, Social Accountability) when the question was also on the agenda of the Council of Europe.³

The reason for this project is not very clear, or at least, it cannot been found explicitly in the IFIP archives. The project was initiated by George Glaser, IFIP Vice-President at the time, and also Chairman of the Activity Development Board. In a circular document of September 1988, 'Initial Project Proposal for an IFIP Code of Ethics,' Hal Sackman explains the provisional scope of the project: 'IFIP rules and bylaws list the aims of IFIP as essentially a) to promote Information Science and Technology; b) to advance international cooperation in the field of Information Processing; c) to stimulate research, development and application of Information Processing in science and human activity; d) to further the dissemination and exchange of information on Information Processing; and e) to encourage education in Information Processing. In addition to the above aims, or as corollaries to the above aims, IFIP also strives to achieve the further challenging goals of a) social responsibility, particularly facilitating the constructive computerisation of developing nations, and b) aspiring to earn recognition as the world leader in international developments in information processing on the basis of IFIP's demonstrated merit and excellence. These two corollaries to IFIP's established aims are major focal points bearing directly on the scope of IFIP ethics.' We could say that it is one of the usual and well recognised goals of many of the codes, namely 'to enhance the profession's reputation and the public trust, and to preserve entrenched professional biases.'⁴

After the New Delhi General Assembly, an ethics survey questionnaire was administered 'to approximately 100 IFIP professionals'⁵

---

² *IFIP-Newsletter* 6(March 1989)1.

³ Herbert MAISL, 'Legal Problems Connected with the Ethics of Data Processing,' Study for the Council of Europe (CJ-PD[79]8), Strasbourg, August 29, 1979. Secretariat Memorandum (CJ-PD[81]8), and the last report (CJ-PD[82]19) with the Minutes of the Meeting (CJ-PD[82]31).

⁴ M.S. FRANKEL, 'Professional Codes: Why, How and With What Impact?' in *Journal of Business Ethics* 8(1989).

⁵ This is the wording of the 'IFIP Ethics Questionnaire Package' of November 28, 1988, which I found back in my own archives. Later, in another presentation, Hal Sackman wrote: 'to 80 national computer societies, IFIP officers and international affiliate organizations worldwide' (See, for instance, H. SACKMAN, 'A Prototype IFIP Code of Ethics Based on Participative International Consensus' in C. DUNLOP and R. KLING

which resulted in a first draft contemplating four major areas of ethics: individual professional ethics, multi-national organisational ethics, international legal informatics ethics and international public policy ethics. This first draft was presented at the IFIP Technical Assembly (TA) and then to the General Assembly in San Francisco, September 1989. The Minutes of the General Assembly stated: 'Mr. Sendov, President, advised that Mr. Sackman had presented his Interim Report to Technical Assembly and it was noted that a great deal of work had been carried out since the Geneva Council Meeting. It was felt that the project should go ahead and Technical Assembly had recommended Mr. Sackman to prepare a revised, shortened version for distribution to a broader audience.' It was also suggested to present it to a larger and more diversified part of the IFIP community. The revised draft was then published in the IFIP-Newsletter for comments and rating.[6]

Intensive discussion took place as soon as the revised version circulated within the national Societies, Technical Committees (TCs), Working Groups (WGs), and Special Interest Groups (SIGs) of IFIP. One cannot say that enthusiasm was shown towards the proposal, and some national Societies were opposed strongly to any international standard on the subject. IFIP Technical Assembly and General Assembly became nervous, asked Technical Committee 9 to stimulate the debate. The Minutes of Buenos-Aires General Assembly, 1990, regarded Dr. Sackman's 'revised preliminary IFIP Code of ethics' as a good basis for discussion. (...) The General Assembly, in order to analyze the acceptance and relevance of this 'Draft IFIP Code of Ethics' as it was now to be called, also asked its Member Societies as well as its organisations, TCs, WGs, SIGs, to discuss this paper in detail. Among others, aspects of education to and enforcement of ethical behaviour should be evaluated in the face of the diverse economic, social and cultural backgrounds.' IFIP General Assembly requested the commitment of Technical Committee 9 in close cooperation with interested TCs and WGs.

The Minutes of the Trondheim Council Meeting in 1992 are clear: 'The status of the 'IFIP Code of Ethics' has been discussed. It appeared that some Member Societies had responded rather negatively to the existing proposal. Technical Committee 9 had therefore agreed to develop a reference frame including some general state-

(eds.), *Computerization and Controversy, Value Conflicts and Social Choices*, San Diego, Academic Press Inc., 1991, 1st edition, p. 698).

[6] *IFIP-Newsletter* 6(December 1989)4.

ments. There would be a discussion on the subject during Congress 92 following which Technical Committee 9 would report to the next Technical Assembly.'

IFIP World Computer Congress 1992, in Madrid programmed an open session. Current IFIP President, Academician Blagovest Sendov, personally chaired that session where no less than 12 panellists debated the opportunity to go further with the proposal. Many IFIP General Assembly representatives and most of the IFIP officers were present. A special brochure had been prepared by WG9.2, with a discussion paper and different codes of national Societies. Moreover, papers expressing concerns about the topic were gathered in one of the streams of the conference and may still be found today in the IFIP World Congress '92 Proceedings.[7]

The statements of the panel were sometimes sharp:

- 'The phrasing smacks rather of motherhood.'
- 'There is no international mechanism to deal with such topics'.
- 'We are doubtful about any pronouncement on ethics which lacks any obvious policy mechanism other than international opinion'.,
- 'We have considerable reservation because of the differences between cultures, traditions, and legal frameworks within the international community, and hence we doubt that a universal code of ethics can be written.'
- 'We consider it necessary to reconsider the draft Code from several perspectives: content, format, process'.

(The Chair, mindful of the arguments, concluded:) 'The time is not ripe to adopt an IFIP international Code!'

Subsequent IFIP Technical Assembly and General Assembly, in Toledo, September 1992, decided to thank the author of the 'draft Code', not to consider it any longer as an 'IFIP Draft Code' and 'to set up an IFIP Ethics Task Group to prepare a document on *Guidelines for Codes of Ethics and Professional Conduct*, closely collaborating with IFIP Member Societies and renowned IT organisations.' Technical Committee 9 was responsible, and asked one of the authors of this paper to chair that Task Group, with some stringent milestones and deadlines including the production of a Handbook within 15 months and a final report to the Technical and General Assemblies in 1994.

[7] 'Ethics of Computing: Information Technology and Responsibility' in *Information Processing 92*, vol. II: *Education and Society*, R. AIKEN (ed.), Proceedings of the IFIP 12th World Computer Congress (Madrid, September 7-11, 1992), Elsevier Science Publishers B.V. (North-Holland), 1992, pp. 344-373.

A New Process of Discussion

What appeared clearly in this process is the deep refusal of an international code. Why? After a long time of reflection, we think one of the main reasons was that it was impossible to find a procedure for enforcement. We shall elaborate on this in the conclusion. Going through the minutes of IFIP meetings and notes of that period, we are convinced the discussion was not first about the content itself, but about the principle of having an international code. Several times, it is repeated that it would be better and sufficient to provide the different national Member Societies with a 'set of guidelines' in a format they could consider for local adaptation.

Among the causes of the failure in the process of the past years is also probably inadequate understanding of the IFIP structure: IFIP is an international organisation made up of national organisations. With the exception of some individual members, who are admitted in recognition of relevant contributions, and a few honorary members awarded for life on the basis of exceptional merit, the membership within IFIP – be it full (one per country), corresponding, or affiliate – is available only to organisations. The 'ethics survey among IFIP professionals' probably neglected that fact, getting the feedback of the real members when the whole affair was nearly completed.

Interestingly, at the same period, a rather similar debate took place within the International Federation of Accountants, but resulted in another way. It issued its first code in 1990 and a revised version in 1992.[8] But a paper in the *Journal of Business Ethics* deeply questioned the capacity of such a code for having a real impact on local constituencies. The analysis of the authors is made essentially on the basis of cultural and socio-economic constraints: 'International professional guidelines are often ethnocentric; they reflect the ethical and cultural standards of the developed countries whose organisations are more influential in writing them...'[9] We are today convinced that many were not far from thinking the same about the Draft IFIP Code.

---

[8] The version which is now available on the Internet is a January 1998 revised version: (http://www.ifac.org/StandardsAndGuidance/Ethics/CodeOfEthicsForProfAccnts.html)

[9] J.R. COHEN, L.W. PANT, and D.J. SHARP, 'Cultural and Socio-economic Constraints on International Codes of Ethics: Lessons from Accounting' in *Journal of Business Ethics* 11(1992), pp. 687-700. This paper is based on the 1990 'IFAC Guideline on Ethics for Professional Accountants'.

The Ethics Task Group decided to start a new process and to get input from all the IFIP Member Societies, and its working bodies namely, its Technical Committees, Working Groups and Special Interest Groups. A brochure for discussion was sent to the 47 Member Societies, 11 TCs, 71 WGs and 2 SIGs.[10] The response was very positive: 30 national Societies, and 17 TCs and WGs were involved in the process at one point or another, collaborating or manifesting interest. Some of them provided the Task Group with additional material or more elaborated answers.

As a result of this cooperative process with the Member Societies, we analysed 31 Codes: 21 from 13 IFIP national Societies, 3 from 2 IFIP Affiliate Members (Regional Societies), 6 from other Computer Societies, and the Draft Code which had been submitted to IFIP previously. The list of the Societies and their Codes is given in Annex.[11]

Other Codes were also received, but have not been included in the analysis because they were restricted to a specific field, such as privacy or health care, or because the status of the Society was not considered to be 'national'. The Task Group was also provided with tentative proposals of oaths for informaticians. Those supplementary documents were as follows:

- Standards of Computer Science Deontology of CITEMA (Centro de la Informática, Telemática y Medios Afines), Spain.
- Health Informaticians' Deontology Code, Greece.
- CPSR (Computer Professionals for Social Responsibility) Proposed Privacy Guidelines for the National Research and Education Network (NREN) and for the National Information Infrastructure (NII), USA.
- Oaths: An Engineer's Hippocratic Oath, An Oath of An Informatician.

[10] IFIP Information Bulletin (January 1994)25. General Assembly of Hamburg (September 1994) admitted new Member Societies which, of course, were not questioned at the time we started the new process. There are now 49 full, associate, and corresponding Members and 11 affiliate (regional or international organizations) Members. See: IFIP Information Bulletin (January 1999)29.

[11] Other IFIP national Member Societies have now adopted a code. This is the case, for instance of the Information Processing Society of Japan (http://www.ipsj.or.jp/english/codeengl.html) enacted since May 20, 1996; or of the Finnish Information Processing Association (http://www.ttlry.fi/): the English translation is available on request at (heidi.lind@ttlry.fi); the Nederlandse Vereniging van Registerinformatici (VRI) updated its own in 1997 (http://www.vri.nl/info/gedraguk.htm); the Hungarian John von Neumann Society (http://www.njszt.iif.hu/) – the text is unfortunately not available in English; the Hong Kong Computer Society (http://www.hkcs.org.hk/ethics.htm); and the

Analysis of the Codes

The Content of the Codes

The first thing which is noticeable resides in the titles of the codes themselves: they are called either 'Codes of Ethics' or 'of Conduct', and more rarely 'Ethical Guidelines' or 'Standards of Conduct'. Sometimes the words ethics and conduct are used together, without really distinguishing the content or considering the rules of conduct as applying the principles as given in the ethical rules. Even when comparing the codes which use only one of the words, no significant content difference can be found between them, inducing confusion and controversy about the meaning of 'ethics' and 'conduct': there are Societies which prefer the wording 'ethics' when applied to the obligations due to the public, and 'conduct' when considering their members as belonging to a 'profession'; but there are others which totally avoid the word 'ethics' since they think that codes must treat only professional matters, leaving to individuals their own appreciation of what ethics may mean! This raises a fundamental question to which we shall come back later, but it shows that the underlying ethical theories are not always made explicit.

It became obvious rather quickly that the codes could be analysed easily according to a rather simple grid of analysis. Most of the rules or of the statements of the codes are formulated along the same pattern: 'X is responsible to Y for Z', where X is the 'Subject', Y the 'Reference', and Z the 'Field of responsibility'.

X is responsible	to Y	for Z
↕	↕	↕
Subject	Reference	Responsibility field

Here are the major findings. Full details are given, along with the codes themselves, and IFIP-GA recommendations, in the 'IFIP Ethics Handbook'.[12]

Malaysian National Computer Confederation (http://www4.jaring.my/mncc/code.htm). Most of the IFIP Member Societies Codes are available at (http://courses.cs.vt.edu/~cs3604/lib/WorldCodes/WorldCodes.html).

[12] See J. BERLEUR and M. D'UDEKEM-GEVERS, 'Codes of Ethics or of Conduct Within IFIP and Other Computer Societies' in J. BERLEUR and K. BRUNNSTEIN (eds.), *Ethics of Computing: Codes, Spaces for Discussion and Law*, London, Chapman and Hall (now Kluwer), 1996, pp. 3-41.

As for the Subject being concerned, most of the time s/he is an 'individual subject' (21 times/31), sometimes specified as a 'computer professional' or a 'voting member', or a 'leader', or a 'teacher'; but on occasion the Subject is an 'institutional subject' like a 'company' or an 'organisation', or a 'computer society', etc. Of course, this question has to be linked to the membership structure of the computer society, since the rules may not have the same characteristics of enforcement depending on the position of the Subject.

Analysis of the rules also gives a very good insight into the References of the rules inside the codes. Members of computer societies recognise their responsibility equally towards the 'public' (23 times/31), the 'organisations', and mainly the 'clients and the users' (25/31), and the 'profession' if not the 'computer society' itself or, sometimes 'oneself' (24/31). Less frequently, within the organisation, a specific responsibility towards the employer, or towards other employees or towards colleagues, is also mentioned (respectively 11, 8 and 9/31). In our opinion, the responsibility towards the clients and the users was to be expected. The recognition of a responsibility towards the public or society as a whole is more remarkable and demonstrates the fact that computers and information technology have influenced attitudes towards our social life.

The responsibility field is, of course, the most developed part within the codes. Five main categories emerge and regroup the different wordings as adopted by the different computer societies:
- respectful general attitude,
- personal (or institutional) qualities: conscientiousness, honesty and positive attitude, competence and efficiency,
- promotion of information privacy and data integrity,
- production and flow of information,
- regulations.

Let us examine these five main categories in more detail:[13]

- *Respectful general attitude* (/30)

    This attitude includes: respect for the interests or rights of the people involved (15), respect for the prestige of the profession

[13] *Ibid.* We give in parentheses the number of codes concerned by the mentioned wording.

(11), respect for the interests or rights of the public (10), and respect for the welfare, health of the public and for the quality of life (10). Sometimes it also includes: respect for the reputation of the computer society (8), respect for the quality of life of the people involved (6), respect for the public in general (6), respect for the environment (6), and respect for the differences of the public (4).

- *Personal (or institutional) qualities, such as conscientiousness, honesty and positive attitude, competence and efficiency* (/30)

  In practice, the terms conscientiousness and honesty are frequently encountered under the expressions acceptance of responsibility (19) and integrity (26). Moreover, appeals to respect for requirements or contracts or agreements (14) and to conscientious work (11) are also frequent. Other topics relating to conscientiousness and honesty are: professionalism (7), credit for work done by others (6), good faith or goodwill (4), concern to meet overall objectives (3), and the courage of one's convictions (1). With regard to the expressions competence and efficiency, two other terms are very common: professional development and training (19) or limitation of work to the field of competence (18). Two others are also worth noting: general competence (13) and effectiveness or work quality (12).

- *Promotion of information privacy and data integrity* (/31)

  Confidentiality (22) is required by nearly all the general codes of the IFIP societies (13/15). Privacy in general (14) and respect for property rights (12) are appealed to quite often. Three other topics, no computer crime, no information piracy or misuse (7), data integrity (6) and data minimisation (2), are less frequent.

- *Production and flow of information* (/31)

  The majority of the codes (23) requires flow of information to involved parties or people. Information to the public (16) is also insisted upon. Half the whole set of codes calls for comprehensive information (14). Several codes also ask for the production of tests, evaluations, results or specifications (7) or for the flow of information from the involved parties or people (7).

- *Attitude towards regulations* (/30)

    Regulations do not appear as a major theme. Less than half the codes requires respect for the code (13), respect for the law (13), and respect for IT and professional standards (12).

    Few codes refer to development of standards (5), of the law (2), or of the code itself (1); some consider sanctions against a breach of the code (9). Regulation of the code itself is often taken into account outside the code, in the procedures.

The Environment of the Codes

An analysis of what we have called the 'environment' of the codes has also been performed. Using the available information, we have tried to compare the sanction levels, the disciplinary procedures, the process of updating the codes, the status of the different computer societies, and their membership structures.

Sanctions and Procedures

The levels of sanctions are generally in four categories: caution or warning, reprimand, suspension, and exclusion or forfeiture, revocation of membership.

Disciplinary procedures normally involve no more than five steps: complaint, investigation (eventually with a suggested reconciliation), hearing process and decision, appeal process, and publication of opinion. Some Societies have also support procedures for members.

Update

The updating of the codes is not done on a fixed date basis: codes may be reviewed 'regularly' or 'when necessary'. Some Societies have no formal mechanism for updating. Other Societies leave it up to the wisdom of a specific committee.

Status and Membership

The status of the Societies varies from a Society 'incorporated by charter granted by the Crown' (BCS: British Computer Society) to a 'registered Society' according to the specific laws of a country. The status surely has an influence on the membership structure, or at least on the process of selection of the 'full members'.

With regard to membership structure, the differences are such that no comparison is worth doing. As stated earlier, we generally find 'individual members', but there are Societies which admit also institutional members (7/18) from education or business, for instance, not always as full members but as affiliates. A distinction is frequently made between 'members entitled to vote' and members 'not entitled to vote'. Within the first group, there is a panoply of denominations: voting, ordinary, regular, individual, full, professional, honorary members, fellows, etc. While within the second, some denominations overlap with the first category, such as fellows, etc. but we also find a wide range of denominations: associates, affiliates, student members, overseas members, etc. We have found Societies which establish no less than nine grades of membership! It is not clear that members are bound by codes in the same manner, although their grades differ. Some Societies require the signature of the members, only if they are 'ordinary members' (AICA: Associazione Italiana per l'Informatica ed il Calcolo Automatico).

Remarks Deriving from the Analysis

The codes show a rather fine convergence as far as their content is concerned. This means that for ethical debates, they offer a basis for discussion on which an agreement may be easily reached. They also offer an already experienced 'framework on ethics' which may help to increase awareness and maintain openness in a dialogue to be deepened. However, the computer-related ethical issues do not seem to be sufficiently elaborated and taken into account. At the most, two out of the five 'responsibility fields' are directly linked with information and communication technology. The others could also be applied to other disciplines and professional associations. Of course, some features may be nuanced when applied in the computer context, such as for instance the 'respectful general attitude towards the interests and rights of involved people' where one could interpret it as determining a kind of 'information right'. We shall see hereafter that the situation is not far from that encountered in the Societies of Civil Engineers.

Codes cannot be too precise, because procedures for enactment are often lengthy. Moreover, they must be short if we want them to be read and applied: long codes are at risk of being inefficient. The different Societies rarely mention that they could offer their members advice in using and interpreting the Code or how to deal with ethical issues at stake. A balance has to be found: some associations have supplemented the code itself with comments, explanatory notes or guidelines to assist members 'in dealing with the various issues contained in the code' or 'helping them in making decisions in their professional work' (ACM: Association for Computing Machinery, USA, Preamble).

Weaknesses must be pointed out. When looking at the list of the national Member Societies, we must recognise the prominent Anglo-Saxon representation, and even the more restricted zone of English influence. Less developed countries outside that zone are clearly underrepresented. Most probably, this is due to different cultural and legal tradition. The Anglo-Saxon world is more open to the role of professional associations and their self-regulation. Continental Europeans, Mediterraneans and Latin Americans are more confident in the law!

The enforcement procedure is all but clear and evident. There are complaint and disciplinary procedures, often rather complicated, but among the 13 IFIP national societies, for instance, only 5 have more than one single sanction, i.e. revocation / exclusion. Otherwise, the commitment of the individuals towards the code is not always an *explicit* condition of membership. These questions of enforcement and sanctions seem more and more crucial, with the current trend toward self-regulation.

Participation of people in drafting provisions seems also rather weak. When we say people, we do not mean only the members of the association or of the national Society, but also the public. Self-regulation must be developed with the participation, as large as possible, of the people concerned – this is a requirement of democracy.[14] There are 'boundaries' between the profession and the society at large. Codes are worded in such a way that they would require the public to be involved in the process of deciding the norms.

'As a member of X, I will contribute to society and human well-being'; 'Members shall in their professional practice have regard to basic human rights and shall avoid any actions that adversely affect such rights'; 'We, members of Y, (...) agree to accept responsibility in

---

[14] J. BERLEUR, 'Self-Regulation and Democracy: Choice and Limits?' in S. FISCHER-HÜBNER, G. QUIRCHMAYR and L. YNGSTRÖM (eds.), *User Identification and Privacy Protection, Applications in Public Administration and Electronic Commerce*, Proceedings of the joint IFIP-WG8.5 and WG9.6 Working Conference, Stockholm 1999, DSV – Department of Computer and Systems Sciences, Stockholm University/Royal Institute of Technology – on behalf of IFIP, Report Series 99-007, ISBN 91-7153-909-3, pp. 1-19.

making (...) decisions consistent with the safety, health and welfare of the public, and to disclose promptly factors that might endanger the public or the environment', etc.

One may easily understand that well-being, basic human rights, health, welfare, environment, etc. could benefit from the participation of those affected, even potentially, if we want the meaning of such 'words' not to be restricted by professional interests.

Finally, as we have already emphasised, the codes examined show too weak a tie to the really emergent hot issues of the profession. We shall come back to this question.

A Short Comparison with Engineering Ethics

The WWW Ethics Center for Engineering and Sciences is one of the centers providing us with course materials and instructional resources, bibliographies, lists of professional societies with their ethical codes and guidelines, research outlines, ethics centers collecting and making information available to all. It is sponsored by the USA National Science Foundation, formerly located at MIT (Massachusetts Institute of Technology), then at the Case Western Reserve University, and now apparently 'independent'.[15] This site is a 'must' in ethics for engineering and science. Above all what we just mentioned, there is an interesting list of 'keywords in science and engineering ethics', subdivided in five categories: general ethical issues, issues in educational and workplace settings, research, ethical courses, and legal. This kind of list is interesting because it represents, as for every classification, a 'state of the art'. It shows where the preoccupation of people in science and engineering could be.

Surprisingly, we found in this list many words which we had found earlier in the codes of IFIP for computer scientists and professionals. Restricting ourselves to the first two mentioned categories, the main words are: privacy and confidentiality (including trade secrets, databases, medical information,...), general ethical issues (ethics and the law, deception, distrust, accountability, ...), safety, competence, environmental issues, fraud, bribes, conflicts of interest, professionalism, copying, ethical responsiveness of organisations (harassment, workplace relationships, communication, gossip ...), diversity (culture, gender, ...).

---

[15] Online Ethics Center for Engineering and Science: (http://ethics.cwru.edu) or (http://www.onlineethics.org)

Two objections may come to mind. First, a list of key words does not give a hierarchy of issues: one item may be encountered in real problems only once whereas the other may be encountered 100 times. Second, many of the issues mentioned are not really linked to science and engineering, as we noted for computing. This last objection shows one of the main difficulties: How can we remain sufficiently general, while being sufficiently specific at the same time? We would not be honest not to add that the WWW Ethics Center also gives a disciplinary index by field of interest, i.e. by subdiscipline in engineering, such as aeronautical, biomedical, civil/structural, chemical, electrical, environmental, materials, mechanical, nuclear, ... engineering. However, on going further, you are not rewarded by discovering substantially new things. So, for example, when looking at the content of the subdiscipline Computer Science, references are given to privacy and cryptography, intellectual property issues, social impacts of computers, safety critical systems, unfair business practices, and other computer science related topics.

A Confusion Between Ethics and Conduct?

Let us come back on the debate about 'ethics' and 'conduct'. We already said that there is a confusion in the titles of the codes, if not in the mind of their authors, since the words are not really differentiated, and that sometimes there are Societies which seem to prefer the wording 'ethics' when applied to the obligations due to the public, and 'conduct' when considering their members as belonging to a 'profession'. We shall see that the IFIP General Assembly will also propose that the codes of ethics be seen as 'mission statements' and the codes of conduct as dealing with a more professional content.

The question is not simple, since IFIP, as an international body, must take into account the different ethical traditions which lead its different Member Societies. It is quite clear that the Anglo-Saxon world refers to consequentialism and utilitarianism, which leads to give to the word 'ethics' a quite different content than in a world where the deontological or Kantian approach is prevalent. Deontology seems to be more normative and has an ethical content which is more decisive.[16]

---

[16] For a short introduction to the different theories of ethics, see for instance: D.G. JOHNSON, *Computer Ethics*, Englewood Cliffs, NJ, Prentice-Hall, 1994 (2nd edition), pp.16-36.

One must admit – and that becomes more obvious than ever today – that self-regulation (to which the codes belong) is motivated by the desire to avoid a greater degree of statutory regulation or to curb government regulation.[17] Ethics in those conditions is becoming more problematic.

So, in order not to fall in the trap of confusion and ambiguity, and to cope with the diversity of theoretical approaches of ethics, we suggest focusing on a more procedural one, as suggested by Jürgen Habermas in his 'Discourse Ethics' or by Marc Maesschalck in his 'Ethics of convictions,' where a procedure for exchanges which recognise the differences but also what could be called an horizon of universalization for the norms is defined.[18] This approach takes longer, but it also means a re-appropriation of everyday life, including the professional life as well as the differences of cultural heritage, in the field of ethics. Applied ethics may cover so many domains and questions that there is a need for getting all the required parties involved. Ethics in the economic and secularised world has taken a predominant position since other discourses seem to have lost their own persuasive meaning, if not their legitimacy. The 'question of meaning' is coming back to the forefront of social life, when there are people willing to confine it in the private sphere. In our age of globalisation, it cannot be examined in another way than on a world-wide scale.

Entrepreneurs are facing day-to-day problems of their firm, including social problems raised out of their aims and aptitudes to cope with. They discover, or re-discover, in daily contacts, their own commitment to create meaningful situations for their employees and workers. If ethics does not seem a major preoccupation among those who govern the world at the level of financial globalisation, the situation is not the same at the level of firms and organisations which face people in search of understanding their own life. The profession may be a level where daily contacts are appreciated. Is it not one of the reasons why ethics is becoming predominant on that scene? However,

---

[17] One may refer to the Press releases at the creation of the Global Business Dialogue, an electronic commerce initiative set up by top executives at 17 companies (see http://www.gbd.org/library/news.htm)

[18] J. HABERMAS, *De l'éthique de la discussion*, Paris, Cerf, 1992. [Orig.: *Erlauterungen zur Diskursethik*, Engl. Transl.: *Justification and Application: Remarks on Discourse Ethics*, Cambridge, MA, The MIT Press, 1993]. M. MAESSCHALCK, *Pour une éthique des convictions. Religion et rationalisation du monde vécu*, Brussels, Publications des Facultés Universitaires Saint-Louis, Coll. Philosophie, 1994, p. 376.

one may call it fair conduct, when another will prefer to speak about ethics. The most important thing is to provide meaningful insights to life in search of understanding, and places where discussion is promoted and open, and where ethical issues are not obliterated.

Recommendations of IFIP General Assembly

On the basis of the work undertaken by its Ethics Task Group, mandated by its General Assembly, IFIP made recommendations to its Member Societies (Hamburg, September 1994). Of course, such recommendations cannot be compulsory, but have to be regarded as the product of an extensive consultation and as resulting in a better appreciation of the historical, cultural, social, political and legal diversity of these Member Societies. The IFIP position has been spelled out as follows: 'IFIP regards as essential that, when wanted and needed, codes of ethics or of conduct should always be developed and adopted within the Member Societies themselves.' This meant that IFIP is not contemplating an IFIP international code.

IFIP recommended that a careful distinction be made between 'Codes of Ethics' and 'Codes of Conduct', since it appeared in the analysis and comments that the former are more often oriented towards the public and the society as a whole, while the latter seem to be related more directly to the 'computing profession'. Codes of ethics could be seen as 'mission statements' of Computer Societies, providing visions and objectives in relation to their public mission and anticipating the issues at stake in a computerised world or in an information society. Codes of Conduct would have to deal with those issues, in the specialised fields of the profession. Moreover, certain authors who have been working on the 'ethics of computing' for some time think that 'the rules of conduct have to reach, beyond the well-structured body of computer scientists, the larger circle of computer users. We must shift from a deontology of informaticians to an objective deontology of informatics under the control of the law.[19]' The question is then raised of the role of the codes in society: do they have to anticipate the law, to supplement it, or be controlled by it? This is explicitly linked to one of the very sensitive questions of today as we already mentioned, namely the question of 'self-regulation'.

[19] H. MAISL, 'Conseil de l'Europe, Protection des données personnelles et déontologie' in *Journal de Réflexion sur l'Informatique*, (Août 1994)31.

As far as the content of the codes is concerned, IFIP suggested, as a first step, an appreciation of the different responsibilities of the 'members' of its Member Societies and of the main rules which are already mentioned in the majority of the existing codes, noting how they could be adapted or included when writing or updating one's own code. It is interesting to stress again that the different existing codes state that the responsibilities of the members extend from the profession itself to the society as a whole. It is not easy to understand how it can be applied realistically, but this is not a reason for avoiding such a statement or provision which could encourage a dialogue between the Societies and the public.

As mentioned earlier, codes of IFIP Member or Affiliate Societies should address computer-specific ethical issues in more depth. IFIP and computer scientists should provide their expertise in dealing with threats and dangers which appear daily in specialised fields. Specific computer-related ethical issues have been listed by the Ethics Task Group, from current case studies, experiences, literature, workshops, or recommendations drawn up by organisations such as the Council of Europe in the domain of 'computer crime' for instance.[20]

The list of computing ethical issues may be long: one may easily imagine issues related to stealing, sabotaging, harassing, etc. Others stress computer abuses, such as alteration, hacking, piracy, introducing viruses, and so on. One cannot ignore all the issues linked to what is handled by the European Commission as a 'Promoting best use, preventing mis-use' Action Plan where the preoccupation is to protect children and human dignity from all illegal or harmful material.[21] Others, in the framework of the Internet and of electronic commerce will stress intellectual property rights, copyright, ownership of data, etc.[22] Other issues have also been examined for a long time in the field of 'Computers and Society' and have been considered as controversial in the computerisation processes. They cannot be forgotten when examining today's computer-related ethical issues, which include

---

[20] Council of Europe, *Computer-Related Crime*, Recommendation N° R(89)9 on computer-related crime and final report to the European Committee on Crime Problems, Strasbourg, 1990.

[21] Action plan on promoting safe use of the Internet:
(http://www2.echo.lu/legal/en/best_use/best_use.html)

[22] J. BERLEUR, 'Final Remarks: Ethics, Self-Regulation and Democracy' in *Ethics of Computing: Codes, Spaces for Discussion and Law*, op. cit.

what was formerly classified as the social, economic, managerial, political, cultural, philosophical consequences of computing.[23]

These lists show that there is still a long way to go in order to reach a satisfying state of the art where these problems will be handled thoroughly. However, it is the duty of computer societies to work on them and propose solutions which will really protect society, as well as the public and the computer profession. Are codes the most suitable way to meet the expectations? This is another question which will not be treated here.

Since the work to be faced is challenging, IFIP has recommended creating 'spaces for discussion' and has established some specific procedures to meet this challenge. To 'create spaces for discussion' on ethical issues is an urgent task for all the constituencies concerned which deal with information and communication technology in order to study specific computer-related issues more deeply. The IFIP General Assembly has proposed different objectives for such spaces including: 'submitting, through the IFIP *Newsletter* for instance, specific case studies, encouraging members to submit their own responses; making available all the up-to-date codes of IFIP national societies, with related pointers to existing documentation for further research; publishing, as foreseen in the European Directive on data protection, significant codes;[24] providing a forum –

---

[23] One may think of:
- the impacts on work organisation and working conditions; employment; changes in qualification and skill needs; the place of women; human resources at work;
- the rationalisation of managerial, professional and technical work; involvement of users namely in systems design; security and reliability;
- the role of ICT in the process of global economy;
- the increasing gap between developed and developing countries;
- the distribution of power; decision-making procedures, centralisation-decentralisation; interactions between the public and computer-using organisations; increased surveillance in automated office; public and private databases; consequences for democracy, privacy and civil liberties;
- the impact on health care, on households, on education, morale and culture;
- the predominant paradigm of instrumental reason; influence on perception of oneself, etc.

See, for instance, J. BERLEUR, A. CLEMENT, T.R.H. SIZER and D. WHITEHOUSE (eds.), *The Information Society: Evolving Landscapes. Report from Namur,* Springer Verlag New York-Heidelberg and Captus University Publications, 1990; R. KLING (ed.), *Computerization and Controversy: Value Conflicts and Social Choices,* San Diego, Academic Press, 2nd ed., 1996; C. HUFF and T. FINHOLT (eds.), *Social Issues in Computing: Putting Computing in its Place,* New York, McGraw-Hill, 1994, p. 726.

[24] See particularly art. 27: 'The Commission may ensure appropriate publicity for the codes which have been approved by the Working Party.' (Directive 95/46/EC of

under the Chairmanship of the IFIP President – where discussion could be raised about harmonising codes of Societies, in order to prevent restrictions in one country being prejudicial to another; participating in international *fora* where similar questions are treated; assisting in the resolution of conflicts which could arise between national codes that are completely different; etc. Therefore, IFIP will collect, compare and help disseminate knowledge on developments in the national Societies. In the case of controversies, it will also advise on the resolution of problems in projects with professionals from countries which have very different codes.'[25]

Moreover, the IFIP General Assembly has requested the establishment of a Special Interest Group (SIG9.2.2: IFIP Framework on Ethics of Computing) which has been accepted by its Technical Committee 9 and WG9.2 and whose role would be to act as a catalyst within IFIP to collect case studies, new codes and comments, and disseminate all relevant information so that the national and regional Member Societies may develop a climate sensitive to ethical questions which could arise world-wide. SIG9.2.2 will regularly inform IFIP about essential achievements and progress in the international discussion, and discuss and suggest solutions for emerging problems.

IFIP does not pretend to have any monopoly in the questions we have discussed. It just assumes its role as the widest international organisation of computer scientists. It knows that others have also started a process of re-assessing their role with regard to ethical issues. There are Universities, for instance, which have established their own 'policies'.[26] There are also other ways to meet the requirements of higher standards of ethics in our society: international guidelines, public policies, legal instruments, etc. The international guidelines may provide statements which act as reference documents or a basis for the development of legal instruments in particular jurisdiction. Public policies may incorporate aspects of acceptable behaviour, practices and standards. Legal instruments are generally the most enforceable, provided they are drafted correctly and the courts are sufficiently qualified to assess the matters brought

---

the European Parliament and of the Council of 24 October 1995 on the protection of individuals with regard to the processing of personal data and on the free movement of such data, Brussels, *Official Journal of the European Communities*, 23.11.95, No L/281/31-50).

[25] See 'Recommendations from IFIP General Assembly, Hamburg September 1994' in *Ethics of Computing: Codes, Spaces for Discussion and Law*, op. cit.

[26] J.W. CORLISS, 'Analysis of Universities Policies', in *ibid*.

before them. We are in favour of the legal instruments, but as said earlier, ethical principles, codes, guidelines, policies, may anticipate and supplement the law. They are the first steps.

Expanding the instances where spaces for discussion are created is a trend to be promoted: ethical questions must be discussed as and when they arise. However, we must also think of coordinating the principles which could emerge from the different convictions, and consider the work of international organisations which have proved their value in the past, such as the Organisation for Economic Cooperation and Development (OECD) or the Council of Europe, when they enacted specific guidelines or the Convention Nr. 108 for the protection of individuals with regard to automatic processing of personal data.[27] The work which has been done in the field of the protection of individual liberties and privacy could be usefully followed up by similarly important work on the ethical issues we are facing today.

Conclusion from our IFIP Experience

The first 'lesson' we could derive from our experience is that to draft an international code, be it European, does not seem to be a realistic goal. The reason which is often given is the impossibility of enforcing it. But there is also another one, which in our view is at least as important. A code has several functions one of which is to make people discuss ethical issues, as long as a time and a place are specifically allocated in the life of the association. This means that to be fruitful the discussion must interpret the rules and implement them in specific political, legal, economic, social, and cultural environments. The role of any international body should, in our view, be to help and support local constituencies in taking their own responsibility.

Moreover, as we have said, codes are very often quite short, between two and four A4 pages long. In a way this is good, because if they were too lengthy, they would be ignored, or if they were too detailed, they would become obsolete very quickly. There is a need

---

[27] Council of Europe, *Explanatory Report on the Convention for the Protection of Individuals with Regard to Automatic Processing of Personal Data*, Convention opened for Signature on 28 January 1981, Strasbourg 1981. OECD, *Recommendation of the Council Concerning Guidelines Governing the Protection of Privacy and Transborder Flows of Personal Data*, 23 September 1980, OECD, Paris, Acts of the Organisation, 1982, Volume 20, pp. 535ff. See also, OECD, *Recommendation of the Council Concerning Guidelines for the Security of Information Systems*, Paris, 1992, OECD/GD (92)190.

then, to permanently elaborate on and discuss emerging new problems and new issues to discern the ethical attitude and behaviour necessary to be adopted. Again, the best place for this is the local constituency. Let us not deprive discussion groups of their right to discuss!

If we had to advise anybody in writing or updating a code we would suggest they:
1. identify the general and specific issues at stake in the profession, but also in the social and economic environment; to try also to define a hierarchy as well as a frequency of the issues;
2. raise some simple and unavoidable questions such as: 'Whom do we serve?' (the public, organisations – mainly clients-users – oneself / the profession / computer society, or...?) 'For what, for what good or benefit do we serve?', 'What is the proper decision-making relationship between our profession and the people we serve, as well as between our profession and the purpose we pursue and strive towards?' We have noticed that some codes have arranged their statements according to the target public (CIPS: Canadian Information Processing Society, BCS: British Computer Society, NZCS: New Zealand Computer Society). Others are still mixing the ethical imperatives and the people being served (ACM: Association for Computing Machinery, USA, IEEE: The Institute of Electrical and Electronics Engineers or GI: Gesellschaft für Informatik, Germany);
3. make clear the ethical principles which support specific code provisions;[28]
4. consider the provisions which may meet the requirements of the situation (1.) and which could meet the questions (2.). We give again here the 'fields of reference' which have been considered by at least one third of the codes we have examined:
    - respect for the interests or rights of the people involved, for the prestige of the profession, for the interests or rights of the public, for the health and welfare of the public, and for the quality of life;
    - conscientiousness and honesty, acceptance of responsibility and integrity, respect for requirements or contracts or

---

[28] The 1992 Code of the ACM (Association for Computing Machinery, USA) makes the distinction between 'General Moral Imperatives' (Title 1), 'More Specific Professional Responsibilities' (Title 2), 'Organizational Leadership Imperatives' (Title 3) and 'Compliance with the Code' (Title 4).

agreements, conscientious work, professional development and training, competence, effectiveness and work quality;
- confidentiality, privacy in general and respect for property rights;
- flow of information to involved parties, and information to the public;
- respect for the code, for the law, and for IT and professional standards.

5. make members of the association participate, as well as the public, in the elaboration of the code or its updating: the larger the participation, the more chance the code has of being accepted;
6. give the code wide publicity: this is a condition *sine qua non* for avoiding the criticism which self-regulation faces today; it also seems to us that publicity is the only way to make the complaint procedure known to different parties; this also means that this complaint procedure must be included somewhere in the code itself or at least that the by-law be indicated and referred to;
7. open a space for discussion where the ethical question is not obliterated and the ethical debate is open and made lively, and where more specialised questions can emerge and have significant consideration;
8. collect cases – we insist real cases as opposed to the fictitious ones which are still circulating either in handbooks or on the web!

In our view, the goal is not to have a code as such, but starting from it, or from another similar document, to stimulate the debate in the association which enacted it. More and more contemporary choices in research and technology raise ethical issues: enhancing the quality of the ethical decision-making seems a target not to miss. Scientists, researchers, engineers and practitioners may help in improving the awareness of hidden issues.

Annex

1. Within IFIP National Member Societies: 21 Codes for 13 National Societies

- ACS (Australian Computer Society, Australia): ACS Code of Ethics
- AICA (Associazione Italiana per l'Informatica ed il Calcolo Automatico, Italy): Codice di Condotta Professionale dei Soci Ordinari AICA
- BCS (British Computer Society, UK): BCS Code of Conduct: Rules of Professional Conduct (1992), BCS Code of Practice (1978)
- CIPS (Canadian Information Processing Society, Canada): CIPS Code of Ethics and Standards of Conduct (1985)
- CSI (Computer Society of India, India): CSI Code of Ethics (1993)
- CSSA (Computer Society of South Africa, South Africa): CSSA Code of Conduct (1988)
- CSZ (Computer Society of Zimbabwe, Zimbabwe): The CSZ Code of Ethics for Institutional Members (1992), The CSZ Code of Ethics for all Individual Members (1992), The CSZ Code of Professional Conduct for Individual Corporate Members (1992), The CSZ Code of Professional Conduct for Registered Consultants (1992), The CSZ Training Accreditation Code of Practice (1992)
- FOCUS (Federation On Computing in the United States, USA)
  - ACM (Association for Computing Machinery, USA): ACM Code of Ethics and Professional Conduct (1992)
  - IEEE (The Institute of Electrical and Electronics Engineers, Inc., USA): IEEE Code of Ethics (1990)
- GI (Gesellschaft für Informatik, Germany): Ethical Guidelines of the GI (1994)
- ICS (Irish Computer Society, Ireland): ICS Code of Professional Conduct (1994)
- NZCS (New Zealand Computer Society, Inc., New Zealand): NZCS Code of Ethics and Professional Conduct (1978)
- SCS (Singapore Computer Society, Singapore): SCS Professional Code of Conduct

2. Within IFIP Affiliate Member Societies: 3 Codes for 2 Regional Societies

- CEPIS (Council of European Professional Informatics Societies, Europe): CEPIS Code of Professional Conduct
- SEARCC (South East Asia Regional Computer Confederation, South East Asia): SEARCC Code of Ethics, and SEARCC General Guidelines for the Preparation of Codes of Ethics for Members

3. From other Computer Societies: 6 Codes for 5 Societies

- ASIS (American Society for Information Science, USA): ASIS Code of Ethics for Information Professionals (current draft 1992)
- CPSR (Computer Professionals for Social Responsibility and Privacy International, International + USA): CPSR Code of Fair Information Practices (privacy)
- JISA (Japan Information Service Industry Association, Japan): JISA Code of Ethics and Professional Conduct
- VRI (Nederlandse Vereniging van Registerinformatici, The Netherlands): VRI Code of Ethics
- IPAK (Information Professional Association of Korea, Korea): IPAK Code of Ethics and IPAK Standards of Conduct

2.3

REFLECTION

2.3.1

TAKING RISKS
AND THE VALUE OF HUMAN LIFE

Göran Möller

In many different situations engineers have to take stand towards what measures should be taken, in the form of investments, regulations or whatever it may be, in order to prevent human lives being lost. What factors influence the level of risk that is acceptable? How does one proceed in deciding what risk-levels are acceptable in a given context?

Taking Risks and the Value of Human Life

In many fields of society, one is confronted with the necessity of deciding what degree of risk taken, in the form of investments, regulations or whatever it may be, in order to prevent human lives being lost. The dilemma can be illustrated with the help of a few examples. What speed limits ought to apply in road traffic? Can the risks of nuclear power be accepted? And if so, how rigorous should the security regulations be in connection with the production of energy from nuclear power stations? What safety-margin ought to be provided in different kinds of constructions? In many of these issues engineers take part in different roles, as constructors, managers and decision-makers.

Road traffic alone claims many thousand victims in Europe every year. At the same time, we know it is possible to considerably lessen this statistic. If sufficiently radical measures were taken, one could reduce the total of fatal traffic accidents by several thousands annually. (Assume, for example, that one brought in a law according to which all cars should be provided with a device preventing them from going faster than seventy kilometres per hour). Yet we are not prepared to take all the measures which would thus lessen the rate of accidents. One of the reasons for this is quite simply that our resources are limited and that we do not want to dedicate more than a certain proportion of them to preventing such fatalities. For we want also to satisfy other needs and desires. We do not want to renounce other values we have in exchange for increased security. Nor do we want our freedom to be constrained to such a high degree. Some of the measures requiring prohibition of or limitation upon dangerous activities do not, in fact, require very great resources, but they imply restrictions upon people's freedom of action. If too many such measures are taken then the private person becomes ultimately so ensnared in a tangle of different ordinances and rules that his or her autonomy comes under threat.

Realistically deciding whether different risks can be accepted or not requires that one take into account certain fundamental facts concerning human life and its conditions. Our human beings are constitutionally fragile and vulnerable. Our life expectation is necessarily limited. We live, in addition, in an environment where various dangers constantly threaten us. Many such dangers belong to human life as such and are therefore impossible to eliminate. From others, however, we can more or less free ourselves. However, even though we can mitigate many forms of suffering and death in various ways, we are not prepared to do it at any price. For we are interested in more than life as such; we want a life of a certain quality. Other values besides human survival have to be considered, therefore. This situation can be brought into relief by considering that if one robbed life of any value by directing all effort and attention to ensuring a continued existence then life itself would no longer be worth living. So we do not want to live in a society where all available resources are used for saving or maintaining life. Nor do we want all actions involving the slightest danger to other people's lives to be forbidden. For that would limit our freedom of action to a completely suffocating degree.

How great then are the risks that can be tolerated in this or that connection? We can not of course allow all risks whatever. So what factors decide the level of risk that is acceptable? It is obvious that the acceptable level of risk varies with circumstances. One such circumstance will be the value attached to the activity causing risk. There must, that is, be tolerable parity between the degree of acceptable risk and the importance of whatever is causing it. It is not so obvious, however, whether, for example, the respective ages of those exposed to some risk have any significance regarding what level of risk is acceptable, or whether a risk is more serious if it bears upon people now living than if it threatens coming generations. The present article aims at discussing some of the ethical problems arising out of decisions where the risk that people may come to grief is such that one can modify it. By the term 'risk' I mean the existence of a probability that human beings may be harmed.

The Legitimisation of Certain Risks

How can one affirm the principle of the high value and inviolability of human life while simultaneously accepting that in many contexts we expose both others and ourselves to mortal risk? This question has been analysed in an interesting way by the American philosopher Charles Fried. As basis for his analysis Fried uses a concept which he calls 'life plan'. To speak of a life plan is to give a summary designation of a person's way of organising his or her life so as to attain the ends desired.[1] Everyone has (hopefully) certain purposes in life in that one wants to realise various values and attain to various goals. These can consist in gaining certain experiences, developing various innate aptitudes or coming to know and be of service to other people. To be able to realise one's goals one has to arrange one's life in line with certain organising principles, since some goals require for their realisation that certain preconditions be fulfilled. Besides this, there can be conflicts between one's different purposes so that some goals cannot be attained without sacrificing some of the others. There has to be, therefore, both an ordering of priorities between one's different values and purposes and a renunciation of those of the latter as would be to the detriment of the whole, being incompatible with other goals considered more worthwhile. For a person to structure what he or she wants to attain in life in this way is, in Fried's terminology, to set up a life plan. It may perhaps seem rather abstract and even exaggerated theoretically to consider one's different personal choices as if they formed part of a kind of plan for

---

[1] C. FRIED, *An Anatomy of Values*, Cambridge, MA, Harvard University Press, 1970, pp. 97ff and 162ff.

one's life. For it is certainly not the case that every time we take a decision we are following a careful, preconceived plan. The expression 'life plan', however, refers in general to a conscious ordering of one's life. One reason why we have to choose and order our priorities with respect to different goals is quite simply that our time is limited. It is granted to scarcely anyone, for example, to be both a skilful footballer and a virtuoso pianist. Moreover, some purposes are directly incompatible with others, besides the fact that certain goals require various types of investment during the earlier stages of life for their achievement.

What has this concept of a life plan to do with our question as to which risks can be accepted? Well, a life plan has to take our mortality into account. Many of the goals we seek, indeed, and which form part of our life plan can never be fulfilled unless we are prepared to take certain genuinely mortal risks. Certainly these risks are for the most part very slight. But they are there, even when realising many very trivial aims. In order to meet our piano teacher or our football trainer, for example, we may need to go through heavy traffic. Many of our daily occupations carry with them in this way a certain risk of losing one's life; but we still do not give them up. We think they are worth the risk. We would see a life-style which avoided everything the least bit dangerous as neurotic.

It is also the case, however, that, by our actions, we often set at risk not just ourselves but other people too. Thus every time we go on the road, we expose our fellow road-users to risk of accident, just as certain industrial processes involve toxic emissions which can be dangerous even to people otherwise unconnected to the industries in question. Indeed it seems almost impossible to have any contact at all with others without exposing them to certain risks. One might, for example, be a carrier of some dangerous disease without oneself knowing it. But how can we morally justify any behaviour endangering the lives of others? It is one thing, indeed, to take risks oneself, but quite another thing to involve others in these.

Charles Fried offers the following rationale for finding it legitimate to expose others to certain risks. Assume that I want to attain a particular goal but cannot do so without exposing both myself and some others to a particular risk. We have already seen that it can be rational to endure a risk for one's own part, namely when the goal of my action is judged sufficiently important in relation to the size of the risk. It remains to justify the risk to which I am exposing the others. For this we need to suppose that these other persons also have certain goals they wish to realise but such as are not possible unless they in turn expose others to certain risks. Now let us imagine that I am prepared to expose myself to risks resulting from these persons' actions on condition that I myself am permitted to expose them to risk in a corresponding way while I am achieving my own goals. Through letting others expose me to certain risks, therefore, I myself obtain the freedom to act in a way involving risks for them. We exchange risks with one another, so to say. Fried calls this forming a 'risk pool'. We contribute to the pool when we ourselves run risks as a result of the actions of others, but we make use of the pool when we expose others to risks.[2] The risk to which we may expose our fellow human beings is limited, all the same. For we are prepared ourselves to renounce certain goals in exchange for a certain measure of security, i.e. in exchange for other people's not exposing us to too great risks to a corresponding degree[3].

A troublesome question is that of how one shall proceed with people who do not wish to 'join' the risk pool. We can imagine someone who does not accept the exchange of risks which takes place in relation to the risk pool. The problem is reminiscent of an argument often used against contract theories of different kinds, to the effect that it is impossible to be bound to a contract without having in some way oneself entered into it. Against this, however, one can argue that a person gives tacit consent to a contract through enjoying some of the contract's advantages. This is John Locke's view, for example.[4] Joining the risk pool could be shown necessary in a similar way. Adapting Locke a little one could describe the tacit consent to the risk pool as follows: everyone exposing another person to a degree of mortal risk (without obtaining that person's express consent) gives thereby his or her tacit consent to being him/herself exposed by others to risks of the same degree.[5] By this argument everyone seems by some kind of tacit consent to be joined to the risk pool.

To understand more clearly the difference between fatalities occurring within the ambit of a risk pool and those that could be

[2] *Ibid.*, pp. 187ff.
[3] *Ibid.*, p. 192.
[4] J. LOCKE, *Two Treatises of Government*, P. LASLETT (ed.), Cambridge, Cambridge University Press, 1960, pp. 365f.
[5] *Ibid.*, p. 366.

classed as the deliberate sacrifice of designated private individuals we can imagine two situations. In one of them we suppose that each one of ten thousand persons is exposed to a mortal risk factor of 0.0001. This entails that one of the ten thousand will probably be killed. We assume also that the cause of this risk is an activity of central importance for the society concerned. In the second situation, by contrast, a definite person is selected to be killed because other persons or society as a whole will in some way benefit from his death. In both these cases the expected total of deaths is the same. What then is the difference between the two situations? Well, in the first case the risk is spread over a large population. Everyone runs a certain risk of being killed and this risk can be legitimised through its being rational for each individual to accept it. In the second case, on the contrary, a certain definite individual is sacrificed so that others shall benefit.[6]

It is important to notice that the risks attaching to the risk pool are limited in character and, besides this, that there is not here any intention to harm anyone. The fatalities which occur are all unintended side effects. The exchange of risks contributing to the risk pool need not, therefore, be incompatible with respect for other people's inviolable dignity. It is common to all the situations envisaged in the following discussion that no harmful intention is involved. It is anyhow only a question of such limited risks that the risk of an individual's coming to grief in any situations of this type is not greater than one in a thousand annually.

Factors Influencing the Degree of Risk Acceptable

Positive Gain from Risk-Bearing Activities

How much value, attached to the activity giving rise to the risk, will of course be decisive in judging what level of risk is acceptable in a given context? A risk is acceptable only if the positive consequences from what causes it are great enough to balance its negative character. There has therefore to be a fair relation between the size of a risk and the importance of what is causing it. For example, we would scarcely consider it permissible for people to devote themselves to

---

[6] R. Nozick, *Anarchy, State, and Utopia*, New York, Basic Books, 1974, pp. 32f.

private or leisure occupations which caused toxic emissions of the same order and amount as we can accept from an economically significant industry. It is therefore necessary to weigh the inconvenience of a risk against the positive consequences of whatever causes the risk.

Further, any judgement as to what measures might reasonably be taken to reduce a given risk must depend on the resources available in the context of the activity concerned. Such a judgement must also take cognisance of other important needs and desires which may be involved. Thus the significance of a given safety measure in relation to other values depends upon the context in which the risk has occurred. A security measure seen as going without saying in a Western European society can thus be seen as an unnecessary and costly luxury in a society having more limited resources. One can also stipulate that the acceptable level of risk for an activity should be seen in relation to the risks arising from other similar activities in the same society. There should, that is, be a certain uniformity in the judgment of different kinds of risks and in the public measures for reducing various such risks. So whether a particular risk is acceptable or not depends upon the context in which it occurs. That is to say that the less resources society has available and the higher the risk level found in other activities then so much higher will be the level of risk acceptable in each case considered.

The Voluntariness of the Risk

In the discussion above of the general principles regarding the legitimisation of risk-bearing activities, a distinction was made between risks to oneself and risks to which other people might be subjected. It emerged from the discussion that, in order to attain a particular goal, we might expose ourselves to greater risks than those to which we can expose others. This conclusion suggests a distinction between voluntary and enforced risks. Voluntary risks are those resulting from such actions as the person who is exposed to the risks has him- or herself chosen to carry out or expressly approved. Enforced risks, by contrast, are those that bear upon a person without them having given any consent to them. The argumentation supports the idea that the less a risk is imposed the greater is the degree of risk that we can accept.

To decide what degree of voluntariness exists in a given situation is of course a complex problem in itself. Important considerations for such a verdict would be questions such as what alternative actions are open to the agent and to what degree he or she is aware of such alternatives. For the most part, we are not free in any absolute sense but rather find ourselves somewhere on a continuous scale between utterly voluntary and completely imposed risks.

Discrimination of Candidates for Exposure to Risk

Is it more important to save some people's lives than others? Ought such qualities among potential victims as age or usefulness to others influence society's efforts at preventing fatalities in particular cases? These questions are of contemporary urgency in many sectors of society, most dramatically perhaps in the health service. There one can be forced, for example, to decide which of two critically injured people shall have access to the only available respirator. The dilemma is operative, however, even in regard to the assessment of more limited risks. Here, though, one has not to decide between concrete individuals. Instead it is a question of whether, and if so then how, one should discriminate between different categories of people when applying risk-reducing measures. For the threat posed by certain risks to different groups of people varies in intensity with each group, just as measures to alleviate a risk can influence these groups in respectively different ways.

One commonly occurring notion is that one should take into account the ages of those whose lives are at risk. According to this point of view, a risk-reducing measure should be the more positively judged the lower be the average age among those whose lives one is protecting.[7] Thus some authors suggest that the criterion one should use when deciding whether a given safety measure should be applied or not ought not to be the number of fatalities likely to be avoided but rather the increase in the sum total of lifetime remaining to all those to be affected by the measure. Another recurring idea is that one should take social utility into account by giving priority to such accident-reducing measures as especially favour those judged of greater use to society.

---

[7] See, for example, D. GOULD, 'Some Lives Cost too Dear' in *New Statesman*, 21 November 1975, and R.M. VEATCH, 'Justice and Valuing Lives' in *Life Span: Values and Life-Extending Technologies*, R.M. VEATCH (ed.), San Francisco, Harper and Row, 1979, p. 218.

As a starting-point for this discussion, I would like to assume the idea that all human beings are of equal worth and entitled to the same fundamental respect. There is no place here to argue for this position. It is, however, in accordance with a Humanistic and a Christian view of human kind. This understanding of things supports the view that society should be impartial as regards which groups of individuals should benefit from different kinds of life-saving and accident-preventing measures. On this basis, everyone in generally similar circumstances ought to have a similar right to benefit from societal efforts to protect the citizens' lives. This implies that society ought not to favour certain categories of people above others when making communal investments promoting life and security. That there can be cause, in certain cases, to protect by special measures people who for various reasons are especially vulnerable to certain phenomena or find it especially difficult to keep clear of certain dangers is quite another matter. It can therefore often be reasonable to give special consideration to protecting such groups as children and the handicapped. Such efforts are then justified by the special needs of these groups and not through their being thought to be of more value than others.

Long-term Consequences

Some types of activity can result in people suffering serious harm long after the activity has ceased. This can entail that people not yet born can come to be exposed, in a more or less distant future, to various consequences of contemporary behaviour. Such types of risks are attached to the storing of radioactive waste or the emission of certain chemicals into the environment. Here we are confronted with a new kind of problem, namely, whether risks to future generations can be justified and how in that case these risks should be evaluated in comparison with risks to people now living.

A fairly uncontroversial starting-point for forming an opinion in these questions might be our conviction that we have moral obligations even towards people as yet unborn.[8] The fact that a person is not yet born cannot make allowable actions which can be dangerous to him or her once he or she has been born. So have we any right at

[8] See, for example, J. RAWLS: '...persons in different generations have duties and obligations to one another just as contemporaries do.' *A Theory of Justice*, Oxford, Oxford University Press, 1972.

all to expose future generations to any risks? The argumentation pursued earlier in this chapter concerning legitimisation of certain risks through an exchange of risks between different persons in a risk pool can scarcely be carried over so as to apply to risks to future generations. There cannot, for one thing, be any mutual exchange of risks between different generations not alive at the same time. Such risks are 'one way'. They necessarily concern the future. We can expose our descendants to risks but they have no possibility of in turn subjecting us to any corresponding risks. The basic preconditions for a risk pool are therefore not fulfilled. One could possibly say that future generations can expose their descendants to risks in the same way as we have done to them and that this could correspond to the reciprocity entailed by the risk pool. We cannot be sure, however, that the people of later times will spread risks into the future of the same order and size as we are doing today. The idea of the risk pool needs to be supplemented, therefore, with arguments of another kind. Such an argument might be that it is not only negative phenomena such as risks, exhausted sources of raw materials and the destruction of the milieu which a generation can leave to those coming after. One also transmits (hopefully) elements of value. Examples of such a positive heritage could be a more profound knowledge within certain areas or a refined and less wasteful technology, not to mention a more complex and profound cultural inheritance. People of the future, therefore, can draw profit and rejoice in the investments made in our time in the same way as we can harvest the fruits of what our fathers and forefathers have achieved. A positive transmission such as this is also directed exclusively towards the future. One can therefore see the positive values which a generation thus transmits as a kind of compensation for the negative effects, such as risks among other things, which it also leaves behind it.

How shall we then evaluate the risks we project into the future? One can scarcely discuss this question in quantitative terms. Yet it would seem rational not to regard them as less serious than those affecting people now living. We ought not, therefore, to allow any purely temporal factor to reduce the level of seriousness assigned to such risks as extend into the future.[9] It should also be noted that many of these risks affecting future human beings are to be reckoned as enforced to a high degree.

[9] See E. ASHBY, 'The Search for an Environmental Ethic' in *The Tanner Lectures of Human Values 1980*, Salt Lake City, University of Utah Press,1980, pp. 18f.

The Catastrophe Scenario

The number of victims of certain kinds of accidents can be very large. An accident can take the form of a catastrophe claiming thousands of victims. Such devastating consequences can occur in connection with some kinds of nuclear disasters or with the bursting of one of the larger type of dams. This situation throws into relief the question as to whether there are any ethically relevant reasons for considering risks involving the possibility of a catastrophe in a fundamentally different way to how one considers risks where the number of victims of any individual accident would be more limited? We assume that the statistically probable number of victims during a given period of time is equally large in either case. So, in other words, are such risks as would entail a great number of fatalities should an accident actually occur more serious than the risks where the number of victims for each accident is more limited but where the accidents occur more frequently instead?

It ought to be clear that we are inclined to apprehend such risks as can cause a catastrophe as more serious than other risks presaging the same statistically expected number of victims per temporal unit. However, can such a difference in our evaluation be justified by any rational argument? One reason why a very limited probability of catastrophe is judged more serious than other types of risk could be quite simply our lacking the power to imagine the proper ordering of quantity where small numbers are concerned. It is difficult for us to grasp, for example, the difference between the two proportions of one in a million and one in a milliard.[10] On the other hand, we can very well understand the difference in quantity between the numbers 1 and 1000. This seems to depend upon our being able to associate them with concrete quantities. This feature of our powers of perception can easily lead us to interpret the risk of one in a milliard that a thousand people will be killed as something *quantitatively* greater than the risk of one in a million that one person shall perish. We are inclined to such an error quite apart from whether or not the catastrophe scenario as such contains any ethically significant aspects. This state of affairs has to be seen as a defect in our ability to interpret certain data and ought not to influence our decisions.

There are, however, certain circumstances which do suggest that

---

[10] L. SJÖBERG, 'Riskupplevelse' in L. SJÖBERG (ed.), *Risk och beslut: Individen inför samhällsriskerna*, Stockholm, Liber förlag, 1982, p. 89.

a situation of catastrophe ought to be assessed in a different way to other accidents. One such circumstance is the fact that the organic structures of society can be seriously damaged by some types of catastrophe. When individual people perish in accidents, it is those close to them who are mainly affected, especially their own families (apart from the victim himself of course). The complicated network constituting a society collapses locally around the person who perishes. This can of course have devastating consequences for those concerned. But seen in a larger perspective the damage is limited as immediately affecting only a very restricted part of society as a whole. The damages to various social organs caused by some catastrophes, on the other hand, are so extensive that these can scarcely be ignored in the context of our discussion. Assume, for example, that a tidal wave or a nuclear disaster should cause mass death and destruction in a country's capital city. Such damage could not be healed without significant setbacks being endured by the country as a whole. For far too much of the network constituting a society would have been destroyed to be able to escape such setbacks. The danger of catastrophes of such enormous extent should therefore be regarded as more serious than the risk of lesser accidents even if the statistical expectation of the number of victims might be equally high in both cases.

Uncertainty in the Assessment of Risk

In some cases a risk can be assessed with great precision. That is to say one can state with a high degree of reliability that the real risk lies within a narrowly defined interval. To play Russian roulette with a known number of bullets in the revolver is a perfect example of this. In other cases, by contrast, it is not possible definitely to assess the probability. There is then uncertainty in the evaluation of the risk itself. Such uncertainty can be called a second order risk, viz. the risk that the real risk is greater than supposed.

Is there any reason then to consider such second order risks when making decisions concerning risks? Is it the case that the fact that the real risk can be greater than has been calculated can be balanced by the fact that it might just as easily be less? Or is it the case, rather, that an increased uncertainty in risk-assessment makes a risk less acceptable than it otherwise would have been? The problem is particularly pressing in connection with the assessment of new technical systems of great complexity. It is often not possible to assess the accident risk in such systems with too great precision. The difficulty consists partly in the need to foresee all possible causes of accidents, partly in the need to assess the probability of simultaneous occurrences of several events collectively necessary for an accident. Since the systems are new there is, besides, only limited evidence from accidents or narrow escapes to work from.

Assume that we have two risk-filled activities which are both assessed as carrying a human fatality risk of p. Both activities are of equal value even in other respects apart from the fact that the uncertainty of the risk assessment is clearly greater in one case. Which of the two activities should be chosen in this situation? It should be clear that increased knowledge as to the level of risk obtaining ought not to lead to one's being less inclined to accept a given risk. On the contrary, it is natural to choose the alternative where the assessment of risk is surest. For, of course, it is more satisfying to base a decision upon tolerably certain knowledge than to build upon more or less arbitrary guesses. A reliable risk assessment implies, therefore, that a higher level of risk can be accepted than would be the case where the assessments were more uncertain.

How Should One Proceed?

How shall one then proceed in order to establish the acceptable risk levels in a given context? I would here like to connect to the Aristotelian tradition in moral philosophy. According to this tradition, being able to adopt a moral position as to how we ought to act in a given situation implies that, from the factual circumstances obtaining and the various consequences which can result from different alternative actions, we can determine what different values and what different duties carry most weight in that situation and should therefore guide our behaviour. Such a weighing up of a situation must be done intuitively. There do not seem to be any simple short cuts here.

Instead, as with cost benefit analyses, of calculating all relevant effects using one common measure, money, one has to resort, in the form of various descriptions of consequences, to giving details of the results of different alternative actions with respect to different kinds of values. To be able to decide whether the overall consequences of a certain activity are acceptable or not one must then weigh these different consequences against one another and relate them to alternative, comparable activities as well as to the available resources. Besides this one must take into account ethically relevant aspects such as, for example, the degree of voluntariness involved.

It was noted earlier that the less the resources society has at its disposal and the higher the risk level of other activities, the higher will be the risk level that can be accepted. There is therefore no kind of neutral or objective guiding value given once for all which can serve as point of departure for an analysis as to whether a certain risk is acceptable or not. The decision will to a high degree depend upon the context in which it is taken. This implies that a given risk judged acceptable in one particular context can very well be considered unacceptably high in another where the resources are greater and/or the risk-level is in general lower. The risk-level obtaining within similar activities in the society concerned is thus an important basis for determining whether a certain risk be acceptable or not. Such a comparison, however, is not sufficient. For there can be reason to question whether the value of human life is considered in a worthy way within these other activities. There can also be legitimate cause for judging it more important to limit certain kinds of risk in comparison with others. For they can differ in various relevant respects. As has been stated earlier, to be able to decide if a given level of risk is acceptable it is also important to consider other relevant factors. Is the risk voluntary or enforced? Is the activity concerned connected with any risk of some devastating catastrophe occurring? The degree of uncertainty in the assessment of the risk can also be relevant.

References

ASHBY, E., 'The Search for an Environmental Ethic', in *The Tanner Lectures of Human Values 1980*, Salt Lake City, University of Utah Press, 1980.
FRIED, C., *An Anatomy of Values*, Cambridge, MA, Harvard University Press, 1970.
LOCKE, J., *Two Treatises of Government*, P. LASLETT (ed.), Cambridge, Cambridge University Press, 1960.
NOZICK, R., *Anarchy, State, and Utopia*, New York, Basic Books, 1974.

SJÖBERG, L., 'Riskupplevelse' in L. SJÖBERG (ed.), *Risk och beslut: Individen inför samhällsriskerna*, Stockholm, Lennart Sjöberg Liber förlag, 1982.
VEATCH, R.M., 'Justice and Valuing Lives' in S.E. RHOADS (ed.), *Valuing Life: Public Policy Dilemmas*, Boulder, Colorado, Westview Press, 1980.

2.3.2

WHO WILL BEAR MORAL RESPONSIBILITY: ENGINEERS OR CORPORATIONS?[1]

José Luis Fernández Fernández

The purpose of this chapter is to answer the question that is the title of this paper in a balanced and impartial manner: both the organisation and the engineer share ethical responsibilities, although in a different and unequal manner. The theoretical basis of the aforementioned thesis is a kind of approach that will be appropriately established in this chapter. This, for example, is what should be understood as the 'ethical dimension' of technical problems, a special place to exercise the profession responsibly; the unexplained but omnipresent Weberian dichotomy between the 'ethics of beliefs' and the 'ethics of responsibility' (this dichotomy should be dialectically transcended in 'responsibility that is encouraged by and rooted in deep beliefs of conscience and/or responsible personal beliefs. In other words, mindful of the results and consequence of one's acts'). And finally, a sincere bet on the institutional-organisational level as an area of ethical display in which the engineer performs his work, together with other colleagues and other professionals. Seen from this perspective, lucidity and critical ability are prerequisites for ethically responsible and technically excellent professional work.

'Technocratism', 'Moralism' and 'Heroism'

In my view, there are three very common ways of improperly approaching the issue of the professional ethics of engineers in an organisation: *'technocratic ideology'*, *'moral'* and *'converting heroism into a myth'*.

---

[1] Original text translated by B. DECEUNINCK.

By *'technocratic ideology'* I mean the approach that attempts to immunise science, techniques and technology from moral questioning. According to this point of view, the engineer is simply a professional, an expert in resolving technical problems. By definition, there is no place for moral judgements or ethical values in these problems.

By *'moral'* I mean the version of ethics that proposes an inappropriate, empty and superficial morality, one that is full of common places and pious exhortations to do good, without taking sufficient account of the actual processes involved in professional practice, or the instrumental restrictions or concrete institutional mediations. As a result, although it often succeeds in articulating well-prepared and haughty discourses, in the end, it is highly simplistic and of little value when decisions need to be taken. Accordingly, the value of these approaches in professional ethics is not only highly questionable; at times, it is actually negative.

The *'conversion of heroism into a myth'* has become popular in public opinion, with disastrous and regrettable results, many of which are well known and documented, in which, in one form or another, engineering professionals are involved. Some of these successes have subsequently formed the basis of articles and books on engineering ethics, the border area of business ethics. (We must acknowledge that, in the end, both are only 'applications'; in other words, different 'species' of the same type: ethics). The case of the Ford Pinto and the no less famous space shuttle Challenger are some of the best-known examples. However, the casuistry could also be added to *ad nauseam* with European examples or with examples from anywhere else in the world.

Indeed, the fact that an engineer, in any of these 'cases' decided to risk his neck (or, at least, risk his job) for the sake of the social good, public health or safety of workers reveals an underlying assumption that *as a matter of principle and without further discussion* professional ethics always requires that sort of 'heroic' behaviour from the engineer. (This has been particularly true in the film industry.) It is obviously worth examining this in greater depth.

Briefly stated, none of the above-mentioned approaches is suitable or compatible with the *conceptual framework* from which I want to write this chapter. *'Technocratic ideology'* because it falls short, with its myopia (and 'self-interested') idea of what science, technical methods and the role of the professional engineer are. The *'moral'* approach is unsuitable because it goes too far and is guilty of the opposite flaw, attempting to go too far with empty statements whose morality is non-viable and ineffective. Finally, I find the *'conversion of heroism into a myth'* questionable in terms of its focus as well as its scope.

I find its focus unsuitable because I do not share the implicit belief that ethics should be understood more as a 'work of titans' than as a daily and habitual practice by ordinary people. I find its scope unsuitable because (apart from a certain negligence of the structural and institutional dimension that corporate ethics requires and conveys), this perspective reduces the engineer's 'ethical-professional agenda' to an excessive extent. By placing so much emphasis on the moral obligation or the lack of moral obligation of whistle blowing (the engineer denouncing the company's 'improper procedures') it neglects other types of obligations and values that are also important in a professional exercise that claims to adhere to ethical parameters, such as, for example, confidentiality or loyalty towards the organisation.

Without a doubt, things are more complex and varied, as we will see in the following pages, which, let us not forget, reflect an attempt to answer the following question: 'Who will bear moral responsibility: engineers or corporations?'

With this in mind, after having delineated in a negative manner the theoretical approach adopted in these pages, we must explicitly state the *conceptual framework* that we will use to address the issue. It is therefore necessary to clarify the thesis that, in my view, is implicitly affirmed in the three approaches criticised above and assumes in an uncritical manner an excessive dichotomisation and an unsuitable compartmentalisation of the technical methods and technology concerning ethics and moral philosophy. This is true, despite the fact that we must recognise that there is indeed an undeniable difference between what we could call *'technical problems'* and those other problems that can be catalogued under the heading of *'ethical problems'*.

The 'Ethical Dimension' of Technical Problems

The following, among others, are *'technical problems'*: 'What is the best way to divert the course of this large river?', 'How can we increase the storage capacity of the hard disk?', 'How should we design the car prototype in order to maximise comfort and speed, while taking account of economic and space restrictions that are required for the future production process?', 'Is it possible, here and now, to contain inflation, lower taxes and reduce the public deficit by two percentage points?', 'When is it necessary to remove a sick patient's appendix?', 'How should such an operation be performed?', 'Is there financial justification for investing so much money in this business project?', 'How many cubic meters of water can this dam support?'...

There are also many *'ethical problems':* hunger; war; the lack of respect for human rights; the demographic explosion; poverty and misery among large parts of the population; drugs; prostitution; lack of solidarity; injustice; ecology; abortion; the cloning of live beings and everything related to bioethics...

'Technical problems' (and in this case, technocracy, which, moreover, is different from 'technocratic ideology', is fully justified) require technical solutions. And it is improper and undesirable to mix things: technical problems require solutions of the same type. Naturally, those who are best able to provide such answers are professionals specialised in each of the distinct areas of knowledge: engineers, economists, doctors, etc. Moreover, 'excellence' in the exercise of the respective occupational activity, is based on the number and quality of the solutions that the professional in question is able to provide, given the technological restrictions and possibilities of each moment and place, regardless of their ethical and religious beliefs or their political options.

'Ethical problems' have a common factor, a family air that identifies them and can be summarised in the abstract and deeper question of *'what should I do?'* This question (we should highlight the verb 'to have to') does not refer only to the 'means' that we should use ('what can I do'?) or in what concrete forum we should use them in order to obtain the result ('which of the possible alternatives do I use to build the bridge?'); does not refer only to what is included within the limits of current legality ('what are we permitted and what are we prohibited from doing?'). Rather, its scope is much wider, because it aims at the ultimate ends and the great utopias of life in society: *'how should we organise ourselves in order to live a fully human life?'*

Nevertheless, just as it was easy to identify those 'professionals' who are supposedly able to resolve technical problems, reality is completely different with regard to ethical problems. Indeed, if, in this case, we ask ourselves who should provide the solution to this problem, we see that we are unable to find any 'expert' who could or who wishes to exclusively devote himself to resolving them. However, given the existence of the said problems and the fact that they are not specific to anyone, they are everyone's problem; and, in a way, we are all 'responsible' for their solution. In this connection, being aware of the magnitude of the problem and the fact that we all feel concerned by this true deficit of humanity is a timid step toward their possible solution.

However, the reality from which we must start is not, in any way, sufficiently defined with the distinction that we just made (which, moreover, is highly artificial and somewhat simplistic) between *'technical problems'* and *'ethical problems'*. We need to introduce a new element into the analysis, that, by integrating both consequences (technical and ethical) in a unitary and more complex whole, can be more fruitful in terms of appropriately addressing professional ethics and therefore, the engineer's moral responsibility. This naturally leads us to what I like to describe as the field of *'the ethical dimension of technical problems'*.

In order to understand what we are speaking about, we would need to stress how the diversion of the large river's course from the aforementioned example could have concomitant effects, which could be positive or negative, in the interests of certain groups. It could also affect the life of people and the ecological equilibrium (for example, in the case of the floods that occurred in China in the spring of 1998). Needless to say, these matters are not 'technical problems' *per se*; they appear during the implementation of a series of 'technical' actions... As a result, they are an empirical conclusion of what should be understood as 'the ethical dimension of the technical problem of diverting the course of a large river' and, as such, *they must be addressed and resolved* as and when technical and economic decisions are taken, if not before. Failure to do so, in other words, to repudiate these issues, has been (and, unfortunately, continues to be) the cause of many disasters that could have been avoided, thus saving the high-price paid in human suffering and in economic resources. This is because this 'ethical dimension of technical problems', which sometimes lacks importance, on other occasions is truly dramatic.

Between What Is Desirable and What Is Possible

We should never forget that techniques, left free to run their course, are not 'neutral', but instead 'blind' to the ultimate objectives and final goals. As a result, the stakes of our actions are high for us if we do not to take this matter seriously.

In this connection, I would like to make two statements that, although they are obvious, are not always taken into account, as they should be, in theoretical debates or in practical decision-making:

1. that not everything related to techniques (including ethically) which is considered desirable at a given time in history can be immediately institutionalised in social reality;
2. and that not everything that is technically feasible at a given moment in time is in itself ethically useful.

The first statement does not create major problems: it has to do with the pace of technical progress (which varies, but is unstoppable) and with the 'vectorial', open and utopian structure of ethics, which, aspiring to what is *'good'* will never achieve it completely, as, by definition, it is *an unattainable predicate.*

The second statement is much deeper and important, especially nowadays, when technological possibilities have exploded, while the results to which said possibilities could lead us are, at the very least, uncertain and, in the worst of cases, could even be counterproductive.

In line with what has been said from this perspective, it is worth making several comments concerning professional ethics in general and engineers' professional ethics in particular. To begin with, beyond the ethical problems that we all share (as human beings affected by the above-mentioned humanity deficit) *there is a specific area, specific to the various professions, to which moral reflection must also be 'applied' without delay.* Secondly, it is not necessary to make huge theoretical efforts to identify this area of problem, since they appear to take the form of *another dimension (and not necessarily the least important) of the customary tasks of professionals.* Finally, *the 'moral responsibility' in the approach to and the solution to this type of problem tends to be amply scattered and is almost always highly dispersed throughout the organisation* in which the practical actions and technical decisions are implemented. As a result, although it will always be necessary to have the assistance of professionals (in this case, engineers) in order to find adequate solutions to these problems, we should not think that they have to be the only valid contacts for this purpose.

We will resume this argument later on in this chapter. For the time being, given that we have just come across the concept of *'moral responsibility'*, we should take some time to clarify it and to briefly describe the structures and the conditions of possibilities that make it possible to assign 'responsibilities' of this type to someone (whether an individual or an organisation).

To speak in this context of 'responsibility' means that we must begin by clarifying the anthropological meaning of the 'moral fact'; in other words, the particular way of being of human beings and that, since ancient times and throughout the history of thought, has defined it as an 'ethical subject', 'being structurally moral', 'working on oneself', or using other, similar expressions, which tend to stress the same particular way of being, which is specific and exclusive to human beings.

This is because it seems quite clear that the 'programming' of our behaviour is qualitatively different from that of other living beings. While animals have schemes that are rigidly pre-determined by nature (through instincts) that 'enable them to behave appropriately' in every situation, we human beings are in another, very different, situation, as our vital structure is more open and undefined. It is clear that we also have instincts and impulses that are genetically determined by nature. However, we cannot say that this component 'conditions' our behaviour in a manner that is as rigid and inflexible as other living beings.

According to the Spanish philosopher Xabier Zubiri, all of the above seems to have to do with a complex evolutionary process, whose final result is what he called the 'hyper-formalisation' observable in the humanised hominid (and not before) and the appearance of 'intelligence through feeling', specific to human beings. Equipped in this way, we human beings can innovate and be creative in our behaviour, going beyond 'animal', mechanical and standardised responses to stimuli received from our surroundings.

However, if this is the case, we are faced with a paradoxical and challenging situation, a genuine 'condition of possibility' of the unavoidable moral dimension of human life. Animals do not reflect, they simply act. Human beings, on the other hand, do reflect. Based on the fragmentary elements that appear to us in every situation, we are able to anticipate a future scenario and to project the consequences of our actions. We are able to 'discount' predictable results, and to calculate the pros and cons. Finally, when we act, we are never completely sure that we are doing the 'right' thing.

As the reader can see, our life is much more difficult but also much richer. Since nature does not tell us 'clearly' how we should behave, we have to 'seek out life'. The various cultures and proposals of moral canons are merely answers and adaptations, of varying complexity and validity, to the specific human condition.

A human being is not born already made and complete; they go through a process of self-creation. The agent of this process is man himself; its means is none other than human action itself (what the Greeks called 'the praxis'); the goal of which is to attain perfection ('excellence'), which is always described as unattainable, utopian and asymptotic.

As a result of all of the above, it is said that man, by 'doing things' is also 'creating himself'. Nevertheless, just as one can get things right and, by doing good things, it is also possible to do well ('good') for oneself; there is always the possibility that the human beings will lose their way or take a less suitable path and, by doing bad things, hurt oneself ('bad'). As such, it is relevant to investigate and discover how we should behave in order to live human life well and make ourselves 'good'.

This, in essence, is the mission of ethics, understood as moral philosophy (Kant's famous *'what should I do?'*). And this, in turn, is the most important task in which each and everyone of us, as human beings, are involved throughout our life: acting based on 'virtue', 'regularly' putting into practice that which ethical theory says is honourable and good. By doing so, we are told that we will succeed in creating in ourselves a good 'character' (a kind of second nature), thereby achieving 'the good life' for ourselves and entering the path to achieving the 'good' above which there is no other greater good *in this world*; i.e. 'full personal realisation', the complete attainment of our potential in a 'successful life', and genuine 'happiness' (or, to express it in classical terms: 'eudemonism').

How Can We Explain the Existence of Moral Responsibility?

The existence of moral responsibility is a fact that cannot be denied – at least according to many lines of thinking in traditional

philosophy which only makes sense, at least, based on the existence of the two following conditions of possibility:
1. that human beings have the ability to 'prefer' one thing to another, as well as the ability to choose, to the extent possible, based on the said preference; and that
2. human beings are not only able to 'justify' their preferred options, they are also able, in passing, to articulate these justifications in action plans that can be used as a model for subsequent, similar occasions.

These two requirements form the central core of what should be understood as the concept of 'moral responsibility'; and point to what has traditionally been studied under the headings of 'freedom' and 'conscience' in the specialised literature. These are problems and issues that we cannot examine thoroughly here. However, it is appropriate to point out that an inappropriate understanding of the phenomenon of freedom – either because it exaggerates what is possible from a kind of 'fantasy of omnipotence' or because it does not offer any margin for manoeuvre in terms of action, from a comfortable and comforting 'feeling of powerlessness' – can distort the analysis of problems and prevent them from being addressed in an appropriate manner.

'Ethical knowledge', 'moral conscience' and so-called 'practical reason' seek to grasp and propose possible objectives useful in themselves, for the subject's action. The basic form is the conscience that calls for action along a determined line, the conscience that prohibits acting in another way, the conscience that requires the application of a given principle of action. As such, it is a 'subjective standard of morality', without which ethics loses its foundation and reason for being. Nevertheless, the said standard, although it is absolutely necessary, is in no way completely sufficient and must be complemented by and rooted in the moral contents that take the form of principles, values and the 'objective standards of morality'.

It is clear that conscience, as the centre of subjectivity, conforms to the theme of the distinct circumstances and vital experiences of the individual. However, at the same time, it is what makes possible each person's specific personal experience. As a result, it is a structure that makes it possible to experience the circumstances, as and when these develop. Re-elaboration, precariousness and lack of absolute certainty about getting things right are a part of the essence of conscience (and as a result, makes possible dogmatism, pluralism, tolerance and even error, which we see in all spheres of what is moral).

In order to speak based on 'moral responsibility', we also need *a certain margin of manoeuvre, with a certain degree of freedom* when taking what we believe are the best decisions ('with a clear conscience'). Although this requirement seems obvious in view of the personal experience of each person, this is relatively often distorted in the analysis, whether to defend the position that denies that freedom is critical to action, or to affirm the presence of a quasi-absolute freedom in every individual or organisational action. Nevertheless, given that none of these extreme approaches adequately explains the phenomenon of freedom when decisions are made, they should be rejected openly, especially if we consider the practical consequences that can result from them and those that, in another place I called 'the fantasy of omnipotence' vis-à-vis the 'feeling of powerlessness'. Moreover, both can serve as an intellectual alibi to legitimise very different points of view regarding morality.

'The fantasy of omnipotence' appeals to both 'heroic deeds' and a 'moral'. This is because it considers that the individual subject (in our case, the engineer) is capable by itself (*'and if he or she can, he or she must'*) of successfully carrying out any action. Nevertheless, this focus, which underrates the existence of the conditionings as determining the action, reflects a complete lack of reality and an extreme ignorance of the formal paths of life within organisations.

At the opposite extreme, 'the feeling of powerlessness', whenever it is necessary to excuse particular responsibilities given objectively undesirable events: 'we did not have any alternative'; 'we cannot change things'; 'I have to do what I am told and try to reach the objectives that are set for me'; 'I complied with my obligation...'

However, decision-making in real life is far removed from both the 'candour' of the first position and the 'cynicism' of the second. This is because there are always levels of free action, together with conditioning of varying rigidity. In addition, this freedom (relative and precarious) is unequally distributed among the various agents and the various organisations.

As I said above, *'conscience' and 'freedom' are premises of modern life* of both people and *companies and organisations*. As a result, 'moral responsibility' to which individuals and organisations may be tributary is, *above all*, based on the respective abilities to act freely. I stress 'above all' because, in addition to this requirement, we must also mention the requirement that springs from the very mission sought by the professional as representative of a 'social role'; in other words, of his particular function in the organisation and in the broader social context.

I am one of those people who think that the question should be answered in the affirmative: organisations too (whether or not in a similar manner) have to justify the decisions that they take (which cannot be reduced to the mere sum of decision-makers taken individually) by reference to a series of values, goals and objectives. We also see that as they act they shape their 'corporate culture', their special character, the 'way of being acquired' which, as is the case with individuals, sets them apart, gives them an identity and enables the environment to act accordingly, based on the economic and moral desirability of their acts.

Engineering as a Profession

The term 'engineer' means so many things and covers a semantic field that is so large that it can even be ambiguous. There are those who wonder whether or not engineering is a 'genuine profession'. Those who are reluctant to consider it in this light are very stubborn, in that they use an extremely rigid definition of the concept that, with a significant effort could be extrapolated beyond the occupations carried out by *ut sic professions*: doctors, lawyers, priests... Nevertheless, if we change the focus and if we broaden the angle of vision, we will agree fairly easily that engineers are genuine professionals; not only due to the type of work that they do, but above all, due to a series of 'attitudes' with regard to how they do it and how they think of themselves.

Assuming the above, and without going into greater detail, let us say that from a sociological perspective, a profession, like a 'social practice', could be loosely defined in the following terms: 'the occupational activity of a group of people, organised in a stable manner; that claims exclusivity of competence (acquired as a result of a long process of theoretical-practical training); that shares a series of specialised knowledge that is of interest to society and that they make available to the latter, charging compensation for their work, thereby earning their living'.

However, if we want to consider the profession of engineer based on the above definition, we would have to do more or less the following: first, we would have to find a common denominator for all 'engineers' in order to identify the goal, *what is 'the good'* that engineering supposedly attempts to do. At the same time, we would have to make clear the added value that engineers contribute and that no other group is able to offer society with the same levels of excellence. If we define in this way what engineers 'do', this will facilitate our efforts to define 'what they should do' (objectives, values and positive proposals), and 'what they should avoid doing' (prohibitions, restrictions and limitations). By proceeding in this manner we are taking a necessary (although insufficient) step to define the scope of their ethical responsibilities.

The Engineer's Moral Responsibility
within the Organisational Context

In companies, the engineer occupies a specific position that involves a series of rules or practices (the description of the position) that are a kind of implicit contract between the professional and the organisation. The said 'implicit contract' is the result of a series of obligations and responsibilities: many of these are explicit; others are customs, uses, implicit expectations (a certain style when doing things)... There are also minimum standards of excellence, that reflect the state of the art at each moment, based on which it is possible to articulate a series of ethical standards and an ideal model of what could described using the concept of 'good engineer', as the person who complies in a satisfactory manner with the implicit and explicit obligations inherent to his professional role.

The following 'direct functions' of the engineer tend to be those referred to in most Codes of Ethics for Engineers: design, manufacture and maintenance of equipment, products, processes and systems or technical services. The most commonly mentioned support functions are consulting, inspection and monitoring that result from the specific professional training of engineers and that tend to be based on the drafting of highly varied technical reports.

As the reader can see, based on these types of efforts directed at self-comprehension, it is possible to arrive at a fairly clear definition of some of the competencies that tend to be attributed more or less exclusively to engineers (and those that are not); what the 'critical points' could be with those that these tend to encounter in carrying out their professional activity; which ethical values are most often violated in the performance of their functions; and finally: what moral responsibility can we attribute to them, in general terms, given that *in addition to 'engineering professionals', they tend to be 'employees of a company'* and, accordingly, perform their services, not in the manner of the 'self employed' decades ago, but rather, within the context of an organisation.

This framework, which has its own dynamic and its own particular objectives (essentially economic), and the fact that the engineer (in the vast majority of cases) is an engineer and works as one within an organisation, results in a series of potential conflicts, as well as a series of moral responsibilities, such as confidentiality and institutional loyalty. This is because, in reality, perhaps one of the features that is most often considered to be understood in all the codes and documents of this type to which I have had access is the fact that, in most cases, the engineer performs his work within an organisation, inside a company structure.

As a result, since they are immersed in organisations (with the exception of those who work as civil servants or those who operate as 'self-employed entrepreneurs'), engineers are not very different from other groups. As a result, although *the moral responsibility of engineers in the decisions that they take is maintained with full effect* (especially since a mistake made by an engineer can have consequences that are more serious that other bad practices in other professions), *it is modified in any way*. We will now examine this point in greater depth.

Is the Engineer Responsible for Deciding
which Risks Can Be Assumed?

There is a starting point that is regularly recognised by engineers regarding the impact that their acts can have on the environment. This recognition at the same time stresses the fact that engineering can involve significant risks for the collective well being or for the health and safety of people.

At this point it is not difficult to affirm that one of the most basic moral obligations for an engineer may be the ability to *understand the ethical dimension of his technical actions;* based on the 'anticipation' of the foreseeable consequences of his actions. If these actions involve higher levels of risk to the health or integrity of people or the environment, they should not be carried out without minimising them as much as possible, or without the prior 'informed consent' of those who could be affected by these actions. In this sense, engineers are the best placed to provide this type of technical information, which concerns the risks and dangers that a process, product, system or concrete experiment can involve, to the public and to the ultimate decision-makers.

Nevertheless, assuming the above, and in line with what we have already stated, we should make a series of clarifications in order to conclude this chapter.

Above all, we should not forget that in a company, the role of engineer (however important this is) is one of many (also important); and that, as a result, the engineer is not the only or perhaps the most important link in the decision-making chain... Because of this, I believe that those approaches that assign engineers a role that is close to that of 'supervisor' of decisions adopted by the organisation's managers are unfortunate. For example, this is the function of internal or external auditors; or another type of professional, who is specifically responsible for attending to inspection and quality. These people are particularly well placed to perform these functions.

Clearly, the company's management, before taking a firm decision regarding a policy or implementing a given procedure must obtain the necessary information (the more strategically important the plan in question, the greater the need to obtain the information). As such, they need to attend to several factors (both historical and economic) and must collect the opinions of many professionals who work inside and outside the organisation (financial analysts, experts in marketing, engineers, etc.). Nevertheless, the ultimate decision is in the hands of the senior executives. It is not in the hands of the aforementioned professionals, nor, of course, is it in the hands of the engineers... In short, as engineers are not paid to manage, they do not have the moral responsibility to do it.

As a result, an engineer can essentially comply with his obligations by producing good work and by providing the company's management with suitable technical reports regarding the implications and risks inherent and foreseeable in the taking of the decisions under consideration. Nevertheless, *in principle,* the engineer is not competent to decide whether or not to assume these risks. There is no reason why the latter, who is actually an expert in accurately identifying and quantifying the risks and costs, is particularly qualified to decide what risks to assume and to what extent, and which risks not to assume. It is necessary to distinguish between the *descriptive* aspect (objective and quantifiable) of the risk and the consequent *normative* aspect (the decision concerning its acceptability). Engineers are extremely competent for the former, but not for the latter.

Needless to say, engineers have a moral responsibility to withstand vigorously anyone who would like to force them to act improperly (fortunately, a highly unlikely event), using their technical knowledge to openly attempt to harm others. In this case, and this applies to any other profession, the principle of *'non-maleficence'*, which has been emphasised since the time of Hypocrates with *'primum non nocere'* ('above all, do not cause harm gratuitously').

In addition, engineers are morally responsible to attempt to do what is good in a positive manner (*'principle of beneficence'*) to the best of their ability, optimising the final results and means used, making use of the 'state of the art' and the concrete circumstances under which the engineer's work is being performed. 'Optimisation', a popular concept in the profession, provides all of the ethical meaning to this technical requirement and involves a pledge in favour of a development that is compatible with its efficiency and the fight against the pilfering of scarce resources and non-renewable energy.

In view of the above, I think that the key to the vault of the entire ethical structure can only consist of trying to maintain three parameters in equilibrium: the pole of *values, personal beliefs* and conscience; the scope of the professional, of *obligations* and *the 'good'* pursued in the practice of engineering; and the institutional-organisational context based on which the technical action must be carried out in a *responsible* manner.

Divided Loyalties and Conflicting Duties

It may be assumed that this equilibrium is not automatic and it is not easy to maintain, whenever one is attempting to be faithful to interests and values that may conflict: personal coherence, the pursuit of the common good and institutional loyalty, etc. It is therefore critical to give a formula that can be applied automatically to resolve *a priori* the innumerable casuistry that we may assume results from the possible conflict between the three poles. Rather, this is something that each person must resolve in each concrete case. The *beliefs* of the conscience itself must serve as a guide throughout the process. However, neither the *ethical standards* emanating from the professional group to which one belongs (the often abused 'ethical codes'), nor, above all, those that can be articulated based on the structures of the organisation itself, which assumes *responsibility for the consequences of its actions,* are unnecessary.

An extreme and paradigmatic example of this is the above-mentioned problem of whistle blowing. There are times when, based on their personal *beliefs*, engineers are amply justified (*and there are times when they are required*) to report his own company for acting against the public good or for seriously endangering human health. This is true in the following circumstances:
1. the engineer is the last recourse and has already exhausted all internal possibilities;
2. in the case of serious, systematic and permanent harm (something that is not incidental or of little importance);
3. it affects third parties that are not sufficiently informed of the risks to which they may be exposed;
4. the engineer has positive and documentary evidence (he or she can prove his claims in court);
5. there is evidence that the engineer will avoid the harm in question by revealing this information.

Assuming a well formed personal conscience with clear guidelines regarding what 'should be' the work of an engineer (something that professional groups do), the only thing that is needed in order never to reach such extreme situations is for the organisational structures to promote ethical behaviour and see to it that immoral acts are not perpetrated. This requires taking ethics seriously. It means avoiding the obsession of 'short-termism' and taking the time to analyse policies, practices, the real *modus operandi*, incentives, controls... In short, it requires a genuine 'cultural' analysis of the ethical climate 'in' the organisation and 'from' the organisation outwards. This is because *acting morally requires people of high moral standing and solid ethical principles, as well as (and above all) the moralisation of structures.*

The first, improving personal quality, is the individual, non-transferable responsibility of each of us. Preparing the company to introduce the 'ethical moment' in their way of working (designing increasingly human and humanising organisational structures) should be carried out voluntarily and spontaneously by the companies themselves and the industrial sectors. Indeed, we could cite some good examples of 'self-regulation'. Nevertheless, in the more than likely case that companies and industries do not satisfy this type of requirement suitably, and in the event that they actually act against the common good, the public authorities, *subsidiarily*, are then authorised and required to pass legislation that is as strict as possible against them, in order to force them to take *responsibility* for the consequences of their actions, as well as for their failure to act.

Conclusion: Between Beliefs and Responsibilities

Consciously or unconsciously, whenever we speak of 'moral responsibility', we are implicitly alluding to Max Weber's well-known distinction between the *'ethics of beliefs'* and the *'ethics of responsibility'*. Naturally, we do not wish to suggest that the 'ethics of beliefs' of conscience must be identical to a lack of responsibility or that the 'ethics of responsibility' should be put on the same level as a lack of beliefs of conscience. This is not the case. The point here is simply to call attention to the difference between acting in one direction by any of those moral maxims. In addition, in order to implement any minimally sensible and realistic approach in professional ethics, it is absolutely crucial to understand the message that lies at the heart of that dichotomy.

Taking an extreme situation, someone who acts based solely on principles and beliefs will have a tendency to adhere to rigid and non-negotiable approaches, to emphasise the existence of duties and obligations that must be fulfilled at all times and in all places, despite the fact that the final results of the action would have to be more or less desirable. In this case we are referring to a new *ethical* version of the well-known Latin saying that asks 'that the law be complied with, even if it destroys the world' (*fiat ius pereat mundus*).

On the contrary (and this is also an extreme case), to act based solely on the 'ethics of responsibility' would require one not to hesitate in putting aside, if this were the case, the principles and dictates of the conscience itself, and adapt, in a somewhat 'betraying' manner, his personal values to the circumstances and to the requirements of an ordinary *'consequentialism'*, using for this purpose means that are not fully compatible with personal beliefs.

In line with all that we have said in this chapter and after having explained the above distinction, we can now answer the question posed at the beginning of this chapter (*Who will bear moral responsibility: engineers or corporations?*) stating the following conclusive thesis as a corollary:

1. Raising the ethical tone of professional action requires 'good people', as well as 'good organisational structures'. Therefore:

1.1. Both the engineer (as a professional and as an individual) and the organisation (as a legal entity) can be charged with the consequences of their actions. As such, there is a *shared responsibility*, although this is clearly *unequal*.

1.2. Both the engineer and the company must *dialectically go beyond the possible dichotomy existing between the 'ethics of beliefs' and the 'ethics of responsibility'*, by articulating an approach that takes account of the need to act based on 'responsibility' that is rooted in clear values and in deep beliefs of conscience, and is based on personal 'beliefs' and organisational values that take account of results and are concerned about the consequences of the acts themselves, as well as the failure to act.

References

APPELBAUM, D. and S.V. LAWTON, *Ethics and the Professions*, Englewood Cliffs, NJ, Prentice Hall, 1990.

DIDIER, C., A. GIREAU-GENEAUX and B. HÉRIARD DUBREUIL, (eds.), *Éthique industrielle. Textes pour un débat*, Brussels, De Boeck Université, 1998.

DOU, A. (ed.), *Evaluación social de la ciencia y de la técnica. Análisis de tendencias*, Madrid, Universidad Pontificia Comillas, 1996.

ESCOLÁ GIL, R., *Deontología para ingenieros*, Pamplona, Eunsa, 1987.

FERNÁNDEZ FERNÁNDEZ, J.L. and A. HORTAL ALONSO (eds.), *Ética de las profesiones*, Madrid, Universidad Pontificia Comillas, 1994.

FERNÁNDEZ FERNÁNDEZ, J.L., *Ética para empresarios y directivos*, 2nd ed. Madrid, Esic, 1996.

JOHNSON, D., *Ethical Issues in Engineering*, Englewood Cliffs, NJ, Prentice Hall, 1991.

MITCHAM, C., *Engineering Ethics Throughout the World: Introduction, Documentation, Commentary and Bibliography*, STS Press, Pennsylvania State University, 1992.

WEBER, M., *La ciencia como profesión*, Madrid, Espasa Calpe, 1992.

3

TECHNOLOGICAL DEVELOPMENT AS A SOCIETAL ISSUE

INTRODUCTION

Göran Collste

Technological development is often described as a predetermined process out of control of human beings. This image is a myth, and indeed a myth that has an ideological function, as this part explains. It goes on to argue that technology is the result of many choices. In the social construction of technology many actors are involved. They often have different interests and different power resources. Thus, technology is neither predetermined nor something that is shaped in isolation separate from society, but rather it mirrors the social conflicts.

A prevailing theme of this part is how to understand technological development as a societal issue. What factors determine the development and implementation of technology? How can citizens, interest groups, consumers, i.e. the users of technology influence the technological development? What is the role of engineers? These are some of the questions raised in this part.

A Different Kind of Technology?

Obviously enough, modern societies are characterised by and, to a large extent, determined by advanced technology. The modern, high-technology society, i.e. the kind of society which has developed since the 1910-20s in Europe, the USA and Japan and since the 1960s and 70s, has been conquering the rest of the world, has several characteristic features. One such feature is its *large scale*. A large proportion of production and trade takes place in huge multinational companies, and services like transportation or medical care are provided in large-scale systems.

Connected to the large-scale character is another feature, the *powerfulness* of the new modes of production. There are now technical means to control nature in ways unbelievable just a generation ago. 'Nature as a human responsibility is surely a *novum* to be pondered in ethical theory', writes Hans Jonas[1].

Another feature of modern society is its *complexity*. Today's society is more complex and complicated than earlier societies. The prime example of this are the devices that are used. These are becoming more and more complicated, and ever more advanced levels of education are needed in order to carry on a profession in this society.

These characteristic features – large scale, powerfulness and complexity – have consequences which are very significant in the area of ethics. We can summarise the consequences using the phrases 'difficult to survey' 'difficult to manage' and 'difficult to understand'.

Do these features of large scale, powerfulness and complexity imply that modern technology is in a different category than pre-modern technology, Peter Davies asks in the chapter *Managing Technology: Some Ethical Considerations for Professional Engineers*. He contests what he considers as three myths concerning our attitudes towards technology. The first myth is that modern technology is just more of the same, the second that technology is neutral and the third that technology is progress. Citing the qualitative new aspects of modern technology, he argues that a 'technocentric' attitude has to give way to new imperatives. He would like to encourage engineers to generate a deeper understanding about the underlying philosophical assumptions of their work. To put these into practice, he asks several questions: 'What does technology free us for?', 'What evidence do we really have that we are able to control technology?', 'What role does technology have in human evolution?' .

New technical systems solve problems and make human life easier in several ways. It is because of these intended effects that there is a market for a system. However, little by little, certain unwanted side effects also start emerging. A problematic feature of the modern high-technological society is that it takes time before unintended side effects of technology become evident. This is, perhaps, most obvious for man-induced climate change. The difficulty to overcome uncertainties and the need to take the scientist's normative presuppositions into consideration to understand the debate on climate change is discussed in the chapter by H.J.M. de Vries, *Objective Science? The Case of Climate Change Models*.

---

1 H. JONAS, *The Imperative of Responsibility*, Chicago/London, The University of Chicago Press, 1984, p. 7.

The new biotechnology provides another illustrative example of the problems encountered in surveying the consequences of new technology and reach unanimous decisions on what to do. As is shown by Philippe Goujon in the chapter *The Case of rDNA Techniques: Ethics and the Complexity of Decision-Making*, ethical aspects were present at the beginning of the discussions in decision-making units, like the European Union, but later on economical and political considerations became decisive for the decisions.

Constructive Technological Assessment (CTA) is a means of enabling public participation in the development and implementation of new technology. However, as Dominique Dieng shows in the chapter *The Hemacard Project: Applying the Constructive Technology Assessment Method to Computerised Health Cards*, CTA must be introduced at the very start of a project in order to influence the system development.

What is the role of the engineers in system development? Do they have any means of influencing technology? Do they want to be actively involved in the decisions? In the chapter *Engineers and the Dialogue on Extendinging Their Horizon of Action*, Eva Senghaas-Knobloch refers to the results of intense dialogues with engineers practising in different areas. Some of the engineers see themselves as victims without any real influence. Others do not care but see themselves as technological heroes. However, some also want to see their task in a wider perspective. One of them says, 'I am a technician but also a father'.

Is the New Technology Possible to Control?

People's freedom and desire to control the future is a precondition of ethics. If we are subject to forces that cannot be influenced, there is no reason to make ethical evaluations and deliberations. A question is whether we today have reached a point at which we are subject to the innate power of the technological system. It is difficult to influence the development at the time when a system is becoming established. The unwanted side effects of a system are sometimes noticed too late, once society has already become dependent on it. By then, we have become locked into a situation from which we cannot escape and there are strong forces behind it. Major investments are required in order for a technological system to be developed and introduced. This means investment in knowledge. Many researchers have devoted themselves, for example, to nuclear research and genetic research, and nuclear power and genetic engineering are the results of their research efforts. Technological systems also require considerable economic investments. Large sums have been invested in the car industry and road building, the nuclear power industry, the pharmaceutical industry and the computer companies, just to take a few examples.

Since such large amounts of resources in both knowledge and money have been invested, there are also powerful interests behind the established technological systems. The researchers and technicians want to continue their work, and those who have put money into the system want to obtain some return on their investments.

Technological change has in various ways impact on society. The limitations of the established neo-classical paradigm in economics to take this impact into consideration is the theme of Luc Soete's article: *The Role of Technological Development in the Economic Process*. The challenge in front of us is, according to Soete, to see technology as an endogenous process, whereby technological change is adapted as a resource for a more human and environmental friendly society.

To take care of the environment, Raoul Weiler makes several proposals in the chapter *Sustainability: A Vision for a New Technical Society?* He says that governments have to promote sustainable development, which could be obtained by reducing the resources used. He also states that this is not incompatible with an increased standard of living.

Who Accepts Responsibility?

The feeling that both society and technology are developing out of control is reinforced by globalisation. As Riccardo Petrella shows in *Globalisation and Ethical Commitment*, the rhetoric of globalisation tells us to 'adapt' to a world of privatisation and competitiveness. As a result it has become even more important, Petrella pleads, that the primacy of politics over finance is restored.

Even if they are difficult to control, there are no hidden, mystical forces behind the introduction of new technologies – they are the result of human intentions and decisions. These decisions have major consequences for people and the natural environment, and they therefore involve ethics.

There are, however, important differences between taking a position on the type of moral problem that we encounter in everyday life, i.e. when we have to make a moral decision as an individual, and the type of decision which contributes to the introduction and development of technological systems. One difference is the far-reaching consequences of the technical systems that make them difficult to survey. Another is that the technological system is the result of a long series of decisions, where many people are involved and influence the outcome. It is therefore usually impossible to pick out any one individual as ultimately responsible.

For example, the discovery made by Watson and Crick in 1953 opened the way for the development of genetic engineering. Can they therefore be held responsible for how their technology is applied today, almost 50 years later? No, not really. They could not even guess at the consequences that would result from their discovery.

To take another example, who was responsible for the expansion of nuclear power? Many technicians were involved in the preparation and construction, and they are, therefore, responsible for the consequences of their efforts. Politicians made the decision to develop nuclear power. They are in turn responsible for the consequences of their decisions. The problem is that when many different people are involved in the decision-making chain, no one takes responsibility – shared responsibility often means no responsibility!

In his article *Globalisation from an Ethical Perspective,* Göran Collste points at the far-reaching consequences of our actions and the need for a global ethic. According to Collste, a universal ethic is possible and indeed necessary, and he formulates some ethical principles that could be used as the basis for a shared global morality.

To go in that direction, the chapters in this part illustrate a need for the formation of cultural and ethical *fora* for dialogue and reflection in a technological society. To prevent technology, in the absence of countervailing forces from setting its own goals, groups like active citizens, consumers, labour unionists and students should join with engineers in discussions on the goals of the global, technological society.

3.1 DESCRIPTION

3.1.1

THE ROLE OF TECHNOLOGICAL DEVELOPMENT IN THE ECONOMIC PROCESS

Luc Soete and Berit Schneider

The field of the economics of technological change has often been described as the 'science' which restricts itself to the positive welfare increasing effects of technological 'progress', implied by the allocation principles of the market mechanism. This old traditional view is now increasingly being brought into question because of fundamental issues concerning the measurement of economic growth and more broadly welfare. This article illustrates the recent attempt of economics to explain the determinants of technological change in an endogenous fashion partly induced by the observation that societal effects in technological change play an increasingly important role. It also examines the implications of the endogenisation of technological development for government policy. It shows that whereas support policies based on traditional economic analysis have tended to create a supply bias of new technologies not adapted to social need; the beneficial effects of technological advance depend on its integration into society. It concludes with the dynamic character of technology assessment that arises from a more coherent picture of the interrelation between technology and society.

Introduction

In revenge for the theft of fire from heaven, it is told, Zeus commanded Hephaestus to make a woman from soil whose beauty would bring misery onto the human race. Hermes gave her gallantry and acuteness, Aphrodite gave her beauty and the gods named her

Pandora: full of poison. When Epimetheus, brother of Prometheus, made Pandora his wife, he received as a gift a jar containing all human ills. Unfortunately, the inevitable happened: the inquiring Pandora opened the jar and all those ills spread over earth. Only 'hope' could not escape the bottle, before it was impulsively shut again. Hope for a better future that makes life worth living for.

The story of Pandora, which enjoys some popularity in scientific philosophy, keeps us since Greek antiquity aware of the fact that, since the 'poison' of knowledge (the theft of fire), the human race has taken over control of its own fate; that 'science' and acquiring knowledge are no exogenous, alien facts, but that they are formed by and made applicable to society; in simple accounting terms, that while the welfare increasing effects of the use of science and technology might be put on the credit side of the balance sheet of humanity, the destructive wars, the scientific and technological errors and accidents of the Bhopal's or Tjernobyl, not to mention the environmental damages, have to be put on the debit side. While current scientific and technological activities are likely to deliver answers and solutions to many of these present-day problems, they are likely to lead to the crystallisation of new problems and questions, including increasingly questions about ethics and the right to intervene in life itself.

The field of the economics of technological change has often been described as the 'science' which restricts itself to the positive welfare increasing effects of technological 'progress', implied by the allocation principles of the market mechanism. This old traditional view is now increasingly being questioned. It raises, as argued in *section 1*, fundamental issues about the measurement of economic growth and more broadly welfare. *Section 2* illustrates the recent attempt of economics to explain the determinants of technological change in an endogenous fashion partly induced by the observation that societal effects in technological change play an increasingly important role. Section 3 and 4 are dedicated to the implications of the endogenisation of technological development for government policy. *Section 3* shows that support policies based on traditional economic analysis tend to create a supply bias of new technologies that are not adapted to social needs, and, therefore, how beneficial effects of technological advance depend on its integration into society. *Section 4* deals with the dynamic character of technology assessment that arises from a more coherent picture of the interrelation between technology and society.

•

If the technological 'black-box' be opened, a multidisciplinary approach to the issue of technological development and innovation is clearly inevitable. As Nathan Rosenberg states it: 'With apologies to Clemenceau it might be said that if technological change is not too important a subject to be left to the economist, it certainly is too diverse a subject to be left to the economist who refuses to step across narrow disciplinary boundaries'.[1]

Economics and Technological 'Progress'

Traditionally, that is since economics adopted its marginalistic approach identified with Alfred Marschall and the 'neo-classical' approach became dominant, economic analysis has reduced technology to an 'exogenous', external factor. The impact of such a given factor on e.g. economic growth can be best described – just like in the case of population growth – in terms of a particular parametrical value: a 'black box, not to be opened except by scientists and engineers'.[2] This awareness about the limitations of the contribution of economics to society would be laudable, if it did not also imply a particular economic vision and interpretation of the contribution of technology to economic development and growth. In this particular vision of economics, technological change is, thanks to the general allocative efficiency characteristics of the market, associated with welfare increasing aspects, i.e. with technological progress.

This is also reflected in the way technological change is generally being measured. From an economic point of view, measurement of technological change is, indeed, generally reduced to those new technologies which have a well-defined economic impact, either in terms of productivity growth or in terms of new product demand. And even with respect to the latter, the methodological problems raised in effectively incorporating new products in the production function framework has generally led to a further reduction of the economic measurement of technological change to productivity growth – typically expressed in terms of some weighted average of labour and capital productivity, called total factor productivity. This

---

[1] N. ROSENBERG, *Perspectives on Technology: 1976*, Cambridge, CUP, 1985, p. 1.
[2] C. FREEMAN, *The Economics of Industrial Innovation*, 2nd ed., London, Frances Pinter, 1982.

measurement issue illustrates well the common perception in economic analyses of technology that technological change and its societal impact can be correctly assessed only in economic terms. It leads to two restrictions in economic analyses of the consequences of technological change.

First, to economists it often comes as a surprise that there are many innovations which have very widespread societal effects, but whose measurable economic effects are small, or at best indirect in terms of macro-economic growth and efficiency. The innovation of an oral contraception device had a major impact on sexual behaviour in the 60s and 70s in most Western countries, giving rise to some fundamental debates about medical and social ethics. Its economic impact was at best indirect through greater participation of women in the labour market. Genetic fingerprinting – a more recent technological advance in bio-technology – is said to be of great importance in forensic medicine, crime detection and the judicial process, especially in cases of rape, assault and murder. It could also have major implications for medical prognosis and life insurance, which will also raise some fundamental questions of medical and social ethics. The emerging information society is having major impacts on new forms of international communication, civil rights and democratic control; its economic impact seems almost unilaterally focused on the new possibilities for substitution of physical economic activities such as commerce, banking, also the workplace for electronic commerce, banking and telework. In this latter case the economic significance of these new forms of electronic, commercial exchange are likely to be large, but the other impacts are probably much more significant. In summary, for many innovations the societal impact may be very great even though the direct measurable economic impact is insignificant.

A second fact which is insufficiently introduced in economic analyses is the difference between innovations which find applications in only one sector and those which effect many or all sectors of the economy. In the technological taxonomies, suggested by a number of authors in the science and technology field (Christopher Freeman, Richard Nelson, Keith Pavitt, Nathan Rosenberg), technological advances are often identified as either 'localised' or 'pervasive' in terms of their impact. Again, an illustrative example might clarify this distinction. The 'float glass' process introduced by Pilkington's in the 1960s was certainly of enormous economic importance for that firm and for the glass industry generally, as it was licensed to almost all the major glass manufacturers in the world over the next few years. However, it has no applications outside the glass industry and its macro-economic significance is, therefore, relatively small. The microprocessor or the computer, by contrast, have found applications in practically every single sector of the economy, with, one suspects, major economic impact on the efficiency and growth performance of the economy.

In other words, economists are not only rarely aware of the societal impact of technological change, but they are also insufficiently aware of the wide variance in economic impact of technological change. The problem has undoubtedly become more severe over the recent period. Most empirical 'economic' studies in this area, in order to at least eliminate some of the more societal technical advances, have limited the analysis to the industrial sector, either as 'purveyor' of technological advances, or as funding such advances. Yet, such an approach becomes increasingly problematic, when dealing with the increasing number of service sectors as major initiators of technological change. A systematic inclusion of service activities means, indeed, that one is increasingly confronted with questions about the actual 'direct' economic impact of such technological advances. In many service sectors the separation between economic 'measurable' impact and 'societal', quality of life impact of technological change is indeed far more difficult to make. We come back to some of these issues in the concluding comments.

Assessing Economic Progress and Endogenous Technological Change

There is increasing recognition, particularly since the late 70s, early 80s, that the traditional economic 'exogenous' approach to technology and the related measurement biases are becoming hindrance, rather than useful conceptual frameworks, within which some of the recent policy concerns about the technology economy interaction can be discussed. Hence the notion of technology has become broadened to the notion of knowledge which puts more emphasis on the human element behind the development of new technology and the human 'embodiment' of essential complementary technical skills.

The economic profession has increasingly started to recognise the fact that such technological knowledge accumulation can be analysed like the accumulation of any other capital good; that economic principles can hence be applied to the 'production' and 'exchange' of knowledge; that in the end knowledge accumulation is endogenous to the economic and social system. Hence, while knowledge has some specific features of its own, it can be 'produced' and used in the production of other goods, even in the production of itself, like any other capital good. It also can be stored and will be subject to depreciation, when skills deteriorate or people no longer use particular knowledge and 'forget'. It might even become obsolete, when new knowledge supersedes and renders it worthless.

However, there are some obvious and fundamental differences with traditional material capital goods. First and foremost the production of knowledge will not take the form of a physical piece of equipment but generally be embedded in some specific 'blueprint' form (a patent, an artefact, a design, a software program, a manuscript, a composition) or in people and even in organisations. In each of these cases there will be so-called positive externalities; the knowledge embodied in such blueprints, people or organisations cannot be fully appropriated, it will, with little cost to the knowledge creator, flow away to others, just as I can share today my ideas with you and I am sure I will continue to acquire some of your knowledge in our conversations and discussions. Knowledge is from this perspective a 'non-rival' good. It can be shared by many people without diminishing in any way the amount available to any one of them. But of course there are costs in acquiring knowledge. A current central theme of economic theory is what is referred to as information asymmetry: the person wanting to buy something from someone who knows more about it obviously suffers from an asymmetry (a lack) of information.

From a purely economic point of view, this explains why markets for the exchange of knowledge are rare and why most firms have preferred to carry out Research and Development in-house rather than contracted out or licensed. Furthermore, it provides a rationale for policies focusing on the importance of investment in knowledge accumulation. Such investments are likely to have high so-called 'social' rates of return, often much higher than the private rate of return. Hence, investment in knowledge, as will be further elaborated on in section 3, cannot be simply left to the market.

It will be obvious that these relatively narrow 'economic' policy concerns do have their far broader societal 'mirror' concerns. First, there has been a general perception that technological developments, particularly the so-called 'new' information and communication technologies, are not contributing to a satisfactory degree to growth in economic and social standards – the term 'productivity puzzle' or 'Solow paradox' has often been used to summarise this problem. Second, there are growing concerns about a division emerging, both within and among nations, between technology - haves and those with limited access to new technologies. Nations are concerned they may be left behind in the pursuit of new markets if they are not on or near the technology frontier; groups within nations worry that, unless the opportunity to master new technologies is available, the broad sweep of technological change may make their skills obsolete. Finally, rapid changes in technology, together with the increasing integration of the international economy, have brought new pressures to bear on the rules and institutions that regulate international co-operation.

With respect to the productivity puzzle they raise the more general issue about the assessment of economic 'progress', as already touched upon in the first section. It could be argued here that the post war period up to the mid seventies – Jean Fourastié's *'Les Trente Glorieuses'* – was a period in which there was general agreement[3] that the quantitative economic data on the post-war increases in production and in productivity did provide a consistent picture of the growth in 'progress' over this same period. As Fourastié put it 'the great hope of the 20th century' has been fulfilled, based on 'the facts of production, consumption, length of working hours, health care and life expectancy'. Over the more recent period, however, and in line with the productivity puzzle evidence, increasing discrepancies are emerging amongst such facts and indicators, but also in the perception, weights allocated to material growth indicators and the many material and non-material 'externalities' of such 'growth'. There is far greater awareness that economic indicators do not measure correctly many of the social and environmental costs of economic growth. Similar arguments can be made with respect to the other concerns raised above.

[3] With of course some exceptions, particularly in the US. For one of the first economic analyses on the difference between material progress and happiness see Tibor Scitovsky's fascinating *The Joyless Economy*.

The result of these arguments is a gradual shift in the way technology is analysed at the economic policy level. It leads, as will be argued in the next two sections, to a significant shift in the purpose and elaboration of technology policy, within which technology assessment does play a central role. The OECD *Sundqvist Report* (1990) is probably the first economic policy report which emphasised so strongly the importance of technological change as a wider process of social change. In the words of the report: 'technology can be defined as a social process which, by meeting real or imagined needs, changes those needs just as it is changed by them. Society is shaped by technical change, and technical change is shaped by society. Technical innovation – sometimes impelled by scientific discovery, at other times induced by demand – stems from within the economic and social system and is not merely an adjustment to transformations brought about by causes outside that system'.[4] In other words, technological change, if it has to have beneficial effects on society will need to be 'embedded', integrated in society. An interesting illustrative example can be found in the lack of development and growth which took place in most East European countries over the last twenty years, despite massive investments in science and technology and higher education.[5] While the lack of economic integration, more specifically the lack of a market separating the technically from the economically feasible, pushed the science and technology system into isolation, the 'market' failure of the science and technology system in East European countries is clearly only one facet of this isolation. In my view, one of the greatest paradoxes of so-called 'socialist' countries' past development has been the total lack of social integration of technological change (lack of safety standards, higher health risks, lack of ergonomic considerations, etc.). Thus, in contrast to the so-called 'capitalist' societies, science and technology became far more imposed on society, and workers in particular, with the resulting lack of efficiency improvements at the shop floor.

This rather obvious point seems to have been least recognised in the purpose and aims of most technology policies, as they have been developed in most developed economies over the last decades. It is to a brief elaboration of the underlying economic motivations for such policies that we turn now.

---

[4] See OECD, *Sundqvist Report* (1990), p. 117.
[5] See GOMULKA, *The Theory of Technological Change and Economic Growth*, London/New York, Routledge, 1990.

'Market Failure' as a Failed Technology Policy

From a traditional economic point of view science and technology policy have been guided by relatively minimalistic questions such as: are there cases of market failure or suboptimality in science or technological effort? Following the argumentation of section 2, there is general agreement amongst economists that market failure is, indeed, one of the intrinsic characteristics of science and technological activities and that underinvestment in research will be the logical outcome of market allocation.[6] The fact that technological advances can be readily copied will deter companies from investing in these, even though a significant advance would lead to enhanced efficiency or performance. Particularly, in cases where technological advances are not well protected by patents and easily copied, plenty of examples of such underinvestment in research can be found. A typical example is agriculture: 'Before the advent of hybrid corn seeds, which cannot be reproduced by farmers, seed companies had little incentive to do R&D on new seeds, since the farmers, after buying a batch, simply could reproduce them themselves. The farmers themselves had little incentive to do such work since each was small and had limited opportunities to gain by having a better crop than a neighbour.'[7] Similar arguments have been raised with respect to many other industries, where scientists and engineers are mobile and where it is hard to keep secret information about the operating characteristics of particular generic designs, or about the properties of certain materials.

On the other hand, once the framework of the perfect atomistic market – as in the case above of farmers – is dropped and some of the more common features of imperfect market competition are introduced,[8] it becomes obvious that market allocation could also lead to overinvestment in research. As illustrated in so many recent popular cases of new major technological breakthroughs (superconductivity, HDTV, etc.), there is often a clustering of research effort involving duplication, which with the vision of hindsight is inherently wasteful. Economies of scale and/or scope that could be achieved through co-ordination will be missed.

[6] Such arguments were already put forward more than thirty years ago in contributions of NELSON (1959) and ARROW (1962).

[7] R. NELSON, *Understanding Technical Change as an Evolutionary Process*, Amsterdam, North Holland, 1987.

[8] As emphasised in some of the more recent contributions in the field of industrial organisation (see e.g. DASGUPTA and STIGLITZ, 1980).

Many of the most well-known authors in the science and technology policy field[9] have discussed in detail the conditions under which these straightforward 'market failures' are more or less likely to occur. It has always been tempting for policy makers to regard them as providing both justification and guidance for governmental actions to complement, substitute for or guide private initiatives. In practice, these arguments have led to the justification for active government support policies in most OECD countries in the technology field, with government support for research, whether in the public or private sector as the central issue at stake. Big national and international Research and Development projects, financed and planned directly by the state, have thus become some of the most dramatic illustrations of 'government failure'. The recent acknowledgement of the co-ordination problems that lead to possible waste and duplication of research efforts has induced a shift towards support for generic technologies and for so-called 'pre-competitive' R&D support.[10] Following the perceived Japanese policy in this field, the US and particularly Europe now all have major programs of such pre-competitive, 'co-ordinated' R&D support.[11] It is too early to describe those new programmes as cases of joint, collaborative government failure.

What is clear, though, is that the narrow market failure approach leads to a massive supply bias in technology policy, exacerbating further the lack of economic, social and even societal integration of new technologies. It could be argued that concerns of policy makers not to interfere with the market process, have made them identify 'pre-competitive' research with non-applicable research! Particularly, in cases of so-called generic technologies the need for support is often far more essential at the application end. Generic technologies can be best described as technologies which are flexible in use: where the process of efficiency gains, improvements and learning to use the technology in a better way are all intrinsically related to the particular application case. Often this more application-focused

---

[9] See e.g. ERGAS (1985), JUSTMAN and TEUBAL (1986), MOWERY and ROSENBERG (1989), PAVITT and WALKER (1976), FREEMAN (1982, 1987), NELSON (1983, 1984, 1987), ROSENBERG (1982, 1989).

[10] For a particularly useful overview of some of the issues involved from a Dutch technology policy making perspective see VAN DIJK and VAN HULST (1988).

[11] For an interesting insight in the different meaning of 'pre-competitive' research in the Japanese and US context, see KODAMA (1988).

research will involve similar uncertainties.[12] Here too, governments have a particular responsibility, going beyond the simple transfer of money, but including public procurement, standard setting and other more application-oriented support. To summarise: there existed a strong concern not to interfere with the market or with anything involving possible 'competitive' applications and with it the justification for technology support in the 'pre-competitive' phase of research. This has led many policy makers to become -- often because of specific national prestige reasons -- 'over-concerned' with the creation of new, potentially pervasive scientific and technological knowledge, at the expense of possible applications, which have fallen outside governments' immediate concern.

The previous point brings to the forefront the implicit 'supply bias' in many science and technology policies, particularly in countries at the technologically leading edge. Such policies have insufficiently paid attention to the capacity of the economic and social system to incorporate technological changes and transformations. At some stage it is possible that new technologies, which have been developed with the support of government sponsored science and technology programmes, rather than being introduced and diffused, further outrun the capacity of the economic system to adapt or generate new demand. An illustration of such an overemphasis of technology supply policy has been the European sponsored megabyte project, which successfully led to the production of so-called megabyte chips by Philips and Siemens. The demand for these chips appeared to be strongly overestimated and, indeed, turned out to be restricted to the relatively limited and fragmented European mobile telephone market.

Endogenous Technology Policy:
Or How to Assess Technology 'Constructively'

The previous policy discussion has focused primarily on the economic context of technology. However, it is clear that the economic feasibility of a new process or product is only one, often decisive, part of the 'societal' integration of technology. Other contexts, i.e. social, ethical and socio-political do play important roles. Once the

---

[12] See also ZIMAN's (1990) contribution to the TEP Competitiveness Conference.

argument is accepted that the creation of technological capabilities does involve a complex, endogenous process of change, negotiated and 'mediated' by society at large, it is obvious that policies in the area of science and technology can not, nor should be limited to the economic 'integration' of technological change, but must include all aspects of its broader societal 'integration'. It is in its broader interaction with society that technological change is adapted and selected and the further realisation of technological progress enhanced. From this perspective, technological renewal is, by definition, a broad concept encompassing research and development, diffusion and imitation of new technologies, as well as the associated social and organisational changes and innovations.

It is, of course, within this broad conceptual framework that (constructive) technology assessment has emerged as an institution and instrument of government policy aimed, on the one hand, at improving the societal integration of technology and, on the other hand, at directing the use and further development of technology. Therefore, technology assessment developed practically completely separated from economics and in particular economic policy. The recognition of the importance of both, economic and broader societal conditions opens up the possibility of a far more coherent and complete conceptual framework for the development of technology policy, which in essence is and should be nothing else than (constructive) technology assessment (TA).

Both the old traditional, and what could be called static, TA issues have their place here, as well as the more dynamic TA concerns following the previous discussion.

Static TA issues have been discussed quite extensively for many years now and could be said to fall in economic language under the heading of the unequal distribution of positive and negative externalities of technological change. As emphasised by authors such as Harvey Brooks, there is an apparent paradox in the distributional impact of technology. The costs or risks of a new technology frequently fall on a limited group of the population, whereas the benefits are widely diffused, often to such an extent that the benefits to any restricted group are barely perceptible even though the aggregate benefit to a large population amounts to considerably more than the total cost to the limited adversely affected group. Examples abound. 'Automation' e.g. benefited consumers of a product by lowering its relative price, but the costs in worker displacement were borne by a small number of people, and may be traumatic. A large electrical generating station may adversely affect the local environment, while providing widely diffused benefits to the population served by the electricity produced. Workers in an unusually dangerous occupation such as mining carry a disproportionate share of the costs associated with the resultant materials, which may have wide benefits throughout a national economy.

This disproportion between costs and benefits can, of course, also work the other way, as in many cases of environmental pollution and emissions. The effluents from a concentrated industrial area such as the Ruhr Valley or the American Great Lakes industrial complex may diffuse acid sulphates over a very large area, which derives little benefit from the industrial activity, but may have its quality of life as well as agricultural productivity seriously degraded thereby.

The issue of sharing costs and benefits of technological change shows, first, how important it is, both from the national and international point of view, to draw up some 'rules of the game' to ensure that adverse effects are less harmful than they would be if everything was left to free competition and, second, to establish such rules fairly early on, before vested interests, acquired privilege and the fierceness of competition jeopardise their compulsory application. The word static is of course inappropriate to describe all such distributional issues. With the increase in the complexity of technology, possible risks not only threaten larger areas, they might also have impacts over a longer period, lasting several generations.

Traditional, static TA should, therefore, be complemented by a more dynamic approach, whereby, as argued above, the broad societal integration of technology, as well as the adaptation of technology to societies' needs, are the central issues at stake. The issue is thus clearly more than one of dynamic externalities. Yet, from a practical economic policy perspective it seems useful, at least in the first instance, to elaborate further upon the externality terminology, rather than the constructive, active TA one. One of the reasons for this is the dynamic character of externalities, which prevents them from being susceptible to definitive once and for all categorisation and makes them more intimately related to particular historical and institutional contexts. As Nelson put it: 'It could be said that technical change is continually tossing up new 'externalities' that must be dealt with in some manner or other. In a regime in which technical advance is occurring and organisational structure is evolving in response to changing patterns of demand and supply, new nonmarket interactions that are not contained adequately by prevailing laws and policies are almost certain to appear, and old ones may disappear. Long-lasting chemical insecticides were not a problem eighty years ago. Horse manure polluted the cities but automotive emissions did not. The canonical 'externality' problem of evolutionary theory is the generation by new technologies of benefits and costs that old institutional structures ignore.

From this perspective, the concept of a 'constructive' or 'socially optimal' way of assessing long term impacts of technological change looses much of its meaning. Occupying a more central place in the technology policy analysis is now the notion that society ought to be engaging in experimentation and that the information and feedback from that experimentation will be of central concern in guiding the present evolution of the economic and technological system. Like in the case of 'market failure', the complexity and subtlety of the dynamic interaction between technology and society suggest that simple normative rules will not be very helpful in the design of 'constructive' technology (assessment) policies. From this perspective, I would argue, as Nelson and Winter (1982) did with respect to the market failure argument, that TA 'policies should focus on problems of dealing with and adjusting to change. It involves in the first instance abandonment of the traditional normative goal of trying to define an optimum and the institutional structure that will achieve it, and an acceptance of the more modest objectives of identifying problems and possible improvements'.[13]

Conclusion

Technological change, as we have argued, is not so much an exogenous 'manna from heaven' factor, superimposed from the outside through the activities of scientists and technologists. Rather, it has to be interpreted as an endogenous process, whereby the produced technological change is continuously being adapted and selected to the broad needs and requirements of society.

---

[13] R. NELSON and S. WINTER, *An Evolutionary Theory of Economic Change*, Cambridge, MA, The Belknap Press of Harvard University Press, 1982.

The endogenity of technology arises, of course, at all levels of technological development. At the level of technology 'creation', technological innovation is not only impelled by scientific discovery, but is also induced by demand. The development of a potential economic idea into new products and processes requires many stages of experimentation, in which market possibilities interact with the original idea. The process of invention and innovation interacts in many loops before attempts to market the product or process are made. There are further interactions and feedback loops between the initial successful marketing of the product and its wider diffusion nationally and globally. The acceptability of a product or process will, of course, also be conditioned by societal attitudes and norms. Thus, in broader terms, technological change stems from within the economic and social system and is not merely an adjustment to transformations brought about by causes outside that system. Societies have a say in the shape technology is likely to take. Hence, the importance of technology assessment for the policy choices which need to be made.

References

ARROW, K., 'Economic Welfare and Allocation of Resources for Invention' in R. Nelson (ed.), *The Rate and Direction of Inventive Activity*, NBER, Princeton University Press, 1962.

DASGUPTA, P. and J. STIGLITZ, 'Industrial Structure and the Nature of Inventive Activity' in *Economic Journal* 99(1980), pp. 266-293.

ERGAS, H., 'Why Do Some Countries Innovate More than Others?', Brussels, Centre for European Policy Studies, 1985.

FREEMAN, C., *The Economics of Industrial Innovation*, 2nd ed., London, Frances Pinter, 1982.

FREEMAN, C., *Technology Policy and Economic Performance: Lessons from Japan*, London, Frances Pinter, 1987.

GOMULKA, S., *The Theory of Technological Change and Economic Growth*, London/New York, Routledge, 1990.

JUSTMAN, M. and M. TEUBAL, 'Innovation Policy in an Open Economy: A Normative Framework for Strategic and Tactical Issues' in *Research Policy* 15(1986), pp. 121-138.

KODAMA, F., 'Innovative Approach to Research Draws Conspiracy Cries from Abroad' in *Japan Economic Journal* (Nov. 26, 1988).

MOWERY, D. and N. ROSENBERG, *Technology and the Pursuit of Economic Growth*, Cambridge, Cambridge University Press, 1989.

NELSON, R., 'The Simple Economics of Basic Scientific Research' in *Journal of Political Economy* (1959)67, pp. 297-306.
NELSON, R., 'Government Support of Technical Progress: Lessons from History' in *Journal of Policy Analysis and Management* 2(1983)4, pp. 499-514.
NELSON, R., *High Technology Policy: A Five Nation Comparison*, Washington and London, American Enterprise Institute for Public Policy Research, 1984.
NELSON, R., *Understanding Technical Change as an Evolutionary Process*, Amsterdam, North Holland, 1987.
NELSON, R., M. PECK and E. KALACHEK, *Technology, Economic Growth and Public Policy*, Washington, DC, Brookings Institution, 1976.
NELSON, R. and S. WINTER, *An Evolutionary Theory of Economic Change*, Cambridge, MA, The Belknap Press of Harvard University Press, 1982.
PAVITT, K. and W. WALKER, 'Government Policies towards Industrial Innovation: An Overview' in *Research Policy* 5(1976), pp. 11-97.
ROSENBERG, N., *Perspectives on Technology: 1976*, Cambridge, Cambridge University Press, 1985.
ROSENBERG, N., *Inside the Black-Box: Technology and Economics*, Cambridge, Cambridge University Press, 1982.
ROSENBERG, N., 'Why Do Firms Do Basic Research (with their Own Money)?' in *Research Policy* (1989).
VAN DIJK, J.W.A. and N. VAN HULST, 'Grondslagen van het Technologiebeleid' in *Economische Statistische Berichten* 21(September 1988).
ZIMAN, J., 'The Restructuring of the Links between Fundamental and Applied Research', paper prepared for the TEP Conference on Technology and Competitiveness, Paris, OECD, June 1990.

3.1.2

GLOBALISATION AND ETHICAL COMMITMENT

The Challenge of the 21st Century

Riccardo Petrella

At first sight, it seems as if the greatest challenge of the 21st century lies in being able to adapt quickly enough to remain abreast of the current trend of globalisation and ahead of the competition. General opinion accepts that globalisation will have far-reaching consequences on economic, social, political and cultural life, not all of which are positive. There is even a certain fatalism regarding the present trends in development, good and bad alike, as inevitable. However, a closer look reveals that a greater challenge lies in simply ensuring that every human being has access to the most basic necessities for life such as access to drinking-water. There is a need to restore the primacy of politics over finance by ensuring the development of democratic structures capable of regulating private enterprise and financial institutions on the new world-wide scale. The author invites us to ask the question: 'Within the present configuration of the world economy, where do the public interest, the common good (...) and respect for the life of others fit in?'

The Dominant Vision

According to the most wide-spread public opinion in political, socio-economic and scientific environments north and south as well as east and west, the globalisation of the economy, of the markets, of companies and capitals constitutes – on a par with the 'technological revolution', which is especially tied up with new information and communication technologies – are the main source of challenges contemporary societies have to face in every single domain at the threshold of the 21st century.

From this point of view the key word is 'adaptation'. It is said that we have to adapt to the present globalisation and the shape it is taking (that is to say, in the context of the increasing and total liberalisation of the markets, of the deregulation of the economy, and of the privatisation of everything that can be privatised, even, for instance, prisons and air...).

It is repeated incessantly that we have to adapt to the new technologies. Those who will not adapt, will be eliminated.

Within the scope of the 'grand technological world revolution', the present globalisation is said to be engaged in burying the 20th century and in giving birth to the third millennium. It is said to be posited as an irreversible constraint, which no one is able to fight, and in the face of which the only possibility is to accept it and to transform it into opportunities for oneself, in a more efficient and profitable way than the others.

Hence the imperative of world-wide competitiveness. This competitiveness, which is the new gospel of the new planetary world of high technology, is elevated to the rank of ineluctable strategic option for every company, city, region, country, State. Hence equally the idea that the new technologies represent the most powerful and efficient instrument in order to secure and reinforce one's worldwide competitiveness.

Realising a remarkable pirouette of self-referential rhetoric, the dominant opinion starts off by positing as dogma the historical inevitability of the present globalisation and the progressive character of the contemporary usage of the 'technological revolutions'. This position, however, is completely wrong.

One wants to deduce from this premise the inescapable rationality of adaptation as the only possible strategy and the sole expression of liberty, and one concludes that the prevalent globalisation of the economy and the prevalent technological revolutions, simply follow history, and are therefore normative objectives, the achievement of which is regarded as the major societal global challenge of our times.

Of course, the dominant opinion is conscious of the particularly negative effects which accompany the present globalisation in the economic, social, political and cultural areas, not to mention the spiritual field, but it holds firm to its belief that, globally and in the long run, the globalisation and the new technologies will yield positive results.

The Two General Prescriptions which Orientate the Dominant Vision Today

First prescription. One has to believe that the present globalisation, the new technologies and the competitiveness of all against all, are not incompatible with social justice, with solidarity, with respect for minorities, with democracy, and with recognition of the other. On the contrary, social justice and solidarity are claimed to adapt to the present globalisation and to the present technological revolutions. Competitiveness is the primary and fundamental condition for insuring social justice and solidarity.

Second prescription. One has to believe that the desire for any generalisation other than the present globalisation – which is illuminated by the principles of the liberalised, deregulated, privatised, competitive capitalistic market economy with a high intensity of more and more so-called 'intelligent' knowledge and technologies – is unrealistic, ideological and demagogical. There must be no thought of a different globalisation. It is impossible, utopianistic. The existing world-wide balance of power will not stand it. Besides, this would mean that one would want to return to either a sterile and suicidal national protectionism, or an unliberal communistic socialism which has already demonstrated its total inefficiency.

Moreover, we are warned not to be naive. There is a genuine world-wide economic war going on. As they radically and relentlessly modify the particulars of the economy, the new technologies contribute to accentuate and accelerate the transformations of professions, of professional abilities, of comparative and competitive advantages, of production methods and localities, of the distribution and consumption of goods and services, and of the organisation and the role of the financial markets. As a consequence the world is profoundly marked by a course which is ever more intent on innovation, competition and the conquest of (other) markets. It is dominated by those who are commercially the most competitive and the most aggressive.

From this viewpoint, each company and each country is obliged to think about insuring and reinforcing its own survival and continuation, by putting its money at the same time on coöperation and on competitiveness. The economic war will not be fought by preaching 'economic peace'. *Si vis pacem para bellum.* To the dominant opinion this ancient Roman proverb applies to the present economy. One has to be strong, and even the strongest, if one wants to let those solutions prevail that are more just in the social and the human field.

The priority then is not to fight the present globalisation, but to humanise it; and the same applies to new technologies. What counts is not to stop them but to make them more human.

The Challenge Lies Elsewhere

An analysis that is closer to the reality of the everyday life of a very wide majority of the world-wide population, shows, on the contrary, that the chief socio-economic, technological and relational challenge does not limit itself to the promotion of globalised and liberalised capitalistic market economy with a human face. It shows that the chief challenge consists in defining and setting to work socio-economic principles, political rules and institutions, as well as applications of science and technology which will allow the 8 billion people who will inhabit this planet in less than 25 years to have access to the most elementary human and social rights. Examples of these rights are, amongst others, access to drinking water, to housing that is worthy of a human being, to sufficient nourishment, to education, to free and adequate means of information and communication, to freedom of opinion, religion and expression, etc.

Let us take water as an example. At this very moment, on an overall world population of 5.8 billion human beings, 1.4 billion human beings have no access to drinking water. This is in spite of all the fine declarations, charters and texts which affirm and maintain that water is the main source of life for the human being and that it represents the most precious natural 'good' for humanity.

In reality, water is becoming a more and more rare and inaccessible 'good'. It is the source of more than 50 local wars amongst neighbouring countries. People not only die for want of water, they also die in order to take away the access to water from other human beings. Water is equally the subject of an increasing number of conflicts between town and country. Moreover, everywhere people increasingly tend to privatise water and to reduce it to a 'marketable good'. It is seen as an important source of overabundant profits in the years to come.

As a consequence of this, how many people will no longer have access to drinking water in 2020? No doubt this will be in the neighbourhood of 2.5 to 3 billion human beings. As a matter of fact, by that time there will be 8 billion of us, and the resources of fresh water will have diminished, unless we have succeeded in reversing the present tendencies.

Is such a reality acceptable for an ethics that wishes to respect human dignity?

*First Proposition: Make Water the First
Common Patrimonial Good of Humanity*

The example of water illustrates quite well the existing relations between natural, geopolitical, economic, cultural and social factors. It illustrates equally the actual functioning of the present economy on a local, continental and world scale. It elucidates in a very pragmatic way the nature and the extent of the present world challenge: how to *be, live and act together* today in a spirit of solidarity with all the populations of the world society and with the future generations – a solidarity which is based on the very concept of *durable developments* everyone claims to support henceforth, which is an altogether fortunate thing.

Water is such a paradigmatic example of the nature of the real world challenge that it justifies my first proposition. In the face of such a challenge, I have the impression that the ethics of engineers should make water into a privileged field of reflection, as a living testimony of the will to face the challenges of the human condition on the threshold of the 21st century. We will have to act in order to enforce the acceptance and the implementation of the principle that in the age of globalisation water must become the first common patrimonial world good. From this viewpoint, it cannot be the object of private appropriation. The access to drinking water must be supported and respected as a basic economic and social right for every human being. Besides, numerous public and private institutions and organisations are already engaged in this direction, be it the FAO, the World Bank, hundreds of local NGOs in the north and south of the planet, or, more recently, the World Water Council, the Fondation de l' Eau (the Water Foundation), the Stockholm International Water Institute (SIWI) and the Global Water Partnership. On top of this, international intergovernmental programmes to the benefit of the access to drinking water have been defined and realised within the scope of the *Décennie de l'Eau* (The Water Decade) (1977-1987), followed by the United

Nations Conference on water in Mar del Plata in 1977. In this very year, a *Congrès Mondial de l'Eau* (World Water Congress) has taken place in Montreal, during which the issue of access to water for everyone has been dealt with once more.

Considerable improvements have been realised, of course. However, the progress remains insufficient and since a few years the situation has been gravely deteriorating especially because of the politico-economical-cultural trend which promotes liberalisation, deregulation and privatisation within the scope of a competitive world market economy.

Although most of the initiatives taken in recent years may have been particularly useful, deserving and extremely valuable, they also have the disadvantage of being more and more marked by a fundamentally techno-economic and market conception of water: water is merely considered as a marketable 'resource' and a marketable 'good'. According to this conception, only market forces and rules of privatisation can insure the objective of access to water for everyone.

If one accepts the principle that water is a common good and a world patrimony, it would be useful to promote a series of socio-economic inquiries, reflections and actions, which would consider water as a concrete reality and a concrete symbol of the common good and of world solidarity.

Second Proposition: Restore the Primacy of Politics over Finance

One of the important obstacles (amongst others) which render particularly precarious every action that would give water the status of a 'patrimonial common good for humanity', or that would vouch for the access to drinking water as a basic economic and social right guaranteed to every human being, is the growing supremacy of the financial powers over the democratically elected political power.

To restore the primacy of politics over finance is the most important socio-politico-economic challenge in our societies, provided, of course, that one is dealing with a context of representative democracy and/or, a fortiori, a context of direct or participative democracy. If, on the contrary, one does not consider it a problem that at this moment the financial markets (and more and more the speculators) determine the priorities of the economic and social politics of the countries, rather than the parliaments, then the most important challenge is obviously situated 'elsewhere'.

Nevertheless, one may wonder to which extent a responsible ethical position can acquiesce in the present situation, which is characterised by:
- the fixation of the priorities of our countries by the financial markets that have no other responsibility than towards themselves,
- the movement of capital on a planetary scale, hence escaping every control ensuing from national and international political and monetary powers,
- the existence of 37 tax paradises all over the world, which, owing to the banker's duty of secrecy, are becoming the legalised places for the criminalisation of the economy (tax evasion, speculation, drug traffic, illegal arms trade, ...).

If only to take the case of Europe, is it not high time for the intellectuals, the scientists, the engineers, the researchers of our universities to take the initiative of opposing what is drummed into the citizens of the monetary and economic Union: that the construction of Europe implies the installation in 2002 of a central Bank which is entirely autonomous and independent of every political power. In this hypothesis, what would be the role of the European Parliament which has been elected through a direct universal suffrage? Why would we keep on electing a parliament that would be powerless to direct and control the central European Bank, which then, in its turn, would become the genuine place of power in the future European Union? That being so, why would we then not elect the 30, 40 or 50 directors of the central European Bank?

These are fundamental questions. Solutions to the challenge to democracy require the correct answers to these questions.

Third Proposition: Devote Oneself to the Installation of a Political Management of World Productivity

In the present globalisation processes certain phenomena are new when compared to the already known processes of internationalisation – and multinationalisation – of trade, companies and capitals:
- the development of structures for producing goods and services (going from the R&D phase to the recycling phase), organised – and more and more connected and integrated – on a world scale. This is particularly possible thanks to the new information, communication and transportation technologies. What is being born is the 'made in and by the world';

- this development is accompanied by a policy of production, distribution and consumption, which is defined and conducted on a world scale. The compared productivity and profitability (between different places, companies and markets) become the criterion for evaluation and decision-making. One witnesses thus the emergence and the consolidation of a world management of productivity/profitability. This way of managing has an increasingly private nature, structured by private organisations and institutions, and relying on the processes of liberalisation, deregulation and privatisation. All of this takes place at the expense of public management on the national level.

What kind of positions do engineers take when confronted with these developments? Are they prepared to participate in the definition and the installation of a public world management of the productivity of the planet. And if so, according to which principles, with what means, in which stages, and in collaboration with which social and economic forces?

Is it not in this roundabout way that one could and should be interested in the politics of science and technology? Why should we consign to science and technology the mission of creating the means for the most competitive private companies and indeed the technologically already most developed countries, to be able to take possession – for the private purpose of productivity and power – of increasingly large sectors (that is to say of market shares) of the power of orientation and control of the allocation of the material and immaterial resources that are available on this planet?

Why should we continue to reduce technology (and technological innovation) to a powerful instrument in the industrial and commercial war, with a view to the conquest of the market and world productivity? What are we to make of the affirmation according to which the companies of all the countries in the world and of whichever proportions, must necessarily, in order to survive, conquer more markets? Is the underlying ideology of the principles of competitiveness and conquest of world productivity compatible with the ethical principles in our traditions? Within the present configuration of the world economy, where do the public interest, the common good, solidarity, love of one's neighbour, respect for the life of others fit in?

*Fourth Proposition: Globalisation
as a New Theme for Ethical Reflection*

I hope that the preceding pages have abundantly brought into the limelight that globalisation is a new, multidimensional process which touches upon numerous spheres of action and calls for ethical reflection. What kind of planet do we want? The whole world is in question, and it is necessary to articulate the questions, the concepts, the collected data, the theories and values that could enlighten our choices as well as their consequences. Different professions and disciplines have to work together for this purpose: engineers as well as economists, anthropologists, historians, literators, sociologists, philosophers, climatologists, jurists, political scientists, etc.

What is especially at stake is to be conscious of the fact that once engineers, researchers, scientists, etc. start treating a phenomenon, they are 'personally committed' because they are social constructs themselves and because, consequently, their act of creativity and their contribution to philosophy (love of truth) are the result of a socially constructed personal commitment. In other words: while studying, analysing and describing globalisation, they smuggle in their conceptions and their choices; they reflect their history, their commitments and their experience.

References

BAUMAN, Z., *Globalization: The Human Consequences*, Cambridge, Polity Press, 1998.
GIDDENS, A., *The Consequences of Modernity*, Cambridge, Polity Press, 1990.
HIRST, P. and G. THOMPSON, *Globalization in Question*, Cambridge, Polity Press, 1996.
KIELY, R. and P. MARFLET (eds.), *Globalisation and the Third World*, London and New York, Routledge, 1998.
PETRELLA, R., *Le Bien Commun. Éloge de la Solidarité*, Brussels, Labor, 1996 and Lausanne, Presses de Page 2, 1996.
PETRELLA, R., Lisbon Group, *Limites à la Compétivité*, Paris, La Découverte, 1995.
SASSEN, S., *Losing Control? Sovereignty in the Age of Globalisation*, New York, Columbia University Press, 1996.
SEN, A., *Poverty and Famines*, Oxford, Clarendon Press, 1981.

3.1.3

GLOBALISATION FROM AN ETHICAL PERSPECTIVE[1]

Göran Collste

Globalisation is a contested conceptualisation of technical, economical, political and cultural processes that taken together are establishing tighter relations between people living in different parts of the world. The world is more and more apprehended as a whole and there is a growing sense of interrelatedness. Globalisation has important consequences for engineering practice and for ethics. Our actions influence increasingly the lives of people and the natural environment at a far distance, which has implications for the question of responsibility and creates a need to work out a global ethics.

The social context of engineering has changed considerable during, say, the last 30 years. For a newly graduated engineer it is not obvious in which country he or she will work. The employer will perhaps send him or her to South East Asia and the company may very well have its head quarter at another continent. Or the new engineer may perhaps him- or herself apply for work in another country than his or her own. The components of the machines he or she is working with are most certainly produced in some far away part of the world.

These observations illustrate aspects of what is often thought of as a new era, called globalisation. This is a much-used concept in present political rhetoric. For some it has a positive value connotation, imply-

---

[1] I thank the anonymous referees of this chapter for valuable and constructive comments.
This chapter is written as part of a project on Globalisation and Social Justice, financed by the Bank of Sweden Tercentenary Foundation

ing broaden minds, international exchange, tourism and mutual understanding across the borders. For others it is a threat, implying injustices and power out of control. But what does it really stand for?

In this article I will discuss the meaning and implications of globalisation. I will focus on the globalisation processes from an ethical point of view.

A Contested Concept

There is as yet no unanimous view on the globalisation processes. Rather, there is a multiple of interpretations. A few examples may illustrate this. According to the Japanese economist Kenichi Ohmae, globalisation is both an inevitable and a blessing process. 'No more than Canute's soldiers can we oppose the tides of the borderless world's ebb and flow of economic activity',[2] Ohmae writes. It is blessing while it will lead to material welfare for everyone. 'In today's borderless economy, the workings of the 'invisible hand' have a reach and strength beyond anything Adam Smith ever could have imagined',[3] Ohmae writes.

A picture of globalisation, that in certain respects is similar to Ohmae's, is drawn by its critics. They also mean that globalisation will imply a transfer of power from nations to financial institutions and corporations but they are, contrary to Ohmae, very critical of this development. It will, they argue, imply a concentration of power and wealth. As one of the critics, D.C. Korten writes, 'The underlying pattern of the institutional transformation being wrought by economic globalisation are persistently in the direction of moving away from people and communities and concentrating it in giant global institutions that have become detached from the human interest.'[4]

There are also observers who deny that there has been a qualitative change of the world economy. The economists Hirst and Thompson argue that the world economy was more integrated at the end of the 19th century than today. What has happened, according to Hirst and Thompson, is a growth in economical co-operation led by the

[2] K. OHMAE, 'Putting Global Logic First' in K. OHMAE (ed.), *The Evolving Global Economy*, Boston, Harvard Business Review Book, 1995, p. 137.

[3] *Ibid.*, p. 129.

[4] D.C. KORTEN, *When the Corporations Rule the World*, London, Earthscan, 1996, p. 226.

nation states. As a consequence, the role of nation states has become more rather than less important. This is concealed by the globalisation rhetoric, which, according to Hirst and Thompson, will have political implications. 'One key effect of the concept of globalisation has been to paralyse radical reforming national strategies', they write.[5]

We Are All Sitting in the Same Boat!

Although, there is as yet no common understanding of the exact meaning or the implications of globalisation, the concept catches a wide spread impression that the world is shrinking and that, in an earlier unknown way, the lives of people living far apart are in many ways interdependent. There is a growing awareness that we are all sitting in the same boat.

'Globalisation can thus be defined as the intensification of worldwide social relations which links distant localities in such a way that local happenings are shaped by events occurring many miles away and vice versa', the sociologist Anthony Giddens writes.[6] Giddens' definition can be seen as a way of relating different kinds of tendencies that have taken place, say, in the last three decades, into an interpretation of a more or less coherent development. One could even make the definition wider, including not only 'social relations' but also economic, environmental and cultural relations. Technology is for better or worse a driving force behind globalisation. On the one hand, new technology tightens the links between people through information networks while, on the other, it poses a threat to the environment and global sustainability.

The Global Information Infrastructure

Information technology (IT), e.g. Internet and e-mail, has been a prerequisite for the globalisation process. The global networks, or the Global Information Infrastructure (GII) as it has been called, have made it easier and cheaper to communicate across the world to the extent that geographical distance in many instances has become irrel-

---

[5] P. HIRST and G. THOMPSON, *Globalisation in Question*, Cambridge, Polity Press, 1996, p. 6.

[6] A. GIDDENS, *The Consequences of Modernity*, Cambridge, Polity Press, 1990, p. 64.

evant. Banks and other financial institutions can move currency from one part of the world to another without any delay in time and without any restrictions. As a consequence of the GII, many corporations now perceive their operation environment to be the world economy. However, there is not only an economical potential in the GII. The possibilities to spread information across the world contain also a democratic potential. Well-informed citizens are a prerequisite for a vital democracy and the GII offers new ways of obtaining information. There is also another democratic potential in the GII. In some countries, the authorities try to control the media and censor critical reports. This becomes more difficult in the era of GII. Thus, if a free flow of information is seen as a condition for democracy, the GII will facilitate the realisation of this condition.

However, there is also room for some question marks when arguing that GII will enhance democracy. One is the, so far, unequal distribution of access to Internet and e-mail. 80% of world's population still lack access to the most basic information technology.[7] Another is the problem of accuracy, reliability and relevance of information. Not all information will help to empower the citizens. On the contrary, there is a risk that the ordinary citizen will get lost in a torrent of information, unable to distinguish relevant information from rubbish. The result will be frustration rather than information.

As a consequence of globalisation politics has become a global phenomenon. People active in NGOs (Non-governmental organisations) can use information technology in campaigns for human rights and environmental protection for example. Still, as a democratic basis, political participation is a local rather than a global phenomenon. The well informed, active citizen will normally engage in political activities related to his or her own neighbourhood or city, rather than in decision-making taking place in global governmental bodies at a far distance. For political participation at the local level, GII does not seem to have much to offer.

Anonymity and Distant Effects

Ethics is reflection on criteria for a right or a wrong action. In a paradigmatic case, person A acts towards person B in a way that benefits

[7] R. KIELY and P. MARFLET (eds.), *Globalisation and the Third World*, London and New York, Routledge, 1998, p. 5.

or hurts B. The moral evaluation of the action presupposes that it is possible to survey its effects. The link between the action and its effects is clear. In the global society, this is no longer the case. In order to illustrate this, I will use a parable from philosopher Derek Parfit. The parable is fictitious and I use it to illustrate a different point than Parfit:

The Harmless Torturers. In the bad old days, each torturer inflicted severe pain on one victim. Things have now changed. Each of the thousand torturers presses a button, thereby turning the switch once on each of the thousand instruments. The victims suffer the same severe pain. But none of the torturers makes any victim's pain perceptibly worse.[8]

This parable can illustrate some aspects of global action. The torturers are anonymous, and there is no individual torturer who causes the victim's pain, because it is the consequence of their collective action. Further, there is a distance between the torturer and the victim. It is possible to imagine the torturers being in a different place or even in a different country from the victim. They never have to see the victims or hear their screams.

Similarly, anonymity and distant effects are two features of actions in the global society. As an illustration we can take transportation habits. Many of us have a choice whether to take the car or bicycle to work. It makes no difference to the environment if you or I choose to take the car. But if everyone makes the same choice, the consequences will be destructive. And there is a distance between the victim and us: the long-term environmental effects may only be revealed to generations as yet unborn.[9] 'Carbon dioxide emissions from the transport sector continue to rise and the available evidence on human health and transport pollutants indicates that critical thresholds have been tripped', John Whitelegg writes.[10]

---

[8] D. PARFIT, *Reasons and Persons*, Oxford, Oxford University Press, 1984, p. 80.

[9] This development of transportation can be illustrated by statistics. Between 1970 and 1990 the number of registered automobiles increased from 250 million to 560 million, the kilometres driven/year increased in the OECD countries for passengers' cars from 2,584 billion to 4,489 billion and for trucks from 666 billion to 1,536 billion. Still ahead of us is a possible demand of cars in for instance China and India. As John Whitelegg points out, consumer habits like the growth in soft drink consumption, for instance, is indicative of trends in transport. The growth in consumption of products of this kind is a major source of the growth in road freight. As a consequence, the conditions for sustainability are not met.

[10] J. WHITELEGG, *Transport for a Sustainable Future: The Case for Europe*, London and New York, Belhaven Press, 1993, pp. 4ff.

This example from transportation can illustrate the new moral situation in the global society. Despite the fact that the effects of the individual choice is negligible, the individual choices – combined with the choices of many other people in similar situations – can still have disastrous effects for people or the environment in the long term.

Global warming and the ozone hole, i.e. the environmental hazards following from many individual actions, can illustrate what Ulrich Beck calls a 'worldrisk society'.[11] These and other global effects lay the ground for a shared insight that the modern society is fragile and that we as world citizens have a common destiny. Global risks serve as a potential for global ethics.

Technological Standardisation
– a Means of Globalisation

Technological standardisation is a prerequisite for increased globalisation. Standards are documented agreements containing technical specifications or other precise criteria to be used as rules, guidelines or definitions. This ensures that material, products, processes and services are fit for their purpose. The format of the credit cards, phone cards and 'smart' cards is one example. Adhering to the standard, which defines such features as an optimal thickness (0,76 mm), means that the cards can be used worldwide.

A main reason for global standardisation is to facilitate international trade. As long as the standards are not harmonised, there are obstacles to the export of products. When international standards are widely used, suppliers can compete on many markets worldwide and customers have a wider choice of compatible offers. Another reason is that often one component, is used in different products. The same bolt can be used in aviation and for agricultural machinery. Then, it is an advantage if the bolt is standardised.

Standardisation is promoted by The International Organization for Standardization (ISO), a worldwide federation of national standards bodies from 130 countries. Thus, ISO is one main agent for technological globalisation.

---

[11] U. BECK, *Was ist Globalisierung?*, Frankfurt a.M., Suhrkamp, 1997.

The standards decided on affect everyone, but who sets the standards? Experts, often connected to industry, have a great influence and in some industries like software computing a few dominant firms have been very influential in setting the standards. In order to attain more user- and environmental friendly technologies, one should strive for more public participation in standard setting.

Another problematic question from a moral point of view is who gains and who loses through a global technical standardisation. Those who believe that trade liberalisation is a blessing for poor countries will argue that everyone is a winner, while those who think that protectionism can help in some situations, will not be equally enthusiastic. They see a standardisation that is decided by powerful global companies as a threat to weaker producers in poor countries.

A Global Political Economy

Engineers are increasingly employed by companies that are acting on a global market. A liberalisation of trade and financial flews, a growth in the number and size of multinational corporations and an international banking system out of control of national governments are some features of the economic globalisation. Of the world's hundred largest economies, fifty are corporations (1996). These companies have plants in numerous countries. This leads to interdependence while the production at one plant, is dependent on the production in another. As a matter of fact, 1/3 of international trade is intra-multinational company trade. The tendency towards globalisation is rooted in the homogenisation of markets, the decreasing costs of transport and communication, the decreasing trade barriers and the pressures from new competitors

Economic globalisation implies technological transfer. Through foreign investments, developing nations can get access to new technology, a necessity for economic growth. However, the wish to attract investors opens the door to political pressures. In order to get favourable conditions, a multinational corporation can threaten to move a plant to another country. Their size and influence create a 'race to the bottom'. This means that in order to attract investments, nations will compete in deregulating, for example, laws for environmental protection or workers rights.

Global integration of international banking is another aspect of economic globalisation. Even here, one will find many different forms. The foreign exchange market was the first to globalise, in the mid-1970s. A deregulation of domestic financial markets and the liberalisation of international capital flows contributed, in combination with computer networks, to an explosive growth in financial marketing as well as to an intertwining of stock markets all over the globe. Inventions of numerous new possibilities to 'swap' currencies and interest rates in order to secure investments, have accelerated the activities of banks and other financial institutions, so called 'non-bank banks', on the world market.[12]

Globalisation seems to have changed the conditions for the nation state in at least two ways. The first is that the states have become more dependent on the multinational corporations. This means that a government that tries to act in a way that threatens the interests of a multinational corporation runs the risk of suffer retaliation. The second is that the role and influence of different kinds of international bodies has grown. Saskia Sassen talks of 'an unbundling of sovereignty'[13] when elements of sovereignty is now being relocated from the state in new transnational private legal regimes and new supranational bodies, like the World Trade Organisation (WTO) and the European Union.

Traditionally, democracy has been developed in the framework of the nation state. Ultimate responsibility and legitimacy is according to the western political tradition in the hands of democratically elected national parliaments. If globalisation implies that more and more issues have to be decided on a level above the nation state, there is a need to redefine democracy at a global level. Different prospects can be envisioned. Either a transformation of sovereignty from national to international bodies will lead to a new democratically anchored rule of law at a global level, or political influence will wither away all together. The not yet realised proposal of the Multilateral Agreement on Investment (MAI) can illustrate the latter alternative. This agreement that was negotiated within the OECD,

---

[12] Daily foreign exchange trades rose from $20 billion per day in 1973 to $1,260 billion per day in 1995. From 1983 to 1993 total cross-border sales and purchases of United States Treasury bonds rose from $30 billion to $500 billion. J. EATWELL, 'The Liberalisation of International Capital Movements: The Impact on Europe, West and East' in *Understanding Globalisation*, Stockholm, A&W, 1998, p. 74.

[13] S. SASSEN, *Losing Control*, New York, Colombia University Press, 1996, pp. 29ff.

prescribed that the signing countries should not be allowed to treat foreign investments different from domestic and that property rights should be protected. It would serve as '...the constitution of a single global economy', according to Renato Ruggiero, the general director of the WTO. The proposal was met by a storm of criticisms and temporarily stopped. The critics argued that the agreement would prevent the signing nations implementing any restrictions on foreign investors motivated by, for instance, environmental protection, human rights or social equality.

The globalisation processes raise some important ethical questions. One concerns the problem to identify the agents, a presupposition for responsibility. Another concerns the possibility to formulate global norms for global action. Is a global ethic possible?

Global Effects – Global Responsibility

When the nuclear plant in Chernobyl broke down in 1986, the discharge of radioactivity reached the northern part of Sweden, 2000 km away. As a result the forests were contaminated and the meat of the elks and the reindeers made unfit for consumption for decades ahead.

This example illustrates, in a distinct way, one aspect of globalisation; the distant effects of large-scale systems. In this example it was a technical system, but as we noticed even large-scale economic systems raise the same question. Who is responsible?

However, in order to discuss responsibility in the era of globalisation we must make clear what responsibility means and why it is important from a moral point of view. Let me sketch a simple model in order to grasp the meaning of responsibility. First we have a moral agent A. A has an *intention* to do something. A performs an *action* in order to realise her intention. Some *consequences* follow from A's action. These consequences can be both the intended ones and not intended. When we say that A is responsible, we mean that A caused the consequences and that A, if necessary, should be able to motivate her action and the consequences following, i.e. they should be able to answer the question Why did you do that? in a reasonable way.

However, some conditions have to be fulfilled to make it meaningful to talk about responsibility. One is that A was *free* to choose and act in line with her intention. If A was forced to perform the action, it is not A but the one who was forcing A, who is responsible for the consequences. Another condition is that A was able to *foresee* the consequences. If consequences follow that were impossible for A to foresee, then A is not responsible for them. However, if A is aware of the fact that he or she can not foresee the consequences but acts anyway, we say that A acts irresponsibly. In this case, it is reasonable to say that A is responsible for the unforeseen consequences of their action.

In real life, this ideal view of responsibility is seldom met. People's freedom to choose and ability to foresee the consequences are restricted in different ways. For instance, an engineer who is employed by multinational company, which is acting in a morally or legally blameworthy way at another part of the globe, may be totally unaware of this fact. In this case, the engineer, as a co-worker in the company, unknowingly is participating in blameworthy collective actions, without having an opportunity to choose any other direction. To avoid this kind of situations, it is important for the employees to seek information about their companies and to establish channels for influence and participation in decision-making.

Why, then, is it important to identify a responsible agent? One answer is that being responsible means being a moral agent, i.e., being a human being and not a robot. Thus, responsibility is a distinguishing mark for humanity. However, it is also for consequential reasons important to identify a responsible agent. If anything goes wrong, the responsible agent is able to change his or her action in the future in order to avoid doing the same mistake again. This is not to say that our moral obligations are restricted to those situations we have created, by intention or neglect. Globalisation implies new channels for information about what is happening in other parts of the world, for instance political repression or famines, and gives us new means to intervene. Thus, globalisation and the decreasing relevance of distance increase this kind of global moral obligation that is not necessarily related to effects of our own actions.

So far, we have assumed that A, i.e. the agent, is an individual. However, A can also be corporate bodies like states, companies and organisations. Even these perform actions that have consequences for human beings. In this case, those individuals who make the decisions to perform the actions are the ones responsible.

However, sometimes it is difficult to discern an agent. One says that 'The market demands' or that 'It was necessary due to globalisation'. But what then does 'the market' or 'globalisation' mean. Is the market an agent? Who is responsible? 'The market' disappears as an agent but the idea of this imaginary agent has influenced the performance of real agents, for instance a government. Then, the government is hiding behind 'the market' or 'globalisation'. It is a way of saying that we, the government, are not responsible because we were forced to act or not to act. However, sometimes for 'market' one should read 'companies'. Then, the meaning is that if a government, for example, introduces legislation with the aim of protecting workers rights, companies acting on a global market will in fact move their investments to another country with more liberal labour laws. In this latter case, it is possible to identify the responsible agent although there is no one to enforce any rules. 'Whoever is free to run away from the locality, is free to run away from the consequences',[14] Zygmunt Bauman notices.

The global financial market is perhaps the best example of a development towards global irresponsibility. Borrowers and lenders are provided with the possibilities of hedges against the risk of interest rate and exchange rate movements, so called derivatives. These new financial inventions are based on insecure future expectations and they are difficult to understand, monitor and control. Responsibility is accordingly disguised in the global economy.

The large-scale dimensions of the global market and the mobility of capital make it difficult to discern the responsible agents. There is a chain of actions from a decision at a stock market to sell the share of a company to a possible effect that a plant, perhaps situated on the other side of the globe, has to be closed down.

The development of large-scale technical systems has similar consequences. Chernobyl is a good example of the problem of fatal consequences of irresponsible performance of technicians. The destructive environmental effects were so distant that the responsible technicians never could take notice of them. The new global risks have no boundedness in space.

In order to overcome the tendency towards global irresponsibility global action is necessary. Different solutions have been proposed. Political scientist David Held argues for a 'cosmopolitan democratic law', i.e. democratic laws agreed upon by the different states but

---

[14] Z. BAUMAN, *Globalisation: The Human Consequences*, Cambridge, Polity Press 1998, pp. 8ff.

valid and implemented on a global level. These laws would include protections of human rights, of decent labour conditions and environmental standards.[15] Regional economic and political integration, like the European Union, could be solutions in the same direction. In order to overcome the anarchical financial market different proposals for a more highly regulated international regime have been presented. The economist James Tobin has proposed a 0,5% tax on foreign exchange transaction in order to dampen speculative international financial movements without preventing long-term investment.[16] Further, there is a need for a global watch, not only – which already is in existence – concerning human rights violations, but also for environmental protection and economic black mailing from the side of large multinational companies.

However, globalisation not only asks for new global institutions, but also, and more fundamental, for a global ethic that can provide criteria for the global institutions.

Towards a Global Ethic?

Globalisation means that people from different parts of the world come closer to each other. As we noticed, an action performed in one part of the world will have consequences far away in another part. According to a recent report from the United Nations we all live in a 'global neighbourhood'. This process creates a need for a global ethic.

When we talk of a global ethic we can mean, descriptively, that there in fact exist common moral norms throughout the world. But we can also claim, normatively, that moral norms are universal. This position then denies the relativity of normative ethics. The argument outlined here does not deny that there exist differences in moral norms in different parts of the world, but insists that there are some universal norms behind these differences, at a more fundamental level of moral thinking.

But is it reasonable to expect the development of a global normative ethic? Are not the differences between cultures too great? As the

[15] D. HELD, *Democracy and the Global Order*, Stanford, Stanford University Press, 1995.

[16] J. TOBIN, 'A Currency Transaction Tax, Why and How' in *Open Economies Review* 7 (1996).

example of different views on child-labour, a commonly used example in arguments for ethical relativism, should illustrate, it is possible to assume an ethical universalism.

In one part of the world, child-labour is considered immoral as an abuse of children. In another part, it is natural to see children as a part of the work force. The first view can be formulated as the norm 'Child-labour is immoral!' and the second view as the norm 'Child-labour is desirable!' How shall we consider this disagreement? Is it a genuine moral disagreement, illustrating the impossibility of a global ethic? I think the answer is no, for the following reasons.

Those who hold the view that child-labour is immoral argue that it is a form of exploitation. Children are unable to defend their rights and should, instead of working, go to school and have some time for play. Child-labour is, according to this view, an obstacle to the natural development of children. Those who see no moral problem in child-labour argues that in a poor society, all members of the family have to earn money in order to survive. In this kind of society, going to school is no alternative because there simply are no schools! The view that children should get the opportunity to play belong, according to this view, to a western way of constructing 'childhood' that is alien to cultures other than the western one.

When we look closer to what, at the first sight, can be apprehended as a moral disagreement, we find that there are other, non-moral, differences or disagreements behind. One difference is the opportunities that children, due to economical and social circumstances, have. When one can offer the children better alternatives, i.e. to go to school and to play, it is immoral to force them to work. However, if, on the other hand, starvation were the alternative to work, surely it would be immoral to prohibit child-labour! Thus, although the conclusions are different, both views can be based on the same value assumption that one should choose the alternative that is to the children's advantage.

Thus, the disagreement concerning child-labour seems to be explained, rather as consequences of socio-economic differences than as deep moral disagreements. If this can be considered as a representative example of culturally based moral disagreements, it seems reasonable to assume that actual differences in moral norms, should not be an obstacle to the possibility of constructing a global ethic.

However, although globalisation creates a need for ethical universalism, and such a universalism is possible to achieve, globalisation may also lead to an unfortunate moral, cultural and social regimentation. Then, cultural norms and modes of behaviour will be imposed from abroad, quelling the traditional ones. MacDonalds hamburger may serve as an illustration of this tendency. If globalisation implies cultural imperialism some important values will be lost. Firstly, local cultures have developed through generations and have, one could claim, as a consequence of this process intrinsic value. As a consequence, their preservation and development is a moral obligation. Secondly, to belong to a culture is for many people a basis of their sense of identity and community. If their local culture is undermined, their lives will become fragmentised and lose meaning. Thirdly, although globalisation implies a need for democratic institutions at a trans-national level, there is also a connection between community and democracy. For many people, democratic participation is only meaningful in relation to the community to which one belongs. Thus, if we want to improve democratic participation, the local community must be protected.

The quest for 'appropriate technology' or 'intermediate technology' is one aspect of, what has been called the communitarian movement. Thus, the argument for appropriate technology goes, new technology should not be forced from outside, but instead be integrated into the technological and cultural tradition of the specific community. Then, important cultural values will be preserved. Only when this happens, can a technological development that will serve the whole community and not function as an alienated enclave take place?

Obviously, globalisation will have different implications for ethics. There is in the era of globalisation on the one hand a growing need for a global ethic, i.e. a shared moral basis. On the other hand there is also a need for a communitarian ethics, i.e. norms that are anchored in moral communities.

Which moral principles should be included in a shared moral basis? This must be a matter for further discussion even though a ground is laid by the UN declaration of human rights. In accordance with the UN declaration, a basic principle for a universal ethic is the respect for human dignity. This principle can also be formulated as an equal right for each individual to a certain quality of life.

The principle of human dignity is abstract and general. However, it can serve as a basis for at least the following three derived principles:

1. A principle of human rights.
   According to this principle, universal human rights (as they, for instance, are elaborated in the UN declaration) should be respected.[17]

2. A principle of equality.
   The principle of human dignity implies a principle of equal dignity. This principle stresses the need to work for a more equal distribution of world's resources. The discussion on the more specific principles of international distributive justice is extensive[18] and includes, e.g. a proposal for a 'global resource tax'.[19]

3. A principle of respect for identity and integrity.
   According to this principle, the uniqueness of individuals and communities should be respected. This implies a respect for the integrity of local communities and traditions while they form the social context for identity-formation.

Of course, the possibilities of implementing these moral principles depend on social and economical circumstances. As Amartya Sen has shown, democratic institutions are a presupposition for the elimination of famines.[20] However, as Sen also argues, even if a free market economy is a prerequisite for economic growth, the protection of human rights and a policy for social equality can not be left to the market but requires political intervention.[21]

Conclusion

The continued evolution towards globalisation seems to be a probable prospect for the future. The driving forces behind are technical and economical. As a consequence nation-states are losing control to

---

[17] See S. SHUTE and S. HURLEY (eds.), *On Human Rights*, The Oxford Amnesty Lectures, Oxford, Basic Books, 1993, for different theoretical justifications of universal human rights.

[18] For a survey, see C. BEITZ, 'International Liberalism and Distributive Justice. A Survey of Recent Thought' in *World Politics* 51(1999)2.

[19] T. POGGE, 'An Egalitarian Law of Peoples' in *Philosophy and Public Affairs* 23(1994).

[20] A. SEN, *Poverty and Famines*, Oxford, Clarendon Press, 1981.

[21] A. SEN, 'Property and Hunger' in *Economics and Philosophy* 4(1988)1, pp. 57-66.

the advantage of market forces and international agencies. This new situation will put new demands on all of us, but not least on engineers who have special possibilities to influence the direction of this development. The far-reaching consequences of our actions put a demand of an equally far-reaching scope of responsibility as well as of new institutions for global governance. In a shrinking world, the need of a universal global ethics is constantly demanding.

References

BECK, U., *Was ist Globalisierung? Irrtümer des Globalismus – Antworten auf Globalisierung*, Frankfurt am Main, Suhrkamp, 1997.
DOWER, N., *World Ethics: The New Agenda*, Edinburgh, Edinburgh University Press, 1998.
GIDDENS, A., *The Consequences of Modernity*, Cambridge, Polity Press, 1990.
HELD, D., *Democracy and the Global Order*, Stanford, Stanford University Press, 1995.
SASSEN, S., *Losing Control? Sovereignty in the Age of Globalisation*, New York, Columbia University Press, 1996.
WATERS, M., *Globalisation*, London and New York, Routledge, 1995.

3.2

EXAMPLES

3.2.1

THE HEMACARD PROJECT

Applying the Constructive Technology Assessment Method to Computerised Health Cards

Dominique Dieng

The Constructive Technology Assessment[1] method aims to integrate the human aspects throughout the development process when building technology. This method was applied to the development of the computerised medical file integrated into a smart card, known as the Hemacard health card. The results of this experiment demonstrated the need to introduce a method like this from the very beginning of the project, as human aspects need to be included in the design of the technology itself. The experiment also highlighted the need to undertake regular assessments throughout the project and to take great care not to neglect the human aspects in the interests of technology and not to be carried away by 'technological delight'.

The purpose of this article[2] is to show, using the example of the Hemacard health card, how a technology can be developed taking into account and integrating both socio-organisational and ethical or legal aspects. Although the *Constructive Technology Assessment* process presupposes a multi-dimensional approach and would also integrate economic aspects, we intend to focus here chiefly on the more social aspects. This does not mean to say that this dimension was underestimated in the Hemacard project, but it was assessed at a macro-economic level.

---

[1] Readers will find a comprehensive description of the approach in N.T. NGUYEN, G. FOUREZ and D. DIENG, *La santé informatisée. Carte santé et questions éthiques*, Brussels, De Boeck Université, 1995.

[2] This research is funded partly by the P.A.I. programme (Interuniversity focal point) – Phase IV, P4/3 programme of the S.S.T.C. (Prime Minister's Services).

We also intend to show through this article that a *Constructive Technology Assessment* process is not always easy to implement. It is a 'forum' for ongoing negotiations and although the engineer risks facing difficulties, 'the game is worth the candle', because if this method is applied correctly, it will result in technology being better suited to users' needs and, ultimately, being more readily accepted. After a brief description of the aims and technical aspects of the project, we will explain the *Constructive Technology Assessment* approach adopted and point out the difficulties which an engineer is likely to face. We will also illustrate our article with examples of negotiations which resulted in modifications to the initial technological project.

The Aims of the Hemacard Project

Hemacard is a research project. The research team's aim was to test the applicability of the Constructive Technology Assessment method on emerging technology. The technology chosen to test this method was that of the microprocessor card applied to a specific area – the health care sector. The initial idea was to provide a *portable minimum medical file* for haematology patients.[3]

The experiment was conducted in the haematology department of *the Hôpital Universitaire de Mont-Godinne* (Mont-Godinne teaching hospital) in Belgium by the P.A.I.[4] Health team of the *Cellule Interfacultaire de Technology Assessment* (CITA – Technology Assessment Interfaculty Unit) of *the Facultés Notre-Dame de la Paix* in Namur, Belgium. The multidisciplinary team, in consultation with the director and one of the doctors in the department, defined other objectives which focused more on ways of improving patient care:
- improving the quality of care,
- improving communication between health professionals, that is among doctors and between doctors and nurses, etc.,
- improving relations between patients and doctors,
- making it easier for patients to move around and travel,
- ultimately, enabling the use of the data gathered for statistical purposes or to trace development curves.

[3] Patients suffering from blood disorders (leukaemia, etc.).

[4] P.A.I: Pôle d'Attraction Interuniversitaire (Interuniversity focal point). Research funded by the S.S.T.C. (Prime Minister's Services – Scientific, technical and cultural services).

Description of the Technology Used

System Functionalities

The interface developed needed to enable the use of the following functionalities:
- patient identification,
- the provision of a complete summary of the medical data required for in-depth patient follow-up,
- maximum assurance in terms of security and privacy,
- the possibility of printing all or part of the medical file,
- and, at a later stage, the possibility of making statistical calculations.

General Technical Specifications

The assessment criteria defined to decide upon the technology which would be used for the experiments concerned card capacity and data security. Having compared the various forms of memory card technology (magnetic cards, optical cards, microprocessor cards, etc.) the technology adopted was that of the smart card. In fact, when the technical assessments had been completed, the development team felt that this technology corresponded best to the required specifications in terms of the security of the information and storage capacity. Smart card technology offered the best combination, offering both a high level of security and a large memory capacity.

Two different types of smart card were selected, depending on the intended holder. The patient card, kept by the patient and containing data from his medical file, used COS 24K card[5] technology. The health professional's card, which authorised access to the contents of the patient card and enabled authorised actions to be carried out – read only, read and write, modifications – consisted of an MCOS card.[6] In addition to the cards, the patient and the doctor were both given a personal code. The access code for the health professional's card defined the valid operations which the health professional was likely to be able to carry out, while that of the patient was intended to reinforce patient identification.

---

[5] COS 24K: erasable card containing just one directory.
[6] MCOS: erasable card containing several directories.

In practical terms, to be able to access the data contained on the patient's card, four elements were required:

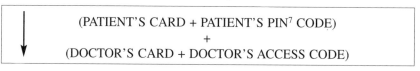

Access to Data on Patient Card

The basic equipment consisted of a PC 386 and two single-slot card readers, one for the patient card and the other for the health professional's card. The equipment was installed in a consultation booth in the day hospital.[8]

Description of Software Applications

The software applications developed needed to cover the following functions:
- management of interaction between the user and the card, that is the user interface, maintenance of writing and reading operations on the card,
- maintenance and saving of data on the computer hard disk,
- management of security functions as regards reading, writing and access to the contents of the card.

Potential Users and Their Clearance

Three potential categories of users had been identified beforehand:
- the patient: that is outpatients and not hospitalised patients. The patient carries the card and consequently part of his medical file. However, he is only authorised to read the contents of the card, and even then only when he is with his doctor. This makes him a 'passive' user,
- the nurse: she is responsible for the patient's medical follow-up and for administering medical treatment and therefore needs to

---

[7] PIN: Personal Identification Number

[8] Mobile hospital unit: outpatients' department where patients do not stay. They come for consultation, receive their treatment (in this case chemotherapy) and then return home the same day.

be able to access the part of the card relating to treatment, but not the entire medical file,
- the doctor (and his assistants[9]): he is the main user of the patient card. He is therefore authorised to read all the data on the card, to write on it, that is to enter new data or modify existing data,
- later on, one of the administrative secretaries had to be involved in the project to create the card and enter administrative data.

Description of the Constructive Technology Assessment Process in the Hemacard Experiment

When preparing the project, the researchers' aim was to develop a health card, the structure of which integrates both the ethical dimension and the legal or even social aspects. They wanted to be able to implement the *Constructive Technology Assessment* method.

During an initial phase, in parallel with the development team which defined its technological choices, the legal aspects of this new technology were examined. Particular emphasis was placed on the rules governing privacy. However, we subsequently came to realise that developing a health card had other legal implications, such as the ownership of the card and its contents (who owns the card? the patient? the doctor?) or the division of responsibilities among health professionals.

A series of ethical questions were also raised by the team ethicist: what could such a medical file mean for the patient? How will the relationship between the doctor and the patient be affected by the use of a health card? Will the patient remain as such for the doctor or will he become a partner? Should the computer screen be directed towards the patient so that he can read the contents of his medical file or not? The ethical aspect of the question of card ownership was also raised: who will be the owner? The hospital? The patient? Who will own the information on the card? The doctor? The patient? Given that the patient holds the card, can he or can he not access all or part of its contents alone? Will he need the assistance of a doctor? These, very briefly, are just some of the main questions which came up as the Hemacard health card was being developed. The development team and the project promoters held numerous discussions to

---

[9] The medical assistant is a young doctor who has almost completed his/her medical training and is in his/her specialisation year.

answer these questions, and in particular to make the necessary choices. The decisions reached contributed towards designing the card and were integrated into the development process and the definition of the specifications. Hence the wish to ensure that the patient was involved, even passively, rather than simply holding the card, resulted in the need to give him an access code. By entering the code the patient is giving a form of tacit consent, authorising the doctor to read the data on the card. Similarly, the need to ensure maximum security for the data and the best possible data confidentiality resulted in the choice of a dual security system. A health professional's card is needed to be able to read the data on the patient card. In addition, the health professional can only access the data on the card in accordance with his clearance as determined by the code entered. The data on the patient card will only be accessible if the patient has given his consent by entering his personal code. However, one exception was provided for: should a patient arrive at the hospital unconscious, an area of the card containing emergency data was accessible without the need to enter the patient's code.

Let us take another example to illustrate the way in which ethical considerations guided the development team's organisational choices. However, before we do so, it is important to point out that the health card is not a whole in itself but part of a technical environment that includes the computer, the card reader, a printer, etc. The development team's discussions with the doctors, the psychologist and the patients highlighted the importance of the patient's involvement in his own health. On the basis of these considerations, they suggested to the promoters that the computer screen should be directed so that the patient can read the screen at the same time as the doctor, it being understood that the doctor would ask the patient if he wanted to see the screen or not. During training sessions particular attention was paid to increasing doctors' awareness of this human dimension of the technology: a screen that comes between the doctor and the patient or a 'facilitator' screen? These few examples illustrate the way in which the technology and the socio-organisational, ethical and legal aspects constantly interact. These social, ethical and organisational aspects provided food for thought throughout the entire project.

The second phase which followed the examination of the social, ethical and legal context consisted, as we mentioned earlier, of defining the specifications with the promoters. Meetings were then held with the other doctors in the department. The computer specialist responsible for the project also met the hospital computer specialist on several occasions. However, in practice this rapidly became the two men's project: the department head and one of the doctors who is a computer enthusiast. Two years after the start of the development phase, individual interviews were organised with the patients, the paramedical staff (senior nurses and the psychologist) and the medical staff (doctor and assistants). The development team was keen to integrate the ethical and psychological dimension and was particularly careful not to upset patients too much, adapting the interview method to the different categories of users. For the medical and paramedical staff, the method adopted was that of the face-to-face interview. For the patients, different techniques were used and adapted, bearing in mind their wishes and their state of health. In fact, haematology patients are physically and psychologically weakened by their illness. Shorter group discussions were organised (two hours rather than the usual three), supplemented by questionnaires to be filled in at home. Individual meetings were organised for patients who had agreed to take part in the evaluation but preferred not to be involved in a group discussion (for instance, because they were afraid that they would have to reveal their health problem or be confronted with those of other people). The aims of the first wave of discussions and meetings were as follows:

- to assess the situation of the socio-relational dimension within the haematology department: patients' experiences, relations between patients, doctors and nurses, etc.,
- to assess the way in which the various categories of people involved (doctors, patients, nurses) perceived *the concept of the health card*.

The second phase of meetings was organised to assess the technology applied. Between these two phases, several information meetings were organised either with the doctors only, with the nursing staff only, or with the entire medical and paramedical team. The purpose of these meetings was to keep them up to date with the project and the way it was developing. As for the patients, they were not involved in this awareness and information campaign because of the state of their health: the promoter did not wish to trouble them with this project until its development was assured. Only when the prototype was operational, some of them – those who were asked by the doctors if they would agree to take part in the experiment – started to be informed. The other patients were informed at a later stage and indirectly, by means of a poster in the waiting room.

We have already mentioned the patients' participation in the assessment of the concept of the health card and, later on, in its technical realization. This also included an ethical dimension. Although strictly speaking this was a non-medical experiment, the development team and the promoter adopted attitudes consistent with a medical experiment because patients were involved. They requested authorisation from the hospital advisory committee to conduct the experiment and undertook to respect the Helsinki Declaration on medical experiments.[10] A contract co-signed by the promoter, the research director and the patient himself was handed to the patient. It included the following clauses: the possibility for the patient to withdraw from the experiment at any time without this affecting his subsequent medical treatment in any way, an undertaking by the team to reproduce precisely the words of the patients questioned and to respect confidentiality.

When the information had been gathered from the various categories of people involved, it emerged that certain changes were necessary to avoid the risk of losing the commitment of doctors who, incidentally, had already proved difficult to motivate. After a few weeks in use, the evaluation showed that the introduction of a new patient into the experiment and therefore the creation of a new card lengthened the consultation period by half an hour. In a department in which doctors already have a very busy schedule, this increase in the length of the consultation was becoming difficult to manage. To overcome this difficulty, it was decided that one of the secretaries in the department would be made responsible for creating new cards and entering the patient's administrative data, a step which had not

---

[10] World Medical Association. The first version dates from 1964 and it was subsequently amended in 1975 (Tokyo), 1983 (Venice) and 1989 (Hong Kong). These texts emphasise the quality of the care provided and the issue of privacy. As regards experiments involving human beings, they stress the scientific quality of the experiment; the competence of the experimenters; the assessment of the risks and benefits linked to the aim pursued and the search for the right balance between them; protection of physical integrity and of privacy; the need to inform the subjects involved of the aims, methods and progress of the experiment, possible risks and anticipated benefits; respect for the autonomy of the subjects in the process of obtaining consent; fairness in the selection of patients and subjects. Cf. in particular M.-L. DELFOSSE, *L'expérimentation médicale sur l'être humain. Construire les normes, construire l'éthique*, Brussels, De Boeck, 1993.

been planned at the outset. Another change brought about by the discussions concerned the positioning of the equipment. Having examined the way the department operated, it emerged that the solution initially proposed risked causing operational difficulties. Another solution was therefore adopted, which involved installing the equipment in one of the consultation booths.

The Lessons Drawn from the Project

The results of the research led us to conclude that the first stage in a technological project like this is to establish a *chart of the people* likely to be involved in the project and to describe their respective roles (even if, at first, the prevailing feeling is that they will not have a role to play). This initial description contributes towards facilitating the audit of the environment in which the project will take place and ensure that no one is forgotten. However, this is not always an easy task. In the Hemacard experiment, one of the mistakes made was to begin to draw up the chart of the people involved from the point of view of the doctor rather than that of the patient who, after all, is the centre of the project and the main person concerned by the health card as he holds it. This is why the role of the administrative secretary had been overshadowed to some extent. The retrospective analysis concluded that the chart of those involved in the haematology department could have been prepared by following the patient's path step by step from the moment he enters the hospital until the moment he leaves.

It is also important to describe the *socio-professional relations* between those involved. This description will make it possible to identify the curbs and driving forces. As we have already mentioned, it is important to remember that a technological development is part of an environment which has its own history and in which people are used to interacting in a certain way. These personal interactions can have consequences for the acceptance of the new technology by those concerned.

This initial audit, which can start with documentary research (for example by analysing the organisation chart of the department), will be supplemented by meetings with representatives of the various categories. It will be implemented *from the very outset of the project*, so that the observations and expectations of those involved can be integrated into the design of the technology. However, this does not mean that this has to be the last stage in their involvement. Throughout the development of the project, those involved will be kept informed of the progress made in building the technological tool and will be consulted at the various phases: before the construction in the preparatory phase, during the test phase, and once the technology has been introduced. They will also be consulted when the engineer has to make a major technological choice.

The need for an *ethical commitment* on the part of the engineer is one point which must be stressed. He has a moral obligation to the people who are to use the technology which he has developed or those who could be affected by it. His commitment will depend on the people involved. Of course, it will not be the same if those involved are patients taking part in a health card project or if they are consumers confronted with genetically modified plants, or groups of people likely to be affected by the development of chemical weapons.

The Hemacard experiment also stressed *the development of the mental processes* within the development team and the way in which the engineer's *perception* of the object he is in the process of constructing is likely to evolve. In the health card experiment we conducted, the development team and the promoter presented the initial concept as a 'minimum medical file' when they referred to the contents of the card. By the end of the project, these contents had become a 'summary of the medical file'. This is a major difference and reflects the changes which occurred during the construction of the project, and in particular the way in which the use of the technology modified people's perception of the initial concept. We also noted another shift in the perception of the health card which emerged as it was being used. At the beginning of the project, the card had been intended for the 'medical follow-up' of the patient. By the end it had become an 'emergency card'. When the doctors presented the card to the patient, they often said how practical it would be in an emergency. Moreover, most of them did not use it at the beginning of the consultation to remind themselves of the patient's history, but simply updated it at the end of the session by entering the modifications and new treatments. These observations clearly show that the gap between theory and practice may be considerable, sometimes giving rise to user dissatisfaction. To some extent this was the case with the Hemacard experiment during which doctors expected to find a minimum medical file but were surprised to discover the 'limited' possibilities of the system, at least compared with their expectations. This is why we lay particular emphasis on the *clarity of the message* given to users by the engineer: the aim of the project must be clearly explained and the engineer must be sure to check that future users have really understood what he means. This assumes that the engineer has a very precise idea of what he is going to develop and has thought about the issue in depth, on the basis of the audit carried out in the initial project phase. This also means that when constructing the technological object, the engineer, the researcher evolves, too. However, even if this evolution is necessary, it is important that the project does not deviate too much from the initial aim and that the engineer refers regularly to the specifications: *the aims must not be modified* (unilaterally at least), *but the means to achieve these aims may change*.

It is also necessary to carry out a test phase with a few users, in order to minimise the risks of subsequent problems. In fact, users often expect a great deal from technology but do not wish to put up with the teething problems. They want the tool they are going to use to have been perfectly finalised. This is why we will also stress the need for the technological object to be finalised before it is put into widespread use, particularly since on a larger scale other problems risk arising. The engineer but also, and above all, the organisational team supporting him, must keep an eye on users' assessments so as to improve the technological object developed and transform the developments made into a success. Why do we put so much emphasis on this point? Because as soon as the user is involved, he is likely to make a substantial personal investment in the project. This can be illustrated by an example: in the Hemacard experiment, the assessments carried out among patients revealed reactions whose importance we had under-estimated. In fact, some of the patients had invested the card with a form of 'power'. This card, this little piece of plastic, was going to save them in an emergency. And yet they were aware of the limited nature of the experiment in terms of both space and time. Nonetheless, for these people, who always have a 'little suitcase, packed and ready to go', at the back of their minds, as they may be admitted to hospital at any time due to their illness, the health card represents security and the assurance that they will be treated very quickly in an emergency, even if they arrive unconscious. Some patients said that this idea made them feel safer, calmer.

We also discovered that the fact that the patient can read the data on the screen at the same time as the doctor contributed towards improving relations between the doctor and the patient, a view which is shared by both parties. Being able to read the screen enabled the patient either to ask questions about information he had not understood or to remember a point he wanted to talk to the doctor about, or to check that the doctor had not forgotten anything . All these are examples which illustrate how technology can affect relations between those involved, a dimension which the engineer will be sure to take into account.

One of the points which is often overlooked concerns the presence of the development team on site throughout the development and test phases, when they will help users having difficulties. The team's presence on site will facilitate not only their integration but also their perception of the tensions and issues involved. It will also facilitate user mobilisation.

Four years passed between the inception of the Hemacard concept and the end of the project. During this period, the project itself evolved, and so did the researchers. We have observed three main categories of changes: changes in the way the researchers perceived the various categories of users, an awareness that they did not involve certain groups enough, such as the patients or the nurses, and socio-technical adaptations.

The socio-technological adaptations include not only those we have just mentioned, but others as well. For example, the fact that at the beginning of the experiment the plan was to use two single-slot readers. Due to an incompatibility between the two readers, the development team, while waiting for this problem to be resolved, had proposed that for a time the doctor and the patient should use the same reader. This initially temporary solution became definitive. This example illustrates the interaction between Man and Technology: technical difficulties bring about changes and users adapt to these changes. The contrary can also be seen and the example, which we gave above regarding the integration of secretaries into the project and the development of new applications, clearly shows how technology adapts to users' needs.

A properly conducted audit will help identify certain difficulties, but will not make it possible to pinpoint them all. This means that it is important to maintain ongoing negotiations regarding technological developments. These, and above all the resultant objects, are one of the guarantees of user acceptance.

Conclusion

All the examples we have given show how the Constructive Technology Assessment method must be part of an ongoing process of negotiation. Negotiations between the engineers and the promoters, negotiations between the promoters and the actors, negotiations between the actors and the engineers. Some of the resistance can be highlighted relatively easily, but other cases, which are more difficult to demonstrate, risk endangering the Constructive Technology Assessment process. This is why we will once again stress the need for an in-depth audit in the project launch phase, so as to make it easier to identify all those involved and define their needs.

The Hemacard experiment also brought to light the ethical, legal, psychological and socio-organisational implications which a technological project may have. All too often these are underestimated, but they should be taken into account throughout the project. The experiment also showed that throughout the construction of the technological tool, new changes are likely to occur in the development process. In some cases, users will adapt to the technology, in other cases, the technology will adapt to the users. There is a permanent balance between the technological requirements and the social requirements.

References

DELFOSSE, M.-L., *Quelques réflexions éthiques sur les cartes santé*, Cahiers de la CITA, S3, Namur, 1993.

CAMBRIOSO, A. and C. LIMOGES, 'La controverse, le processus-clé de l'évaluation sociale des technologies' in *Analyse évaluative et évaluation sociale des technologies*, Québec, Cahiers de l'ACFAS, 1988.

KUSTERS, B., C. LOBET-MARIS and N.T. NGUYEN, 'Some Methodological Issues in Information T.A. – Two Cases Studies' in *Third European Congress on Technology Assessment: Post Congress Workshop*, Copenhagen, 1992; reprint in Cahier de la CITA T.A. 2.

MENKES, J., 'The Role of Technology Assessment in Decision Making-Process' in *International Symposium on the Role of T.A. in Decision Making-Process*, Bonn, 1982.

MOATTI, J.-P., 'L'expérience américaine de l'évaluation technologique aux U.S.A.' in *Culture Technique* (juin 1983)10.

NGUYEN, N.T., G. FOUREZ and D. DIENG, *La santé informatisée: carte santé et questions éthiques*, Brussels, De Boeck Université, 1995.

NGUYEN, N.T., *Bio-éthique et Technology Assessment, (Contrôler la science ?)*, De Boeck-Wesmael, 10/1990.

QVORTRUP, L. et al. (eds.), 'Social Experiments with Information Technology and the Challenges of Innovation', a Report from the FAST Programme of the CEC, Dordrecht, 1987.

VAN BOXSEL, J.A.M., 'The Relevance of Technology Dynamics for the Practice of Constructive Technology Assessment' in *Advanced Training Course on Technology Assessment Methodology*, Mol, 14-15 October, 1993.

3.2.2

THE CASE OF rDNA TECHNIQUES

Ethics and the Complexities of Decision-Making

Philippe Goujon

The 1970s rapidly became a focus for public interest and excitement. All areas of biology were influenced by the revolution in technical developments and in particular by the rDNA technique. This technology has been, and still is, used to create recombinant DNA from a variety of viral, animal and bacterial sources. There is serious concern that some of these artificial recombinant DNA molecules could prove biologically hazardous. This article is about the adventure of the regulation of the use of DNA technique, research and work. It is written fundamentally from an historical perspective, the events described taking place over a ten-year period, which covers the birth and the early years of modern biotechnology. In focusing on regulation, and in particular on the European public and science policies, the intention is to study how societies learn to digest new knowledge and to manage its inherent risks and consequences. Taking into account that this learning process is a multidimensional one and that the debates about biotechnology were from the start international, we will show that the process of regulating a new technology is very complex. We will also reveal how ethical and moral intentions can be confronted, in the debates and negotiating processes, with other interests, such as personal, sectoral, national, international, economical, legal and political interests.

Ethics and genetics, these two terms are often associated, often without really knowing how to define the first, nor even if our questions should concern research activities or their application. Ethics is a global conception of existence, it results from 'a philosophical reflection enabling man to find his place in relation to himself and enabling him to apprehend the society of which he is part.'[1]

In focusing on public policy and regulation, in particular at European level, on the new techniques in genetic engineering during the 1970s, we will study how societies learn to regulate the arrival of new technological knowledge and possibilities and to manage their consequences. Taking into account that this learning process is multidimensional and that the debates about biotechnology were from the start international and multi-sectoral, we will show to what point the processes designed to regulate a new technology become complex. We will show that moral and ethical intentions are often at issue, and that allowance must be made for other factors such as personal, sectoral, national, international, economic, legal and political interests.

We will restrict ourselves to a purely historical perspective, with the aim of describing the process of regulating rDNA techniques at the highest levels (the main American, British and European agencies or scientific bodies which determine policy.)

Recognition of the Problem

Biotechnology has been praised to the skies, high hopes placed in a better knowledge of DNA,[2] key to a better world, but also creating anxieties about the use of these often esoteric techniques in the public. Following the discovery of the double helix by Watson and Crick, the genetic code was explained. After the work of Avery, and thanks to the work of Lederberg and many other scientists, the field of bacterial genetics gradually expanded.

Several decades later, Stanley Cohen and Herbert Boyer at Stanfield published[3] and patented their use of restriction enzymes with bacterial plasmids for the basic 'cut and stick' operation. They took

---

[1] F. SÉRUSELAT, 'Vers une approche des définitions du concept de l'éthique' in *La lettre éthique* 5(1990).

[2] Deoxyribonucleic acid: a molecule having the structure of a double helix and representing the chemical basis of heredity. Present in chromosomes and in mitochondria and chloroplasts

[3] S.N. COHEN, A.C.Y. CHANG, H.W. BOYER and R.B. HELLING, 'Construction of Biologically Functional Bacterial Plasmids *in vitro*' in *Proc. Natl. Acad. Sci.* 70(1973), pp. 3240-3244.

two organisms incapable of joining in nature, isolated a segment of the DNA of each of them using chemical scalpels called restriction enzymes, and combined the two fragments of material in a plasmid that was then introduced in a host cell. This cell absorbed the plasmid and began to replicate it indefinitely, thus generating identical copies of the new chimera. Soon known by the name of genetic engineering, this operation marked the dawn of the age of genetic recombination and the 'mastery of heredity'.

A financier – and ex-biochemist – named Robert Swanson contacted Boyer and asked him whether their technique could enable the creation of an organism producing proteins foreign to its constitution. Boyer replied in the affirmative. He borrowed $500 and joined Swanson to set up a company to exploit the potentialities of the new techno-science; he named the company 'Genentech'. On 15 August 1977, the scientists at Genentech succeeded in inserting the gene of somatostatin in the genome of live bacteria which subsequently produced it. On 24 August 1978, they managed to have insulin produced by the *E. coli* bacteria. Many other experiments followed. Dozens of biotechnology companies were founded. Thanks to the talents of molecular biologists, genetic recombination became common practice and the number of human genes inserted into bacteria and other simple organisms grew rapidly.

The implications regarding the public policy of this new technology, its potentialities and the biological revolution it initiated did not affect all the departments of government at the same time nor in the same way. The reactions were heterogeneous and their harmonisation a hard-won victory. However, the inventions could not be wiped out nor their discovery denied. The question remained: could they, or should they, be controlled? This was the issue at the origin of the Asilomar conference, which was to mark the beginning of a long process to regulate modern biotechnology.

Asilomar (California, 1975) and Its Consequences

At the start of this revolution, some scientists, aware of the potential risks, organised in February 1974 a conference on the biological risks. This hardly drew much attention from the scientific community or the media, but it did stimulate thought.

In June 1973, the annual meeting of the Gordon Conference on Nucleic Acids was held in New Hampton (USA) and was mainly devoted to the question of the risks involved in recombinant DNA (rDNA) research and techniques. The joint chairmen of the conference, Maxime Singer and Dieter Soll, drafted a letter[4] to the National Academy of Sciences and the Institute of Health requesting the creation of an advisory committee to evaluate the biological risks of rDNA research and recommend appropriate action. The letter was published in *'Science'*. In reaction to this letter, the National Academy of Sciences announced that Paul Berg would head the advisory committee. The 11 members, all involved in rDNA research, were aware of the extremely rapid development of the research and techniques and of the shared misgivings about the potential risks. Their report was published in the journal *'Science'* on 26 July 1974[5] and, almost simultaneously (though slightly abridged), in the scientific journal *'Nature'*.[6]

The Asilomar Conference

In response to this letter, a congress was organised. It was held in Asilomar, California, in February 1975: one hundred and forty biologists from seventeen countries took part to outline the risks for the environment and human health that might result from experiments with recombinant DNA. This congress received considerable coverage from the world press.

At the most factual level, this congress represented a meeting of invited scientists during which eminent specialists discussed the risks that might be associated with work and techniques involving recombinant DNA and the ways of containing or of reducing these conjectural risks. The scientific press remarked that most of the participants were quick to return to their work, and opposed any regulation of their research. When the congress was well in its stride and seemed in all probability to be heading towards granting the moratorium requested by the eleven specialists in their letter of 26 July 1974, an invited lawyer read a statement on the legal responsibilities

---

[4] M.F. SINGER and D. SOLL, 'DNA hybrid molecules' in *Science* 181(1973), p. 1114.
[5] P. BERG et al., 'Potential Biohazards of Recombinant DNA Molecules' in *Science* 185(1974), p. 303.
[6] In view of its historic importance, the text of Berg's letter is given in the annexe at the end of this chapter

of researchers responsible for a biohazard. The last speaker, Professor Harold Green from George Washington University Law School, captured the full attention of the participants with a communication entitled *'Some conventional aspects of the legislation and the way in which they are likely to affect you in the form of, say, a damages and interest suit for several million dollars'.*[7] The fear of being involved as the defendant with a very substantial financial risk at stake would soon lead to where more 'altruistic' considerations had failed.

The following day, during the closing session, the researchers adopted a safety programme. This programme emerged from numerous discussions on the different levels of risk in order to classify experiments and on the corresponding physical confinement of potential hazards to be required. Amongst the more constructive ideas – emphasised by British participants such as Sydney Brenner – was the concept of 'biological containment': the use of micro-organism disabled in ways which would limit their ability to survive or reproduce outside the contained vessel and special conditions provided in the experiment. This line of argument was the starting-point for a great deal of risk assessment research over the following years – pratically all of it reassuring, but always limited by the logical impossibility of proving the opposite.

Consequences

The Asilomar conference started a fundamental debate on the control of science and technology and extended it to all fields of life sciences and technology. Scientists were often shocked by the popular misconceptions of the issues and by the violent attacks to which they were subjected; especially from the major ecological movements. However, some of them agreed with the critics and continued to demand safety measures – or even a total moratorium – on all research with recombinant DNA. The upshot of all these factors was an animated public debate involving scientists facing the wrath of public interest groups or local politicians who were often ill informed.

Public concern in the USA reached a climax in 1976-1977, a period corresponding to the introduction in Congress of proposals to regulate rDNA research. During the same period, the National Institute of

---

[7] Michaël ROGERS, 'The Pandora's Box Congress', *Rolling Stone* (19 June 1975), p. 77.

Health (NIH), under its director, Donald Frederickson, set about developing a regulatory framework for conducting such research under the NIH Recombinant DNA Advisory Committee (NIHRAC). The first version of this regulatory framework was published by the Federal Register on 7 July 1976.

The American Society of Microbiology (ASM) played an important part in the gradual development of a structured, balanced response from the scientific community. It enacted a recommendation approved and widely distributed in May 1977 which emphasised the required scientific and technical skills and placed all responsibility for regulatory measures on the Department of Health, Education and Welfare (HEW). An advisory committee composed of lawyers, and representatives with an appropriate technical expertise was to be set up. The recommendation likewise stressed delegation of responsibility to local committees at the level of the institutions involved in research into recombinant DNA to, at the same time, the experts at the institutions and the representatives of the public. This recommendation proposed exempting low-risk experiments from regulation, in order to preserve a certain flexibility and re-evaluate regulation in light of the experience. All these points will hold good not only in the discussions over the years ahead, but also in other countries and national laws.

During 1977, the scientists' concerns were communicated to the public and to the teams of the Congressmen concerned. Some amendments were made to the first bills, integrating advice from scientists. Information indicating non-dissemination or unwanted effects had considerable influence on the course of events.

In September 1977, Senator Adlai Stevenson wrote that most of the legislations under review were ill-equipped for attaining their objective, that is, the protection of the public, without however curbing research. He declared his intention to explore the use of existing laws in order to regulate research on recombinant DNA. In November 1977, the ASM expressed its concern with regard to the apparently unreasoned haste of attempts to establish a legislation to regulate research on recombinant DNA without first consulting the qualified scientists and medical experts. It underlined the need to make allowance for the fact that the first allegations concerning rDNA research were characterised by an uncontrolled imagination and, very often, by overstated assertions by persons not having a knowledge of infectious diseases. The ASM stressed the urgent need for a minimum provisional legislation to extend the regulatory frame to include all rDNA activities regardless of their sources of funding.

Throughout late 1977 and 1978, the ASM continued to work closely alongside the Congress committee, and the prospect of any federal legislation faded. However, this success was not self-evident. A national conference on 'Recombinant DNA and the Federal Government' was held in October 1980 with presentations by federal officials from 17 Agencies, by congressmen and their advisors in charge of legislative activities in the field, and by lawyers from Washington expert in such problems.

Lessons from the US Experience

The US experience, after Asilomar, was of great importance and constituted an example of an open dialogue between scientific and political circles. One lesson that might be learnt is that Congress was able to react promptly once public health was concerned. Another lesson is that the legislative process is adequate in the sense that it is sufficiently slow to prevent ill-adapted, hurriedly-passed legislation. A third lesson is the Congress' attention to the scientific community and its capacity to modify its opinion once presented with new, well-founded arguments. This does not, of course, mean that the legislation is dead and buried. There are still several Congress committees that continue to hold hearings concerning NIH activities and research involving rDNA. There is also still a specialist medical press that continues to follow the problem. Finally, if nothing else, the mere excitement surrounding the problem will sustain the topicality of the question of whether or not there should be any government involvement in rDNA research and technology. These three factors might explain why Congress continued to show interest in the issue.

This adapted regulation may be connected with the absence, during this period, of problems as regards genetic research and engineering, with the initial economic successes that resulted therefrom, with the prospect of a considerable potential market and – also – with the leading position in the scientific and economic fields of biotechnology that the USA wishes to defend. The US experience gives a lesson on how to manage the interface between science, technology and society while keeping to transparency, democracy and method, a lesson which will, consequently, be generally accepted and effective.

The British Reaction

The Ashby Working Party

When 'the Berg letter' appeared in 1974, British scientists were the most affected in Europe, because of the number of programmes and research centres concerned. This research was financed by the universities or by the Research Councils, so the Department of Education and Science were able to simply call a halt to all work involving recombinant DNA techniques until such time as the government had set a regulatory frame in place.

In July 1974, the Ashby Working Party was set up to decide whether research should be continued. Its report, published in December 1974,[8] recommended the pursuit of research provided appropriate precautions were taken, in particular concerning biological confinement. The Ashby report especially focused upon the concept of 'biological containment', by crippling the plasmid vector and its bacterial host. As a result of this prompt reporting, the concepts developed by the Ashby Working Party were used by the UK and other scientists in the February Asilomar conference.

Reply of the British Government and the Creation of GMAG

The government replied by setting up, under the direction of Sir Robin Williams, another Working Party in August 1975 to prepare a code of practice to regulate all activities involving genetic manipulation. This committee gave its answer in 12 months and recommended the creation of a supervisory authority for genetic manipulation and a regulatory framework. Consequently, the Genetic Manipulation Advisory Group (GMAG) was set up by the Department of Education and Science: it held its first meeting in January 1977. Its members included eight scientists and medical experts, four 'public interest' representatives, four trade unions representatives, four workers interests and two representatives of management – one nominated by the Confederation of British Industry, the other by the Committee of University Vice-Chancellors and Principals.

The concept of 'public interest' representatives was an innovation and allowed a more efficient communication with the public

---

[8] HMSO, *Report of the Working Party on the Practice of Genetic Manipulation*, Williams Working Party, Comnd. 6600, Her Majesty's Stationery Office, London, 1974.

although the GMAG meetings remained private. The GMAG followed the advice of the Williams Report with regard to a regulatory framework for analysis of the files concerning work on rDNA, concerning in particular four levels of physical confinement from the lowest level up to the strictest confinement level. This scheme was abandoned in 1979 and replaced by a risk assessment scheme. The largest confinement level was above the restriction levels for the bottom confinement level laid down by the NIH.

The publication of the Williams Report and the creation of the GMAG allowed scientific work to resume in the United Kingdom, where some ten level-III containment facilities were constructed. At the request of the GMAG, the Medical Research Council (MRC) also financed training courses for biological safety officers at Porton Down, the government microbiological research centre (now Centre for Applied Microbiology and Research). Following the introduction of the evaluation scheme in 1979, most experiments have since been recategorised in confinement level I in which a mere 'good microbiological practice' is required.

The 1970s saw increasing demand in many countries for the improvement of health and safety at work. A growing awareness of the risks influenced the debate around rDNA. In 1974, the United Kingdom adopted an important law: the Health and Safety at Work Act, giving more extensive powers to the government's Safety Executives (HSE) and the Factory Inspectorate. Further specific legal regulations followed under this act, requiring that all establishments (including the ministry and research institutes) set up local safety committees. Regulations concerning genetic and operational manipulations were likewise introduced (SI 1978 No. 752), operative from 1 August 1978. These latter demanded that 'persons should not carry on genetic manipulation unless they have previously notified the Health and Safety Executive and the Genetic Manipulation Advisory Group.' The GMAG did not give its opinion until the proposals had been discussed at local biological safety committee level.

The GMAG's development of a risk evaluation scheme in 1979, implemented since the 1980s, calmed scientific nerves (reassured the scientists) regarding the possibility of overzealous safety committees, useless delays and an excess of transparency damaging economic and commercial interests and distorting national and international industrial and economic competition.

Since the United Kingdom did not have a hard and fast description of 'good microbiological practice', a policy memo was put together for 'Guidelines for Microbiological Safety' by the scientists themselves in a 'Joint Co-ordinating Committee for the Implementation of Safe Practices in Microbiology'. The GMAG accepted this in July 1980. The risk assessment scheme resulted in most work being recategorised to either category I or GMP. In general categorisation of experiments was carried out by the local biological safety committee. From 1980 onwards, it was only in cases of uncertainty, or where work in categories II, III or IV was envisaged, that GMAG was asked to advise on a case by case basis.

The Regulatory Frameworks of the GMAG

The UK Williams Report and the GMAG code of practice (which differs from that of the NIH on essential points dealing with classification procedures and containment methods) came out before publication of the NIH guidelines (June 1976). As a result those European countries involved in research with rDNA decided initially to adopt the regulatory frameworks of the GMAG. In general, the lack of a legislative working structure such as the Health and Safety at Work Act made the introduction of legislation to cover genetic engineering all the more complex. This absence also explains the fact that, since the regulatory framework of the NIH was introduced with the lowest confinement requirements and with a codified system of categorisation, almost all the main European countries decided to adopt these. However, the implementation methods and standards vary greatly.

The recommendations of the US NIHRAC and the UK GMAG had enormous initial influence. Because they were discussed principally in scientific circles, the natural internationalism of science and its culture of critical scrutiny limited the extent to which these early guidelines were mere expressions of national interest. Such was the background of the first European legislative initiative by the European Commission.

The European Commission's Reaction:
From a Directive to a Recommendation

If the regulation, development and use of a new technology are complex at State level, they reach an even greater order of complexity when a community of several States is concerned, as is the case in the European Community. The complex sectoral make-up (the Directorate General Divisions, each with its own professional bias and sectoral interests to defend) of the European Commission, the problem of the co-ordination and harmonisation of its actions and the heterogeneity of national and corporate interests, the distance from national politics and from citizens and local communities and in general the multi-institutional structure (Commission, Parliament, Council) compound this complexity. If we bear in mind that biotechnology is essentially multidisciplinary, multi-sectoral both scientifically and as regards its multinational and multi-international applications and implications (and in consequence could not be the monopoly of any one specific Directorate General (DG)),[9] we begin to see the difficulties of putting in place a European regulation that is sufficiently homogeneous and binding or influential for all the Member States to respect.

The Situation of the 1980s

Let us attempt to take stock of the 1980s.

The expert opinion of the state of play in connection with genetic engineering as regards Commission services is to be found in DGXII, the base of a team of scientists recruited to carry out research in biological safety under the Euratom treaty who could identify (and interact with) competent external scientists.

As its main hopes rested on a proposal for a Community-wide research programme, the DGXII 'biological team' followed closely the international debate on the safety of rDNA research. On 21 January 1977, they organised a meeting of the chairmen of the national committees charged with the control of rDNA research. The purpose of the meeting was to study the way in which the Commission might contribute towards the compulsory strand of the guidelines in respect of this type of research such as it existed among the Member States and how the Commission might be placed to promote its harmonisation. The first objective was particularly important for indus-

---

[9] The main Directorate-Generals concerned by biotechnology that we may cite in this article are: DGIII: internal market and industrial affairs, IV: competition, V: employment, social affairs and education, VI: agriculture and rural development, XI: environment, nuclear safety and civil protection, XII: science, research and development, XIII: telecommunications, the information industry and innovation (in 1989, education and consumer protection became separate services, split off from DGV and XI respectively).

trials. The resulting opinion was that the Commission should put forward a Directive demanding the setting up of national control committees, defining their terms of reference, and promoting harmonisation of national guidelines.

DGXII (the Directorate-General for Science, Research and Development), in consultation with the Scientific and Technical Research Committee (CREST), the European Science Foundation (ESF) and other sources of scientific advice, proceeded, throughout the following year, to the formulation of a 'Proposal for a Council Directive establishing safety measures against the conjectural risks associated with recombinant DNA work'.[10] The legal base of this proposal was Article 235.[11] The Commission put this before the Council on 5 December 1978. The preamble stressed, in positive terms, the value of the pure and applied science, and the need to combine protection of the person, the conservation of food reserves and the environment, and rDNA research. It was clear about the international character and epidemiology of the conjectural risks, and about the fact that a delay in the development of research of work among the Member States could affect their scientific and technical competitiveness. The rapid evolution of science, the need to consider local circumstances and the need to maintain scientific industrial secrecy and intellectual property were also acknowledged. Since this proposal, both aspects – *the ethical and industrial/economic interests* – are now in evidence. The definition of work on recombinant DNA was identical to that for genetic manipulation in the implementation regulation in the United Kingdom.

In substance, the proposed Directive required a preliminary notification to, and authorisation from, the national authorities before commencement of any activity involving recombinant DNA. The national authorities would have to develop a categorisation system and inform the Commission accordingly, while the latter would publish them. The Member States were supposed to submit to the Commission the list of authorisations granted at the end of each year, with a covering general report on their experiments and problems.

[10] European Commission, Proposal ... *Official Journal of the European Communities*, 301, pp. 5-7.

[11] 'If any action by the Community appears necessary to achieve, in the functioning of the Common Market, one of the aims of the Community in cases where this Treaty has not provided for the requisite powers of action, the Council, acting by means of a unanimous vote on a proposal of the Commission and after the Assembly has been consulted, shall enact the appropriate provisions.'

Article V of the Directive provides for revision where necessary, at regular intervals not to exceed a period of two years, thus introducing a measure of flexibility. As this proposed Directive became the subject of debate in the European Parliament, the Commission team, scientists and authorities at DGXII level were alerted by the swing of opinion in the USA in 1978 towards non-legislation. The American NIHRAC and British GMAG continued to gain experience. The scientific debate pressed on, and consensus took shape around the recognition that certain of the initial fears had been exaggerated. The Director of the NIH, Don Frederickson, visited Gunther Schuster, Director-General of DGXII in 1978, to discuss the lessons of the US experience. He emphasised the desirability of avoiding fixed statutory controls.

The Parliament had commenced its scrutiny and was adding amendments more rigidly specifying containment requirements. Inspired, however, by experiences in the US and the UK, the Commission – on Schuster's advice – decided in 1980 to withdraw their proposed Directive and replace it with a proposal for a Council Recommendation.[12] In substance, this non-binding proposed recommendation was that Member States adopt laws, regulations and administrative provisions requiring notification – not authorisation – of recombinant DNA work. A whole body of advice coming from outside sources (The European Science Foundation (ESF), the ESF Liaison Committee on Recombinant DNA research,[13] The European Molecular Biology Organisation, the North Atlantic Assembly, The Council of Europe, The Economic and Social Committee...) concur in emphasising the economic importance of the new discipline of biotechnology, the absence of observed problems, the fact that work using rDNA did not present *per se* any new biological risks and that research should continue, providing certain minimum confinement procedures were respected.

During a meeting with the Scientific and Technical Research Committee (CREST)[14] in September 1978, the Commission recognised the importance of rDNA technology for the understanding of the structures of genetic functions which, in the long term, might well

---

[12] European Commission, Draft Council Recommendation concerning the registration of recombinant DNA work, Com (80), Vol. 467, 8 July 1980.

[13] This committee brought together the representatives of the national rDNA committees.

[14] The European Committee of Member State Officials – advising on matters of research, science and technology.

revolutionise certain methods of agricultural and industrial production. The risk associated with rDNA work was *'particularly conjectural and controllable'*. Reference was made to experiments and other considerations encouraging the idea that man and his environment, having survived the continual flux of information between species, could be considered to be relatively tolerant to any new form of recombinant DNA.

The examples of regulation in the USA and the UK were cited with, for the USA, mention of the rules that should, in future, become more supple and, for the United Kingdom, that the current code of good practice was applied on a voluntary basis, but with the understanding that the health and safety inspectors had wide powers to enforce implementation of the recommended precautions. The other Member States were presented as similarly preparing a regulatory system. The national committees in France, in the Netherlands, in Denmark and in Belgium had been assigned the task of indexing the work in progress and analysing the proposals for research. Apart from the United Kingdom, only one Member State, the Netherlands, clearly declared its intention to introduce a legislation to govern rDNA research. The Commission was also careful to stress the importance of reports and analyses from the ESF's *ad hoc* committee on rDNA research and from the Standing Committee on Recombinant DNA of the European Molecular Biology Organisation (EMBO) in the elaboration, and for the adoption, of regulatory systems at Member-State level.

Despite the declining evaluation and the real importance of the conjectural risks, the Commission put forward six reasons for national legislation:
1. The seriousness of the risks.
2. The expansion of rDNA work.
3. The transnational nature of the risks.
4. Research in the laboratories of private companies (with the risk that, in the absence of the relevant legal provisions, private laboratories and industry would not follow the same rules as those for the public sector).
5. The need to establish harmony between the Member States (so as to avoid disparities and a concentration of activities at the most permissive sites).

---

[15] The ESC is a consultative body statutorily involved in all Community legislation, and representing the social partners: business, consumers, trade unions.

6. The great value of the legislation on rDNA technology (rDNA was presented as an ideal way of arriving at compatibility between the legislation and the development of modern technology and to prepare a first base for the provisions that, inevitably, must be taken to protect man against his own inventions).

The 1978 Proposal for a Council Directive on rDNA Work

The draft Community Directive was presented to the Council in December 1978, but the first official response came from the Economic and Social Committee (ESC), a committee representing management and labour. Their report, delivered in July 1979, gave their evaluation, which dwelt on the declining evaluation of risks and on the absence of any specific problems. It stressed that the introduction of safety measures was not in itself proof of danger. Industry and agriculture, on the other hand, would benefit from the application of the new technology and, in general, had doubts as to the usefulness of the Directive. However, this evaluation supposed a continuous *self-discipline* on the part of the scientific community that could not be guaranteed. Certain countries (the Netherlands) took the view that legislation might help to reduce the latent suspicion among the population, stressing that if legislation were adopted, it would need to be adaptable to rapid change. The Economic and Social Committee[15] ESC report nonetheless backed the Commission draft directive, and proposed that ESC itself hold a public hearing with the Commission to consider scientific opinion as well as the opinions of the groups of unions, industries, agriculture and the general interest.

The Commission responded to the shift of opinion (which was at that time general)[16] by replacing its draft Directive with a proposal for a Council Recommendation in June 1980.[17]

Divided Opinion on the Adoption of the Recommendation

During the debate on the Commission proposal for a Council recommendation, rather than a Directive, for the control of rDNA work,

[16] A more and more favourable view of biotechnology in conjunction with growing confidence in the fact that it presented far less risk than initially thought was inevitable (opinion shared in particular by the EUROPEAN Science foundation, the European Molecular Biology Laboratory, the ESF Liaison Committee on recombinant DNA Research, the North Atlantic Assembly and above all by the Council of Europe.)

[17] EUROPEAN COMMISSION, *Draft Council Recommendation, concerning the Registration of recombinant DNA work*, COM (80), Vol. 467, 8 July 1980.

opinion in the European Parliament – was largely split. Subsequent to its hearing of May 1981, the Economic and Social Committee continued to support a Directive, thereby clashing with the Commission. At the European Parliament, the Rapporteur was Domenico Ceravolo, an Italian communist. His report found that, even if the risks were due only to a set of hypothetical events, this did not constitute a justification for thinking that they were any less significant nor any less valid. The conjectural risks could not be disregarded, for no appropriate criterion was available for their evaluation.[18] However, the liberals and conservatives, in this period, representing a majority in the Parliament, supported the Commission's proposals mindful that an excessively mandatory legislation could stunt the growth of the European biotechnological industry.[19] The proposal for a recommendation was approved by the Parliament in early 1982[20] and adopted by the Council in June the same year.

In October 1984, the Committee of Ministers of the Council of Europe adopted almost the same text as the recommendation for their Member States with a rather greater degree of flexibility for Member States, who could decide freely as to the risk categories necessitating a notification (because, it was argued, the biological risk had been overestimated). It was also stressed that studies would continue on the ethical questions.

With the adoption of Recommendation 82/472, recommending national notification systems in respect of work on recombinant DNA, the debate on regulation in Europe died down for some years. The recommendation came as a rational response to uncertainties. It authorised those Member States engaged in large-scale research activities having the corresponding monitoring systems in place to develop them and to adapt them to the perceived requirements. Aware of public feelings, the scientists involved co-operated spontaneously with the national authorities. International harmonisation developed via the usual scientific networks and international authorities such as those already mentioned.

---

[18] This remains a problem for all attempts at 'risk assessment' and underlines the basic limits inherent in this approach: to evaluate the risks implied by new technologies, reference is possible only to the risks implied by an existing technique similar, even if only remotely, to the new technique.

[19] J. BECKER, 'Recombination DNA Research: EEC Dispute' in *Nature*, 24(December 1981)31.

[20] Council Recommendation (472 EEC): *Concerning the Registration of Work Involving Recombinant DNA*, 1982.

The discussions that led to the recommendation of 1982 were followed by a period of some years of relative calm, at least as regards initiatives for the regulation of biotechnology. As pointed out in the previous description of communications from the Commission in 1983, 1984 and 1985, the general feeling in the departments responsible for various sectoral products was that the applications of the new biotechnology, linked to the sectors, did not pose an insurmountable problem and could be dealt with by the sectoral legislations.

An Uncertain Future

The situation evolved however as the large-scale production biotechnological installations and the demand for dissemination of genetically modified organisms multiplied. Public interest and attention was stimulated again and again by a considerable journalistic coverage centring on:
1. The scientific discoveries and their impact on knowledge
2. The economic implications and potential
3. The ecological and industrial risks stressed by the ecological groups
4. The ethical aspects (screening the population, in *vitro* fertilisation, the problem of interfering with nature, the danger represented by scientists playing sorcerer's apprentice). The World Council of Churches (WCC) drafted and published a rather hostile report containing, according to certain commission' members, glaring scientific inaccuracies.[21]

The entry of ecological interests into the political debate on biotechnology was one of the more notable developments of the 1980s; at the same time, the public authorities at national and Community level interpreted their general responsibility for the protection of the environment in relation to the challenges of the new processes and products resulting from biotechnology.

The creation of the European Biotechnology Co-ordination Group (EBCG)[22] in 1985 by the main European sectoral federations, not on an

---

[21] WORLD COUNCIL OF CHURCHES, *Biotechnology: Its Challenges to The Church and to the World*, Report by WCC Subunit on Church and Society, WCC Geneva, 1989.

[22] Member European sectoral federations of the EBCG in 1991: Association of Microbial Food Enzyme Producers (AMFEP), Conseil Européen des Fédérations de l'Industrie Chimique (CEFIC), Confédération des Industries AgroAlimentaires, Comité Européen des Obteneurs des Variétés Végétales (COMASSO), European Federation of Pharmaceutical Industries' Association (EFPIA), Fédération Européenné de la

industrial initiative but at the request of Etienne Davignon (then Vice-President of the EEC) to dispose of an intersectoral conciliation structure for the initiatives of the Commission of the European Communities in biotechnology, met the need for a representation of the bio-industries capable of expression via a trans-sectoral organ of communication.[23] This body was also useful for a dialogue with the European or international authorities (e.g., the Organisation for Economic Co-operation and Development (OECD)) or with foreign bio-industrials (USA, Japan, etc. ...).

To support the action for the regulation of biotechnological research and development, the Commission established internal structures under the authority of the 'Biotechnology Steering Committee' (BSC), the secretariat of which was provided by the Consultation Unit (CUBE). The Biotechnology Regulation Interservice Committee was instituted at Community level in July 1985 under DGXI and DGIII. Its task was to facilitate consultation between the services responsible for the preparation of draft Directives; currently, these concern the confined use of genetically modified micro-organisms[24] and information for the consumers.[25] These Directives, which followed the Council Recommendation for the regulation of rDNA, entail a different 'story'.

It should be remembered that this was, and still is, a more complex situation with, in the Commission in particular, the problem of harmonisation due to the conflict of interests between the DGs. It should further be pointed out that pressure from ecologists increased the more the genetic engineering techniques left the laboratory and gave rise to industrial applications. Although the new

---

Santé Animale (FEDESA), Green Industry Biotechnology Platform (GIBIP), International Group of National Associations of Agrochemical Manufacturers (GIFAP). Member national bio-industrial associations of the EBCG since April 1989: Associazione Nazionale per lo Sviluppo delle Biotecnologie (Italy) (ASSOBIOTEC), Bioindustry Association (United Kingdom) (BIA), Bioresearch Ireland (Ireland) (BRI), Foreningen af Bioteknologiske Industrier i Danmark (Denmark) (FBID), Groupe Belge de la Co-ordination de la Bio-industrie (Belgium) (GBCB), Nederlandse Industriële en Agrarische Biotechnologie Associatie (Netherlands) (NIABA), Organisation Nationale Interprofessionelle des Bio-Industries (France) (ORGANIBIO).

[23] Interesting to note that the European Council of Chemical Industry Federations (CEFIC) gave the EBCG a chairman and a secretary and defrayed the costs of the secretariat.

[24] Directive 90/220 of 23 April 1990 and Regulation No. 258/97 of 27 January 1997.

[25] Regulation of 27 January 1997 No 258/97 and No. 1139/98.

techniques could legitimately also claim the description of clean technology, the interaction in Europe with ecological movements was more a collision than an accommodation. At Parliament level, this was felt by a marked attention to the Greens. The Greens, because of their political gains in the late 1980s, spurred other political parties on to a new strategic will (to win back lost votes) to show their own 'green orientation and inclinations'. A restrictive approach towards the new technology struck them as a popular and easy course of action.

With this coincidence of popular fears, political interests and opportunist or conservative bureaucracies, protests from the scientific community were few and frequently ignored, even when they came from Nobel Prize winners. The OECD report, stressing that there was no scientific basis in favour of a specific legislation for rDNA, was noted for its prestige and authority, in support of just such legislation. DGXII lost, for a while, its fight for influence in the Commission. Its proposal to offer scientific advice was vigorously rejected and counter-attacked. The opinion of safety specialists from the EFB was aggressively rejected by the Director General of DGXI.

A similar reaction of rejection greeted the suggestion (at the Biotechnology Regulation Inter-service Committee (BRIC) level[26]) that the details of a rapidly evolving field should be worked out by technical experts in the Standards Committees. DGXI represented the head of the line as regards biotechnological legislation, but not as regards the standards. Consequently, the technical details of scope, a central problem in the US debates, were defined in annex to each of the biotechnological directives – 90/219 (confined use) and 90/220 (dissemination) – in terms specific to the comprehension of the science legislators during the 1980s, as modified by the experts chosen by the ministers for the environment, who removed these annexes from the field of the procedures of the 'Committees for the adaptation of the technical programme'. The consequences in terms of cost, delay and dispute dominated the debates on regulation during the 1980s.

---

[26] In July 1985 the BSC agreed to establish the Biotechnology Regulation Inter-service Committee BRIC: the functions of it in particular were defined to be to review the guidelines for rDNA research, to review the regulations applied to commercial applications of biotechnology and to determine whether current regulations adequately dealt with the risks that might be introduced by biotechnology.

European industry had sought to establish a communications network for the expression of bio-industrial interests. However, this failed to take on the task of expressing these interests energetically. This failure became evident with the disappearance of the EBCG. This was, on the one hand, largely brought about by the fact that each sector was jealous of its independence, making it unbearable that one of them should try to gain supremacy. On the other hand, certain federations had the feeling that their freedom of action in their own section had been reduced (a clear sign of insufficient consultation). Finally, these situations resulted in divergent views among the federations, some wishing to strengthen the structure, others to weaken it, which led to the rift. Nonetheless, following this setback, spring 1989 saw the emergence of the Senior Advising Group on Biotechnology (SAGB) in the European Council of Chemical Industry Federations (CEFIC). The SAGB[27] provided an industrial forum allowing debate on aspects of Community policy in the matter of biotechnology, with the aim of promoting a climate of support for biotechnology. The SAGB gradually led to the development and structuring of the dialogue between the bio-industry and the EEC, enabling account to be taken at Community level of the urgent need to redirect Community policy on biotechnology with a view to competitiveness. Continuing to defend stubbornly their main sector interests, the federations displayed a conservatism similar to that encountered within the Commission.

In the early 1990s, the general situation evolved with the realisation, even at Commission level, of the consequences of the failure with inter-services co-ordinations. At the request of President Delors in 1990, the Secretary-General initiated the constitution of the Biotechnology Co-ordination Committee (BCC) and upheld and developed the central role of the BCC within the Commission services. The Commission acted as a brake on the autonomous behaviour of the individual DGs and developed de facto a greater degree of horizontal transparency in the Commission. A greater degree of transparency developed vis-à-vis the outside. Round table discussions with industry and a more general spread of non-governmental representatives of interests became a new characteristic of the activities of the BCC. The communication of 1991 announced that the European Standards Committee was to be charged with the development of standards for biotechnology.

---

[27] Members of the SAGB in 1991: Ferruzzi, Hoechst, Hoffmann-La Roche, ICI plc, Monsanto Europe sa, Rhône-Poulenc, Sandoz Pharma Ltd., Unilever plc.

The fact remains, however, that the hostility towards this new technology was far from spent. If a DG had been thwarted at the BCC, a telephone call or a fax could soon mean a letter from a member of the European Parliament to the secretary-general. Nor was there any lack of groups of activist organisations to bring the arguments to the public sphere.

The balance between the pro- and anti-biotechnology camps still remains delicate. In spite of the advances made during recent years, regulations on biotechnology and, in particular, the legislation concerning genetically modified organisms, still remain problematic. The adopted directives are still far from coming to a broad consensus, and require further improvement. Considering the stakes (economic, political and ethical), the complexity and the multiplicity of the forces involved, the constant evolution of the field and the antagonistic pressures, one thing is certain: the debate is far from finished, and the perfect regulation is yet to be invented.

Conclusion: Lessons and Reflections

If we limit ourselves to the story here under review, that of the regulation of recombinant DNA techniques, what lessons can we learn? Begun by the scientists, the debate on the regulation of rDNA research and technology may be considered as a genuine prototype for all attempts to strike a balance between the need to control and legislate and the development of a new technology. This debate laid the foundations for the provisions that have been made to protect man against his own technical successes. During the debate, the reasoned communication and attention among the various partners and the co-operation, from the outset, between the scientists involved with the authorities – this with salutary initiatives – were decisive in the success chalked up in the USA and in the United Kingdom or throughout Europe.

These factors allow the regulation to be characterised by a flexibility enabling it to adapt to extremely rapid evolution, the regulatory process avoiding, by its complexity, unduly hasty decisions that might be detrimental to scientific research, industrial and economic competitiveness.

More specifically, with regard to the different regulation procedures the NIH system of safeguards relied to a greater extent on biological containment, the UK on physical. The American guidelines were written in more detail than the UK recommendations and code of practice. The NIH administrative system was designed to cover research funded by the NIH, the UK system would address all laboratories under the aegis of a national central advisory committee. In both cases a consensus was reached on the fact that the inherent risks attached to the new biotechnological techniques had been exaggerated.

A major feature of the debates in the US was the progressive development of a well-organised, articulate and balanced response by the scientific community. The leading role was played by the American Society of Microbiology (ASM), but many other professional associations of Biological and medical sciences joined with ASM in a broad alliance, through semiformal linkages via their executive officers, and widespread networks capable of providing rapid responses.

The US experience in the post Asilomar period was of significance, as a successful example of open dialogue between the scientific and political communities. The successes can be related to the flawless safety record of genetic engineering in the US over the following years; and to the position of scientific and economic leadership in biotechnology which the US maintained. More generally and importantly, the US experience provided, for scientists in all fields and legislature everywhere, an object lesson in how to manage the interface between science and society in a way that was democratic and transparent.

For the UK, Gibson describes the Joint Co-ordinating Committee's action, by the scientific community, as 'almost exceptional in the rDNA debate' and emphasises the role of individual scientific contributions. In particular, Sydney Brenner first gave – in July 1978 – the GMAG the initial concept of the risk evaluation scheme that proved so successful. It was introduced in March 1979 and revised in January 1980. Local biological safety committees were able to operate with ease the risk assessment scheme, which was sufficiently flexible to allow medical or scientific information to be introduced when reaching a decision. The scheme led to the majority of work being reclassified in level I. The development of a new strain of *E. coli* by S. Brenner allowed GMAG to incorporate the concept of biological confinement in its risk evaluation scheme with more confidence.

With regard to the European process, it is clear that lessons were drawn from the American and British process, in particular the necessity of listening to the scientists. Nevertheless, it is equally clear that this historical outline emphasises the complexity of the European process, which means that the politicking was and is more intense than in the US. Many factors render obscure the legislative and other action of the Community institutions, shielding them from effective democratic scrutiny, and limiting their transparency: the multi-institutional complexity (Commission, Parliament, Council, etc.) of the machinery; the distance from national politics where Brussels bureaucracy is a convenient scapegoat for nationally unpopular measures and above all from citizens and local communities; and of course the inescapable diversity of Europe's languages and cultures. This lack of transparency means that on complex subjects, only a sustained and determined effort of communication can ensure that all parties with relevant interests and knowledge have the opportunity to participate in preparing proposal and decisions. Complexity without transparency allows, even encourages, the pursuit of individual and institutional self-interest, whilst within the Commission communications are not so easy. Each DG has its own professional bias, which creates real barriers between DGs. Thus in DGIII, legislation was essential to creating a common market for food products, pharmaceuticals etc. Similarly for DG XI the control of chemical products for the protection of human health and the environment was a major challenge. The culture of DGXII, especially in the earlier decades, was scientific in its sympathies and roots. They were reluctant legislators in 1978, and were glad to retire from such matters in the mid-80s. Global trends, economic pressure, the natural internationalism of science, its perceived relevance to economic competitiveness, the increasingly expensive and specialised character of research led to rapidly expanding biotechnology R&D programmes at European level in the later 80s. The pressures of managing these increasing resources with a static or declining complement of staff forced DGXII to focus on the politics of winning these heavier budgets, and on managing the selection and administration of vast numbers of projects. These pressures further diminished the appetite for inter-DG arguments over legislation as in themselves; they caused outside pressures to increase.

Overall, the extreme complexity of regulatory procedures at the level of the major institutions must be underlined and particularly where community bodies are concerned. Equally, the difficulty of taking into account the ethical issues at stake in the midst of such a complex situation must be realised. With regard to Parliament, the difficulties are just as considerable. The fact that the biotechnological field represents esoteric techniques and that the members of the European parliament have in general little time to dedicate to studying dossiers such as those relating to biotechnology means that in practice such dossiers are left to a rapporteur. The basis for formulating the parliamentary opinion on legislation relating biotechnology was therefore typically a narrow one; in an area which shared (with nuclear energy) the most concentrated attention of the 'Green' fraction of the Parliament. Moreover, even MEPs not of this fraction, were in many countries acutely conscious in the late 1980s that the major political parties were losing ground to the green movements; and to recapture these votes were anxious to demonstrate their own green credentials.

Against this coincidence of popular fears, political self-interest and bureaucratic opportunism, the voice of scientific protest were few, feeble and disregarded. DG XII lost the arguments inside the commission and had at the critical moments no interested allies.

This example shows, now that a post-modern current is carrying an increasingly widespread suspicion of the advances of techno-science, that a regulation adapted to the feats of reason and their implications is really difficult to find and that ethical issues rapidly disappear under the weight of political, and above all, economic pressures. This is just as true whether it be in the USA or Great Britain or Europe. The issue at stake which had most bearing on legislation was related, not so much to the ethical issues, but had more to do with the economic and political situation, with an underlying fear of not being competitive enough in this newly emerging and promising area of biotechnology. Should we therefore be pessimistic? No, but it remains beyond doubt, however, that the pluridimensionality, plurisectorality and plurinationality of the scientific and technological fields and of their applications and implications cause difficulties for all attempts at regulation taking any meaningful account of the social, political, economic and ethical risks and problems, in particular as is shown with brilliant clarity by the turn this story takes once industrial and economic interests swing into action.

One Last Point to Be Noted:

Underlying current practice is the assumption that a more or less authoritative appeal to science could help the process of consensus formation on the possible effects of the use of genetically modified organisms. This confidence is problematic for two reasons.

Firstly, we have to deal with a trans-scientific problem, that is a problem that can be stated in the language of science but cannot be solved within the language of science.

Secondly, our current knowledge does not provide us with the means to predict the ecological long-term effects of releasing organisms into the environment. So it is beyond the competence of the scientific system to answer such a question, although it is precisely this assumed competence which normally forms the basis for an authoritative appeal. In fact, science would not ask itself such a question since there is no method to make this question researchable. Reasoned statements on the subject matter cannot go beyond theoretical speculation. The reason for an appeal to science is solely policy motivated: we would like to have the answer to this question for achieving a manageable code of practice.

We can reconstruct two kinds of answers which science has given us so far.

The first answer came from one branch of science, where most scientists were biotechnologists, molecular biologists or microbiologists. They answer the question by acknowledging the trans-scientific problem and stating that the development of a test protocol for identifying the risks attached to any particular organism would be an unachievable task. However, at the same time they argued that this is irrelevant knowledge since we can rely on the experience with traditional plant breeding practices, which, in their opinion, differs insignificantly from the practice of genetic engineering – only in so far as that we now know exactly what kind of new genes we are introducing.

Ecologists on the other hand played down the trans-scientific issues, by saying that they could develop precisely the type of knowledge policy makers asked for by doing research on so called microcosms needed to make predictions possible in terms of quantitative risk assessment.

From a policy perspective both answers are unsatisfactory, because a biotechnologist cannot address the problems in terms of safety or in terms of risk. They just rhetorically state that it could be an acceptable risk (by appealing to the fact that we already have accepted the risks associated with conventional plant breeding). This does not give us an informed opinion on how to regulate individual cases, nor does it address the issue of a precautionary approach. Ecologists, on the other hand, underestimate the difficulties of a trans-scientific issue: the promise of providing a quantitative risk assessment in the course of microcosm-based experiments, and without conducting field experiments, cannot be fulfilled in the foreseeable future. Only if one fully appreciates the trans-scientific issue, does one see the dilemma for policy: allowing uncontrolled field experiments might involve unknown environmental impacts. To impose too many constraints on these experiments, however, would mean that we will never gain knowledge about the behaviour of GMOs.

Supposing the scientists and other experts had been listened to, the appeal to science for policy would have been made without reflecting sufficiently on the trans-scientific issue underlying a scientific controversy. Science reduced this issue by translating it into a question of relevancy, to which both molecular biologists and ecologists came up with unsatisfactory answers. Policy has to be engaged in science to look for answers concerning perceived risks but it cannot make a legitimate appeal to a science, which does not resolve the relevancy question.

So if there is one lesson to be retained from this historical study, it is precisely the fact that practically the only people truly qualified to dialogue with the legislators (whether that be in the USA, Great Britain or in Europe) are the scientists themselves. This should cause us to stop and think about how possible it really is in such a context to take into account the ethical considerations and the problems of democracy with regard to the difficulties involved in taking decisions linked to the regulation and the integration of science and scientific techniques.

References

BECKER, J., 'Recombination DNA Research: EEC dispute' in *Nature* 24(December 1981)31.
BERG, P. et al., 'Potential Biohazards of Recombinant DNA Molecules' in

Science 185(1974), p. 303.

COHEN, S.N., A.C.Y. CHANG, H.W. BOYER and R.B. HELLING, 'Construction of Biologically Functional Bacterial Plasmids *in vitro*' in *Proc. Natl. Acad. Sci.* 70(1973), pp. 3240-3244.

Council Recommendation (472 EEC): *Concerning the Registration of Work Involving Recombinant DNA*, 1982.

EUROPEAN COMMISSION, 'Draft Council Recommendation Concerning the Registration of Recombinant DNA Work', *Com* (80), Vol. 467, 8 July 1980.

HMSO, *Report of the Working Party on the Practice of Genetic Manipulation*, Williams Working Party, Comnd. 6600, Her Majesty's Stationery Office, London, 1974.

ROGERS, M., 'The Pandora's Box Congress' in *Rolling Stone* (19 June 1975), p. 77.

SÉRUSELAT, F., 'Vers une approche des définitions du concept de l'éthique', *La lettre éthique* 5(1990).

SINGER, M.F. and D. SOLL, 'DNA hybrid molecules' in *Science* 181(1973), p. 1114.

WORLD COUNCIL OF CHURCHES, *Biotechnology: Its Challenges to the Church and to the World*, Report by WCC Subunit on Church and Society, WCC Geneva, 1989.

List of Acronyms

ASM   (American Society of Microbiology)
BCC   (Biotechnology Co-ordination Committee)
BRIC   (Biotechnology Regulation Inter-service Committee)
BSC   (Biotechnology Steering Committee)
CEFIC   (the European Council of Chemical Industry Federations)
CREST   (Scientific and Technical Research Committee)
CUBE   (Consultation Unit)
DG   (Directorate General)
EBCG   (European Biotechnology Co-ordination Group)
EMBO   (European Molecular Biology Organisation)
ESC   (Economic and Social Committee)
ESF   (European Science Foundation)
GMAG   (Genetic Manipulation Advisory Group)
HEW   (Health, Education and Welfare)
HSE   (Health and Safety Executive)
MRC   (Medical Research Council)
NIH   (National Institute of Health)
NIHRAC (NIH Recombinant DNA Advisory Committee)

OECD   (Organisation for Economic Co-operation and Development)
rDNA   (recombinant DNA)
SAGB   (Senior Advising Group on Biotechnology)
WCC   (World Council of Churches)

Annex

The Berg Letter

Potential Biohazards of Recombinant DNA Molecules

Recent advances in techniques for the isolation and rejoining of segments of DNA now permit construction of biologically active recombinant DNA molecules in vitro. For example, DNA restriction endonucleases, which generate DNA fragments containing cohesive ends especially suitable for rejoining have been used to create new types of biologically functional bacterial plasmids carrying antibiotic resistance markers and to link Xenopus laevis ribosomal DNA to DNA from a bacterial plasmid. This latter recombinant plasmid has been shown to replicate stably in Escherichia coli, where it synthesises RNA that is complementary to X. laevis ribosomal DNA. Similarly, fragments of Drosophila chromosomal DNA have been incorporated into both plasmid and bacteriophage DNAs to yield hybrid molecules that can infect and replicate in E. coli.
Several groups of scientists are now planning to use this technology to create recombinant DNAs from a variety of other viral, animal and bacterial sources. Although such experiments are likely to facilitate the solution of theoretical and practical biological problems, they would also result in the creation of novel types of infectious DNA elements whose biological properties cannot be completely predicted in advance. There is serious concern that some of these artificial recombinant DNA molecules could prove biologically hazardous. One potential hazard in current experiments derives from the need to use a bacterium like E. coli to clone the recombinant DNA molecules and to amplify their number. Strains of E. coli commonly reside in the human intestinal tract, and they are capable of exchanging genetic information with other types of bacteria, some of which are pathogenic to man. Thus, new DNA elements introduced into E. coli might possibly become widely disseminated among human, bacterial, plant, or animal populations with unpredictable effects.
Concern for these emerging capabilities was raised by scientists attending the 1973 Gordon Research Conference on Nucleic Acids, who requested that the National Academy of Sciences give consideration to these matters. The endorsement of the Academy of Life Sciences of the National Research Council on this matter propose the following recommendations.

First, and most important, that until the potential hazards of such recombinant DNA molecules have been better evaluated or until adequate methods are developed for preventing their spread, scientists throughout the world join with the members of this committee in voluntarily deferring the following types of experiments.
- Type 1: Construction of new, autonomously replicating bacterial plasmids that might result in the introduction of growth determinants for antibiotic resistance or bacterial toxin formation into bacterial strains that do not at present carry such determinants; or construction of new bacterial plasmids containing combinations of resistance to clinically useful antibiotics unless plasmids containing such combinations of antibiotic resistance determinants already exist in nature.
- Type 2: Linkage of all or segments of the DNA from oncogenic [cancer-inducing] or other animal viruses to automatically replicating DNA elements such as bacterial plasmids or other viral DNA. Such recombinant DNA molecules might be more easily disseminated to bacterial populations in humans and other species, and thus possibly increase the incidence of cancer or other diseases.
Second, plans to link fragments of animal DNA to bacterial plasmid DNA or bacteriophage DNA should be carefully weighed in light of the fact that many types of animal cell DNA contain sequences common to RNA tumour viruses. Since joining of any foreign DNA to a DNA replication system creates new recombinant DNA molecules whose biological properties cannot be predicted with certainty, such experiments should not be undertaken lightly.
Third, the director of the National Institute of Health is requested to give immediate consideration to establishing an advisory committee charged with (i) overseeing an experimental program to evaluate the potential biological and ecological hazards of the above type of recombinant DNA molecules; (ii) developing procedures which will minimise the spread of such molecules within human and other populations; and (iii) devising guidelines to be followed by investigators working with potentially hazardous recombinant DNA molecules.
Fourth, an international meeting of involved scientists from all over the world should be convened early in the coming year to review scientific progress in this area and to further discuss appropriate ways to deal with the potential biohazards of recombinant DNA molecules.
The above recommendations are made with the realisation (i) that our concern is based on judgements of potential rather than demonstrated risk since there are few available experimental data on the hazards of such DNA molecules and (ii) that adherence to our major recommendations will entail postponement or possibly abandonment of certain types of scientifically worthwhile experiments. Moreover, we are aware of many theoretical and practical difficulties involved in evaluating the human hazards of such recombinant DNA molecules. Nonetheless, our concern for the possible unfortunate consequences of indiscriminate application of these techniques motivates us to urge all scientists working in

this area to join us in agreeing not to initiate experiments of types 1 and 2 above until attempts have been made to evaluate the hazards and some resolution of the outstanding questions has been achieved.

Paul Berg, Chairman
David Baltimore
Herbert W. Boyer
Stanley N. Cohen
Ronald W. Davis
David S. Hogness
Daniel Nathans
Richard Roblin
James D. Watson
Sherman Weissman
Norton D. Zinder

Committee on Recombinant DNA
Molecules Assembly of Life Sciences,
National Research Council,
National Academy of Sciences,
Washington DC *20418*

26 July 1974

3.2.3

OBJECTIVE SCIENCE ?

The Case of Climate Change Models

H.J.M. de Vries

Man-induced climate change is a complex issue full of uncertainties. In this paper, I will attempt to make explicit some of the values and ethical questions in the so-called 'climate debate'. This is done by unravelling the various layers of complexity; establishing methods of integration and using cultural theory to distinguish various perceptions of reality.
Examples on the basis of the TARGETS *and the* IMAGE*-model are used as illustrations. My conclusion is that we have to learn to make the ethical component in individual and collective actions explicit, if we are to find the integrity and mutual respect needed to guide mankind into an equitable, sustainable and efficient new millennium.*

The threat of human-induced climate change has rapidly become one of the most prominent environmental issues on the global agenda.[1] It is intricately related to the debate about sustainable development which started in the 1980s with the publication of the report Our common future of the World Commission on Environment and Development.[2] To assess the causes and consequences of human-induced climate change, the Intergovernmental Panel on Climate Change (IPCC) established in 1988 three working groups. The first surveyed the scientific evidence on climate change. The second assessed environmental and socio-economic impacts of climate

---

[1] J. JÄGER and T. O'RIORDAN, 'The History of Climate Change Science and Politics' in T. O'RIORDAN and J. JÄGER (eds.), *Politics of Climate Change – A European Perspective*, London and New York, Routledge, 1996, pp. 1-31.
[2] WCED (World Commission on Environment and Development), *Our Common Future*, Oxford, Oxford University Press, 1987.

change. The third explored response strategies. The three working groups published reports in 1990 and 1996.[3]

Climate change refers to the anticipated change in climate as a consequence of anthropogenic emissions of gases which enhance the natural greenhouse-effect. It can best be discussed in the so-called Pressure-State-Impact-Response (PSIR) framework.[4] Observations show a steady increase in the atmospheric concentrations of gases which are known to trap heat, most notably carbon dioxide (CO_2), methane (CH_4) and dinitrogen oxide (N_2O). The resulting radiative forcing will cause a rise in global temperature. This will have an array of impacts on the global environment, among them changes in regional climate regimes and a rise in sea level. To these changes mankind may respond basically in two ways: adaptation in a variety of ways and mitigation by emission reductions. A system-oriented representation of the problem guarantees the necessary integrated approach – or assessment – of the problems and the solutions and can be operationalized in more or less formal 'scientific' models. This essay is largely based on work within the Bureau for Environmental Assessment of the National Institute of Public Health and the Environment (RIVM) in The Netherlands, especially around the TARGETS- and IMAGE-models.[5]

The IMAGE 2 model consists of three fully linked systems of models:
- The Energy-Industry System (EIS);
- The Terrestrial Environment System (TES); and
- The Atmosphere-Ocean System (AOS).

The Energy-Industry System (EIS) computes the emissions of greenhouse gases in 13 world regions. The energy-related emissions

---

[3] IPCC (Intergovernmental Panel on Climate Change), *Climate Change: The IPCC Scientific Assessment*, J.T. HOUGHTON, G.J. JENKINS, and J.J. EPHRAUMS (eds.), Cambridge, Cambridge University Press, 1990; IPCC, *Climate Change 1995: The Science of Climate Change*, Cambridge, Cambridge University Press, 1996; IPCC, *Climate Change 1995: Impacts, Adaptations and Mitigation of Climate Change: Scientific-Technical Analysis*, Cambridge, Cambridge University Press, 1996; IPCC, *Climate Change 1995: Economic and Social Dimensions of Climate Change*, Cambridge, Cambridge University Press, 1996.

[4] J. ROTMANS and H.J.M. DE VRIES (eds.), *Perspectives on Global Change: The TARGETS Approach*, Cambridge, Cambridge University Press, 1997.

[5] *Ibid.*; J. ALCAMO (ed.), *Integrated Modelling of Global Climate Change: IMAGE 02*, Dordrecht/Boston/London, Kluwer Academic Press, 1994; J. ALCAMO, R. LEEMANS and E. KREILEMAN (eds.), *Global Change Scenarios of the 21st Century*, Oxford, Elsevier Science, 1998.

are based on the Targets Image Energy Regional (TIMER) simulation.[6] The model is a systems dynamics model with investment decisions into energy efficiency, electricity generation and energy supply based on anticipated demand, relative costs c.q. prices and institutional and informational delays. The model uses 13 world regions and 5 economic sectors. Technological change and fuel prices dynamics influences energy-intensity, fuel substitution and penetration of non-fossil option such as solar electricity and biomass-based fuels. The objective of the Terrestrial Environment System (TES) is to simulate global land-use and land-cover changes and their effect on emissions of greenhouse gases and ozone precursors, and on carbon fluxes between the biosphere and the atmosphere. This sub-system can be used to evaluate the effectiveness of land use policies for controlling the build-up of greenhouse gases, to assess the land consequences of large-scale use of biofuels, to evaluate the impact of climate change on global ecosystems and agriculture, and to investigate the effects of population, economic, and technological trends on changing global land cover. Figure 1 gives an overview of the IMAGE-model which shows the PSIR-loop.

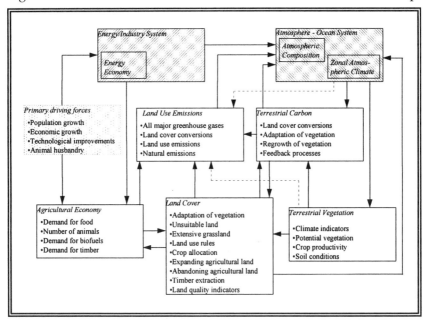

Figure 1. Overview of the IMAGE 2.1 model

[6] B. DE VRIES, J. BOLLEN, L. BOUWMAN, M. DEN ELZEN, M. JANSSEN, E. KREILEMAN and R. LEEMANS, 'Greenhouse-Gas Emissions in an Equity-, Environment- and Service-

All three IPCC working groups used models, in the sense of more or less formal representations of [parts of] reality – are these models 'scientific'? In terms of practical validation – comparison of simulated results with observational data – and conceptual validation – scrutinising the concepts and theoretical laws of the system under consideration and its dynamic behaviour – one may have doubts. There are still large uncertainties, controversies and areas of ignorance.[7] Observational data are inherently limited although their size, reliability and relevance are increasing. Conceptually the models show still known and unknown deficiencies – as can be understood from the evolution history of these models. The uncertainties usually show up in diverging interpretations of past events: various plausible but sometimes contradictory explanations of phenomena can be constructed.

Because of these validation problems, one might conclude it is better to postpone any policy-oriented conclusions until our scientific understanding of the causes and consequences of climate change has improved – if ever. However, in view of the rapid changes which are going on, it is a dire necessity to explore possible futures, in order to be able to anticipate and adjust if desired objectives – such as prosperity and efficiency but also sustainability and equity – are threatened. This requires, that climate change modelling is seen as an ongoing process of learning and communication with interactive use of models. It also requires that uncertainties and controversies are made visible so that underlying views about how the world functions and is or should be managed become an explicit part of scientific exploration. In the remainder of this essay I will discuss these issues in more detail and provide examples to illustrate their relevance for the ethical aspects of the climate change debate.

The 'Objectivity' of the Science of Climate Change

The science, the causes and the policies related to man-induced climate change are about long-term, complex issues which extend

Oriented World: An Image-Based Scenario for the Next Century' in *Technological Forecasting and Social Change* 63(2000)2-3.

[7] J. VAN DER SLUIJS, *Anchoring Amid Uncertainty: On the Management of Uncertainties in Risk Assessment of Anthropogenic Climate Change*, unpublished Ph.D. diss., Utrecht, University of Utrecht, 1997; DE VRIES, BOLLEN, BOUWMAN, DEN ELZEN, JANSSEN, KREILEMAN and LEEMANS, *Greenhouse-Gas Emissions in an Equity-, Environment- and Service-Oriented World*.

across several spatial and temporal scales and include many different scientific disciplines. In climate change research, however, there is only limited room for repeated and controlled experiments and hence part of our knowledge is necessarily and principally 'weak' in the sense of not strictly falsifiable.[8] This holds even more for the complex human dimensions of climate change. In cases of 'weak' knowledge, bounded rationality and biased interpretation and expectation play an important role. There is also a tendency to ignore uncertainties to support the rationality of the world-view one adheres to and to misappreciate the extent of our ignorance.[9] This can be illustrated with Figure 2 which presents a spectrum of levels of reality of which scientists construct and use models – or mental maps – in their attempt to understand and manage the surrounding world. One may associate these levels with increasing degrees of freedom of their elements or view them as representing an ascending order of complexity, intentionality and consciousness.[10]

The first level consists of physical reservoirs. Model variables, at this level, usually have an explicit and formal correspondence with real-world observable phenomena. Modern science and technology have been successful in this realm by combining formal analysis with controlled experiments. This is reflected in the analogies and metaphors that are used: the planetary system as a celestial clock, the heart as a mechanical pump, the psyche as a steam engine, and the brain as a computer. Increasingly, it is realised that such metaphors may be as much a reflection of our own particular way of defining and acquiring knowledge of reality as of reality itself.

The next level maps the behavioural and informational structures, which govern human interference in the underlying physical

---

[8] H.J.M. DE VRIES, *Sustainable Resource Use – An Inquiry into Modelling and Planning*, unpublished Ph.D. diss., Groningen, University of Groningen, 1989.

[9] J. MORECROFT and J. STERMAN, 'Modelling for Learning' in Special Issue of *European Journal of Operations Research* 59(1992)1; S. FUNTOWICZ and J. RAVETZ, *Uncertainty and Quality in Science for Policy*, Dordrecht, Kluwer Academic Publishers, 1990.

[10] Such an arrangement has often been proposed. Teilhard de Chardin (1963) advances complexity and interiorisation as a third dimension alongside the infinitely large and the infinitesimally small and introduces the notion of noosphere. Daly (1973) uses a hierarchist spectrum of values, ranging from worldly means to ultimate spiritual goals. Several others: the spectrum from 'low' to 'high' human needs proposed by Maslow; the fusing of values and self-image (e.g. Harman 1993); the existence of three cerebral layers cf. Vroon (1989); and the Eastern chakra doctrine. See also Simon (1989) on the question whether complexity is partly 'in the eye of the beholder'.

environment. It is usually described with relationships based on correlation analysis of a limited sample of data from observations in social sciences. Often, models from the physical sciences have been employed as analogues for the construction of hypotheses.[11] More recently, behaviour has been introduced more explicitly in the form of information-dependent sets of rules which prescribe actors in response to changes in their [model] environment.[12] Many of the controversies around social-scientific theories arise from the considerable diversity and variety of behaviour that is constantly evolving – a process of which these theories are themselves a part. Universally valid laws such as those of the natural sciences have yet to be discovered.

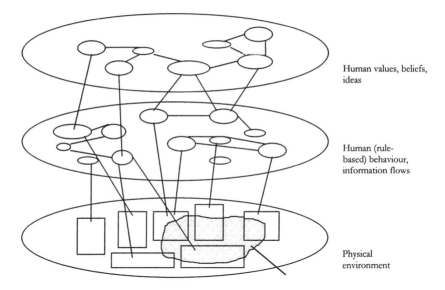

Figure 2. The stratification of [models of] reality

The third level comprises values, beliefs and ideas that reflect and rationalise people's behaviour. At this level the behavioural rules

[11] DE VRIES, *Sustainable Resource Use – An Inquiry into Modelling and Planning.*

[12] The focus is here on human actors because of the anthropocentric tendency of most analyses. In fact a continuous spectrum is proposed within which for instance animals in an ecosystem, given their potentially rich range of behaviour, can be located in the middle level. It is my view that deriving human metaphors from ecology and not from mechanics is an important step.

and information flows of the second level are evaluated from a higher and broader point of view.[13] Ethical questions are mostly addressed at this level. Generally speaking, this level is merely included in models in the form of response variables chosen ex ante and as part of scenario analyses. Many of the normative dimensions and decision-making processes are not, or only in highly simplified and implicit ways, included in quantitative models.[14]

These three levels of complexity are of course a greatly simplified representation of a continuous spectrum. Its use, however, may help to communicate that 'strong' science, generating statements on the basis of controlled experiments, only covers a limited domain of the physical environment and an even smaller part of the levels of behaviour and values. It also stimulates explicit discussion of the concepts and methods – as well as their differences – used in the natural sciences and in the social sciences. Integrated assessment, not only 'horizontal' but also 'vertical', may gradually restore the connection. In the last decades, the search for new methods and approaches to bridge the gap has intensified. Approaches such as simulation games and policy exercises, are promising complements.[15] System dynamics, applied general equilibrium and actor-oriented models, cellular automata and genetic algorithms are some of the tools that have been applied more recently.[16]

The IPCC itself faced the difficulties arising from assessments at various levels of complexity. In 1988 the IPCC established three working groups to assess the scientific evidence, the possible impacts and potential response strategies. Despite serious scientific shortcomings – such as the initially misrepresented role of aerosols – and controversies

---

[13] E. JANTSCH, *The Self-Organizing Universe – Scientific and Human Implications of the Emerging Paradigm of Evolution*, Oxford, Pergamon Press, 1980.

[14] The behavioural models at the second level often incorporate values and beliefs – such as the notion of the rationally optimizing individual consumer. To the extent that the limited validity of such models is not acknowledged, one may speak of model-incorporated ideology.

[15] B. DE VRIES, 'Susclime : A Simulation Game on Population and Development in a Climate-Constrained World' in *Simulation and Gaming* 29(1998)2, pp. 216-237.

[16] F. KREBS and H. BOSSEL, 'Emergent Value Orientation in Self-Organization of an Animal' in *Ecological Modelling* 96(1997), pp. 143-164; M.A. JANSSEN, *Modelling Global Change – The Art of Integrated Assessment Modelling*, Cheltenham, UK, Northampton, MA, USA, Edward Elgar Publishers, 1998; E. VAN DAALEN, W. THISSEN, and M. BERK, *The Delft Process: Experiences with a Dialogue Between Policy Makers and Global Modellers. Global Change Scenarios of the 21st century – Results from the Image 2.1 Model*, J. ALCAMO, R. LEEMANS and E. KREILEMAN (eds.), Oxford, Elsevier Science Ltd., 1998, pp. 193-234.

– about for instance the role of clouds – the report of the first working group was rather widely accepted as 'scientific' and hence detached from political bias and from the series of abnormal climate events across the globe. According to Jäger and O'Riordan, the other working groups:

> ' ... were far less successful, partly because it is undesirable to separate adjustment from impacts, and also to isolate analysis from policy prescriptions. The fact that the Russians chaired WG2 and the Americans chaired WG3 was also politically dictated, and virtually guaranteed that both groups would be ineffective in informing the all important negotiation process. It is also a sad reflection on the state of the social sciences at present, that their practitioners cannot produce a coherent view of what causes climate change in terms of human needs and wants and associated economic and technological 'drivers', what should be done about these, and what would be the social, political and economic consequences. Compared with the consensus-oriented format of the natural scientists, the social scientists have behaved in a more disorganised and non-credible manner.'[17]

Obviously, the ecological and social scientists had to struggle which many more uncertainties and conflicting interpretations of the scarce and scattered data.[18]

Cultural Perspectives: Interpreting Uncertainty

Obviously, the deficiencies in our knowledge about causes and consequences of anthropogenic emissions of greenhouse-effect enhancing gases leaves room for divergent interpretations of the past and projections of the future. The 'cultural theory'[19] can be used as a framework to recognise that knowledge is, at least to some degree, socially constructed. It combines insights from cultural anthropology

---

[17] J. Jäger and T. O'Riordan, 'The History of Climate Change Science and Politics.'

[18] Some social scientists were quick to remind natural scientists that the 'strength' of their knowledge is partly a matter of perception: examination of how 'technical' knowledge about climate change is created by scientists and communicated to policy-makers reveals that at least part of the scientific consensus is socially constructed (Jasanoff and Wynne 1998).

[19] M. Thompson, R. Ellis and A. Wildawsky, *Cultural Theory*, Boulder, Westview Press, 1990; M. Schwartz and M. Thompson, *Divided We Stand: Redefining Politics, Technology and Social Choice*, New York, Harvester Wheatsheaf, 1990.

and ecology in distinguishing cultural perspectives, based on the degree to which individuals behave and feel themselves part of a larger group of individuals with whom they share values and beliefs (the 'group' axis) and the extent to which individuals are subjected to role prescriptions within a larger structural entity (the 'grid' axis).[20] The resulting four perspectives are related to their position along these two axes: the hierarchist (high on both), the individualist (low on both), the egalitarian (high in 'group', low in 'grid') and the fatalist (low in 'group', high in 'grid').[21] Whereas the hierarchist places the emphasis on control and expertise in order to guarantee stability within a world of limits, the individualist is convinced that the world has inherent stability and abundance. The egalitarian emphasises the fragility of nature and the probability of irreversible destruction. The fatalist experiences the world as determined by pure chance.

The interpretation of past events and the anticipation of the future are both filtered through the diverse rationalities which are used by adherents of the four myths to make their lives liveable. One should bear in mind that people seldom express these paradigms in their extreme form – nor should one give in to the temptation to make caricatures. The four paradigms interact dynamically. Excessive hierarchism leads to bureaucratisation – which then collapses in liberalisation and privatisation processes. Excessive individualism leads to the marginalisation of the less successful people who then resort to fatalism, which in turn may feed egalitarian movements in a process of radicalisation. Fundamentalism, nationalism and also environmental radicalism can be partially understood against this background.

Cultural theory offers an interesting framework for an analysis of the climate change controversies and risks (Table 1). Seen from the *individualist* perspective, the past is full of human success stories, the future full of promise. The debate about 'limits to growth' is seen as

---

[20] In anthropology there is reference to work done by Douglas (1970) in which peoples' belief systems are interpreted as reflections of social relationships. Holling's work (1986) on the stability of ecosystems and the cycles within them has been a contribution from ecology.

[21] The fifth world-view, that of the hermit, is beyond the area defined by these two axes. Cultural Theory has similarities with three recent perspectives on economic developments used by the Dutch Central Planning Agency: those of coordination, equilibrium and free markets (CPB 1993). See also Van Asselt and Rotmans (1996), WRR (1994) and De Vries (1996).

exhibiting a lack of trust in the ingenuity of humans and in the resilience of nature. It is precisely the existence of limitations that challenges people to surpass them as we have seen so often in the past. Besides, humans have an enormous capacity for adaptation. Individualists may, however, come into conflict with their own convictions. Increasing population density increases the pressure for coordination. As soon as successful new areas are exploited and new markets are conquered, the necessity of regulations and a loss of exclusiveness emerge. Other confrontations occur when the rules of the game are threatened or social limits to growth appear.

The *hierarchist* world view is expressed by administrators, managers and engineers who have confidence in the results of measurements and risk calculations and assume the need for good management of both the societal process and the environmental system. On the basis of what experts say about the constraints and the costs and benefits, an optimum technical-economic course is determined. Seen through the eyes of the hierarchist, the world is perpetually in need of management: chaos and anarchy are just around the corner. This is the world of bureaucrats and politicians who understand the need for and the subtleties of institutions, committees, and negotiations. The legitimacy of big government and big business, and their definition of what is valued and what matters, are paramount. They too, however, face the shortcomings and contradictions of their view. The hierarchical organisation is intent on securing its own continuity, hence it is not easily mobilised for policies outside accepted risks or challenging the status quo. The inherent delays and buffers resulting from procedures and consultations may have a stabilising effect but they also tend to paralyse commitment and creativity, causing cynicism and loss of credibility. Also, in their struggle for greater power, organisations come across limits in the form of corruption, suspicion and indifference. In this way sudden movement can sometimes occur in favour of the individualist or egalitarian perspective, sometimes via decline or fatalism.

The third world-view is the *egalitarian* one which is characterised by a strong group feeling, with communitarian aims and values and a lack of structured authority. Nature is perceived as a fragile, connected, complex whole. Emphasis is placed on our limited knowledge of the earth. The basic premise is equity between man and nature and among people, as expressed in the image of people

	individualist	hierarchist	egalitarian
world view idea of nature	skill-controlled cornucopia	isomorphic nature	accountable
myth of nature	nature benign perverse/tolerant	nature	nature ephemeral
concept of human nature	self-seeking	sinful	born good, malleable
management style driving force	growth [of opportunities]	stability [of social system]	equity and equality [in man-nature system]
type of management	adaptive	control	preventive
desired system properties	exploitability	controllability	sustainability
attitude to nature	laissez-faire	regulatory	attentive
attitude towards humans	channel rather than change, to avoid market disortions	restrict behaviour avoid loss of control	change behaviour values, to avoid catastrophe
attitude to needs/resources	expand resource base	rational allocation of resources	need-reducing strategy
economic growth	preferred: aim to create personal wealth	preferred within conditions: aim to avoid social collapse	not preferred/not a primary goal
attitude towards risk	risk-seeking	risk-accepting, institutionalisation and formal methods	risk-aversive/risk reduction

Table 1. Characteristics of cultural perspectives (based on Thompson, et al. 1990)

as stewards or partners. Issues related to justice, poverty and security are seen as important aspects of a basic strategy that aims to achieve sustainability. A key issue is a fair distribution of benefits and costs, both between rich and poor – the context for socialism – and between us and posterity – the context for many environmental groups. Usually egalitarians will insist on preventive measures, from a risk-averse position. As with the other world-views, egalitarians face limitations in maintaining their world-view. They always run the risk of individualism and hierarchy arising within their ranks. They may need a future with environmental disasters and a present peopled by acquisitive and powerful enemies in order to keep the group together. The solidarity and dogmatism of such radicalisation is often accompanied by a loss of creativity and effectiveness.

Many of the world's poor are *fatalist* in their societal outlook. They try to cope with everyday life's contingencies within the limited freedom of manoeuvre that they have. People and nature are both capricious, and all one can do is to make the best out of it. To them, one may presume, climate change only gets significance if they are actually hit by the consequences. Fatalists are part of the game together with the other perspectives because they provide legitimacy for the hierarchists – where they may become members of the lower ranks of the welfare society, because they are held out the promise of success by the individualists, and because they are potential converts for the egalitarians.

Uncertainties and Climate Change

The previously introduced schemes allow an ordered discussion of the various uncertainties and issues regarding the threat of human-induced climate change, according to the classification in Table 2. Issues are ranked according to their place in the PSIR-chain and the complexity levels of the underlying [sub]system. Many of them are clouded by uncertainties and controversies, which give rise to divergent or even antagonistic interpretations of what happened and expectations of what might happen. Initially, the focus has been on the S1 cell, that is, on the physics of greenhouse-gas dynamics in the atmosphere. Gradually, it has encompassed other compartments of the biosphere and possible impacts on agriculture (cells I1 and I2). As the possible threat of rising greenhouse-gas concentrations became more widely acknowledged, attention shifted to the pressure side in the form of emissions scenarios (cells P1 and increasingly P2). Policy proposals were framed in the context of the IPCC which raised all kinds of equity and cost issues – the human response (cells R2 and R3). In each cell I have indicated, with a few keywords, some of the dominant questions to be investigated. Obviously, this categorisation is based on rather crude definitions and distinctions and real-world dynamics makes the issues within cells highly interdependent.

At the physico-chemical level there are several near-certainties about climate change at the globally aggregated level and in the medium term of say 50 years. However, there are still major uncertainties which not necessarily soon diminish with more research. For example, the amount of carbon which is taken up from the atmosphere by the terrestrial biosphere and/or the oceans, is still an uncertain quantity which allows for various descriptions with different results, and different policy recommendations.[22] There are similar uncertainties in the nitrogen-fluxes,[23] the complex interplay of biopheric responses,[24] the role of clouds and aerosols etc. The actual changes at the regional and local level, the importance of biotic and

level 3	idea of fairness: mitigation costs burden sharing	economic [in]equity; socio-cultural & political factors environmental security & refugees.	idea of fairness: damages and adaptation costs	constraints on population & economic growth?
level 2	scenarios: population, economy and energy, including fertility & mortality determinations; cost-efficiency analysis	population & economy dynamics stock characteristics, ageing, technology. Fossil-fuel resource base	impacts on food and health, on biodiversity; benefit analysis	economic instruments & & barriers stimulation & transfer (efficiency, renewables)
level 1	scenarios: greenhouse-gas emissions	climate dynamics: clouds, aerosols/SO_2 biospheric feedbacks, climate variability & extremes	changes in land use & land cover; biodiversity, desertification etc	ecosystem response
	pressure P	*state S*	*Impact I*	*response R*

Table 2. *Classification of climate change related uncertainties and issues according to the PSIR-chain and the level of complexity*

[22] D. SCHIMEL, I.G. ENTING, M. HEIMANN, T.M.L. WIGLEY, D. RAYNAUD, D. ALVES and U. SIEGENTHALER, 'CO_2 and the Carbon Cycle' in *Climate Change 1994*, IPCC, Cambridge, Cambridge University Press, 1995; D. SCHIMEL, 'The Carbon Equation' in *Nature* 393(1998), pp. 208-09.
[23] M. DEN ELZEN, H.W. BEUSEN and J. ROTMANS, 'An Integrated Modeling Approach to Global Carbon and Nitrogen Cycles: Balancing Their Budgets' in *Global Biogeochemical Cycles* 11(1997)2, pp. 191-215.
[24] ALCAMO, LEEMANS and KREILEMAN, eds. *Global Change Scenarios of the 21st Century.*

abiotic feedbacks in the longer term and the changes in more refined indicators such as the distribution of weather extremes are also surrounded with large uncertainties.[25] These uncertainties are occasionally used by stakeholders to postpone policy measures or to prophecy doomsday.

On the pressure side, there are the huge uncertainties regarding future emissions of greenhouse-gases. The IS92 IPCC anthropogenic CO_2-emission scenarios cover a range from 5 to 35 Gton carbon in 2100; the new IS99 scenarios to be published in the course of 1999 again cover only a slightly smaller range. Non-CO_2-emission estimates are compounding the uncertainties and add, in the case of sulphur-dioxide emissions, a regional climate impact component. Of course, the emission scenarios reflect the many uncertainties at the higher level of societal dynamics which tend to be wider and more difficult to assess.

The development of regional population is relatively certain over the next few decades because of the population inertia. In the longer term, however, major uncertainties exist with regard to the determinants of fertility, morbidity and mortality and the consequent population growth. The development of regional economies and their interaction is more uncertain. Increasingly it is recognised that the large income disparities which evolve in, for instance, the widely used IS92-scenario – 5.5 billion people in India and Africa with an income of about 7000 1990\$/cap/yr[26] against an average income in the OECD of over 110.000 1990\$/cap/yr in 2100 – may be unacceptable as well as unfeasible. Insights and expectations are again changing, and may keep doing so. On top of this, and usually hardly accounted for in models, is the complex interplay between population, economy and environment.[27] For instance, very skewed income distributions and inefficient and insufficient infrastructure, often in combination with institutional inertia and corruption, may impede economic productivity and growth. The large and rapid

---

[25] Cf. note 3; E.J. BARRON, 'Climate Models: How Reliable are their Predictions?' in *Consequences – The Nature and Implications of Environmental Change* 1(1995)3, pp. 16-24.

[26] Dollars per capita per year

[27] L.W. MACKELLAR, W. LUTZ, A.J. MCMICHAEL and A. SUKRHE, *Population and Climate Change*, Vol. 1 of S. RAYNER and E.L. MALONE (eds.), *Human Choice and Climate Change*, Columbus, Ohio, Battelle Press, 1998, pp. 1-87.

changes in the age distribution give rise to both 'greying' and 'greening'; it is unclear how economies will accommodate to it.[28]

One of the other and rather crucial parts is the energy system. It is an intermediate between population and economic developments on the one hand and greenhouse-gas emissions on the other. A large unknown here is the coupling between monetary and physical flows. As the nature of economic output changes, the 'economic transition' from the agricultural to the industrial, and later on, the service economy takes place. It has been associated with 'dematerialisation' and 'ecological restructuring' but more research is needed.[29] In combination with other factors, energy demand projections may widely differ even for the same population and economic growth projection. Further unknowns enter the models in converting energy demand into greenhouse-gas emissions: at what costs can additional fossil fuels be brought to the marketplace and what determines their mutual substitution; how will the cost of non-fossil options develop over time and how fast can they penetrate the energy market?[30] For the less developed regions, uncertainties are even larger and of a more conceptual and structural nature. Finally, most models have no or a highly simplified feedback from the energy system onto the economy, and take not or hardly into account how capital and land requirements for new energy supply technologies feed back onto the [regional] economies through interest rates, food prices and the like.[31]

---

[28] H. RAE, *The World in 2020 – Power, Culture and Prosperity: A Vision of the Future*, London, Harper Collins Publishers, 1995.

[29] R. AYRES and L. AYRES, *Industrial Ecology – Towards Closing the Materials Cycle*, Cheltenham, Edward Elgar, 1997. For example, a recent study found a marked difference in energy-intensity between market- and government services, which would imply different energy use patterns depending on the role of government (Jespersen 1999). Another issue are the degree to which the informal economy is included in the statistical data; the use of Purchasing Power Parity (PPP) corrected welfare is one of the proposed adjustments.

[30] M. GRUBB, 'Technologies, Energy Systems, and the Timing of CO_2 Emissions Abatement: An Overview of Economic Issues' in *Energy Policy* 25(1996), pp. 159-72; DE VRIES, BOLLEN, BOUWMAN, DEN ELZEN, JANSSEN, KREILEMAN and LEEMANS, 'Greenhouse-Gas Emissions in an Equity-, Environment- and Service-Oriented World: An Image-Based Scenario for the Next Century' in *Technological Forecasting and Social Change* 63 (2000) 2-3.

[31] Often, at least in most official projections, it is assumed that the equilibrating forces of markets will lead to the necessary adjustments in a timely and acceptable manner. This cannot be taken for granted in such regions as India with its scarce capital and land resources or the Middle East and CIS with their large dependency on oil and gas exports.

On the impact and response side, there are only a few regional in-depth analyses of the consequences of changes in certain climate parameters on agriculture and health. Even a large integrated assessment model like the IMAGE 2.1 model deals in a rather crude way with changes in crop yields, farmer response and ecosystem changes.[32]

One of the consequences is that any analysis of emission reduction, damage and adaptation costs is only meaningful for a well-defined reference or baseline scenario – which itself has to be chosen out of a whole set of scenarios of unknown plausibility. From here on uncertainties then propagate throughout the discussions on adequate response policies. No wonder that policy-makers are confronted with widely different views on what should be done where and when, and that many proposals reflect in various ways the different world-views and management styles discussed in the previous paragraph.

Table 3 illustrates this point for the four IPCC-scenarios to be presented in the course of 1999. These scenarios are members of four scenario families: 1-scenarios corresponding to outward-oriented worlds with increasing globalisation and liberalisation and 2-scenarios to more inward-oriented worlds with more cultural pluralism, whereas A-scenarios reflect a small and B-scenarios a large concern about environmental and equity issues. In line with these scenarios, the A1 and B1 scenarios have low population and high economic growth and fast technology development and transfer. The A2 and B2 scenarios, on the other hand, picture a world in which population growth is higher and economic growth lower due to less regional interactions and preservation of more traditional or locally oriented values and policies. Table 3 suggests the links with the previously discussed perspectives: whereas the A1 world reflects the market- and high-tech-orientation of the individualist, the B1 and possibly even more the B2 worlds would be driven by egalitarian values. The A2 world could easily support a fatalist business-as-usual outlook.

Behind these uncertainties and the ensuing controversies lurks a deeper, conceptual question. The dominant framework to model human behaviour in relation to climate change is neo-classical economics in its mathematically refined present-day form. Humans are

---

[32] ALCAMO, LEEMANS and KREILEMAN (eds.), *Global Change Scenarios of the 21st Century*, Oxford, Elsevier Science, 1998.

conceived of in this framework as maximisers of discounted individual utility with perfect foresight, although various refinements have been proposed and implemented. In fact, this simplification is pivotal to most economic analyses. Yet, it is increasingly challenged by novel approaches in which the diversity of human behaviour and in human interference with the environment is simulated in much more detail. Including spatial aspects has given rise to powerful tools to improve our understanding of the interplay between physical and human aspects of change, using amongst others systems dynamics and cellular automata.[33] The insights of environmental psychology have been introduced in a model which simulates human behaviour as an interplay between the four modes of behav-

	A1	B1	A2	B2
Probability of CC	Medium to large A2>A1>B2	Smallest	Highest	Small to medium A1>B2>B1
Attitude towards CC	Economic growth has priority	Other environm. concerns have priority; no-regrets and side-benefits	Muddling through-not an issue / fatalist	Local environm. concerns; incapable to solve global environm. issues
Actions if CC is large	Barricades; adapt on your own; mitigate tensions by compensating poor	Formulate and implement global CC policy (too late but feasible), including adaption	Not feasible: world community incapacitated; cope on your own	Some regions suffer; cope on your own, but (ineffective) attempts to share the burdens
Actions if CC is small	Proud: 'You see, ours is the best of all possible worlds'	Proud: 'You see: the high-tech and SD-orientation did it'	'Well, this time we were lucky... so far'	'Well, we were lucky... but our local concerns may have helped
Attitude towards CC policy	Reluctant, although rich enough to make compensatory gestures	Reluctant, if it slows down the narrowing of the N-S income gap	No expectations: impossible anyway	Low expectation full of good intentions, but they fail

Table 3. Climate Change (CC) in the four IPCC 1999 scenarios.

[33] T.S. FIDDAMAN, *Feedback Complexity in Integrated Climate-Economy Models*, unpublished Ph.D. diss., Report D-4681, Cambridge, MIT System Dynamics Group, 1997; P.M. ALLEN, *Evolutionary Complex Systems: Models of Technology Change*, Amsterdam Conference on Developments in Technology Studies, Evolutionary Economics and Chaos Theory, Amsterdam, 1993; G. ENGELEN, R. WHITE, I. ULJEE, and P. DRAZAN, 'Using Cellular Automata for Integrated Modelling of Socio-Environmental Systems' in *Environmental Monitoring and Assessment* 34(1995), pp. 203-14.

iour, thus broadening the notion of 'rational human behaviour'. My view is that these attempts are the first real efforts towards integrated assessment, in which complex human systems are not simply modelled as isomorphic to some well-researched physico-chemical systems.

The Battle of Perspectives

Much of the indicated uncertainties should not be viewed as a statistical artefacts, but instead be traced to different interpretations of reality. The use of different perspectives or world-views to derive consistent qualitative and quantitative projections has been advocated to deal explicitly with the underlying value orientation.[34] This approach has been applied in a simplified fashion in the TARGETS-model with the aim of opening up the discussion on sustainable development.[35] A more refined approach has been applied to the problem of climate change.[36]

A simple dynamic model has been constructed of the economy (investment decisions and the degree of emission reductions) and the climate system. The three different cultural perspectives on how the world functions and is managed are implemented in the form of diverging assumptions on climate sensitivity, technological development, and mitigation and damage costs. The model is then run to construct three 'utopian' futures, that is, worlds in which the decision-making agents have the correct view of how the world functions (world-view) and manage it accordingly (management style). The different model assumptions and decision rules lead to three different 'utopian' time paths for CO_2-emissions and tempera-

---

[34] DE VRIES, *Sustainable Resource Use – An Inquiry into Modelling and Planning*; S. RAYNER, 'A Cultural Perspective on the Structure and Implementation of Global Environmental Agreements' in *Evaluation Review* 15(1991)1, pp. 75-102; SCHWARTZ and THOMPSON, *Divided We Stand*; THOMPSON, ELLIS, and WILDAWSKY, *Cultural Theory*; M.E. COLBY, 'Environmental Management in Development: The Evolution of Paradigms' in *Ecological Economics* 3(1991), pp. 193-213; WRR (Wetenschappelijke Raad voor het Regeringsbeleid), 'Duurzame risico's: Een blijvend gegeven', Den Haag, Report n°4, Staatsuitgeverij, 1994.

[35] ROTMANS and DE VRIES (eds.), *Perspectives on Global Change: The TARGETS Approach*, Cambridge, CUP, 1997.

[36] M.A. JANSSEN and H.J.M. DE VRIES, 'The Battle of Perspectives: A Multi-Agents Model with Adaptive Responses to Climate Change' in *Ecological Economics* 26(1998), pp. 43-65.

ture change (Figure 3). In the utopia of the hierarchists, fossil fuels are phased out in the longer run to stabilise temperature change at an acceptable level. In the utopia of the individualists, with higher economic growth, the limited efforts to reduce energy intensity cause CO_2-emissions to soar to over 40 GtC in 2100. However, the insensitive nature of the climate system leads to a temperature increase of only 0.5°C. In the utopia of the egalitarians, lower economic growth, a strong policy on energy efficiency and an early phase-out of fossil fuels leads to much lower CO_2-emissions, yet warming exceeds 1°C because the climate system is assumed to be quite sensitive to human interference.

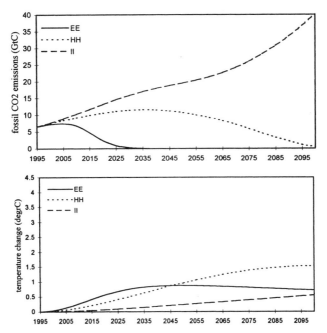

Figure 3. Projections of fossil CO_2-emissions and temperature change in three utopias (Egalitarian (EE), Hierarchist (HH) and Individualist (II)).

Using the more realistic assumption that agents do not have perfect knowledge and that their behaviour is biased by their worldview, a possible adaptive response to climate change can be incorporated in the model simulation by assuming that agents change world-view if their expectation does not match their observation within a certain margin and if another world-view gives a better match. In changing world-view their management style changes as well. Three new experiments result in the emission pathways and temperature profiles shown in Figure 4. Emissions begin to bifurcate in the middle of next century. In the individualist world, emissions increase sharply after 2040 because agents discover that climate change does not happen and switch to the individualist perspectives. If the egalitarian world-view turns out to be correct, the reverse happens. In this way, the adaptive capacity of human actors in the system is taken into account. This allows exploration of a more complex reality, which adds variety but makes validation more difficult.

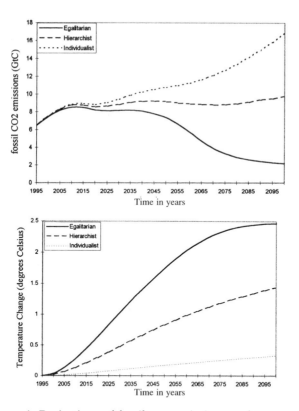

Figure 4. Projections of fossil CO_2-emissions and temperature change if the adaptive response by changes in world-view are taken into account (Egalitarian (EE), Hierarchist (HH) and Individualist (II)).

Ethics Dimensions of the Climate Change Debate

As has become clear from the previous considerations, the climate change issue has unavoidably ethical aspects. A global common – the atmosphere – has been accumulated with greenhouse-effect enhancing gases over time by people. Some peoples have been doing this over centuries, in the form of deforestation and fossil fuel combustion amongst others. Other peoples have hardly begun to emit these gases but are expected to have rapidly rising emissions in the next decades. An enhanced greenhouse-effect will have impacts which are highly uncertain, at least at the regional level, and hence peoples will be affected by these impacts in quite different ways. Hence, along the whole chain there are issues of intergenerational and intragenerational distribution of welfare and well-being.

The issue can be formalised in a simple way by stating that the world community has to agree on what degree of climate change is judged to be acceptable, given uncertainties and adaptation options. From this an upper bound on the cumulated emission of greenhouse-gases can be derived; these can be translated in a series of emission time-paths. Next, scenarios can be made to assess the anticipated exceedence of these emission caps. The sacrifices to avoid this exceedence – referred to as the burden – have to be shared.

As yet, there is no agreement on what an acceptable risk is but the discussions within the IPCC suggest that the world community should aim for stabilisation of the equivalent greenhouse-gas concentration in the range of 450 to 550 ppmv – which roughly coincides with cumulated carbon emissions in the order of 600 to 1000 10^9 ton carbon equivalent.[37] For comparison: humanity has emitted some 300 10^9 ton carbon equivalent between 1900 and 1990. To estimate how much carbon humanity will emit over the next centuries is at least as difficult, as has been discussed in the previous paragraph.

---

[37] R. SWART, M. BERK, M. JANSSEN, E. KREILEMAN, and R. LEEMANS, 'The Safe Landing Approach: Risks and Trade-Offs in Climate Change' in *Global Change Scenarios of the 21st century*, J. ALCAMO, R. LEEMANS and E. KREILEMAN (eds.), Oxford, Elsevier Science Ltd., 1998, pp. 193-234.

Suppose the stakeholders in this debate agree upon the acceptable risk and the anticipated exceedence, they have to find ways to share the burden. In the last few years, various approaches to burden sharing have been offered.[38] One principle is called grandfathering: the present situation – quantified in such indicators as ton carbon emitted per year per person or per unit of economic output – is a legitimate starting point to allocate the costs to bring down emissions. It is associated with the sovereignty of nations. Another principle is that every person on earth has the same emission right, over his or her lifetime, for instance. A third principle says that the ability to pay – or the welfare loss per unit of the costs – provides an allocation key. Obviously, each of these principles may have very different outcomes for the measures and policies to be taken in the regions in the world. The situation becomes more complex if one also introduces past contributions to expected climate change – indicated by for instance the rise in CO_2-concentrations or average temperature.[39] A further consideration is that one should aim for global efficiency in the sense that measures should be taken where they are cheapest. Such an over-all cost-efficiency framework has led to an abundance of analyses on emission permit trading.[40] Finally, there are good arguments to include the anticipated damage into the equation which leads to cost-benefit analyses by counting the avoided damage as benefits.

As one may expect, the various approaches cannot be separated from the participants' perceived interests or, broader, their perspective. They also are intertwined with one's expectations about the future (cf. Table 3). The only robust conclusion can be that the negotiations about how to deal with impending human-induced climate change will be an ongoing process for the decades to come – with the Kyoto protocol one out of several initial and important steps.

---

[38] L. RINGIUS, A. TORVANGER and B. HOLTSMARK, 'Can Multi-Criteria Rules Fairly Distribute Climate Burdens? OECD Results From Three Burden Sharing Rules' in *Energy Policy* 26(1998)10, pp. 777-93.

[39] DE VRIES, BOLLEN, BOUWMAN, DEN ELZEN, JANSSEN, KREILEMAN and LEEMANS, 'Greenhouse-Gas Emissions in an Equity-, Environment- and Service-Oriented World: An Image-Based Scenario for the Next Century.'

[40] P.A. SCHULTZ and J.F. KASTING, 'Optimal Reductions in CO_2 Emissions' in *Energy Policy* 5(1997), pp. 491-500.

Conclusions

In the past few years the threat of climate change as a consequence of large and increasing emissions of greenhouse gases, notably carbon dioxide (CO_2) has become widely acknowledged. Climate change models have a rather poor quality in terms of formal validation as required in the natural sciences. Yet, they can in our view be useful tools to assess the interactions between humans and their environment if the emphasis is on interactive, process-oriented construction and use and if novel approaches such as development of interactive, perspective-based modelling approaches and its use in policy dialogues. This is the more important as the climate change debate is full of normative issues which can easily be obscured by implicit assumptions or model-incorporated ideologies.

References

ALCAMO, J. (ed.), *Integrated Modelling of Global Climate Change: IMAGE 02*, Dordrecht/Boston/London, Kluwer Academic Press, 1994.

ALCAMO, J., R. LEEMANS and E. KREILEMAN (eds.), *Global Change Scenarios of the 21st Century*, Oxford, Elsevier Science, 1998.

ALLEN, P.M., *Evolutionary Complex Systems: Models of Technology Change*, Amsterdam Conference on Developments in Technology Studies, Evolutionary Economics and Chaos Theory, Amsterdam, 1993.

AYRES, R. and L. AYRES, *Industrial Ecology – Towards Closing the Materials Cycle*, Cheltenham, Edward Elgar, 1997.

BARRON, E.J. 'Climate Models: How Reliable are their Predictions?' in *Consequences – The Nature and Implications of Environmental Change* 1(1995)3, pp. 16-24.

COLBY, M.E., 'Environmental Management in Development: The Evolution of Paradigms' in *Ecological Economics* 3(1991), pp. 193-213.

DALY, H., *Toward a Steady-State Economy*, San Fransisco, Freeman and Company, 1973.

DE VRIES, B., 'Contouren van een duurzame toekomst' in R. WEILER and P. GIMENO (eds.), *Ontwikkeling en Duurzaamheid*, Brussels, VUB Press, 1996, pp. 31-73.

DE VRIES, B., 'Susclime : A Simulation Game on Population and Development in a Climate-Constrained World.' *Simulation and Gaming* 29(1998)2, pp. 216-37.

DE VRIES, B., J. BOLLEN, L. BOUWMAN, M. DEN ELZEN, M. JANSSEN, E. KREILEMAN and R. LEEMANS, 'Greenhouse-Gas Emissions in an

Equity-, Environment- and Service-Oriented World: An Image-Based Scenario for the Next Century' in *Technological Forecasting and Social Change* 63(2000)2-3.

DE VRIES, H.J.M., *Sustainable Resource Use – An Inquiry into Modelling and Planning*, unpublished Ph.D. diss., Groningen, University of Groningen, 1989.

DEN ELZEN, M., H.W. BEUSEN and J. ROTMANS, 'An Integrated Modelling Approach to Global Carbon and Nitrogen Cycles: Balancing Their Budgets' in *Global Biogeochemical Cycles* 11(1997)2, pp. 191-215.

DOUGLAS, M., *Natural Symbols*, London, Pelican Books, 1973.

ENGELEN, G., R. WHITE, I. ULJEE, and P. DRAZAN, 'Using Cellular Automata for Integrated Modelling of Socio-Environmental Systems' in *Environmental Monitoring and Assessment* 34(1995), pp. 203-214.

FIDDAMAN, T.S., *Feedback Complexity in Integrated Climate-Economy Models*, Ph.D. diss., Report D-4681, Cambridge, MIT System Dynamics Group, 1997.

FUNTOWICZ, S. and J. RAVETZ, *Uncertainty and Quality in Science for Policy*, Dordrecht, Kluwer Academic Publishers, 1990.

GRUBB, M., 'Technologies, Energy Systems, and the Timing of CO_2 Emissions Abatement: An Overview of Economic Issues' in *Energy Policy* 25(1996), pp. 159-172.

HARMAN, W., *Global Mind Change – The Promise of the Last Years of The Twentieth Century*, San Fransisco, Knowledge Systems Inc., Institute of Noetic Sciences, 1993.

IPCC (Intergovernmental Panel on Climate Change), *Climate Change: The IPCC Scientific Assessment*. J.T. HOUGHTON, G.J. JENKINS and J.J. EPHRAUMS (eds.), Cambridge, Cambridge University Press, 1990.

IPCC, *Climate Change 1995: Economic and Social Dimensions of Climate Change*, Cambridge, Cambridge University Press, 1996.

IPCC, *Climate Change 1995: Impacts, Adaptations and Mitigation of Climate Change: Scientific-Technical Analysis*, Cambridge, Cambridge University Press, 1996.

IPCC, *Climate Change 1995: The Science of Climate Change*, Cambridge, Cambridge University Press, 1996.

JÄGER, J. and T. O'RIORDAN, 'The History of Climate Change Science and Politics' in T. O'RIORDAN and J. JÄGER (eds.), *Politics of Climate Change – A European Perspective*, London, and New York, Routledge, 1996, pp. 1-31.

JANSSEN, M.A. and H.J.M. DE VRIES, 'The Battle of Perspectives: A Multi-Agents Model with Adaptive Responses to Climate Change' in *Ecological Economics* 26(1998), pp. 43-65.

JANSSEN, M.A., 'Optimization of a Non-linear Dynamical System for

Global Climate Change' in *European Journal of Operational Research* 99(1997), pp. 322-335.

JANSSEN, M.A., *Modelling Global Change – The Art of Integrated Assessment Modelling*, Cheltenham, UK, Northampton, MA, USA, Edward Elgar Publishers, 1998.

JANTSCH, E., *The Self-Organizing Universe – Scientific and Human Implications of the Emerging Paradigm of Evolution*, Oxford, Pergamon Press, 1980.

JASANOFF, S. and B. WYNNE, *Science and Decisionmaking*, Vol. 1 of S. RAYNER and E.L. MALONE (eds.), *Human Choice and Climate Change*, Columbus, Ohio, Battelle Press, 1998, pp. 1-87.

JESPERSEN, J., 'Reconciling Environment and Employment by Switching from Goods to Services? A Review of Danish Experience' in *European Environment* 9(1999).

KREBS, F. and H. BOSSEL, 'Emergent Value Orientation in Self-Organization of an Animal' in *Ecological Modelling* 96(1997), pp. 143-164.

MACKELLAR, L.W, W. LUTZ, A.J. MCMICHAEL and A. SUKRHE, *Population and Climate Change*, Vol. 1 of S. RAYNER and E.L. MALONE (eds.), *Human Choice and Climate Change*, Columbus, Ohio, Battelle Press, 1998, pp. 1-87.

MORECROFT, J. and J. STERMAN, 'Modelling for Learning', Special Issue of *European Journal of Operations Research* 59(1992)1.

RAE, H., *The World in 2020 – Power, Culture and Prosperity: A Vision of the Future*, London, Harper Collins Publishers, 1995.

RAYNER, S., 'A Cultural Perspective on the Structure and Implementation of Global Environmental Agreements' in *Evaluation Review* 15(1991)1, pp. 75-102.

RINGIUS, L., A. TORVANGER and B. HOLTSMARK, 'Can Multi-Criteria Rules Fairly Distribute Climate Burdens? OECD Results From Three Burden Sharing Rules' in *Energy Policy* 26(1998)10, pp. 777-793.

ROTMANS, J. and H.J.M. DE VRIES (eds.), *Perspectives on Global Change: The TARGETS Approach*, Cambridge, Cambridge University Press, 1997.

SCHIMEL, D., I.G. ENTING, M. HEIMANN, T.M.L. WIGLEY, D. RAYNAUD, D. ALVES and U. SIEGENTHALER, 'CO_2 and the Carbon Cycle' in *Climate Change 1994*, IPCC, Cambridge, Cambridge University Press, 1995.

SCHIMEL, D.S., 'The Carbon Equation' in *Nature* 393(1998), pp. 208-209.

SCHULTZ, P.A. and J.F. KASTING, 'Optimal Reductions in CO_2 Emissions' in *Energy Policy* 25(1997), pp. 491-500.

SCHWARTZ, M. and M. THOMPSON, *Divided We Stand: Redefining Politics, Technology and Social Choice*, New York, Harvester Wheatsheaf, 1990.

SIMON, H.A., *The Sciences of the Artificial*, Cambridge, MA, The MIT Press, 1969.

SWART, R., M. BERK, M. JANSSEN, E. KREILEMAN and R. LEEMANS, 'The Safe

Landing Approach: Risks and Trade-Offs in Climate Change' in J. ALCAMO, R. LEEMANS and E. KREILEMAN (eds.), *Global Change Scenarios of the 21st century – Results from the Image 2.1 Model*, Oxford, Elsevier Science Ltd., 1998, pp. 193-234.

THOMPSON, M., R. ELLIS, and A. WILDAWSKY, *Cultural Theory*, Boulder, Westview Press, 1990.

VAN ASSELT, M.B.A. and J. ROTMANS, 'Uncertainty in Perspective: A Cultural Perspective Based Approach' in *Global Environmental Change* 6(1996)2, pp. 121-58.

VAN DAALEN, E., W. THISSEN and M. BERK, 'The Delft Process: Experiences with a Dialogue between Policy Makers and Global Modellers' in J. ALCAMO, R. LEEMANS and E. KREILEMAN (eds.), *Global Change Scenarios of the 21st century – Results from the Image 2.1 Model*, Oxford, Elsevier Science Ltd., 1998, pp. 193-234.

VAN DER SLUIJS, J., *Anchoring amid Uncertainty: On the Management of Uncertainties in Risk Assessment of Anthropogenic Climate Change*, Ph.D. diss., Utrecht, University of Utrecht, 1997.

VROON, P., 'Een oud brein in een nieuwe wereld' in *Informatie en Informatiebeleid* 7(1989)4, pp. 19-29.

WCED (World Commission on Environment and Development), *Our Common Future*, Oxford, Oxford University Press, 1987.

WRR (Wetenschappelijke Raad voor het Regeringsbeleid), 'Duurzame risico's: Een blijvend gegeven', Den Haag: Report n°4, Staatsuitgeverij, 1994.

3.2.4

SUSTAINABILITY

A Vision for a New Technical Society?

Raoul Weiler

The concept of Sustainable Development was introduced some 15 years ago. The concern about the evolution of our planet is steadily increasing and has been the subject of several publications. The first report to the Club of Rome Limits to Growth *(1972) is to be considered as a real milestone, drawing the attention of the public to the impact of human activity on the possible depletion of the natural resources and on deterioration of the environment of the planet. Later publications e.g. the Brundtland Report (1987), and several conferences of the United Nations (Rio 1992, etc.) have contributed to a growing acceptance of the vision of sustainable development. Today several governments are setting up institutions and programmes in search for the local application of sustainability.*
In the mean time, technological innovations in different domains have shown numerous and realistic examples which prove feasible considerable savings in resources and energy use and consequently in the protection of the environment. The concepts of 'Factor Four' (Wuppertal Institute for Climate, Environment and Energy and The Club of Rome) and even of 'Factor Ten' (Rocky Mountain Institute) were introduced, meaning that a standard of living twice as high could be attained with half the resources, or even less.
The engineering approach for the design of buildings, cars, consumer products, industrial and agriculture production processes, transportation infrastructure, etc. have to be reconsidered. A long range vision has to take the place of the present consumer driven, fast profit generating industrial and economic system. The consequences on the environment and climate call for a critical reflection about the way of living of the industrialised society.

Rebound effects can substantially reduce the results obtained by technological innovations. Therefore, a wide acceptance and understanding of the objectives of sustainable development by the different populations should be obtained.

The societal aspects of engineering are on the way to becoming of highest importance; the ethical side of the profession of the engineer, technologist and designer is becoming a leading edge in our industrial society.

Emergence of the Concept of Sustainability

The concept of sustainability has its origin in the World Conservation Union's study of sustainable resource use (International Union for the Conservation of Nature and Natural Resources, IUCN, 1981). From there the concept made its way into the World Commission for Environment and Development, the Brundtland Commission.[1] In the well known report – the Brundtland Report – of this commission, 'Our Common Future', a general definition has been given:

> Humanity has the ability to make development sustainable – to ensure that it meets the needs of the present without compromising the ability of future generations to meet their own needs. The concept of sustainable development does imply limits – not absolute limits but limitations imposed by the present state of technology and social organisation on environmental resources and the ability of the biosphere to absorb the effects of human activities. But technology and social organisation can be both managed and improved to make way for a new era of economic growth.[2]

This definition has been subject to much criticism, but nevertheless it is still a very fruitful concept. Recent world conferences on the status of the planet earth have referred to the concept of sustainability, be it meetings on climate change, on habitat, on demography: all of them are concerned about solutions in view of long term evolution of mankind on our limited planet. In the mean time national programmes are on the way to being discussed or set up for the implementation of sustainability, which are directly or indirectly concerned about: greenhouse phenomena, world food availability and

---

[1] THE WORLD COMMISSION ON ENVIRONMENT AND DEVELOPMENT (Chair: G.H. BRUNDTLAND) *Our Common Future*, Oxford, Oxford University Press, 1987.

[2] *Ibid.*, p. 8.

agriculture, transport and urban development, about water supply and health or many others. Finally, it is worth recalling the first report to the Club of Rome Limits to Growth,[3] in which, in very clear terms the limited character of our planet has been described with the help of a mathematical simulation technics. Twenty years later, the publication of Beyond the Limits (1992) by the same authors, has reactualised the topic of limited resources and once again underlined the necessity of such a concept of sustainability.

The implementation of a sustainable industrial society remains an enormous challenge to the engineer, the economist, the politician and the entire civil society. The second half of the twentieth century is doubtless an era of unprecedented technological progress.

Engineering Ethics

The purpose of this paper is to situate the potentialities of technological innovation in the process of setting up a sustainable society. It is a way for engineers to use their capabilities in a way that is inspired by ethical considerations world-wide. For this reason, the concepts of Factor Four first described by Prof. E.U. von Weizsäcker of the Wuppertal Institute for Climate, Energy, Environment and of Factor Ten of the Rocky Mountain Institute, are used to illustrate, by means of numerous examples, the feasibility of a technological innovation approach. However this does not mean, at all, that the other aspects of the problem of sustainable growth are reduced to this technological approach. On the contrary, aspects such as social equity, the participation of the civil society through available or still to be installed democratic channels and institutions, the elaboration and the implementation of new accounting systems taking into consideration the use of nature and environment, etc., are an integral part of the success and are of great importance to reach, in a foreseeable future, a sustainable society.

In a more general and philosophical approach, inspired by the philosophy on ecology, it is pointed out by M. Carley and I. Christie in their book *Managing Sustainable Development*,[4] that:

---

[3] D.H. MEADOWS, D.L. MEADOWS, J. RANDERS and W.W. BERHENS III, *The Limits to Growth: A Report for The Club of Rome's Project on the Predicament of Mankind*. New York, Universe Books, 1972.

[4] M. CARLEY and I. CHRISTIE, *Managing Sustainable Development*, London, Earthscan, 1997.

'The holistic approach recognises the inescapability of integrating an ethical dimension into science of global environmental change.'

In the context of the present analysis, technology should explicitly be added to science as well as resource use to global environment. The holistic character refers to the global approach of sustainability as is mentioned above. As to environmental and engineering ethics as well, four categories or viewpoints can be distinguished: the technocentric, the biocentric, the managerial and the communalist.

The *technocentric* view is characterised by a resource-exploitative and growth-oriented approach. It tends towards a wholly instrumental approach to nature and is resolutely anthropocentric. It places faith in the capacity of technology to intervene in nature for own profit and to substitute man-made capital for natural resources where required.

The *biocentric* view is characterised by a preservationist anti-growth attitude and stands at the opposite of the technocentric view. Moral rights are conferred on other species and humans are required to respect the intrinsic value in all nature and live in harmony with it.

The *managerial* view is characterised as a resource-conservationist and oriented to sustainable growth.

The *communalist* which is resource-preservationist and oriented to limited or zero growth.

None of the technocentric or biocentric positions are compatible with sustainable development. The technocentric attitude does not consider the consequences of growth on resource availability and environmental effects; the biocentric one appears to be impracticable in relation with the world-wide industrialisation process and the aspiration of most of the populations to a better life and human development.

The managerial and communalist approaches represent intermediate attitudes compared to the first two ones. Both are associated with a possible human 'stewardship' of nature, with different emphasis on the extent of the necessity of growth. If one recognises the need for sustainable growth to improve human development especially in the developing world and the need for a policy change in the sense of a sustainable resource use (steady-state or steady-flow) in the industrial world's consumption and production pattern, then the intermediate views are applicable. The *Factor Four* and *Ten* approaches do fit within and contribute very well to a resource-preservationist philosophy combined with a limited sustainable growth, resulting in a possible and feasible 'stewardship' of our planetary situation.

Referring to the work of Prof. E.U. von Weizsäcker of the Wuppertal Institute,[5] the efficiency revolution which is part of the concept of *Factor Four* and *Ten* has to be introduced on large scale basis into our society and especially in the engineering practices. Undoubtedly, behind this efficiency concept stands a planetary ecological reality and necessity. Numerous reports from, for example, The Worldwatch Institute, The World Resources Institute, The United Nations, The Club of Rome, Wuppertal Institute, indicate more and more that the industrial society is evolving towards the natural boundaries of the planet. Evidently it is recommended to avoid, in the long run, these boundaries. If not, humanity could face a situation, expressed in the following way:

> 'If we fail to change course soon enough and the collision occurs, nature will survive the event somehow. Humanity will not.'

Among the challenges expressed at the Earth Summit in Rio de Janeiro in 1992, sustainable development was elevated to one of the three urgent main themes. In fact, it became a reference action point. The concept of possesses the ability to bring a considerable contribution to meet this challenge. Environmental protection has been an uphill struggle from the beginning on and will remain so for many years to come. The main reason is that environmental protection is seen as an economic sacrifice. On one hand society began to realise that a healthy environment was a social asset worth paying a price for it. On the other hand, some entrepreneurs started to accept that preventive pollution intervention, still accompanied with costs, appeared to be, for several reasons, economically attractive. However, in the long run the pollution intensity related to the consumer level of our industrial societies of the North and to 6 billion people world wide is not maintainable. This affluence represents a resource use easily a factor 5, often a factor 20 higher of the developed countries when compared with the developing ones. It means that today's global consumption rates are clearly unsustainable. The pol-

---

[5] E.U. VON WEIZSÄCKER, A.B. LOVINS and L.H. LOVINS, *Factor Four: Doubling Wealth, Halving Resource Use*, London, Earthscan, 1997.

lution paradigm pushes the rich countries towards what may be called a physically impossible goal: 'zero'-emission factories and machines. Classical pollution control is unavailable to 80% of the world and 'zero'-emissions are unattainable for 98%.

The message of the engineering approach tries to get out of the pollution dilemma as a solution for the industrial society. It proposes an increasing productivity with the consequence of decreasing resource use and there by decreased pollution emission.

Dematerialisation and Immaterialisation

Among the primary prerequisites to reach a sustainable society the technological feasibility should be proven. It seems almost impossible to envisage that a society would turn the clock back and be ready to accept a life style with reduced technological facilities. Nevertheless the question remains to find out in how far a given society is ready to work in the direction of sustainability. The degree of urgency for action about sustainability increases every day, hence, not more than about one half of the next century remains for technological innovation and its translation into daily life. Keeping in mind that the amortisation of investments for infrastructure for transportation, energy production facilities, buildings, and so on takes periods of thirty years or more, then the need for clear concepts about sustainability and its technological feasibility arises now or in the very near future. The scientific and engineering community stands in this matter before a tremendous societal and ethical challenge which it cannot afford to ignore

Dematerialisation of products and production processes by the given factor represents the reduction of materials use and fossil energy consumption and consequently an environmental protection benefit is directly related to it.

Immaterialisation of production processes and professional activities does allow a further step towards extreme reduced material use and energy consumption. Within the EU the focus Information Society Technology (IST) research programme is positioned to enable sustainable development (SD).

Both action lines are situated in the engineering concept indicating the capacity and the feasibility to contribute substantially to a sustainable society. The leitmotiv of *Factor Four* is clearly a message about technological feasibility: 'doubling wealth while halving resource use' and *Factor Ten* suggests an even higher technological potential.

The driving force of this approach is definitely technology driven. The ethical contribution relies on the view that engineering activities are dominantly oriented towards the future generations. It is interesting to look at this point at what philosophers of technology have thought about the pace and the significance of technology in the western society. In the early fifties and sixties, technology and economic expansion did not leave much room for much attention about possible side effects. Resource depletion, environmental impact, were not considered as issues. Industrial expansion had a total free hand and restrictions from civil society were unimaginable. Critical and quite alarming analyses about the industrial activities made by philosophers of technology like Jacques Ellul (1960, 1977) and Langdon Winner (1977), and the observations on the state of the nature and the environment as a consequence of human activity, done by scientists like Rachel Carson (1962), did not immediately draw society's attention. Their early but pessimistic analysis expresses the view that an industrial and market driven society is entirely unsustainable. The recent and new framework of sustainable development has created another vision and reference point for the future generations.

'Rebound' Effect

In the fields where, due to technological innovations, substantial progress has been made in the direction of dematerialisation or immaterialisation, the benefits do not seem to have reached their initial objective. A rebound effect appears and annihilates the potential benefits. Reduced energy cost leads to more energy consumption and the net result is, at least at this stage, no progress in resources reduction. The rebound effect indicates the difficulty and complexity of implementation of the proposed solutions. A conscious and deliberate choice by the international community will have to be taken, if any result is to be reached.

Besides the above mentioned action lines, other conditions in terms of social equity and democratic imbedding also have to accompany the process towards sustainability. These boundary conditions will also determine the final outcome of the implementation of sustainable development.

Examples of Productivity Increase through Engineering

It is worthwhile to have a brief look on some examples of realisations of a factor four in different domains of technology. In the excellent book by E.U. von Weizsäcker, A.B. Lovins and L.H. Lovins,[6] which is used in the present survey, a short description of some 50 cases illustrates largely the possibilities and potentialities which technology innovation can offer.

Energy Productivity

Rational use of energy represents a widespread expression for the necessity of better use of these resources. The notion of energy productivity expresses a redirection of technological progress. Increasing energy productivity by a factor of four is proposed as a new standard and the focus is put on examples where a straightforward potential of energy efficiency is available.

One of the most telling examples with high energy saving potential is the ordinary family car. The Rocky Mountain Institute (RMI) has designed such a car. It was given the name of *hypercar*. Summarising the different analyses and even models developed by important and smaller car manufacturers leads to the following conclusion: a low energy consuming car with acceptable performances and size (4 passengers) is a feasible object. Prototyping models have been constructed and tested. Decreasing the weight by choosing carbon fiber technology and powered by a hybrid electric propulsion system, such a car would end up with a fuel use of about 2,0-2,5l/100km. According to the authors, forecasts indicate that by 2005 most cars in the showrooms will be of this ultralight hybrid type. Moreover, these cars will not only save 80-95 per cent of fuel and have cleaner exhaust output but are of a higher technological level than the ones we know today.

Several other examples can be taken from the domain of housing and building construction. Countries like Sweden have adopted

[6] *Ibid.*

thermal insulation standards (50-60 kWh/m2/y) which compared to other countries (e.g. Germany, 200 kWh/m2/y) are lower by a factor four. Such constructions have been built and are normal looking houses.

In the field of electrical household equipment such as refrigerators, electrical bulbs, washing machines, etc. a factor four in energy efficiency has been reached within a decade. Denmark is the country mentioned as an example where such an energy efficiency level has been set forward as the industrial standard. Meanwhile similar figures have been obtained in the US.

Also, with office equipment, considerable energy savings can be obtained e.g. a portable PC uses about one tenth of energy compared to an average desktop device. Power management in a PC or other computers, on software basis, allow considerable additional energy savings.

The food chain production, electrical devices (fans, pumps, motors), air-conditioning are shown to have large potential for increasing energy efficiency levels.

In all cases increased investment costs for technological renewal or materials have been calculated and compared with fuel costs and appear to be within more than acceptable margins.

Material Productivity

The productivity of using material resources has up to now not been a usual concept. We owe the concept and definition mostly to the work of Friederich Schmidt-Bleek, until recently, of the Wuppertal Institute's Division for Material Flows and Eco-Restructuring. It concerns the material inputs per service unit (MIPS) and is a way of estimating for a product or service the quantity of materials that must be moved about somewhere on the planet. It deals with all parts of material involved in the manufacturing of a product or service regardless on which location or continent it has been originated. *Material productivity* is the reduction of the amount of MIPS. The concept of MIPS is wider than longevity and durability, so a product with a high longevity does not necessarily show a good performance. In fact MIPS refers to the total life cycle of a product. As a rule of thumb, the increase of materials productivity is recommended to be higher than a factor four, rather much closer to a factor ten. The Product Life Institute in Geneva has developed strategies for optimising

resource efficiency. As an example, some of these rules are reproduced here:
- leasing instead of selling, wherein the manufacturer's interest lies in durability;
- extended product liability, which could induce manufacturers to guarantee low-pollution use and easy reuse or disposal;
- joint ownership or use, which would require fewer products for the same amount of service;
- remanufacturing: preserving a stable frame of a product after use and replacing only worn-out parts;
- product design optimised for durability, remanufacturing and recycling.

The total concept of material efficiency has to do, in a cumulative way, with: 'how much with how little'.

A few examples will illustrate the practical side of the MIPS concept.

The hypercar, once again, appears to be an excellent example. To build a usual car, about 1.520 tons of material are moved in sequential processes of metal mining, refining, shipping, plastic and glass manufacturing. In addition for the catalytic converter, about 2.000 tons are moved to extract the platinum. The hypercar will weigh about three times less than an current model. E.g. a four- to five-passenger US car of the year 1994 weighs some 1.439kg and an equivalent hypercar only 521kg. The structural mass will shift from steel and other metals to polymer composite materials. Surprisingly the need for more polymer production would be limited to some 3% of the actual year production, for an expansion of an order of magnitude for the production of the composite material and a several-hundredfold production of carbon fiber. The new technology of the engine reduces considerable material flow and spare parts as well. Recycling the hypercar would mean completely new installations, the older ones being in use for another decade or more and do not constitute a real threat to the concept.

Other examples vary from the use of electronic carriers instead of print outs; durable office furniture with specific design for easy remanufacturing or recycling; the use of steel instead of concrete for pylons carrying 110kv electricity.

In the field of agriculture the inefficient use of water in dry regions is well known and very frequent. Subsurface irrigation where drip lines buried about 20-25 cm deep in the soil emit small amounts water immediately in the plant root zone. An overall efficiency increase of a factor 1,8 to 2,4 of water use in an Arizona desert (US) is quite a performance. Side benefits such as less herbicides and fertiliser use are to be added to this result.

Water use can be reduced in manufacturing e.g. paper and board fabrication, production of razor blades, pens and microchips. Water use in the cotton production has been reduced by some 80%. Additionally, the reduction of water use in manufacturing processes is accompanied with a decreasing energy (heating) consumption. In-house water consumption can be reduced from 300 to 110 liters water per person per day (US average).

Other potential material flow reductions are encountered in: recycling materials used for packaging all kinds of goods, rehabilitating existing buildings instead of demolishing them, reuse of bottles, cans and large containers. Rehabilitating wood as a construction material, being more than one quarter less energy intensive compared to concrete.

Transport Productivity

The transport of material goods and individual mobility have considerably shaped industrial society. The impact on natural resource use and on the environment are far from being negligible. Data and communication highways represent powerful symbols of recent technological progress. The potential increase of *transport productivity* lies far beyond a factor four in the domain of long-distance electronic communication technology. Two interesting cases are at this point to be mentioned:
- substituting electronic mail for posted letters;
- substituting video-conferencing for business meetings.

Numerous examples of these technologies are today commonly used in business as well as by private persons and families. When the MIPS method is used for the calculation of the resource savings, then, for a trans-Atlantic business trip and a 6-hour video-conference, a factor of roughly a hundredfold emerges. Such an outcome has to be taken very cautiously from the point of view of personal contacts as well as from the point of view of the material involved, nevertheless a factor four is certainly achievable.

In this respect the adoption of a policy of the real price of transport, e.g. *make transport prices tell the ecological truth* is definitely the right direction to move into. As to common business practice, the *just in time* principle would come into severe problems when all costs have been taken into account.

Another striking example of physical transport involved in a simple dairy product namely strawberry yoghurt in Germany. A study showed the complexity of the transportation routes of the different materials involved in the production of a simple strawberry cup. The location of the manufacturer being Stuttgart, the travel distance, on average, by the yoghurt, its ingredients and packing is 3.500km; and another 4.500km have to be added for the transportation of the supplies to the suppliers of the dairy manufacturer. A reduction of a factor four in transportation productivity, through a more locally organised production process, is readily obtained.

Increasing the capacity of existing railways with the help of electronic control systems would make it possible to increase the frequency of the traffic without decreasing the safety. Along the same thinking pattern, railway transportation consumes fewer resources than road and air transportation. Today with conventional trains the limit of competition compared with other means is, in terms of travel time, about 400 km. A new technology (called Pendolino, invented by Italian engineers), will make it possible to increase travel speed on conventional rails and therefore increase the radius with low investment cost and no additional environmental damage. Other technology approaches are under way of taking advantage of low energy use and low investment costs. The *Factor Four* level could be achieved with these innovations.

Public city transportation can be successfully implemented provided that intelligent urban master plans are conceived. A well-known example is the city of Curibita in Brazil, where by now, 70% of the population of a city of about 4,8 million persons today (1,6 million in 1964) uses public transportation. Generally speaking, urban innovation has the potential of an environmental and energy friendly outcome. From the point of view of *Factor Four* it is within reach for a large number of cases.

This brief survey on productivity and efficiency gives a realistic view on the potential of technological innovations where a *Factor Four* or more have been realised or are on the edge of realisation. The different examples in the fields of energy productivity, of material productivity with the MIPS indicator as a guide line and of the transport productivity concern all the activities of daily life. It represents an optimistic approach which enables a new perspective for mankind.

Recommendations and Tentative Conclusions

Sustainable development is a new framework which possesses a number of potential technological answers for our world community for the next half century. A few conclusions follow from the above analyses:
- sustainability is in the meantime largely promoted in the world community. It concerns in the first place the industrial societies of the Northern hemisphere. The argument often expressed by which sustainable development should be set up in the first place in developing regions, especially in the South (because of the demographic development there), can not be taken seriously;
- sustainability should be implemented in all industrial sectors and in all new products designed and brought to the market. Considerations about the life cycle and the recycling capacity of parts have to be conceived in the early stages of design;
- the technological feasibility of the efficiency and productivity revolution described by the concepts of *Factor Four* and *Factor Ten*, illustrated above by numerous examples, is largely proven;
- governments should increase their efforts to stimulate industry to develop sustainable products and efficient manufacturing processes. The promotion and introduction of standard criteria to be applied by the design and construction of public equipment belongs to the action field of public authorities. The concept of MIPS (material inputs per service unit) should be largely used in the public domain;
- in the domains of urbanisation, transportation, the re-use of buildings, etc, local governments have an important role to play. Long range solutions which guarantee a low pollution effect and MIPS index should be retained;
- engineers involved in the design of new products and goods should be fully aware of the necessity of the approach of sustainability and the related efficiency. An important role has to be played by all universities and institutions concerned with the education and training of engineers and technicians. The focus on sustainability should explicitly be present in all courses about technology.

The conclusions are numerous and widely diversified and indicate the importance of the framework of sustainability. The way out for the future of the planet with the doubled population in the next century is to be seen in an efficiency revolution of resource use. This perspective is in accordance with the original definition of sustainability of the Brundtland Report. The solutions proposed by the vision of *Factor Four* and *Factor Ten* appear to be feasible and therefore realistic. They are part of an optimistic view which relies on the capacity of mankind to benefit from technological innovation. At the same time and as a consequence, society will become more technical than it is already now. *Rebound* effects belong to possible reactions of the society and reducing largely the benefits of engineering innovations. Therefore it appears to be necessary that the framework and objectives of a sustainable society can count on a broad acceptance and understanding by the population. Besides the engineering aspect of sustainable society, the importance for social and economic consequences should also be explained to the public by national governments and leaders of international institutions. Considering all eventualities, the remaining time being short, it is becoming urgent to act in consequence.

References

BROWN, L. et al., *State of the World 1998: A Worldwatch Institute Report towards a Sustainable Society*, London, Earthscan, 1998.

CARLEY, M. and I. CHRISTIE, *Managing Sustainable Development*, London, Eartscan, 1997.

MEADOWS, D.H., D.L. MEADOWS, J. RANDERS and W.W. BERHENS III, *The Limits to Growth, A Report for The Club of Rome's Project on the Predicament of Mankind*, New York, Universe Books, 1972.

THE WORLD COMMISSION ON ENVIRONMENT AND DEVELOPMENT (Chair: G.H. BRUNDTLAND), *Our Common Future*, Oxford, Oxford University Press, 1987.

VAN DIEREN, W., *Taking Nature into Account – Towards a Sustainable National Income, A Report to the Club of Rome*, New York, Springer Verlag, 1995.

VON WEIZSÄCKER, E.U., A.B. LOVINS and L.H. LOVINS, *Factor Four. Doubling Wealth, Halving Resource Use*, London, Earthscan, 1997.

WEILER, R. and D. HOLEMANS (eds.), *De leefbaarheid op aarde. Global Change: voor welke toekomst?*, Leuven, Garant en TI-K VIV, 1997.

REFLECTION

3.3.1

ENGINEERS AND THE DIALOGUE ON EXTENDING THEIR HORIZON OF ACTION

Awareness of Responsibility as a Claim to Competence and as Moral Behaviour

Eva Senghaas-Knobloch[1]

The concept of responsibility of engineers refers to two dimensions: the dimension of professional competence and the dimension of values and moral norms. The consciousness of responsibility, too, relates to these two dimensions: first as an awareness about one's own degree of power to shape the technical world and, second, as the moral evaluation of one's own practice.
This contribution summarises the results of intense dialogues with engineers practicing in different technological fields and professional positions. Consulting, in contrast to R&D activities, seems to be the professional context which obliges engineers and technical experts to develop the most comprehensive awareness of values, interests and needs involved in their projects, particularly when related to the technical infrastructure of the modern society. Engineers in such contexts are open to using new methods of participation and technology assessment. A sense of responsibility with respect to both professional competence and moral sensitivity can put into practice in professional life if adequate structures for participation are instituted officially.

A look back over history clearly demonstrates that there are two dimensions to the concept of responsibility in professional behaviour. One of these dimensions concerns the issue of authority or

---

[1] Original text translated by B. DECEUNINCK.

competence, both in the sense of the competence we allocate to ourselves and that which is allocated to us by others. As technology gained significance, a new social group developed, made up of technicians, engineers and scientists gathering together around the guiding principle of technology. In Prussia and the German Reich, it was this group of technical and scientific experts who understood how to successfully acquire competence for issues of technical regulation from the state – and in terms of professional politics that meant above all from the legally trained civil servants. In the context of the development of the special regulation method applied in Germany, the self-allocation of competence meant the legitimisation of the authority to act on the basis of technical and scientific rather than legal knowledge. Awareness of responsibility as a claim to competence brings with it a claim to professional autonomy. Acquiring a certain degree of freedom in terms of the competence to act involves taking responsibility for the effects and consequences of this action. This is genuine political responsibility. As members of a community – in this case the community of scientific and technical specialists – people bear a political responsibility.

The other dimension of responsibility concerns moral rules, values and standards. This involves the human capacity to divide himself (or herself) into two entities, one acting, the other observing, which also justifies endeavours to 'straighten things out with oneself' as far as possible.[2] Some of the moral rules and commandments are laid down in the rules of law, with the most fundamental of them being enshrined in particular in the code of Human Rights, as emphasised by Ropohl.[3] However, history has also shown that no complete concurrence between rules of law and moral rules is possible in real communities.

The two dimensions of the concept of responsibility also imply a twofold dimensioning of awareness of responsibility. Awareness of responsibility expresses both the awareness of the scale of one's own power to act and the appraisal of one's own behaviour measured against moral yardsticks,[4] which may leave open the issue of the psycho-dynamic basis for this appraisal, be this the internalisation of

---

[2] See Arendt's interpretation of Socrates' words: 'it is better to suffer wrong than to do wrong' in her discourse on collective responsibility, see Arendt 1987, pp. 44ff.

[3] G. ROPOHL, *Ethik und Technikbewertung*, Frankfurt a.M, Suhrkamp, 1996, pp. 320ff.

[4] Hoff 1995, pp. 58ff. also presented this context for the integration of moral control and moral theories.

a punishing instance or the feeling of relation and obligation. By considering these two dimensions of responsibility and the awareness of responsibility, it is possible, taking an ethical point of view, to consider both the social context and the subjectiveness of subjectivity of those who act. When looking back in history at the problem of the first major threats posed by new industrial technology, typical fields of conflict relating to the profession of engineer can now be described in relation to the two dimensions of responsibility and awareness of responsibility.

The historical problems of the steam boiler explosions and the subsequent initiatives in the field of engineering in terms of the formulation of their own design rules and monitoring indicate an awareness of responsibility that focuses on *functional responsibility*. The safe functioning of technology in the twofold sense of reliable, but also danger-free functioning has from the outset belonged to the area of responsibility which technicians and engineers consider themselves obliged to accept. *Functional responsibility* corresponds to 'internal responsibility'[5] as responsibility for the quality of the product and the correctness of business relations as well as the profession's responsibility to maintain group standards. In this respect, the function of the 'state-of-the-art' as a yardstick may be considered an attempt, in view of the dynamics of the technical progress unleashed, to obtain an equivalent to the old guild rules, insofar as these concerned quality in terms of the functional safety of products. As the further development of products, or invention and optimisation are inherent in the technical ideal pursued by the engineering sciences, quality can only be ensured by the rapid generalisation of technical rules of procedure for the time being acknowledged as a yardstick. However, precisely because of its desired function of ensuring operational safety, technical standardisation in the sense of the standardisation of responsibility comprises stipulations which are unwelcome to technical constructors and users and which above all constitute obstacles for innovators. A *conflict* arises here between the professional interest in safety of action thanks to the *technical standard* achieved and the ideal of constant *further development*.[6]

---

[5] W.C. ZIMMERLI, 'Wandelt sich die Verantwortung mit dem technischen Wandel?' in H. LENK and G. ROPOHL (eds.), *Technik und Ethik*, Stuttgart, Reclam, 1987, 92-111, pp. 101ff.

[6] E. SENGHAAS-KNOBLOCH and B. VOLMERG, *Technischer Fortschritt und Verantwortungsbewußtsein*, Opladen, Westdeutscher Verlag, 1990, pp. 19f.

Secondly, the historical example of the steam explosions highlights the special relationship between business and engineers. Basically, engineers' professional activity relies on the economic interest of new technical products and new processes. The severity of the conflicts that result from this can be illustrated historically by the example of the petition from steamship owners in January 1841 against the American law of 1838. The petition criticises the legal clause 'according to which an accident involving fatalities should be considered prima facie proof of negligence. The argument was that if Congress upheld the view that steamship travel is too dangerous for public safety, it would be fairer and more honest to ban it completely.'[7] The regulations aimed at preventing explosions were put across as an attack on steamship travel as a whole. For engineers, whose professional existence depends on economic investment, the tension between an independent interest in development and construction in the field of technology and its economic exploitation becomes clear. From what point can a construction be considered perfected and safe and its application in practice be justified? An area of conflict arises here in terms of the awareness of responsibility between the professional *claim to the practical application of creative technical ideas* and the professional *ideal of technical functional safety* according to autonomous standards.

Thirdly, the example shows that in an age of science-led industrial technology, it is extraordinarily difficult to distinguish between internal and external responsibility. The purchasers and users of technical products or systems are often not the same person. Apart from the customer, there are also those who use the technology acquired by the customer, whether they are passengers on steamships, workers in a factory, or those using infrastructure facilities, such as the postal network. If the effects of technical artefacts are also considered from the point of view of third parties, other endangering aspects become clear, above and beyond the problem of accidents and risks, which are related to the chain of social effects. First and foremost, there is the problem of the 'redundancy' of possibilities for subsistence which have existed hitherto, or in other words professional positions and problems relating to the devaluation of professional skills and experience. In any case, the

---

[7] J.G. BURKE, 'Kesselexplosionen und bundesstaatliche Gewalt in den USA' in K. HAUSEN and R. RÜRUP (eds.), *Moderne Technikgeschichte*, Cologne, Gütersloh, S., 1975, 314-336, p. 314.

widely differing critical trends indicate that engineers may see themselves as being embroiled in a conflict between their *professional claim to innovation*, which requires socio-economic momentum, and the *professional ideal of 'good technology'* as a culturally and socially appropriate means for a good life, which may run entirely counter to this socio-economic momentum.

From the historical point of view it becomes clear that engineers' professional practice has, *from the outset*, been characterised by areas of conflict involving ambivalent quests: under certain circumstances the ambition for *safety of action* is contrary to the ambition for constant *further development*. The claim to the *practical application* of new technical ideas can come into conflict with the ideal of autonomously determined *functional safety* and the ideal of *good technology* as a means of achieving socially professed aims can come into conflict with the perceived socio-economic technical momentum. In all areas of conflict, the issue of responsible action is linked to the subject image of one's own power to act or the scope of action one grants oneself.

Awareness of Responsibility and New Threats

In the second half of the 20th century, in addition to the well-known dangers which scientists and technical experts have had to tackle (technical safety problems and the costs in social terms owing to the successful increase in technical and economic efficiency), two other areas of danger have also made themselves felt: the damage done to the natural foundations of life and the threat to the human self-image. The threat to the natural foundations of life is a classic example of the problem of the unintended and often unforeseen consequences of technical innovation.

As regards atomic and space research, as early as the late 1950s Hannah Arendt described the related problem in philosophical terms in her book *Vita Activa*, when she pointed out that research and development organised along scientific and technical lines no longer adopted a theoretical attitude towards nature in terms of observation or making, but a method that corresponds to acting between people. Since scientists no longer simply take from nature the materials needed to explore and contemplate its inner laws and also no longer use this material to make finite things, which as far as anyone can judge can be returned to the cycle of nature, they themselves create 'natural processes which without men would not exist', they start new unprecedented processes whose outcome remains 'uncertain and unpredictable',[8] they initiate processes 'which can no longer be reversed and generate forces which are not provided for in the natural order of things'. Now, Hannah Arendt considers unpredictability and irreversibility as typical characteristics of the activity of human trading. According to Arendt, when trading, people enter into relationships with one another. The characteristics referred to have their own place in the web of relationships between people and also have their counterparts. By making mutual promises to one another, people can to some extent put an end to the unpredictable nature of affairs between them, and the consequences of the irreversibility of actions which have occurred between people are limited by the power of forgiveness between people. If the characteristics of unforeseeability and irreversibility are transferred from the area of relationships between human beings to their relationship to nature, then that radical new problem area which has been characterised by the term risk society since the 1980s comes into being.[9] People take risks consciously, without being able to gain an overall view of their full extent.[10]

The second area of new dangers is linked to the problem that technical objects which people use in their everyday lives change not only their world but also their subjectivity. Taking a long-term view, engineering historians such as White point to the complexity

---

[8] H. ARENDT, *The Human Condition: A Study of the Central Dilemmas Facing Modern Man*, New York, Garden City, 1959, p. 208.

[9] The imperative aspect of the risk society is vividly expressed, for example, by Hubert Markl, scientist and President of the Max-Planck Association, in an interview with the newspaper Die Welt on 1 July 1996, when he said in response to a question about the quality of the German science policy: 'There is an awareness in all parties that scientific ability and innovative research are decisive for competitiveness' and went on to say that the international market is the only yardstick for the development trend in research and technology and for justifiable national regulations. Basically, he said that: 'Of course, one can take the view that risks cannot be considered enough in advance, and the results are all the safer, and we can successfully produce to a particularly high standard of safety. This would indeed be right in theory, if we lived in a world without competition. But when similar companies in England can bring the same products onto the market one or two years earlier, then later production not only means the loss of one or two years, but also the loss of the market ...'

[10] See SENGHAAS-KNOBLOCH 1993.

and the ramifications of the effects of technological change.[11] Radical social changes which occurred over many centuries with the technology of times gone by are now possible in a far shorter space of time owing to the systematically planned power of new technologies to exert an effect. Whereas in former times changes came about over many generations, these days a single generation of people has to cope with many radical technical changes which intrude into their lives and self-image. Developments in information and communication technology, reproductive medicine and biotechnology are driven forward with the aim of pushing back the spacio-temporal limits of human existence as far as possible. In the field of research into artificial intelligence, the equivalent quest has found its most radical expression in the controversial model of a common 'post human reason'.[12]

The increased power of the 'new collective practice' in technical development to intervene changes the situation as regards responsibility, but it also changes the awareness of responsibility among technical and scientific experts. Depending on the place in which they exercise their particular professional activity, they no longer see themselves not only as players in certain technical developments, for which they consider themselves responsible, but also as parties involved in a process relating to society as a whole, in which they play a part in the development trend, but without being able to determine it or wanting to accept responsibility for it. In this situation, since the 1970s discussions among engineers, in particular in the field of VDI but also among other specialist groups, on the problems of assessing and structuring technology have intensified. In this way, they are reacting to the citizens' initiatives and new social movements which have formed against certain technical projects or lines of development. Against this background, in the Federal Republic of Germany in the 1980s various dialogue projects were set up, in which an attempt was made to hold a constructive debate

---

[11] L. WHITE, 'Technikfolgen-Abschätzung aus der Sicht eines Historikers' in M. DIERKES, T. PETERMANN and V. VON THIENEN (eds.), *Technik und Parlament. Technikfolgen-Abschätzung: Konzepte, Erfahrungen, Chancen*, Berlin, Sigma, 1986, pp. 47-72.
White shows how the development from coal fire pans to fireplaces has affected not only comunication between social classes and levels, but also the decline in child mortality and has thereby led to the possibility of new, more intensive relations between parents and children.

[12] MARZ 1993 analysed this model from discussions among researchers in the field of artificial intelligence.

between technical specialists and technical laymen on the aims of and objections to technical developments.

One decisive factor here is that in the discussion of the problems of responsibility in the field of engineering, engineers consider the specific characteristics of the various areas of technology and the various professional field of practice. Below the problems which arose in the dialogues with engineers in the field of information and communication technology,[13] are compared with those in the field of waste technology.[14]

Whereas information and communication technology encompasses a broad pallet ranging from the principles of microelectronics to the problems of the organisation of work and is a field of technology in which technical development in general is being driven forward, waste technology deals precisely with the consequences of the development of the industrial society. Waste technology aims to provide an answer to the initially unforeseen consequences of socio-technological development and has to solve the practical problems of waste storage and the precautions to be taken against toxic emissions. Those working in the field of waste technology are confronted daily with the unintended consequences of technical development and the socio-economic prerequisites of their own technical responses. Whereas information and communication technology affects mainly social norms, waste technology is above all a matter of norms or values of freedom from physical harm and of health.

In modern societies, waste problems are typically matters for the social infrastructure, which has hitherto usually been state-run but is now increasingly being privatised. There are points of contact here with the field of information and communication technology in the infrastructure networks of the postal service. Other areas which belong to the field of infrastructure are those in which civil engineers work, for example the field of drinking water reservoirs or river dams. Ekardt[15] rightly points out that engineers who work in this field are to some extent working in an intermediary area, where civil society and the state overlap and the necessary foresighted reflection in the future situation as regards the community's requirements

---

[13] See SENGHAAS-KNOBLOCH and VOLMERG 1990.
[14] See VOGELSANG 1998, WERTHEBACH 1998.
[15] H.-P. EKARDT, 'Ingenieurverantwortung in der Infrastruktur-entwicklung – neu beleuchtet im Lichte des Civil Society-Diskurses' in E.H. HOFF and L. LAPPE (eds.), *Verantwortung im Arbeitsleben*, Heidelberg, Asanger, 1995, pp. 144-161.

makes comprehensive, carefully weighed judgements a prerequisite for the successful fulfilment of tasks.

Consequently the professional areas in which engineers operate are to a greater or lesser extent linked to the practical needs of people. Those working in the field of research into gallium arsenide, to develop the bases for more efficient chips, have far less to do with the practical users of technology than those, for example, who provide customer advisory services for certain products, plants and procedures. The actual professional position and the way this position is perceived influence the view of the problem and individual responsibility that will be taken. This will be explained below.[16]

Engineer's Professional Practice and Self-reflexion

Anyone wishing to talk to engineers about their possibilities regarding the socially acceptable and ecological structuring of technology takes it for granted that they see themselves as players.[17] However, the extent to which this assumption is correct must in itself be the subject of discussion. It is difficult to address those who see themselves less as engineers and more as dependent employees like everyone else on the subject of their own possibilities of exerting an influence. In a trade union working party of engineers involved in various tasks concerning the detail development, adjustment and testing of operating systems and computer programs, we came across an attitude that rejected the concept of making any active contribution to the outcome of technical development. This attitude was upheld by the engineers' perception of themselves as the object and victim of technical rationalisation rather than as responsible actors. The need for self-protection which had brought this group of engineers together in the trade union working party in the first place weighed more heavily than the interest in efficient and comfortable information-technology systems as working materials that correspond to the professional engineering ideal. This emerges from the following comments made by an engineer on the issue of sufficiently good working materials:

---

[16] The engineers' comments used are taken from the reports on the discussions with groups of engineers in the field of information and communication technology and the participants in the discussion at the *Ingenieurakademie Nord* on the problems of waste.

[17] See the detailed interpretations of VOLMERG 1990.

> 'I am totally against the idea of looking at things the way the company does, so to speak. I would like to take an example... When a bottleneck occurs on the machine or on the terminals, and you demand more and better terminals, it doesn't get you anywhere. I've done that myself, idiot that I was. I've called for more development systems, and they were purchased immediately. That was a backwards step. Eventually we had to keep going without a break. It didn't ease the pressure at all.'

With the integration of the engineer's ideal for good technology into the interests of employee protection, this group of engineers was overtaxed. People who see themselves as being without any influence cannot assume any responsibility, either within their own firm or for that firm's products. With regard to the field of conflict referred to above between the engineer's quest for good technology as a means to attain social objectives and the socio-economic momentum, it can be seen here that the experience of socio-economic momentum has resulted in the surrender – or at least the repression – of the professional ideal of good technology. The awareness of responsibility has become focused on the role of the dependant employee.

However, there are also other, quite different groups of engineers, also committed trade unionists, who have not abandoned their claim to innovation. Their trade union commitment goes hand in hand with in-depth reflection on the problems of technological development, in which they see themselves involved as actors. In a group of trade unionist engineers, who are working on the development of new digital messaging systems, very far-reaching risk analyses were undertaken in this sense in the field of information and communication technology. This brought with it the harsh hypothesis that the new communications technology might kill (living) communication, is linked to risks of social isolation and, when used in certain ways, also implies an increased security risk. At the same time, however, the subjective side of the innovation was discussed as pleasure in innovation. The following sequence gives an impression about the ambivalence.

> 'Engineers think up an awful lot of marvellous ideas. It is fun thinking up new things. They don't consider whether we actually need whatever it is they have thought of. It's fun, and then they try to convince other people that they need it, but in reality they don't. And companies want to produce and sell thing, and that coincides

with the engineers' interests. And so the company can integrate the engineers' interests... Of course they are partially responsible, because they provide the ideas...'

Engineers involved in development can be aware of their own delight in innovation. People who are aware of their own influence in technical development also see their own responsibility. In language, this takes expression in the manner of speaking in the third person, thereby distancing oneself. But how can the efforts to achieve good technology, that is technology which is acceptable to people and to society, be brought into line with delight in innovation, while retaining an awareness of responsibility?

People who see no chance of integrating their technical skills with areas of social competence feel a deep conflict. It is not unusual to find differing positions being adopted in one person so as to be able to place the emphasis now on one quest, now on another. Apart from their work as a developer, they can also be involved in works councils, or in citizens' initiatives. Those who experience this as a painful division tend to devalue one aspect or the other. In the group of development engineers who are committed trade unionists referred to above, the extremely sarcastic phrase of the 'engineer/man interface' was used and displayed a contrafactual tendency to deny social competence or present it as insignificant and underdeveloped.

One dramatic means of approaching the two opposing ambitions can be seen in an unhappy consciousness of an inner division, as expressed in the following comments from a development engineer:

> 'As an engineer I say 'yes', but as a human being I say 'no' to information and communication technology.'

There is another, third attitude, which should be mentioned here, which is expressed among young, successful engineers who work in research and development departments, in industry or in research institutes, often in managerial positions. In groups which were linked to this area, ideals were formulated that provide professional inspiration. The following comment is a reflection of this:

> 'I believe that in each of us there is a wish to achieve more somewhere. It may be a sportsman, who would like to be better, faster, than the one in front of him ... I also believe that somewhere or other we have a claim – and here of course the values and aims of each generation are redefined – that things should be simpler, eas-

ier, better for us. And to achieve that, we create aids. And, as has been said, we technicians try to improve these aids still further, to make them more efficient, and we try, quite simply, to find really elegant, noble solutions, which are the pinnacle somewhere, at least for the moment.'

Those at the leading edge of technical development who presses on with this do not see any room here for an awareness of responsibility in the sense of moral awareness. Inevitably, every constructive act is also a destructive act as regards that which already exists, according to the following comment from an engineer in workshop programming:

'Yes, when I think about it, I cannot imagine any field of activity which does not have its negative aspects as well. And in that case, either I despair, and I pack it all in and jump in the river, or – but even that doesn't solve the problem. And we live with the dilemma that whatever we do, we find out that we are inflicting damage on ourselves, or on our national economy, or as citizens on this planet... as a result of our existence, our actions. And so the question is, how bad is the damage.'

From this point of view, there is no point in dealing in advance with possible negative aspects of our own conduct. People who get involved in that would perhaps have to despair. In any case, they would lose their position at the leading edge of development. It is precisely those who want to influence technical development from the inside who fear a loss of power, when their own actions are thoroughly considered.

So in the various professional fields of engineering, various ways are suggested of how to deal with the two sides of the awareness of responsibility. They include deliberate distancing from the role of the active engineer, a division into a distancing and an identifying attitude to the professional role and 'over' identification with the professional role. None of these ways succeeds in integrating the two dimensions of the awareness of responsibility in the sense of reflected power to act.

However, the dialogues also reveal possibilities for integration. Such possibilities become clear, on the one hand, in a point of view that considers both individual and collective action against a wider background. On the other hand, chances for integration become visible when the issue is raised as to how good decisions can be reached when establishing technological standards in legal and technical terms. If the one case involves attitudes with which aspects of the problems of technical assessment can be accepted in an appropriate manner by an individual, the other case involves the question of how political communities should organise their decision-making process regarding standards and norms.

The Extended Horizon of Action

It is not by chance that there are often engineers in the field of consultancy who tackle the problem of the integration of the technical and moral aspects of their actions in particular. As consultants, they work in a social field in which contradictory forces are at work and have to be offset. Company and business consultancy takes many forms. Business consultancy is a branch of industry in its own right, with its own professional code for consultant engineers. In addition to this private-sector consultancy, there is also consultancy promoted by the state, for example in connection with the instruments for the state promotion of technology and projects aimed at putting technology into practice at VDI centres, and finally the manufacturers of technical systems also have an interest in increasing sales of their systems by providing suitable customer services and developing special adaptations.

In this situation, personal attitudes and values which are important to consultants, are an indispensable part of professional activity. In dialogues and discussions there is a chance to reflect on them and to disclose moral dilemmas between various values. For example, one consultant made this comment during the discussion on the problem of waste:

> 'I receive an enquiry from a potential customer. I have to submit an offer and fulfil certain framework conditions. And then (...) my ideas do not correspond to those of the customer. Now I have thought about how I deal with this in daily life (...) usually I simply try to explain my own moral concepts as regards this project or as regards the area to be dealt with, to the customer. So I would like to put forward my own moral concepts and at the same time question my own moral concepts and convictions in a discussion process. However it doesn't have to be just the customer. This can also happen within the project team (...) All in all, I don't have a patent

recipe on how to act in an environmentally sensible manner, because for one thing, for example, there are also situations where concerns about the environment can arise with certain orders. And you can actually lose orders and this could affect the employees. Perhaps you could also accept an order like this and then exercise a certain function while working on the order to guide it in this direction. For me, the only thing that is really important is that I can look myself in the eye in the morning without a bad conscience.'

Clearly, in the context of consultancy, the way in which the consultant plays his part is important. Saving face and being able to look himself in the eye in the morning without a bad conscience are two ways of expressing the fact that a broader horizon is being suggested precisely in the field of consultancy. Short-term economic interests with a view to one's own company or the customer's company have to be weighed against the chance of being perceived as a good consultant by customers and colleagues, even at a later date. In addition, the value of the advice is assessed by people who were not present during actual consultation process. Their possible needs and interests have to be taken into account and weighed up when any decisions are made. The consultancy situation is organised in social terms in such a way that the needs, interests and values of everyone who may be affected by the consultancy and the decisions it may lead to must be considered either in actual fact or in theory. This social aspect of professional consultancy can therefore also provide clues on ways in which the various fields of engineering can succeed in integrating technical and moral aspects.

> 'It's a matter of strengthening the individuals, particularly by means of appropriate framework conditions, so that they can adopt a longer-term view and a wide scope of values is present. I am a technician, and within the scope of what politicians give me as a framework (...) and a father who has brought children into the world, and who naturally worries.'

This reflection was done by an engineer in the context of a dialogue-project (see above) on the waste problem, revealing the possibility of becoming involved in the professional role in a way which neither loses sight of technical, political nor moral areas of competence.

The other path to integration concerns the way in which political communities have established technical regulations, i.e. which areas of competence, needs and interests are included in the decisions.

Just as the assessment of the consequences of technology is naturally not undertaken first and foremost by technicians, but by people with other areas of competence, engineers express criticisms of the effects of certain legal or political standards that they have to follow from the point of view of their professional praxis. However, this is obviously not simply a retort to demands and the assessment of the consequences of technology. Rather, the suggestion put forward during the dialogue, for example, for a 'planning permission process for laws similar to the planning permission for a disposal site' indicates that one's own professional activities include experiences which appear suitable for application in other contexts. A disposal site has to be planned for a long period. Those involved know, or perhaps they have learnt through previous mistakes, that the horizon of time, but also of values and interests, have to be sufficiently broad. In fact, the suggestion for a planning permission procedure for laws fails to recognise that in general the legislative procedure is preceded by a lengthy process of consultation with associations and interest groups. However, the results achieved through this process are clearly still not considered appropriate. In any case, the reason for the proposal is to include committed citizens in the assessment process. This therefore forms a new policy of establishing regulations. The varied chains of effects resulting from standards can no longer be overseen by individual groups of specialists. The plan to include active citizens makes it possible for specialists to perceive their professional responsibility more clearly.

Awareness of responsibility can unfold in its twofold dimension as a claim to competence and to morals wherever it is possible to take into account feelings of citizens' solidarity and feelings of social commitment in the process of assessment and decision–making. This is what the engineer succeeded in doing, which expressed his awareness of his responsibility as follows: 'I am a technician, and I am also a father.' However, this is no longer a matter of the responsibility of engineers, but of engineers' political awareness of their responsibility as citizens. Engineers' responsibility can clearly no longer be fulfilled if their political responsibility as a citizen is not perceived at the same time, whether at local or at global level. It is a matter of dealing at the same time with practical projects and the framework conditions within which these are organised.

References

ARENDT, H., 'Collective Responsibility' in J.W. BERNAUER (ed.), *Amor Mundi: Explorations in the Faith and Thought of Hannah Arendt*, Boston/Dordrecht/Lancaster, Martinus Nijhoff, 1987, pp. 43-50.

ARENDT, H., *The Human Condition. A Study of the Central Dilemmas Facing Modern Man*, New York, Garden City, 1959.

BURKE, J.G., 'Kesselexplosionen und bundesstaatliche Gewalt in den USA' in K. HAUSEN and R. RÜRUP (eds.), *Moderne Technikgeschichte*, Cologne, Gütersloh, 1975, pp. 314-336.

EKARDT, H.-P., 'Ingenieurverantwortung in der Infrastrukturentwicklung – neu beleuchtet im Lichte des Civil Society-Diskurses' in E.H. HOFF and L. LAPPE (eds.), *Verantwortung im Arbeitsleben*, Heidelberg, Asanger, 1995, pp. 144-161.

HOFF, E.-H., 'Berufliche Verantwortung' in E.-H. HOFF and L. LAPPE (eds.), *Verantwortung im Arbeitsleben*, Heidelberg, Asanger, 1995, pp. 46-63.

MARKL, H., 'Wir Deutsche sind heute zu risikofürchtig', interview in *Die Welt*, 1 Juli 1996, p. 9.

MARZ, L., 'Mensch, Maschine, Moderne. Zur diskursiven Karriere der "posthumanen Vernunft"', Paper des Wissenschaftszentrums Berlin für Sozialforschung, 1993, FSII93-107.

ROPOHL, G., *Ethik und Technikbewertung*, Frankfurt a.M., Suhrkamp, 1996.

SENGHAAS-KNOBLOCH, E., 'Lust und Unlust am technischen Fortschritt' in *Informatikforum* 7(1993)1-2, pp. 15-22.

SENGHAAS-KNOBLOCH, E. and B. VOLMERG, *Technischer Fortschritt und Verantwortungsbewußtsein*, Opladen, Westdeutscher Verlag, 1990.

VOLMERG, B., 'Die Moral des Ingenieurs und die Ethik technischen Handelns – sozial-psychologische Überlegungen zur Gestaltbarkeit der Technik durch den Wandel kultureller Werte' (1990) in E. SENGHAAS-KNOBLOCH and B. VOLMERG, 1992, pp. 111-123.

VOLMERG, B. and E. SENGHAAS-KNOBLOCH, *Technikgestaltung und Verantwortung. Bausteine für eine neue Praxis*, Opladen, Westdeutscher Verlag, 1992.

WHITE, L., 'Technikfolgen-Abschätzung aus der Sicht eines Historikers' in M. DIERKES, T. PETERMANN, V. VON THIENEN (eds.), *Technik und Parlament. Technikfolgen-Abschätzung: Konzepte, Erfahrungen, Chancen*, Berlin, Sigma, 1986, pp. 47-72.

ZIMMERLI, W.C., 'Wandelt sich die Verantwortung mit dem technischen Wandel?' in H. LENK and G. ROPOHL (eds.), *Technik und Ethik*, Stuttgart, Reclam, 1987, pp. 92-111.

3.3.2

MANAGING TECHNOLOGY

Some Ethical Considerations for Professional Engineers

Peter W.F. Davies

'I suppose it is tempting, if the only tool you have is a hammer, to treat everything as if it were a nail.'

Maslow[1]

This chapter critiques three well known, but still powerful 'technology myths': (1) Modern technology is just more of the same; (2) Technology is neutral; (3) Technology is progress. By reflecting on these, engineers are encouraged to generate a deeper understanding about the underlying philosophical assumptions of their work, a key requirement for any true professional.

Introduction

One of the traditional characteristics of any 'profession' is that the truly professional person thinks reflectively and deeply about the nature and meaning of their work; for the engineer this requires an examination of *The Philosophy of Technology*. Such literature can be heavy to read, so I aim to introduce the reader to philosophical matters on technology by way of examining three popular 'technology myths'. Myths usually gain their mythic status by having a strong *ideological* content; in other words, enduring myths are those that are simple to grasp, appear to be naturally 'obvious', are vague (sometimes deliberately so), have a small element of truth, and confuse the general with the specific. Concerning technology, there are I believe

[1] A.M. MASLOW, *The Psychology of Science: A Reconnaissance*, New York, Harper and Row, 1966, pp. 15-16.

three particularly powerful myths: (1) Modern technology is just more of the same; (2) Technology is Neutral; (3) Technology is Progress. These three highly questionable assumptions fester beneath much technology decision-making, and have a direct bearing on whether engineers can be considered to be 'acting professionally', or not. They therefore need careful scrutiny.

A Briefest Possible Philosophical Landscape of Technology

Before examining the three myths, we need some sort of 'map' with which to ethically approach *technology*. Technology may be usefully considered to be manifested in four modes: the physical objects themselves (computers, cars), knowledge (rules, theories, know-how), activity (design, making), and volition (intended use).[2]

Furthermore, technology can be variously viewed as the extension of human physical & mental skills, as the manifestation of a certain attitude towards the world, as a form of power (especially over the environment), as maximising rationally efficient action, as a way of salvation, as a revealing of the truth hidden in nature, and more besides. For this chapter I define technology simply as *one particular form of expression of human creativity*; it therefore reflects back to us clues as to our ethical development so far, and suggests what we need to do next.

Modern Technology is Just More of the Same

Consider this quote: 'The high-tech phenomenon is not new, even though the financial world treat it as fascinating as the discovery of a new comet. Once the wheel was high tech'.[3] The ideological nature of the argument is clear: surely the only difference with 21st century modern technology is that we are just enabled to do things faster, travel further, see more magnified, communicate more easily and instantly over greater distances, and so on. Wenk's argument certainly seems impeccable at first sight; any feelings of unease about

---

[2] C. MITCHAM, *Thinking Through Technology: The Pathway between Engineering and Philosophy*, Chicago, University of Chicago Press, 1994.

[3] E. WENK, *Tradeoffs: Imperatives of Choice in a High-Tech World*, Baltimore, John Hopkins Press, 1986, p. 112.

technology should be put down to our not having adapted fast enough to technological change; and that is to do with us and our control of technology, not to do with technology having become somehow qualitatively different.

Hans Jonas is one philosopher of technology who counters this by giving five reasons why *modern* technology should be considered a special case,[4] and hence why our traditional ethical notions need developing:

(i) The inherent ambivalence of technological action: even when technology is used only for the most honourable of motives, its ambivalent nature has a threatening side which often only becomes apparent many years later, (e.g. DDT in chemical fertilisers, now banned). So the notion of attempting to guide technology, even by the most noble of values, may be doomed to failure. Perhaps what we label *mere side-effects* are in fact a central part to all technology, and a clear sign that we are relating to technology in an inadequate way.

(ii) The momentum of technology: there appears to be a rule that *'what technology can do, it must do'*. If we can genetically engineer plants and humans, why not? Do we not have an ethical duty to develop potentially beneficial possibilities? In everyday ethics, *'can'* does not necessarily imply *'ought'*; we have now reversed this. It is a dangerous reversal because continually succumbing to the 'technological imperative' gives technology a momentum all of its own. Modern technological development appears to have few cultural restraints, and there is little evidence to suggest that technology's own momentum is somehow driving it towards evolving into a more humanised version of itself.

(iii) The magnitude (or scale) of technology: this effect is in both space and time. Many uses of technology tend to become major, (consider for example the historical expansion of the 'horseless carriage' (motor-car), and the 'digital adding machine' (computer)). The sheer magnitude implies that moral significance is achieved, and suddenly we find ourselves in what Jonas terms an *ethical novum*. Our responsibility has grown exponentially with the power of technology.

(iv) The global power of technology: the first three points add up to this one. Humankind, through technology, is now a major planetary force. We cannot go back on this. Decisions about technology affect

---

[4] H. JONAS, *The Imperative of Responsibility: In Search of an Ethics for the Technological Age*, Chicago, University of Chicago Press, 1984.

all human, vegetable, mineral and animal life, and now even the biosphere. This is forcing humankind to consider itself as a *'trustee' of the whole of creation*.

(v) *The apocalyptic potential of technology:* one could argue that every species eventually reaches the stage where it has the power to destroy itself. The human race, through technology, reached this stage with the atomic bomb, causing Einstein to note that 'everything had changed except our way of thinking'.

I would therefore argue, in agreement with Jonas, that the *quantitative* exponential expansion of technology now places it in a *qualitatively* different category. This is due to technology's huge power and pervasiveness, coupled with the human race's less-than-perfect track record in handling power. Moreover, with power comes responsibility, and it is by no means clear that our traditional notions of responsibility (nor our organisational control mechanisms) are adequate in a global technological society.

If we accept that today's technology means a qualitatively different situation from *all* previous relationships between humankind and technology (and the environment), then the notions that *Technology is Neutral*, and *Technology is Progress* need even more careful scrutiny. Any gap between the reality of technology, and our conceptions of it, needs closing fast.

Technology is Neutral

The idea that *technology = neutral* was philosophically disproved as far back as the 1930s, but it persists due to its ideological nature. After all, is it not obvious? I can use a hammer constructively (and ethically) to build a house for shelter, or I can use it destructively (and unethically) and kill someone. Technology therefore must be a neutral tool; technology is not to blame, but the people who use it wrongly; guns don't kill, people do (so the argument goes).

But is it this simple? First, the fact that technology can be used in good and bad ways does not make it neutral, but ambivalent. Second, technological neutrality is a dangerous myth because it fools us into inadequately considering the complexity of technology, and the organisational supporting structures that grow up around it. It encourages a simplistic *technology-fix* mentality, technical solutions to social problems (e.g., high-rise housing). Such mentality also underpins genetic engineering. Third, the appeal of the *technology =*

neutral myth is that it *does* push the onus of responsibility onto human beings. It is true, we should *not* hide behind a determinist view of an autonomous technology, where our only responsibility is to 'adapt' to the so-called 'technological imperative'. However, basing this truism on some vague notion of technological neutrality, is wholly inadequate.

Consider then in what ways technology might *not* be neutral (i.e., ways in which technology is value-laden, as opposed to value-free). In a world of limited financial resources, some-body (or institution) has the power to decide to develop large-scale technology 'A', rather than technology 'B'; it is therefore in their *interest*, and commensurate with their *power*, to do so. For years, sustainable energy technologies such as wave, tide and wind power have lost out to nuclear energy. The technology we (as users) have, represents not the interest of users and society as a whole (even though we buy them), but the balance of power and vested interests of politics, business and key individuals who control the resources. Our technology is therefore laden with their values, (such as big is beautiful, and centralised is more controllable and efficient). Products are also an expression of someone's preferences, cultural values and worldview; they are designed for a *purpose* which embodies the designer's values, (and not least with the aim of meeting sales targets). Moreover the benefits of technologies are often unequally distributed in favour of certain sections of society, further undermining the neutrality myth. This can be seen through the characteristics patterns (such as hierarchies of control, growth of expertism, dependency, and alienation of the user), which go hand in hand with technology.

Even when intended uses of technology become unexpected emergent ones, and even when we cannot prove the links between cause and effect in the society-culture-technology circle, each technology (and potential technology) carries with it a force which still says: 'Do something with me; respond to me'. (The television begs: 'switch me on'). Our responses may be yes, no, resignation, admiration from a distance, or whatever, but these still belie technology's underlying value-ladenness.

Where does this leave us? The instrumental value alone of technology is sufficient to counter the *technology = neutral* myth. Moreover, technology's ambivalence (hiding behind talk of 'neutrality') at least implies caution, but its continuing appeal lies in our need to believe that technology serves us, rather than the other way round.

If we persuade ourselves that technology is neutral, then we humans at least *feel* we are in control, (for good or ill). Experience however shows us that we only accept the true non-neutral nature of technology too late, when it has become deeply embedded in society and its adverse side-effects can no longer be ignored. *If* it was a simple neutral tool, we could easily root out technologies that were later discovered to be undesirable. This is not the case, and hence the importance of considering the third myth.

Technology is 'Progress'

The world is largely ruled by ideas, whether true or false, and the specific idea of *progress*, has become almost like a secular faith. Nisbet is convinced that the idea of progress 'has done more good over a twenty-five hundred-year period, led to more creativeness in more spheres, and given more strength to human hope than any other idea in history'.[5] If we do not believe that we are progressing, then a major locus of meaning and purpose is lost. Who would not feel threatened by arguments that *our* Age was no better than previous ones? Naturally then, we hold on tenaciously to this general belief in 'progress'.

The notion though that progress comes primarily *through technology* is a more recent belief of the last two hundred years. Like most ideologies it is dependent on its simple appeal, its confusion of language, and its generalisation of the specific. We react strongly to those who question it, labelling them 'Luddite' or 'Greenies'. Such scapegoating is a sure sign that we have an ideological attachment to technology, and that we need to do some careful self-questioning. Western culture, profoundly affected by the course of technology, does not easily accept weaknesses in the *Technology = Progress* argument. But I mention three of them.

First we have to be careful with the word 'progress'. There are several types of progress; for example: technical, intellectual, artistic, social and moral. That there has been technical progress is undoubtedly true, intellectual progress probably, artistic and social progress questionable, and moral progress highly contentious. But above all it is the continuing confusion of technical progress with ultimate social and moral progress which has given most power to the *Technology = Progress* myth.

[5] R.J. NISBET, *History of the Idea of Progress*, London, Heinemann; Dordrecht, D. Reidel, 1980, p. 8.

Second, the way we tell our own history is mainly in 'progress mode'. History, in particular the history of technology, is written from the point of view of how-we-got-to-where-we-are-now, and then extrapolated into the future. This encourages the *Technology = Progress* cultural mentality, and a determinist view of history and technology; the Stone Age 'inevitably' led to the Bronze Age and then on to the Iron Age, The Industrial Revolution, and the Computer Age. We like to tell our historical story as one of continuing cultural (i.e. now technological) advance. Not to do so is to accept that history is cyclical, or in terminal decline, and this would deeply undermine human self-understanding, meaning and purpose. Presumably, if we cannot argue that we have improved our world, then we have to admit that Progress-through-Technology has failed.

Third, in the West we are deeply influenced by early Christian teleological thought-forms. In his book *The City of God* Augustine (4th century) cites all the essential ingredients of the Western Idea of progress: cumulative material and spiritual advancement, historical reform rooted in social awareness, conflict driving history forward, history as a series of epochs, and finally some future golden age of peace, justice, freedom and security. These ideas waned in medieval times; however in the Reformation (1600-1750) they were revived but this time in the form of the Protestant Work Ethic which was itself set in the context of the Industrial Revolution and the role of technological innovation and prowess. Here the link between technology and progress is first made, and given religious, social and political sanction. The idea of Technology = Progress reached its zenith by the end of the 19th century. It then took a severe battering in 1914 as not only was technology developed specifically to slaughter millions on the battlefield, but also as the sheer horror and pointlessness of the bloodshed became apparent. This was then followed by the Great Depression, the Holocaust, World War II, the atomic bomb, and a host of tyrannical dictatorships. The assumption that human nature could be changed via technologically provided freedom and abundance, collapsed; but it did not die. It has now re-emerged with a new set of prophets and a new supporting language (global village, post-industrial society, the Computer Age, global communications and post-modern Man, the Age of Aquarius!). As an engineer, would you accept that the one thing you learn from history is that we never learn anything from history?

The above points may explain *why* the *technology = progress* myth is so resilient, but also gives clues as to why it is a dubious myth. Its basis is the 'philosophy of materialism', (i.e., material abundance leads to social progress and higher moral virtue). However, it would appear that material progress has come with some spiritual losses. Yes, technology has made work safer and more pleasant, but its meaning has been lost. Yes we have more information, but are we any better informed and able to identify the significant and relevant issues from amongst the mass? Yes there are spectacular technical achievements, but huge distributive inequalities remain. The promise of technology captures extremely well the notion of progress, but the multitude of technological *means* is now obscuring the *ends*, and it is to these we now turn.

Special Issues for Engineering Professionals

Suppose then that I have managed to convince you of the inadequacy and danger of these three widely held myths. Assume you accept that modern technology *is* qualitatively different from all previous technology, and that its power and pervasiveness require a completely new understanding of responsibility. Assume also that you accept that technology is *not* neutral or value-free; that all technologies developed are deeply imbued with certain worldviews and values, leading to uncertain emergent outcomes and a skewed distribution of benefits. Assume finally that you are highly sceptical of the notion of automatic progress via the next technological invention; you strongly suspect that utopia may *not* after all be just around the corner as soon as we have cracked the issues of genetic engineering, nano-technology, and a cure for cancer. So what? Where does it all lead? There are, I believe, three special interlinked ethical issues here, on which any truly 'professional' engineer should 'have a view'.

First, *what does technology free us for?* We are perhaps much clearer on what it is supposed to free us *from*, (mainly boring, unpleasant work). What technology is supposed to free us *for* is, I assume, *'the good life'* (more leisure time) and time to pursue 'higher' things. But here we note a profound contradiction. For surely if this were so, then the current millions of unwaged Europeans should be hailed as the first-fruits of the dawning of the new Technological Age of

Leisure: instead they have low social status, are given a below-subsistence package and told to go and find work again. Meanwhile as most people in paid employment well know, work hours get longer and longer. I suggest then that a major ethical issue here is the *meaning, definition, expectations and distribution of work*. As an engineer, how does your work relate to this?

Second, *what evidence do we really have that we are able to control technology?* In other words, *by what criteria can we say: '"yes", we are controlling technology!'*?[6] Consider the motor-car. Did the inventor of the 'horseless-carriage' really envisage the establishing of two new global industries (petro-chemicals and car-manufacturing), the annual toll of deaths and injuries, the change to out-of-town commuting lifestyles, the city-smog and the holes in the ozone layer? What proof have we that we are in control of motor-car technology? Or is the concept of the motor-car fundamentally flawed? Or perhaps a technical Pandora's box of motor-car technology has been opened and has somehow autonomously run away with itself? Is anyone to blame? *If* no one is to blame, then is no one in control? If no one is in control, then does technology have a life all of its own?

I suggest that to *prove* we control technology, we need to demonstrate conclusively that we are able to exercise with *equal* ease and legitimacy both a 'yes' and a 'no' vote to various technologies. If we never exercise a 'no' vote, then we have to accept by default that we humans can do nothing except adapt to whatever technology throws at us. Perhaps engineers need to exercise what Jacques Ellul calls *the ethics of non-power*,[7] saying 'no' to developing certain technologies. In this respect discussion of the special responsibility of the engineering community is only just beginning. Perhaps engineers should promote the idea of *Technology Tribunals* as a form of Technology Assessment.[8]

---

[6] H. SKOLIMOWSKI, 'A New Social Philosophy as Technology Assessment' in DURBIN (ed.), 1982, pp. 130-141.

[7] J. ELLUL, 'The Ethics of Nonpower' in KRANZBERG (ed.), 1980, pp. 204-212.

[8] See Fig. 1; DAVIES, 1992, pp. 307-329.

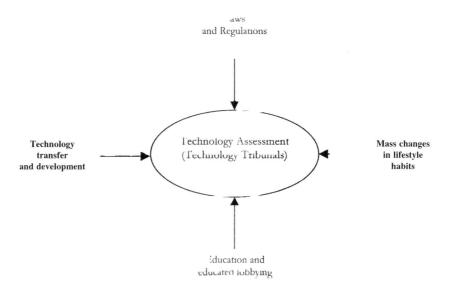

Figure 1

The *third* issue for engineers concerns a fundamental clash in beliefs about whether we need *more*, or *less*, technology. Both camps would agree that we have to work much more *with* nature (as opposed to powering *over* it), but there is disagreement as to *how*. The debate surrounds the question: *which technologies might deprive me of something essential for my human-ness?* In other words *what role does technology have in human evolution?* With the possibility of pregnant men, head transplants and genetically engineered 'nice' people maybe less than 50 years away, technology forces a reconsideration of the human. We have not even yet come to terms with the challenge to our humanity by computers (Are humans no more than mere meat machines? Can computers think? Is denying 'rights' to computers nothing more than sheer protein chauvinism?[9]). For some, the problem is that we have developed technology *too much*; we have changed the world enough and now is the time to step back and try and understand and rectify what we have done. For others we need *more* technology, not less, and the idea of cyborg Man and Woman is to be welcomed as a speeding up of our own evolution

[9] C. MITCHAM and A. HUNING (eds.), *Philosophy and Technology*, Vol. 2: *Information Technology and Computers in Theory and Practice* (Vol. 90 in the Boston Series in the Philosophy of Science), Dordrecht, D. Reidel, 1986.

which indeed is deemed vital for our survival; technology here is a gift which must be developed as fast as possible so we can claim our own evolutionary global destiny. Resolving this issue of technology's role in human evolution is a crucial debate for any professional engineering institution.

Conclusions

In *Cybergrace: The Search for God in the Digital World*, Cobb[10] suggests the gap between machine and spirit is closing. Computers are displaying 'emergent' (creative) properties that in the past have been attributed to the 'hand of God'; i.e., machines, the tools and creations of the engineering profession, have spirituality.

At this point perhaps you might feel like handing everything over to the theologians, but I appeal to you not to do so. This is not because theologians are paying me to protect them against accusations of a conspiracy theory(!), but because such a move would encourage the cult of expertism which can all too easily divert us from doing the hard professional thinking for ourselves. Engineers have a responsibility to articulate in both word and deed, an ethical position in relation to this everyday phenomenon we call 'technology'. The three 'technology myths' examined are intended to provide a vehicle to do just this.

What ultimately is at stake in our technology management is the survival or otherwise of the human race. My gut feeling is that in attempting to take charge of our own destiny by replacing Providence with the Modern Technological Project, we are *less* able to grasp the essential, due to the *multitude of technical means* that continually cry out to be used by us. The essence of engineering practice is to set us free, so maybe the concept of freedom is the most useful starting point to judge the engineering contribution to technological development. Perhaps engineers should always bear in mind that being human and free means communicating face-to-face in a local community, at least as much as being globally plugged in; being human and free means being meaningfully engaged with the technology we have, at least as much as being 'unburdened' by abundant

---

[10] J. COBB, *Cybergrace: The Search for God in the Digital World*, New York, Random House (Crown Publishers), 1998.

technical gadgetry; being human and free means setting limits and saying 'no', equally as easily as saying 'yes' to the latest technological offering.

References

COBB, J. *Cybergrace: The Search for God in the Digital World*, New York, Random House (Crown Publishers), 1998.

DAVIES, P.W.F., *The Contribution of the Philosophy of Technology to the Management of Technology*, unpublished Ph.D. diss., Brunel University with Henley Management College, 1992.

DURBIN, P.T. (ed.), *Research in Philosophy and Technology*, Vol. 5, Greenwich, Connecticut, JAI Press, 1982.

ELLUL, J., 'The Ethics of Nonpower' in M. KRANZBERG (ed.), *Ethics in an Age of Pervasive Technology*, Boulder, Colorado, Westview Press, 1980, pp. 204-212.

JONAS, H., *The Imperative of Responsibility: In Search of an Ethics for the Technological Age*, Chicago, University of Chicago Press, 1984.

KRANZBERG, M. (ed.), *Ethics in an Age of Pervasive Technology*, Boulder, Colorado, Westview Press, 1980.

MASLOW, A.M., *The Psychology of Science: A Reconnaissance*, New York, Harper and Row, 1966.

MITCHAM, C., *Thinking Through Technology: The Pathway between Engineering and Philosophy*, Chicago, University of Chicago Press, 1994.

MITCHAM, C. and A. HUNING (eds.), *Philosophy and Technology*, Vol. 2: *Information Technology and Computers in Theory and Practice* (Vol. 90 in the Boston Series in the Philosophy of Science), Dordrecht, D. Reidel, 1986.

NISBET, R.J., *History of the Idea of Progress*, London, Heinemann; Dordrecht, D. Reidel, 1980.

SKOLIMOWSKI, H., 'A New Social Philosophy as Technology Assessment' in P.T. DURBIN (ed.), *Research in Philosophy and Technology*, Vol. 5, Greenwich, Connecticut, JAI Press, 1982, pp.130-141.

WENK, E., *Tradeoffs: Imperatives of Choice in a High-Tech World*, Baltimore, The John Hopkins Press, 1986.

EPILOGUE

*Philippe Goujon, Bertrand Hériard Dubreuil,
Jean Marie Lhôte, Emmanuel Thévenin, Michel Veys*

This handbook inherits from the European philosophical traditions, which have analysed the influence of modern 'Technik' on contemporary culture. But to highlight the moral dilemmas that face engineers, the editorial committee has asked social scientists to summarise the ethical issues foreseen from their discipline and professional engineers to provide problematic examples. By doing so, this handbook has tried to specify some of the mediations between the moral agents and the consequences of the technological system that they are operating in. This epilogue wants to leave room for reflection by showing the magnitude of the field opened up by this book, in terms of research into the ties between science, technology and societies, in terms of engineers taking care of these ties in their professional life and in terms of the skills to be conveyed by those who train them.

A Problematic Situation

Throughout its successes technical progress has accumulated so much power and assumed such proportions that many regard it as the modern-day equivalent of ancient tragedy. For Heidegger, for instance, 'Technik' is the fate that relentlessly drags along man in a process of total rationalisation in which modernity is accomplished and annihilated. The Greeks feared that human power would provoke the anger of the gods if it exceeded a certain limit: Prometheus was punished for stealing their fire; the transgression of the boundary unleashes divine vengeance. As such, modernity seemed to have shattered all boundaries: nothing stands opposed to the unlimited growth of human competence. Consequently, there is a growing anxiety in our conscience with regard to the means technical progress places at our disposal. For reasons that appear to be more mythical than real, this anxiety has crystallized out on to, among others, nuclear energy. This has conferred an increased responsibility on the actors in the techno-scientific universe with individuals having to assume responsibility for decisions that entail grave human, economic and social consequences. These decisions are all the more difficult to take since the same actors find themselves at the point where various spheres overlap (existential and institutional, economical and political, administrative or judicial), each frequently imposing its own constraints and pressures. The anxiety comes from the particularity of these actors who are at one and the same time endowed with conscience and confronted with the monumental challenges of modern 'Technik'. The construction of the handbook underlines their difficulties and the proposed examples illustrate a painful complexity.

This painful complexity strains the European culture itself. The latter carries a radical suspicion which, in the history of thought and in the history of humanity, weighs upon the modern programme and upon the belief that reason is the positive agent in both the management of the city and in the control of nature. The beginning of the 20th century finally presents us with the sad privilege of a double worldwide confrontation of societies with a sophisticated modernity and with the creation of totalitarianisms, whose relentless character is directly linked to the rational mechanisms of technical management.

From these difficult experiences proceeded a crisis of modernity where reign derision or even Nihilism, a quoting and pastiche of the preceding styles, a multiplicity of 'dogmas' which establishes and separates, flight to and refuge in the past as an unlimited reservoir of pseudo-references. This movement which is sometimes referred to as 'postmodernity', brings into doubt the concepts of rationality and questions the beneficial nature of technical progress. Eventually, the Western claim to universality which relativises every experience and all cultures, may wipe out the awareness of the very future which the unlimited development of technical skills have made possible and which have given this very development its 'raison-d'être'.

In a world that questions everything, 'Technik' becomes – or tends to become – the ultimate reference point for a pragmatic truth, in spite of the suspicions it comes under and the criticism to which it is subjected. 'Technik', not 'Science', because the latter remains abstract, whereas the former imposes itself concretely and orientates the sciences towards remunerative and spectacular applications. The totality of the production apparatus pulls the sciences towards efficiency; the sciences end up coinciding with technology, in the global sense this word has acquired under Anglo-Saxon influence. This has resulted in the decline of the question of truth in the name of an immediate and superficial efficiency.

The teaching of technology does not even transmit the value of objectivity entailed by the teaching of sciences. It only carries out an act of faith in the advent of new technologies: this indicates a double blindness if technological innovation is tied up with the invisible process that animates market economy, and if, in addition, techno-scientific knowledge cultivates the oblivion of ultimate meaning and sense. Technology is an assembly of means and intentions, which surpasses each of its constituents: it is a way of thinking, of acting, of transforming the world, and therefore technology cannot be dissociated from a vision of society. The question posed by this philosophical analysis – which has a European background – is to know whether technology will, by definition, due to its own difficult to regulate growth become a factor of dehumanisation, in the same way that, as an extension of nature, it has become a factor of denaturisation.

However, these questions cannot be asked simply at such a general level.

That is why this handbook tries to specify some of the mediations between the moral agent and the consequences of the technological system that they are operating in. Certainly, the use of technical methods has become systematic and it is important to face up to the consequences of these systems on the economy, on politics and on world ecology. Since technological systems are social systems, people maintain them. Since the systems compete with each other, they allow people to choose between them, even if the scale of any technological system generates consequences of considerable significance.

This is the reason why the third chapter in this book has summarised a large field of research. Will industries ever be able to assume all ecological constraints? Is sustainable development possible? How far are scientific and technological developments arbitrary and to what extent can they be submitted to democratic regulations? Is the ideology of engineering definitely tied up with the ideology of progress, or can engineers broaden their culture by means of other criteria?

On the other hand, if people maintain technological systems, it is very important to start by knowing those people, listening to them, and situating them globally; the first chapter of this book presents this first field of research. Who are these engineers, trained to construct, to manage technological institutions? What are their history and their identity? To what degree are they free agents? What are their values and their moral problems? To what extent do those moral dilemmas translate or not larger societal problems? How much wisdom do they display to solve them? What is the strength of this wisdom, what are its limits?

Technological institutions compete with each other: all kinds of business undertakings, forms of government, engineering societies. What kind of logic dominates the mezzo level? The economic logic of business undertakings? The political logic of territories, in terms of monopoly, market share or the right to enter the technological systems? Or the simple logic of the systems themselves that are programmed with a view to their own growth.

Finding Mediations

The book here opens up a third field of research. The sheer size and complexity of technological organisations, the exponential growth of their powers, the fact that humankind is led by these factors to forget about the order of necessity, raises questions unparalleled in their sweeping scale. Must we use all means that the techno-scientific complex puts at our disposal? Is everything permitted when and if it is possible? Who makes the decisions and on what basis? Can they still profess the neutrality of Science and the pragmatism of technology, which leads to a reliance on external determinism (social demands, market forces or political strategies)? These are the ethical and moral questions one may put to institutional actors, but equally to people and to societies when our very civilization is at stake.

How then should the role of these intermediary bodies be considered? How can they regulate the technological evolution? Does the competition between them give rise to liberty or to a flight forward? Generally speaking, can one assign a moral authority to technological institutions in the same way one has granted them a juristic personality? How to differentiate between the concept of moral responsibility and the concept of civil liability?

The way in which these problems are settled on the intermediary level has a direct influence on the macro level. The superior level itself, however, is taken care of by giant institutions: multinationals whose budget is superior to that of certain states; international pressure groups manipulating public opinion; disproportionate worldwide technological systems, which specifically assure the transport of information, of goods and of services; international law protecting industrial properties and respect for the prevailing standards. On a world scale, technological systems are getting themselves organised in a much faster and efficient way than the political world.

The democratic systems that tend to be the norms in politics try to be so in economy. It still has to learn to get out of the technical fatality, and first of all, it has to foster understanding that the technical destiny is situated *within and not outside us*. Societies, technological institutions and their actors must impose on themselves the limits that the technological transgression may entail: to prefer peace to violence, conviviality to anonymity, ecological equilibrium to display of power... In other words, to choose the technical means that are adapted to the ethical goals they pursue and, for that, to explore at each level their own room for liberty.

Technology has acquired so much power in our modern society that it may give the impression of being fully autonomous. It has become indispensable for many of our contemporaries. It is courted by the political and the economic world to such an extent that it mobilises considerable sums. In our culture, it has become a fully-fledged variable, which leads men and women to look for technological solutions to political, economic and social problems, so that all aspects of social life may be instrumentalised one after the other.

In a word, technology may become a drug as it penetrates our body and soul. That is why it poses problems in the heart of every society, of every institution, of every reflection. It poses moral problems, political problems and problems with respect to society, none of which can be ignored any longer by the actors of the technological world, by engineers and particularly by future engineers. Hence the importance of an ethical manual for today's engineers and their teachers, but which will be equally to the advantage of all those who reflect on the problems that are posed by technological development.

Without pretending to supply answers, this book tries to make ethical reflection possible by putting the technological actor, their conscience and position in the centre together with the proceedings, the localities, the levels and the difficulties involved in decision-making. The objective is not to provide rules or dogmas, even less to determine moral answers. Ethics is never to be found within the answers but in the dynamic movement of debate and inquiry before the real action, and on the frontier which separates our subjective existence from the constraining exteriority. From the pedagogical point of view, engineering ethics cannot conform to a supposedly objective knowledge, particularly because – to resume one of Wittgenstein's thoughts – 'ethics can only be transcendental'.

Room for Liberty within the Heart of a Conviction

The constant evolution of technological systems could lead one to believe that a more or less natural and inevitable dynamic of progress regulates the phenomenon. Technical developments would be seen to occur more or less autonomously within the confines of its world. In other words, it could be assumed that technical progress advances under its own self-generated momentum.

For the individuals who maintain or depend on technological systems, they create a fallacious impression of autonomy. This book tries to refute this vision of the world following three lines of arguments that reflect our convictions.

Out of Realism

This book has provided the opportunity to highlight the competitive environment in which the different technical developments are always implicated. If the first factor of tension appears to be economic, we must add that the technical debate cannot avoid the social and political dimensions. If we neglect to do this, the whole issue risks being relegated outside the decision-making field altogether.

To put it another way, the technical actors can situate themselves in two ways with regard to their actions. They can choose to limit their observations to the purely technical field and decide not to take into account the interactions in other areas. However, this is not a realistic attitude to take, as the chapters written by specialists in the social sciences have attempted to show. Neglect of these other areas springs from an idealising of the technical process.

On the contrary, they can decide to pick out the main interactions which their actions will have with other fields. These can then be taken into account in the decision-making process. In this second context, an ethical question arises: realism assumes a global perception of the world. So the choice of the technical means used to contribute to its development automatically raises the global question of the final goals and the coherence of the decisions.

A Liberty to Be Negotiated

The realism mentioned above demands, in fact, of the engineer, as of the other social actors, to be present in the world. They must be able to gauge the ties between the possible choices and the current or future reality. It is in the negotiation of this future that they will find room for manoeuvre, whose distinguishing feature is to be a space shared by several actors.

Even if technology were to become autonomous, it would have to find the means to become self-regulatory. Further, it would have to find public spaces in which technicians could discuss the social problems they encounter and find the political means to make sure that power would be divided and counterbalanced. It would also have to find the means for economic regulation in order to put its goods and its services at everyone's disposal. Finally, it would have to find the means not to exhaust the planet's resources.

Realism evokes an ethics of responsibility; to the latter also – and perhaps above all – one has to add an ethics of conviction, as ethics must not let itself be resolved into economy or sociology, nor reduced to a self-justification. The fundamental thesis, which runs through this book, consists in maintaining that technological development is far too important to be confided solely to technicians. At the very least, they have to worry about humanity as a whole and to assume responsibility as human beings, which means that they have to choose between those possibilities that constitute the horizons of the world.

An Educational Conviction

Technologies cannot exist without technicians; technical workers often profess it openly, but hide their personal responsibility behind that of the institution where they work. Even engineers feel constrained by the internal development of technical means and their ignorance of economic and social forces. They do not realise their own power.

This book has been written with a very definite aim: to provide an education for the exercise of technological power. Engineers first of all have to become conscious of their power, of its strengths and of its limits. Then they have to become conscious of the fact that they have been confided this power in the name of a technological expertise. Finally they have to enlarge the notion of expertise to include the political, the economic and the cultural fields which are intimately intermingled.

This companion of European contributions would like to stir up in its readers a consciousness by assuring them that in spite of pressure and determinism there is, and there always will be, room for liberty. Throughout the descriptions, the examples and their philosophical analyses, the idea gets reinforced that to use this available liberty means taking a decision of conscience.

The three levels that give structure to this work – micro, mezzo, macro – tend to lead the engineers or the future engineers to shake off their neutralist conception of sciences and their pragmatic view of technologies. In the same movement these actors are invited to establish a possible correlation between accepted or dominant norms and their own human activity for which these norms should serve as a regulating device. Practice must dedicate itself to being understood as an existential way of life of a human being belonging to the world and open to the construction of meaning.

Technologies cannot but limit themselves to the game that is played with their rules – if they develop freely. Beyond this, the game causes vertigo. Being fully conscious of this fact, the technological actors can maintain a relation of responsibility with their experience, they can pass from knowledge to conscience, from awakening to a life open to the world of meaning. To play with the tension between norm and experience, between responsibility, conviction and reality, is much more liberating than to rid oneself of norms that are guided by an all encompassing illusionary truth.

Toward New Social Contracts

Can one claim that a work of this kind suffices to get a clear conscience and to believe that the difficulties will be overcome before long? Of course not, but it can be of assistance in getting out of the confusion. Such a confusion was expressed in all innocence by the organisers of a prestigious conference organised in the spring of 1999 by the National Council of Engineers and Scientists of France (Conseil National des Ingénieurs et Scientifiques de France) and the Society of Civil Engineers of France (Société des Ingénieurs Civils de France) on the occasion of the one hundred and fiftieth anniversary of the latter. The themes of the activities reveal the contradiction in which engineers find themselves at the end of this millenium.

- First theme: 'How to master the dynamism of progress in order to offer the best possible service to the consumer and the citizen?'
- Second theme: 'The industrial imperative of France.'

The program does not offer a synthesis and the organisers are perfectly right: it would be a vain as well as a purely academic exercise. How to draw up a synthesis between the mastering of dynamism and the imperatives of the growth of that same dynamism? Between a service to be rendered and a competition to be won? Between attention for the other and a cultivated aggressiveness? Between prudence and recklessness?

Nevertheless, one has to observe that not so long ago the first term of the contradiction would not even have been expressed; the evolution of our wits is a sure thing even if the imperatives of growth still impose their laws. This companion participates in this growing consciousness. Today it introduces the reflections of engineers, social scientists and philosophers when humanity finds itself confronted with troubling situations like genetic manipulations, nuclear energy or unforeseen capers of the international stock exchange. Up to now engineers saw themselves working rather in the margin of these great interrogations, leaving it to the doctors, the researchers and the financiers to deal with the vertigo involved. Many were satisfied to function as serving-hatches, at best being more preoccupied with their social responsibilities within the company than with the effects induced by their ever so reassuring technical competence.

Is it possible that a serving-hatch begins to think? For some forty-odd years this somewhat preposterous idea seems to emerge. Since technologies imply all its actors, a serving-hatch that starts to think runs the risk of causing great amazement. Does one clearly calculate all the consequences?

If technical training incorporates these tendencies, changes will accelerate. Teaching technology at secondary school level is a big step forward. Note in passing that the increasing use of the term technology in accordance with Anglo-Saxon usage is symptomatic of an attitude change. Technical development presupposed the use of external means; technology brings in the idea of a system within which the person finds themselves and which they seek to regulate. There remains the question of the meaning of these systems, which is what this book attempts to do.

Even if the study of ethical questions is not yet included on technology courses as such in Europe as it has been recently in the USA, the change of vocabulary and the change in teaching methods it implies could well be introduced progressively. All that remains is to ask about the meaning of these systems, which is what this handbook sets out to do.

Doubtless the answers history will give to these questions will in no way be those one would expect. This work may be read by future generations with the same incredulous curiosity with which they may read moral treatises of the 17th century. However, this need not deter us. In those domains in which thought tries to understand the world to find its liberty, the important thing is the space of discussion opened by this book and the new social contracts that the latter seeks to establish.

Editor's note: For a discussion of the use of the different terms: 'technology', 'technique' and 'technik', see the general introduction under the section entitled *Terminological Warnings*, pp. 10-11.

POSTSCRIPT

THE ACHIEVEMENT OF 'TECHNOLOGY AND ETHICS'

A Perspective from the United States

Carl Mitcham

The achievement of the *Technology and Ethics: A European Quest for Responsible Engineering*, which aspires to serve as a European engineering ethics handbook, may be assessed against the background of what has been done previously in the United States. It may also be considered in relation to the basic goals of engineering ethics and engineering ethics education. On both counts, *Technology and Ethics* constitutes a major advance in the engineering ethics field.

Engineering Ethics in the United States

Until quite recently, engineers and engineering educators in the United States have been the primary promoters of engineering ethics education. In form, their approach has been quite individualist – that is, the publication of individually authored texts or the anthologizing of previously published work by individuals. In content, engineering ethics texts in the U.S. have largely rested on what may be described as a distinction between internalist and externalist ethics, with the North American focus being almost exclusively internalist. Internalist engineering ethics emphasizes issues of conduct internal to the profession and generally avoids questions of public policy and politics. Pedagogically, this has led to a heavy use of case studies and scenarios. Any attempt to place engineering in a larger social or philosophical context has been considered social or philosophical criticism, and delegated to others. Indeed, the very phrase 'technology and ethics' is seldom found in engineering ethics discourse, precisely because of the way it breaches the boundaries of the engineering profession.

The historical and sociological background of North American developments in engineering ethics education rests on developments in engineering ethics itself, which may be summarily sketched as taking place in four phases. During the initial phase there were no explicitly formulated ethics codes. By the fourth phase, such codes had not only been formulated, they emphasized public responsibility, although this responsibility remained adjudicated in internalist terms.

Phase One: Implicit Ethics

Professional engineering societies arose in the U.S. during the mid- to late-1800s, at the culmination of the Industrial Revolution. These societies typified what, following Alexis de Tocqueville, the great French observer of *Democracy in America* (2 vols., 1835 and 1840), may be termed intermediate associations.[1] That is, in highly decentralized, individualist nations such as the U.S., persons tend to band together in multiple and diverse associations intermediate between them and the state. Such intermediate associations are exemplified by churches, political parties, fraternal lodges, clubs – and professional associations of engineers. In the U.S., particularly, these intermediate associations were the forerunners of what are today known as non-governmental organizations (NGOs).

In the initial phase of their development, these professional engineering associations or NGOs – fragmented by technical (and social) boundaries into civil, mechanical, electrical, and other forms of engineering – included ethics only implicitly. In their early years neither the American Society of Civil Engineers (ASCE, founded 1852), the American Society of Mechanical Engineers (ASME, founded 1880), nor the American Institute of Electrical Engineers (AIEE, founded 1884) had codes of ethics. Instead, each society promoted an ethos of professional behavior intermixed with technical knowledge and expectations concerning professional etiquette, all communicated primarily through apprenticeship and example.

The NGO character of these professional associations deserves to be emphasized. Today the ASCE and the ASME, for instance, each represent-

---

[1] See, e.g., A. DE TOCQUEVILLE, *Democracy in America*, vol. I, chap. 12.

ing approximately 125,000 members, are both corporate organizations; each has a national headquarters run by permanent staff and a director, who is in turn appointed by an elected board. The primary mission of such societies, to quote the current (2000) ASCE strategic plan, is 'to provide essential value to [its] members, their careers, [its] partners, and society through developing leadership, advancing technology, advocating lifelong learning, and promoting the profession.'[2] As organizations with strictly voluntary membership, however, neither the ASCE nor the ASME has any official or legal power over the professional practice of engineering, neither technically nor ethically.

Furthermore, in the U.S. engineering is not a highly regulated profession. Until the 20th century many engineers were self-educated or became engineers simply through apprenticeship. Today one is commonly denominated a 'professional' engineer in four overlapping senses: as someone employed by an engineering firm, as one who has earned an academic degree in engineering, as a member of a professional engineering association, and as a licensed professional engineer. Licensed professional engineers or PEs, who are registered by the individual states, constitute less than 5% of the 1.5 million individuals who call themselves engineers.[3] Only in the last case, that of PEs, do legally enforceable codes of conduct exist even today.

Phase Two: Ethics as Loyalty

Historically, however, as technical knowledge became increasingly rationalized in its various semi-autonomous spheres (following what Jürgen Habermas has identified as a key element in the logic of modernity), ethics too became an issue for explicit elaboration. During this second phase of professional engineering development, occupying the initial third of the 20th century, professional ethics codes were first explicitly formulated, primarily as a means to promote professional development and prestige.[4]

---

[2] See 'The Next Strategic Plan: Building ASCE's Future,' available at (www.asce.org).

[3] The procedure for becoming a PE is as follows: a person must have an engineering degree, passed the Fundamentals of Engineering exam, had four years experience working under PE supervision, and then passed the Principles and Practice of Engineering exam. The two exams are prepared by the National Council of Examiners for Engineering and Surveying (NCEES).

[4] The best single source for information on this history is E.T. LAYTON Jr., *The Revolt of the Engineers: Social Responsibility and the American Engineering Profession*, 2nd ed. Baltimore, MD, Johns Hopkins University Press, 1986, which was first published in 1971, and focuses on the years 1900-1945.

Primary examples are the codes of ethics of the AIEE, adopted 1912, of the ASCE and of the ASME, both of which were adopted in 1914. Each of these three codes was less than a page in length and stressed that 'the engineer should consider the protection of a client's or employer's interests his first professional obligation' (to quote the AIEE code) or required the engineer to act simply 'as a faithful agent or trustee' (ASCE language). Paradoxically, because of the pride of place given to business interests and company loyalty, these initial codes often had the effect of undermining rather than promoting professional autonomy. Professional engineering adopted a kind of self-imposed tutelage to its most immediate employers.

One reason for such self-imposed tutelage was no doubt the military origins of engineering as a profession. The first explicitly denominated engineers were members of the military corps of engineers. In fact, in the U.S. the first engineering degrees were granted by the U.S. Military Academy at West Point (founded 1802). In the military a primary obligation is obedience to authority, which is easily translated into civilian terms as company loyalty.[5]

Phase Three: Public Safety, Health, and Welfare

Phase three of ethics code development began after World War II, as engineers became increasingly aware of the social impact of their work and of their corresponding social responsibilities. The key characteristic of this period was the codified rise to prominence of a new principle recognizing the importance of public safety, health, and welfare, as can be documented by developments in three acronym-denominated professional organizations: ECPD-ABET-AAES, NSPE, and IEEE.

The ECPD-ABET-AAES Case. In 1947 the Engineers Council for Professional Development (ECPD) – founded in 1932 as an organization of organizations (not of individuals), and charged in part to develop an ethics code acceptable to its constituent engineering societies – adopted an ethics code that made it a leading duty for engineers 'to interest [themselves] in public welfare' and to 'have due regard for the safety of life and health of the public.' Revised in 1963, 1974, and

[5] For a slightly more developed argument concerning the military background of the modern engineering profession, see C. MITCHAM, 'Engineering Ethics in Historical Perspective and as an Imperative in Design,' in id. *Thinking Ethics in Technology: Hennebach Lectures and Papers*, 1995-1996 Golden, CO, Colorado School of Mines, 1997, especially pp. 125-126.

1977, this code eventually formulated the first of seven 'fundamental canons' as follows: 'Engineers shall hold paramount the safety, health and welfare of the public in the performance of their professional duties.'

In 1980 the educational component of the ECPD was restructured into the Accreditation Board for Engineering and Technology (ABET), an agency that now certifies all engineering degree programs in the U.S. ABET assumed the final ECPD revision of its code, along with an extended 'Suggested Guidelines for Use with the Fundamental Canons of Ethics.' In this form the ABET code has had some influence on engineering education, insofar as ABET has increasingly stressed the importance of engineering ethics in the engineering curriculum.

For instance, the new ABET 2000 accreditation criteria include eleven outcomes for graduate engineers, one of which explicitly calls for 'an understanding of professional and ethical responsibility.' At least one and maybe two other outcomes may also be interpreted as ethics-related.[6] This makes professional ethics a substantial component of engineering education. At the same time ABET shies away from any explicit recommendations about the specific content or pedagogy of engineering ethics instruction, leaving these questions instead to the faculty of specific programs.

The restructuring of ECPD educational activities into ABET took place parallel with the restructuring of ECPD interdisciplinary professional development activities into a new American Association of Engineering Societies (AAES). One of the perennial problems of professional engineering development in the United States has been a fragmentation in the professional engineering community, a dispersal of social power that dilutes public influence. Unlike ABET, the new AAES did not assume the ECPD code, but in 1984 officially adopted its own 'Model Guide for Professional Conduct,' which seeks to provide a unifying framework for all existing disciplinary codes. This AAES guide, in revision, has likewise progressively stressed the importance of safety, health, and public welfare.

The NSPE Case. A second illustration of the post-World War II emergence of the importance of social responsibility in engineering ethics is the code developed by the National Society of Professional Engineers (NSPE). Like the EPCD, one of the original objectives of the

---

[6] ENGINEERING ACCREDITATION COMMISSION, *Criteria for Accrediting Engineering Programs*, Baltimore, MD, ABET, November 1, 1999. Available at (www.abet.org).

trans-disciplinary NSPE, founded 1934, was 'the establishment and maintenance of high ethical standards and practices.'[7] Unlike the ECPD, the NSPE is an NGO of some 60,000 individuals, all of whom are PEs. According to its mission statement, the NSPE 'promotes the ethical and competent practice of engineering, advocates licensure, and enhances the image and well-being of its members.'[8]

Although an ethics code was proposed as early as 1935, none was formally adopted until 1946, when the NSPE endorsed the new EPCD code even before the EPCD formally did so. With the 1963 revision of the EPCD code, however, the NSPE moved to create its own code. The evolution of this distinctly NSPE code led by 1981 to the adoption of a short list of 'Fundamental Canons,' the first of which is to 'Hold paramount the safety, health and welfare of the public.'

The IEEE Case. Still a third example of the rise in social responsibility characteristic of United States engineering ethics codes in the second half of the 20th century can be found in the Institute of Electrical and Electronics Engineers (IEEE) – which emerged in 1963 from the unification of the AIEE and the Institute of Radio Engineers (IRE, founded 1912) and is today the largest professional engineering ngo in the world, with over 350,000 members.

In the early 1970s the IEEE undertook to write a new code of ethics. In the preamble to its initial code of 1974 the IEEE declared that

> Engineers affect the quality of life for all people in our complex technological society. In the pursuit of their profession, therefore, it is vital that engineers conduct their work in an ethical manner so that they merit the confidence of colleagues, employers, clients and the public.

The fourth article of the code itself specified that IEEE members have a responsibility to 'protect the safety, health and welfare of the public' and even to 'speak out against abuses in those areas affecting the public interest.'

In 1990, following strenuous debate in the late 1980s about the way to properly amend the code, the code was simplified and public responsibility was elevated to the first of ten principles. IEEE members committed themselves 'to accept responsibility in making

---

[7] For history of the NSPE, see P.H. ROBBINS, *Building for Professional Growth: A History of the National Society of Professional Engineers, 1934-1984*, Washington, DC, NSPE, 1984, with a special chapter on 'Ethics,' pp. 94-110.

engineering decisions consistent with the safety, health and welfare of the public, and to disclose promptly factors that might endanger the public or the environment.'

Phase Four: Ethics Education

By this time, however, engineering ethics development had already entered a fourth phase. During its first three phases, engineering ethics development – and, in consequence, engineering ethics education – had been primarily the prerogative of professional societies. There had been little in the way of the explicit promotion of engineering ethics education at the university level, and virtually no reflection on the social dimensions of engineering practice or the full scope and philosophical foundations of engineering ethics. North American developments to this point had been piecemeal and pragmatic, in accord with the larger character of U.S. intellectual culture. It has also been decidedly internalist.

Beginning in the 1970s, however, individual engineers, especially professors of engineering, became more involved, often working in tandem with academic philosophers. This new phase was stimulated by a series of widely publicized cases perceived as examples of engineering negligence or improper subordination to economic interests, and by federal funding for engineering ethics research.

Among the leading cases were a series of catastrophic DC-10 airliner disasters traceable to questionable engineering designs,[9] an instance of whistle blowing during construction of the San Francisco Bay Area Rapid Transit (BART) system,[10] and poor design on the Ford Pinto automobile that contributed to a number of fatal accidents.[11] The nuclear meltdown at Three-Mile Island, the rise of the environmental and consumer protection movements, general protests

---

[8] See NSPE Strategic Plan, available at (www.nspe.org).

[9] See M. CRUD and L. MAY, *Professional Responsibility for Harmful Actions*, Dubuque, IO, Kendall/Hunt, 1984; and J.H. FIELDER and D. BIRSCH, (eds.), *The DC-10 Case: A Study in Applied Ethics, Technology, and Society*, Albany, NY, State University of New York Press, 1992.

[10] See R.M. ANDERSON, R. PERRUCCI, D.E. SCHENDEL, and L.E. TRACHTMAN, *Divided Loyalties: Whistle-Blowing at BART*, West Lafayette, IN, Purdue Research Foundations, 1980.

[11] See D. BIRSCH and J.H. FIELDER, (eds.), *The Ford Pinto Case: A Study in Applied Ethics, Business, and Technology*, Albany, NY, State University of New York Press, 1994. See also R. NADER's *Unsafe at Any Speed: The Designed-in Dangers of the American Automobile*, New York, Grossman, 1965, which makes a more general argument.

against authority stimulated by the Vietnam War and Watergate, were other factors contributing to the emergence of a perceived need to take engineering ethics into the classroom.[12] Indeed, such factors contributed not just to the emergence of engineering ethics but of a more general ethics of technology – including especially biomedical and environmental ethics as two major components of a broad applied ethics movement.[13]

The publication of three major engineering ethics textbooks between 1980 and 1983 marked the emergence of this new phase. All were supported, in different ways, by new federal governmental grant programs that in turn reflected the new social sense of urgency and questioning of science and technology.

The first was a two-volume collection co-edited by Albert Flores and Robert J. Baum on *Ethical Problems in Engineering* (1980).[14] This work was part of a 'National Project on Philosophy and Engineering Ethics,' directed by Baum at the Center for the Study of Human Dimensions in Science and Technology of Rensselaer Polytechnic Institute (from which has grown into the RPI Department of Science and Technology Studies).[15] The project was a three-year effort that sought to broaden the discussion of engineering ethics. Funded by a

---

[12] One hyperbolic presentation of the spirit of the times is G. MARINE's *America the Raped: The Engineering Mentality and the Devastation of a Continent*, New York, Simon and Schuster, 1969.

[13] The literature on biomedical and environmental ethics is so extensive as to defy ready citation, as is the literature on the applied ethics movement. In fact, the first and third of these denominated fields even have their own encyclopaedias: W.T. REICH, (ed.), *Encyclopaedia of Bioethics*, 2nd ed., 5 vols., New York, Simon and Schuster Macmillan, 1995; the 1st ed., 4 vols., appeared in 1978; and R. CHADWICK, (ed.), *Encyclopaedia of Applied Ethics*, 4 vols., San Diego, Academic Press, 1998. The second encyclopaedia includes a useful entry on 'Science and Engineering Ethics, Overview' (vol. 4, pp. 9-28) by R. SPIER, co-founding editor of the journal *Science and Engineering Ethics* (1995-present). But for one recent introduction to the applied ethics movement as a whole, especially insofar as it relates to teaching engineering ethics, see M. DAVIS, *Ethics and the University*, New York, Routledge, 1999, chap. 1, 'The Ethics Boom, Philosophy, and the University,' pp. 3-21.

[14] A. FLORES and R.J. BAUM, (eds.), *Ethical Problems in Engineering*, 2nd ed.: vol. 1 (FLORES, ed.), *Readings*; vol. 2 (BAUM, ed.), *Cases* (Troy, NY, Center for the Study of the Human Dimensions of Science and Technology, Rensselaer Polytechnic Institute, 1980). Although titled the second edition, the first edition seems to have been a privately circulated working draft.

[15] For a brief introductory announcement of the project, see A. FLORES, 'National Project on Engineering Ethics to Bring Together Engineers, Philosophers,' *Professional Engineer* (NSPE) 47(August 1977)8, pp. 26-27.

major grant from the National Endowment for the Humanities, the project supported just over a dozen two-person teams of philosophers and engineers to examine value issues of engineering skills and activities. Team projects included the preparation of case studies on selected ethics problems, curricular development, and the drafting of recommendations for professional engineering societies. For the next decade the Flores-Baum collection was the single best source of materials; indeed, it remains an important reference even today.[16]

The second textbook was computer engineer Stephen H. Unger's *Controlling Technology: Ethics and the Responsible Engineer* (1982).[17] Unger had participated in the Baum project, and received further funding from the new Ethics and Values in Science and Technology (EVIST) program at the U.S. National Science Foundation. (The creation of EVIST in 1972, with Baum as its first director, is further witness to the critical spirit of the times.) His involvement, on behalf of IEEE, in investigating the situation surrounding the BART whistle blowers and his association with the creation of a special section within IEEE to examine the Social Implications of Technology, made Unger an influential presence in the profession.[18] Unger's book, which surveys cases, argues the importance of professional ethics and for most vigorous ethics activities on the part of professional societies in support of practicing engineers, and puts forth his own model ethics code, remains the single best undergraduate textbook in the U.S.

Still a third important textbook creation from this period was the result of a collaboration between philosopher Mike Martin and engineer Roland Schzinger, *Ethics in Engineering* (1983).[19] The Martin and

---

[16] For another important overview of engineering ethics education at this point in time, one that remains insightful, see R.J. BAUM's *Ethics and Engineering Curricula*, Hastings-on-Hudson, NY, The Hastings Center, 1980, which was one of a series of studies on 'The Teaching of Ethics.'

[17] S.H. UNGER, *Controlling Technology: Ethics and the Responsible Engineer*, New York, Holt, Rinehart and Winston, 1982. Unfortunately, the publisher priced this book out of the market, failed to keep it in print, and then resisted efforts to reprint. As a result it was not until the second, thoroughly revised edition, (New York, John Wiley, 1994) that the book became widely available.

[18] Unger's report on the BART case is included in his book, as is an account of the establishment of the Committee on Social Implications of Technology (CSIT), which evolved into the Society on Social Implications of Technology (SSIT) – the IEEE being composed of a large number of special interest 'societies' – and publishes the quarterly IEEE *Technology and Society Magazine* (1980-present).

[19] M.W. MARTIN and R. SCHINZINGER, *Ethics in Engineering*, New York, McGraw-Hill, 1983; 2nd ed., 1989; 3rd ed., 1996.

Schzinger volume, now in its third edition, takes an approach representative of the Anglo-American analytic tradition in philosophy; that is, it adopts a mixed utilitarian and rights-based ethical perspective and presents ethics as dependent on critical moral reasoning. At the same time, it makes a provocative argument for engineering as social experimentation, and seeks to draw out some of the implications, although in a non-radical manner. It is probably the most widely used text in the field.

Although there have been important additions to the engineering ethics textbook literature since,[20] the approach established by Flores-Baum, Unger, and Martin-Schinzinger has remained the standard in the U.S. Indeed, subsequent texts have largely institutionalized their internalist and individualist focus, using a mix of analytic ethics and case studies with some modest introduction of social implications.[21] Although there have been a number of collaborative conferences,[22] and textbooks are now complemented by videos[23] and even a well developed web-based 'On Line Ethics Center,'[24] collaboration and interdisciplinarity remains modest if not marginal.

Beyond Engineering Ethics in the U.S.

Technology and Ethics thus offers a new departure in engineering ethics on at least two counts: form and content. In regard to form, its

---

[20] See especially the following: D.G. JOHNSON, (ed.), *Ethical Issues in Engineering*, Englewood Cliffs, NJ, Prentice Hall, 1991; C.E. HARRIS, Jr., M.S. PRITCHARD, and M.J. RABINS, *Engineering Ethics: Concepts and Cases*, Belmont, CA, Wadsworth, 1995, 2nd ed. 2000; and C. WHITBECK, *Ethics in Engineering Practice and Research*, New York, Cambridge University Press, 1998. Three modular texts, as part of general introductions to engineering, further institutionalised this approach: in the 'E Source series,' C.B. FLEDDERMANN's *Engineering Ethics*, Upper Saddle River, NJ, Prentice Hall, 1999; in the 'Engineer's Toolkit series,' C. MITCHAM and R. SHANNON DUVAL's *Engineering Ethics*, Upper Saddle River, NJ, Prentice Hall, 2000.

[21] For one early report that can be interpreted to confirm this judgement, see M.S. FRANKEL, (ed.), *Science, Engineering and Ethics: State-of-the-Art and Future Directions*, Washington, DC, American Association for the Advancement of Science, 1988, which summarises a AAAS Workshop and Symposium held in February 1988.

[22] Probably the most important two conferences were organised by the Center for the Study of Ethics in the Professions at the Illinois Institute of Technology, with grant support from both NSF and NEH: 'Workshop on Ethical Issues in Engineering' (1979) and 'Beyond Whistleblowing: Defining Engineers' Responsibilities' (1982). Reports and proceedings are available on both conferences from the Center.

[23] The two best are 'Gilbane Gold' (from NSPE) and 'Testing Water . . . and Ethics' (from the Institute for Professional Practice).

[24] Available at (www.onlineethics.org).

collaborative character – with contributions from engineers and engineering educators throughout Europe – is much more extensive than anything that has been attempted in the United States. Although small groups of scholars have been made to break out of the individualist approach typical of u.s. educational practice, never before have so many different scholars worked together across national and disciplinary boundaries to produce such a systematic reflection.

In regard to content, this is the only engineering ethics textbook to undertake a broad contextualization of the engineering experience. The introduction provides appropriately pluralistic descriptions of the nature of ethics, of engineering, and argues for the possibility of an engineering ethics. Then in three major parts, the text moves from considerations of (1) individual issues of engineers in technical institutions, through (2) technical systems and technical decision making, to (3) technical development as a social issue. Thus *Technology and Ethics* provides a broader perspective on the engineering profession than has yet been achieved in any other engineering ethics volume. Moreover, the tripartite structure of each major section – historical and sociological description, case studies, and philosophical reflection – creates an interdisciplinary matrix that is unparalleled by any previous effort.

In short, in comparison with previous engineering ethics work in the u.s., *Technology and Ethics* constitutes a distinct achievement. One of the special strengths of this European Engineering Ethics Handbook is surely its program for dealing in substantive ways with the ideal of public responsibility, which in the u.s. tends to be left solely in the hands of the cowboy whistle blower.

The Goals of Engineering Ethics

Independent of what has been achieved in the u.s., one may also assess *Technology and Ethics* in relation to the goals of engineering ethics in itself – and especially of engineering ethics education. Such goals may be defined in both narrow and broad terms, intensifying moral consciousness in the profession or placing the profession itself in a more expansive ethical landscape.

Introducing Moral Consciousness in the Profession

In narrow terms, the goal of ethics is simply to introduce the moral conscience into technical education. Professional ethics is an attempt to bring technical knowledge – which, as the Greeks already noted especially with regard to medicine, can be used to help or to harm – under moral authority.

From its inception engineering, like medicine, has been heir to a general presumption that it serves human benefit. Such a presumption nevertheless calls for particular and personal application. If engineers as individuals or in professional associations fail to bring moral standards to bear on their practice, they reduce themselves to what, in the U.S., would be called 'hired guns,' whose powers are easily subverted by ends that ignore or oppose human benefit. That evils sometimes masquerade as goods, that technical actions may have unintended consequences, or that in specific instances what constitutes true human benefit may be highly contestable are all insufficient warrants for opting out of reflective ethical analysis.

In the contemporary world, of course, the adoption of substantive moral ideals is properly subordinate to that ethical pluralism which is the correlate of a Euro-American commitment to personal liberty and democratic politics. In such a social order, neither engineering nor any other technical profession will be allowed to impose technocratic ideals.[25] Physicians are not free to impose their interpretations of health, nor lawyers their understandings of justice.

Given the delimitations of technocratic life, each technical profession has obligations to make its particular technical ideals and any specific recommendations that emerge from technical practice as fully transparent as possible. Moreover, without necessarily providing substantive ideals, engineering ethics may nevertheless provide procedural guidance in accord with democratic principles. Just as it can be categorically stated that the medical researcher must respect human rights and seek the free and informed consent of any subjects of human experimentation, so there are boundary conditions on engineering practice with regard, for example, to honesty that it would have been desirable to spell out more forcefully than *Technology and Ethics* has seen fit to do. Indeed, it would have been a

---

[25] Two useful presentations of issues related to technocracy as it has manifested itself in the U.S. context are W.E. AKIN, *Technocracy and the American Dream: The Technocrat Movement, 1900-1941*, Berkeley, University of California Press, 1977; and F. FISCHER, *Technocracy and the Politics of Expertise*, Newbury Park, CA, Sage, 1990.

noble achievement for the broad international and interdisciplinary collaborative that developed this European Engineering Ethics Handbook to have put forth in clear and concise language a consensus or model code of ethics for engineers. Such a code might well introduce moral consciousness into the engineering profession in an even more forceful manner than the present Handbook will be able to do.

Placing the Profession in an Ethical Context

In broader terms, more than simply introducing moral consciousness into the profession, engineering ethics may also be conceived as having the goal of compensating for a certain restriction of focus and understanding. It is remarkable that as the humanities and the social sciences have increasingly become narrow specializations, it is the technical professions that have increasingly called for the assertion of a general education component into technical curricula. In the U.S. the establishment of medical humanities programs within medical colleges is the most obvious case in point. Increasing efforts to promote the teaching of communications skills, the social sciences, and the humanities in colleges of engineering is perhaps another.

The problem that such concerns seek to address has a significant history. Writing in the early 1850s, in the wake of the Industrial Revolution, in what has become in English a classic defense of 'The Idea of a University' against attempts to reduce all knowledge to direct practical benefit, John Henry Newman admitted that although technological knowledge 'aimed low, (...) it has fulfilled its aim.' The aim of Francis Bacon's technological reform of the sciences 'was the increase of physical enjoyment and social comfort; and most wonderfully, most awfully has he fulfilled his conception and his design. Almost day by day have we fresh shoots, and buds, and blossoms, which are to ripen into fruit, on that magical tree of Knowledge which he planted, and to which none of us perhaps, except the very poor, but owes, if not his present life, at least his daily food, his health, and general well-being' (Discourse V, sec. 8).

According to Newman, however, there also exists a higher aim which, although more difficult to achieve, must for this very reason be protected. Over a hundred years prior to contemporary recognition of the empirical risks of technological progress, Newman argued that the form of education on which this dynamism was based, itself constituted a threat and danger to truly human achievement.

At the core of technological progress is a division of labor that undermines comprehensive reflection. The technically specialized individual contributes more effectively to the accumulation of national wealth, but in the process becomes more and more degraded as a rational being. Human power is inversely proportional to narrowness of mind and purpose.

Newman's vision of the university is as that social institution which has as its aim the determined exercise of the mind as a whole. This is what he terms a liberal education – which is not the same thing as political liberalism. The liberal or comprehensive and liberating culture of the mind is achieved not through the amassing of facts or information, but through understanding the relations between different types of information. Intellectual virtue is not a passive acquisition of data, but an active assessment. As he summarizes, 'That only is true enlargement of mind which is the power of viewing many things at once as one whole, (...) of understanding their respective values, and determining their mutual dependence' (Discourse VI, section 6).

Newman further considers the relation between liberal education and professional skill. As he acknowledges, it is reasonable to ask 'what is the real worth in the market of [an education that] does not teach us definitely how to advance our manufactures, or to improve our lands, or to better our civil economy; or again [since] it does not at once make this man a lawyer, that an engineer, and that a surgeon; [nor] lead to discoveries in chemistry, astronomy, geology, magnetism, and science of every kind' (Discourse VII, section 2).

Newman's response is on two levels. On an abstract, not to say metaphysical level, Newman answers that 'a liberal education is truly and fully useful, though it be not a professional, education' simply because it is good. The good is the foundation of the useful, not vice versa.

On a more concrete or pedagogical level, Newman answers that liberal education introduces a balance into professional life that is not to be had in any other way. By learning 'to think and to reason and to compare and to discriminate and to analyze,' by developing some refinement of taste, judgment, and sharpened mental vision, one will be enabled to practice law, politics, medicine, or engineering without becoming simply a lawyer, politician, physician, or engineer. In a university setting students of technical specializations 'will just know where [they] and [their sciences] stand,' and engineers or other professionals will 'be kept from extravagance by the very rivalry of other studies, [from which they will have] gained (...) a special illumination and largeness of mind and freedom and self-possession, [so that they treat their] own in consequence with a philosophy and a resource, which belongs not to the study itself, but to [their] liberal education' (Discourse VII, section 6).

Newman's notion of the 'gentleman' engineer (which is what the above description characterizes) is limited on at least two counts: its exclusive focus on British men and its ontology of intellectual virtue. Women and diverse cultural traditions now find strong representation in the engineering profession; and intellectual virtue is not dependent, as he argued, on education in the classics. Nevertheless, its concrete observation that the professions must be protected from 'single vision' (William Blake) or a fanaticism of technique is even more to the point in a professional culture such as ours, with a specialization that exceeds anything dreamed of in Victorian England.

Toward a Duty Plus Respicare

In this context, it is remarkable the extent to which the approach to engineering ethics being advanced by *Technology and Ethics* may well serve to meet precisely the challenge Newman identified. Individual confinement within narrow specializations increases social power but thwarts the personal development of intellectual virtue. The multi-dimensional, interdisciplinary ethical reflection embodied in the European Engineering Ethics Handbook constitutes an enlargement of ethics to include business, management, and public policy issues all treated from diverse historical, social science, and philosophical perspectives. Indeed, it does this in such a manner that professional engineers may 'be kept from extravagance by the very rivalry of other studies.' From such counterpoint they may gain 'a special illumination and largeness of mind' so that they achieve a truly humanistic and philosophical purchase on their technical studies.

On another occasion, it has been argued that the most general formulation of engineering ethics and, *mutatis mutandis*, engineering ethics education, may be properly summarized in the form of a duty plus respicare, to take more into account.[26] Engineering ethics has undergone a progressive enlargement from military obedience through company loyalty to social responsibility. Extending that successive taking of more things into account or under consideration, it is reasonable to argue for the enlargement of engineering ethics education to encompass not just professional codes of conduct but broad historical, managerial, public policy, social science, and humanities reflection on the context and role of engineering. Such is one interpretation of the thrust of *Technology and Ethics*.

One weakness that may still be identified in the program outlined by the European Engineering Ethics Handbook is that it makes only minimal efforts to build bridges to other forms of interdisciplinary reflection. Applied ethics is found not only in interdisciplinary engineering ethics, but also in biomedical ethics, environmental ethics, and other fields. Interdisciplinarity is itself a focus of interdisciplinary research in arguments between multidisciplinarity, cross-disciplinarity, trans-disciplinarity – and even anti-disciplinarity.[27] Some self-critical reflection on such alternative approaches to interdisciplinarity might well deepen the remarkable achievement that *Technology and Ethics* remains.

Conclusion

Technology and Ethics: A European Quest for Responsible Engineering is a signal achievement in engineering ethics and in the development of engineering ethics education materials. It clearly complements work in the United States by encouraging engineering ethics to awake from its internalist slumbers and by deepening engineering ethics reflection through a truly unique, interdisciplinary, international, and collaborative effort. It further expands the goal of engineering ethics to the point where it may become a significant component of general, liberating education for engineers. Although the ultimate

---

[26] C. MITCHAM, 'Engineering Design Research and Social Responsibility,' in K.S. SHRADER-FRECHETTE, *Ethics of Scientific Research*, Lanham, MD, Rowman and Littlefield, 1994, pp. 153-168.

[27] Two studies that measurably advance research on interdisciplinarity are humanities professor J. THOMPSON KLEIN's *Interdisciplinarity: History, Theory, and Practice*, Detroit, MI, Wayne State University Press, 1990 and engineering professor S.J. KLINE's *Conceptual Foundations for Multidisciplinary Thinking*, Stanford, CA, Stanford University Press, 1995.

influence of this European Engineering Ethics Handbook cannot be determined short of its entry into history through publication and use, one can only hope that the promises it contains are widely realized.

BIBLIOGRAPHY

ACHTERHUIS, H. (ed.), *De maat van de techniek*, Baarn, Ambo, 1992.
ACHTERHUIS, H. et al., *Van stoommachine tot cyborg. Denken over techniek in de nieuwe wereld*, Amsterdam, Ambo/Anthos, 1997.
AGERSNAP T., 'Consensus Conferences for Technology Assessment', paper read at the conference 'Technology and Democracy', Copenhagen, 1992.
AHLSTRÖM, G., *Engineers and Industrial Growth*, London, Croom Helm, 1982.
ALCAMO, J. (ed.), *Integrated Modelling of Global Climate Change: IMAGE 02*, Dordrecht/Boston/London, Kluwer Academic Press, 1994.
ALCAMO, J., R. LEEMANS and E. KREILEMAN (eds.), *Global Change Scenarios of the 21st Century*, Oxford, Elsevier Science, 1998.
ALLEN, P.M., *Evolutionary Complex Systems: Models of Technology Change*, Amsterdam Conference on Developments in Technology Studies, Evolutionary Economics and Chaos Theory, Amsterdam, 1993.
ALMOND, B. (ed.), *Introducing Applied Ethics*, Oxford, Blackwell, 1995.
ALMOND, B., *Exploring Ethics: A Traveller's Tale*, Oxford, Blackwell, 1998.
ALMOND, B., *Exploring Philosophy: The Philosophical Quest*, 2nd ed., Oxford, Blackwell, 1995.
ALPERN, K.D., 'Engineers as Moral Heroes' in V. WEIL (ed.), *Beyond Whistleblowing: Defining Engineers' Responsibilities*, Chicago, Illinois Institute of Technology, 1982.
ALTHAM, J.E.J., 'The Ethics of Risk', *Proceedings of the Aristotelian Society* 84(1984), pp. 15-29.
APPELBAUM, D. and S.V. LAWTON, *Ethics and the Professions*, Englewood Cliffs, NJ, Prentice Hall, 1990.
ARENDT, H., 'Collective Responsibility' in J.W. BERNAUER (ed.), *Amor Mundi: Explorations in the Faith and Thought of Hannah Arendt*, Boston/Dordrecht/Lancaster, Martinus Nijhoff, 1987, pp. 43-50.
ARENDT, H., *The Human Condition: A study of the Central Dilemmas Facing Modern Man*, New York, Garden City, 1959.
ARROW, K., 'Economic Welfare and Allocation of Resources for Invention' in R. NELSON (ed.), The *Rate and Direction of Inventive Activity*, Princeton, Princeton University Press, 1962.
ASHBY, E., The Search for an Environmental Ethic' in *The Tanner Lectures on Human Values*, Salt Lake City, University of Utah Press, 1980.

AUCLAIR, A., *Les ingénieurs et l'équipement de la France. Eugène Flachat (1802-1873)*, Le Creusot-Montceau les Mines, Écomusée de la Communauté Urbaine, 1999.
AVISON, D.E. and A.T. WOOD-HARPER, *Multiview: An Exploration in Information Systems Development*, Henley-on-Thames, Alfred Waller, 1990.
AVISON, D.E., 'What is IS?', An inaugural lecture delivered at the University of Southampton, 3 November 1994, 1995.
AYRES, R. and L. AYRES, *Industrial Ecology – Towards Closing the Materials Cycle*, Cheltenham, Edward Elgar, 1997.
BAART, A., c.s., *Werkschrift moreel beraad in kerken. Een nadere begripsbepaling*, [Paper on moral deliberation in churches. A further conceptualisation], Driebergen, MCKS, 1990.
BARBER, B., *Démocratie forte*, Collection Gouvernances Démocratiques, Paris, Desclée de Brouwer, 1997.
BARBER, B.R., 'Liberal Democracy and the Costs of Consent' in N.L. ROSENBLUM (ed.), *Liberalism and the Moral Life*, Cambridge, MA, Harvard University Press, 1989, pp. 54-68.
BARBOUR, I., *Ethics in an Age of Technology*, San Francisco, Harper, 1993.
BARRON, E.J. 'Climate Models: How Reliable Are their Predictions?' in *Consequences – The Nature and Implications of Environmental Change* 1(1995)3, pp. 16-24.
BASSAND, M. and P. ROSSEL, 'Swissmetro pour repenser notre vision de la Suisse', Lausanne, Polyrama EPFL, 1994, N° 96.
BAUDET, J.C., *Les ingénieurs belges, de la machine à vapeur à l'an 2000*, Brussels, APPS Ed., 1986.
BAUMAN, Z., 'The Risks of the "Risikogesellschaft"', lecture delivered on 2 April 2 1996, University for Agriculture, Wageningen, The Netherlands (published in Dutch in Z. BAUMAN, *Leven met veranderlijkheid, verscheidenheid en onzekerheid*, Amsterdam, 1998).
BAUMAN, Z., *Globalization: The Human Consequences*, Cambridge, Polity Press, 1998.
BAYERTZ, K. (ed.), *Verantwortung, Prinzip oder Problem*, Darmstadt, Wissenschaftliche Buchgesellschaft, 1995.
BECK, U., *Was ist Globalisierung? Irrtümer des Globalismus – Antworten auf Globalisierung*, Frankfurt a.M., Suhrkamp, 1997.
BECKER, G., 'Event Analysis and Regulation: Are We Able to Discover Organisational Factors?' in A. HALE, B. WILPERT and M. FREITAG (eds.), *After the Event, from Accident to Organisational Learning*, Oxford, Pergamon, Elsevier Science, 1997, Chapter C.3, pp. 197-215.
BECKER, J., 'Recombination DNA Research: EEC Dispute' in *Nature*, 24(December 1981)31.
BEDER, S., 'Engineers, Ethics and Etiquette' in *New Scientist* (25 September 1993), pp. 36-41.

BELL, R., *Les péchés capitaux de la haute technologie*, Paris, Seuil, 1998.
BERG, P. et al., 'Potential Biohazards of Recombinant DNA Molecules' in *Science* 185(1974), p. 303.
BERGER, P.L. and B. BERGER, *Sociology*, New York, Basic Books, 1972.
BERGERON, F. and L. RAYMOND, 'Planning of Information Systems to Gain a Competitive Edge' in *Journal of Small Business Management* 30(1992)1, pp. 21-26.
BERLEUR, J. and Klaus BRUNNSTEIN (eds.), *Ethics of Computing: Codes, Spaces for Discussion and Law*, London, Chapman & Hall, 1996.
BERLEUR, J. and Diane WHITEHOUSE (eds.), *An Ethical Global Information Society: Culture and Democracy Revisited*, London, Chapman & Hall, 1997.
BERNER, B. and U. MELLSTRÖM, 'Looking for Mister Engineer: Understanding Masculinity and Technology at Two *Fin de Siècles*' in B. BERNER (ed.), *Gendered Practices*, Stockholm, Almqvist and Wiksell, 1997, pp. 39-68.
BERNER, B., 'Explaining Exclusion: Women and Swedish Engineering Education from the 1890s to the 1920s' in *History and Technology* 14(1997), pp. 7-29.
BERNER, B., 'Professional or Wage Worker? Engineers and Economic Transformation in Sweden' in P. MEIKSINS and C. SMITH (eds.), *Engineering Labour. Technical Workers in Comparative Perspective*, London, Verso, 1996.
BERNER, B., 'Women Engineers and the Transformation of the Engineering Profession in Sweden Today' in Shirley GORENSTEIN (ed.) *Research in Science and Technologie: Gender and Work* (Knowledge and Society Series, 12), Greenwich, CN, JIA, 2000
BERNER, B., *Perpetuum Mobile? Teknikens utmaningar och historiens gång [Perpetuum Mobile? Technology's Challenges and the Course of History]*, Lund, Arkiv, 1999.
BIJKER, W., 'Life After Constructivism' in *Science, Technology and Human Values*, (1993), pp. 113-138.
BIJKER, W.E., T.P. HUGHES and T.J. PINCH (eds.), *The Social Construction of Technological Systems: New Directions in the Sociology and History of Technology*, Cambridge, MA, MIT Press, 1987.
BOLTANSKI, L. and L. THÉVENOT, *De la justification, les économies de la grandeur*, Paris, Gallimard, 1991.
BONNAFOUS, A., 'Transport et environnement: comment valoriser et maîtriser les effets externes?' in *Economie et statistiques* (1992)258-259.
BOUDON, R., *Widersprüche sozialen Handelns*, Darmstadt/Neuwied, Luchterhand, 1979.
BOVY, P., 'Le Swissmetro en neuf questions', Lausanne, Polyrama EPFL, n° 96, 1994.

BRATT, I., *Mot rädslan [Against Fear]*, Stockholm, Carlssons, 1988.
BRION, R., 'La querelle des ingénieurs en Belgique ' in A. GRELON (dir.), *Les ingénieurs de la crise*, Paris, Éditions de l'École des Hautes Études en Sciences Sociales, 1986, pp. 255-70.
BROWN, L. et al., *State of the World 1998: A Worldwatch Institute Report towards a Sustainable Society*, London, Earthscan, 1998.
BUCCIARELLI, L.L., *Designing Engineers*, Cambridge, MA, MIT Press, 1994.
BUGARINI, F., 'Ingegneri, Architetti, Geometri: La lunga marcia delle professionni tecniche' in W. TOUSIJN (ed.), *Le libere professioni in Italia*. Bologna, Il mulino, 1987, pp. 305-335.
BUNDESANSTALT FÜR ARBEIT [Federal Labour Institute], *Information for Employers and Workers in Engineering Professions*, 25, 1996.
BURKE, J.G., 'Kesselexplosionen und bundesstaatliche Gewalt in den USA' in K. HAUSEN and R. RÜRUP (eds.), *Moderne Technikgeschichte*, Gütersloh, S., 1975, pp. 314-336.
BURN, J.M., 'Information Systems Strategies and the Management of Organisational Change – A Strategic Alignment Model' in *Journal of Information Technology* 8(1993), pp. 205-216.
BUSINESS INTELLIGENCE, *IT and Corporate Transformation*, Business Intelligence, 1995.
BYNUM, T.W., 'Computer Ethics in the Computer Science Curriculum', in T.W. BYNUM, W. MANER, and J.L. FODOR (eds.), *Teaching Computer Ethics*, Research Center on Computing and Society, 1992.
BYNUM, T.W., 'The Development of Computer Ethics as a Philosophical Field of Study' in *Australian Journal of Professional and Applied Ethics* 1(July 1999)1, pp. 1-29.
CALAME, P. and A. TALMANT, *L'État au cœur*, Collection Gouvernances Démocratiques, Paris, Desclée de Brouwer, 1997.
CALAME, P., *Mission possible*, Collection Culture de Paix, Paris, Desclée de Brouwer, 1995.
CALAME, P., *Un territoire pour l'homme*, Collection Territoires et Société, La Tour d'Aigues, Éditions de l'Aube, 1994.
CAMBRIOSO, A. and C. LIMOGES, 'La controverse, le processus-clé de l'évaluation sociale des technologies' in *Analyse évaluative et évaluation sociale des technologies*, Cahiers de l'ACFAS, Québec, 1988.
CARLEY, M. and I. CHRISTIE, *Managing Sustainable Development*, London, Earthscan, 1997.
CEDRE, *Le défi régional de la grande vitesse*, Paris, Syros-alternative, 1992.
CEMT (European Conference of Transport Ministers), *Tendances du transport européen et besoins en infrastructures*, Paris, CEMT, 1995.
CENTRE FOR EMPLOYMENT MEDIATION, CENTRE FOR LABOUR MARKET INFORMATION, *Annual Report of the ZAV 1997*, No. 26/1998.
CHECKLAND, P.B., *Systems Thinking, Systems Practice*, New York, Wiley, 1981.

COBB, J., *Cybergrace: The Search for God in the Digital World*, New York, Random House Crown Publishers, 1998.
COHEN, S.N., A.C.Y. CHANG, H.W. BOYER and R.B. HELLING, 'Construction of Biologically Functional Bacterial Plasmids *in vitro*' in *Proc. Natl. Acad. Sci.* 70(1973), pp. 3240-3244.
COLBY, M.E., 'Environmental Management in Development: The Evolution of Paradigms' in *Ecological Economics* 3(1991), pp. 193-213.
Collste, G., ETHICS AND INFORMATION TECHNOLOGY, NEW DELHI, NEW ACADEMIC PUBLISHER, 1998.
COUNCIL RECOMMENDATION (472 EEC), *Concerning the Registration of Work Involving Recombinant DNA*, 1982.
DALY, H., *Toward a Steady-State Economy*, San Fransisco, Freeman and Company, 1973.
DASGUPTA, P. and J. STIGLITZ, 'Industrial Structure and the Nature of Inventive Activity' in *Economic Journal* 99(1980), pp. 266-293.
DAVIES, P.W.F., *The Contribution of the Philosophy of Technology to the Management of Technology*, Ph.D. diss., Brunel University with Henley Management College, 1992.
DAVIES, P.W.F. and John QUINN (eds.), *Ethics and Empowerment*, London, MacMillian, 1997.
DAVIES, P.W.F. (ed.), *Current Issues in Business Ethics*, London, Routledge, 1997.
DAVIS, M., 'Thinking Like an Engineer: The Place of a Code of Ethics in the Practice of a Profession' in *Philosophy and Public Affairs* (1991), pp. 150-167.
DAY, C.R., *Les écoles d'arts et métiers, XIXe-XXe siècles*, Paris, Belin, 1991.
DE KERCKHOVE, D. (ed.), *Connected Intelligence: The Arrival of the Web Society*, Toronto, Somerville House Publishing, 1997.
DE LURDES RODRIGUES, M., *Os Engenheiros em Portugal*, Oeiras, Celta Editoria, 1999.
DEMARCO, T. and T. LISTER, *Peopleware*, New York, Dorset House Publishing, 1987.
DE VRIES, B., 'Contouren van een duurzame toekomst' in R. WEILER and P. GIMENO (eds.), *Ontwikkeling en Duurzaamheid*, Brussels, VUB Press, 1996, pp. 31-73.
DE VRIES, B., 'Susclime: A Simulation Game on Population and Development in a Climate-Constrained World' in *Simulation and Gaming* 29(1998)2, pp. 216-237.
DE VRIES, B., J. BOLLEN, L. BOUWMAN, M. DEN ELZEN, M. JANSSEN, E. KREILEMAN and R. LEEMANS, 'Greenhouse-Gas Emissions in an Equity-, Environment- and Service-Oriented World: An Image-Based Scenario for the Next Century' in *Technological Forecasting and Social Change* 63(2000)2-3.

DE VRIES, H.J.M., *Sustainable Resource Use – An Inquiry into Modelling and Planning*, Ph.D. diss., Groningen, University of Groningen, 1989.
DELFOSSE, M.-L., 'Quelques réflexions éthiques sur les cartes santé', *Cahiers de la CITA*, S3, Namur, 1993.
DEN ELZEN, M., H.W. BEUSEN and J. ROTMANS, 'An Integrated Modeling Approach to Global Carbon and Nitrogen Cycles: Balancing Their Budgets' in *Global Biogeochemical Cycles* 11(1997)2, pp. 191-215.
DESSAUER, F., *Philosophie der Technik*, Bonn, Cohen, 1927.
DEUTEN, J.J., A. RIP and J. JELSMA, 'Societal Embedment and Product Creation Management' in *Technology Analysis and Strategic Management* 9(1997)2, pp. 219-236.
DIDIER, C., A. GIREAU-GENEAUX., B. HÉRIARD DUBREUIL (eds.), *Éthique industrielle. Textes pour un débat*, Paris/Brussels, De Boeck Université, 1998.
DORN BROSE, E., *The Politics of Technological Change in Prussia*, Princeton, NJ, Princeton University Press, 1993.
DOU, A. (ed.), *Evaluación social de la ciencia y de la técnica. Análisis de tendencias*, Madrid, Universidad Pontificia Comillas, 1996.
DOUGLAS, M., *Natural Symbols*, London, Pelican Books, 1973.
DOWER, N., *World Ethics: The New Agenda*, Edinburgh, Edinburgh University Press, 1998.
DURBIN, P.T. (ed.), *Research in Philosophy and Technology*, Vol. 5, Greenwich, CT, JAI Press, 1982.
DURBIN, P.T. and C. MITCHAM, *Research in Philosophy and Technology: An Annual Compilation of Research*, London, 1980.
DURBIN, P.T. and F. RAPP, *Philosophy and Technology*, Dordrecht, Reidel, 1983.
DURBIN, P.T., 'Introduction: Some Questions for Philosophy of Technology' in P.T. DURBIN and F. RAPP, *Philosophy and Technology*, Dordrecht, Reidel, 1983, pp. 1-14.
DWYER, T., 'Industrial Safety Engineering: Challenges of the Future' in *Accident Analysis and Prevention* 23(1992), pp. 265-273.
EARL, M.J., 'Experiences in Strategic Information Systems Planning' in *MIS Quarterly* (March 1993).
EKARDT, H.-P., 'Ingenieurverantwortung in der Infrastrukturentwicklung – neu beleuchtet im Lichte des Civil Society-Diskurses' in E.H. HOFF and L. LAPPE (eds.), *Verantwortung im Arbeitsleben*, Heidelberg, Asanger, 1995, pp. 144-161.
ELLUL, J., 'The Ethics of Nonpower' in M. KRANZBERG (ed.), *Ethics in an Age of Pervasive Technology*, Boulder, CO, Westview Press, 1980, pp. 204-212.
ELLUL, J., *La technique ou l'enjeu du siècle*, Paris, Colin, 1954.

ENGELEN, G., R. WHITE, I. ULJEE and P. DRAZAN, 'Using Cellular Automata for Integrated Modelling of Socio-Environmental Systems' in *Environmental Monitoring and Assessment* 34(1995), pp. 203-214.

ERGAS, H., 'Why Do Some Countries Innovate More than Others?', Brussels, Centre for European Policy Studies, 1985.

ESCOLÁ GIL, R., *Deontología para ingenieros*, Pamplona, Eunsa, 1987.

EUROPEAN COMMISSION, 'Draft Council Recommendation Concerning the Registration of Recombinant DNA Work', *Com* (80), Vol. 467, 8 July 1980.

FARBEY, B., F. LAND and D. TARGETT, *How to Assess Your IT Investment*, Oxford, Butterworth Heinemann, 1993.

FERNÁNDEZ FERNÁNDEZ, J.L., and A. HORTAL ALONSO (eds.), *Ética de las profesiones*, Madrid, Universidad Pontifícia Comillas, 1994.

FERNÁNDEZ FERNÁNDEZ, J.L., *Ética para empresarios y directivos*, 2nd ed., Madrid, Esic, 1996.

FIDDAMAN, T.S., *Feedback Complexity in Integrated Climate-Economy Models*, Ph.D. diss., Report D-4681, Cambridge, MIT System Dynamics Group, 1997.

FIDLER, C.S., and S. ROGERSON, *Strategic Management Support Systems*, London, Pitman Publishing, 1996.

FINON, D., *L'échec des surgénérateurs: autopsie d'un grand programme*, Grenoble, Presses Universitaires de Grenoble, 1988.

FISSCHER, O.A.M., M.C. POT and A.H.J. NIJHOF, 'Kwaliteitszorg: op weg naar een volwassen organisatie' in C.T. HOGENHUIS and D.G.A. KOELEGA (eds.), *Technologie als levenskunst. Visies op instrumenten voor inclusieve technologie-ontwikkeling*, Kampen, Kok, 1996, pp. 197-216.

FLORES, A. and R.J. BAUM (eds.), *Ethical Problems in Engineering*, Troy, NY, Rensselaer Polytechnic Institute, 1979, 2 vols.

FLORES, A., 'Designing for Safety: Organisational Influences on Engineers' in V. WEIL, (ed.), *Beyond Whistleblowing: Defining Engineers' Responsibilities*, Chicago, Illinois Institute of Technology, 1982, pp. 153-167.

FLORMAN, S.C., *The Existential Pleasures of Engineering*, New York, St. Martin's Press, 1976.

FLORMAN, S.C., *The Introspective Engineer*, New York, St. Martin's Press, 1996.

FLYNN, D.J. and P.A. HEPBURN, 'Strategic Planning for Information Systems – A Case Study of a Metropolitan Council' in *European Journal of Information Systems* 3(1994)3, pp. 207-217.

FORCHT, K.A. and D.S. THOMAS, 'Information Compilation and Disbursement: Moral, Legal and Ethical Considerations' in *Information Management and Computer Security* 12(1994)2, pp. 23-28.

FORESTER, T., 'Megatrends or Megamistakes? What ever Happened to the Information Society?' in *The Information Society* 18(1992), pp. 133-146.

FOUREZ, G., *Nos savoirs sur nos savoirs, un lexique d'épistémologie*, Brussels, De Boeck Université, 1997.
FOUREZ, G., 'Constructivism and Ethical Justification' in M. LAROCHELLE, N. BEDNARZ and J. GARRISON (eds.), *Constructivism and Education*, Cambridge, Cambridge University Press, 1998, pp. 123-115.
FOUREZ, G., 'Le Technology Assessment, nouveau paradigme éthique' in J. PLANTIER (ed.), *La démocratie à l'épreuve du changement technique*, Paris, l'Harmattan, 1996.
FOUREZ, G., 'Scientific and Technological Literacy as a Social Practice' in *Social Studies of Science* 27(1997), pp. 903-936.
FOUREZ, G., *Liberation Ethics*, Philadelphia, Temple University Press, 1982.
FRANKEL, M.S., 'Professional Codes: Why, How and With What Impact?' in *Journal of Business Ethics* 8(1989), pp. 109-115.
FRANKENA, W.K., *Ethics*, Englewood Cliffs, NJ, Prentice Hall, 1963.
FREEMAN, C., *Technology Policy and Economic Performance: Lessons from Japan*, London, Frances Pinter, 1987.
FREEMAN, C., The *Economics of Industrial Innovation*, 2nd ed., London, Frances Pinter, 1982.
FRIED, C., *An Anatomy of Values*, Cambridge, MA, Harvard University Press, 1970.
FUNTOWICZ, S. and J. RAVETZ, *Uncertainty and Quality in Science for Policy*, Dordrecht, Kluwer Academic Publishers, 1990.
GAUDIN, T., *L'aménagement du territoire vu de 2100*, Collection Territoires et Société, La Tour d'Aigues, Éditions de l'Aube, 1994.
GERT, B., *The Moral Rules*, New York, Harper and Row, 1966.
GIDDENS, A., *The Consequences of Modernity*, Cambridge, Polity Press, 1990.
GILLE, B., *Histoire des techniques*, Paris, Gallimard, 1978.
GILLE, B., *Les ingénieurs de la Renaissance*, Paris, Seuil, 1964.
GILLIGAN, C., *In a Different Voice: Psychological Theory and Women's Development*, Cambridge, MA, Harvard University Press, 1982, 2nd rev. ed., 1993.
GODARD, O. (ed.), *Le principe de précaution dans la conduite des affaires humaines*, Paris, Éditions de la Maison des Sciences de l'Homme, 1997.
GOLDSTEIN, J., *Die Technik*, Frankfurt a.M., Rütten und Loening, 1912.
GOMULKA, S., *The Theory of Technological Change and Economic Growth*, London/ New York, Routledge, 1990.
GOTTERBARN, D., 'The Use and Abuse of Computer Ethics' in T.W. BYNUM, W. MANER and J.L. FODOR, (eds.), *Teaching Computer Ethics*, New Haven, CT, Research Center on Computing and Society, Southern Connecticut State University, 1992, pp. 73-83.
GOTTERBARN, D., K. MILLER and S. ROGERSON, 'Software Engineering Code of Ethics' in *Communications of the ACM* 40(November 1997)11, pp. 110-118.

GOUJON, P., *From Biotechnology to Genomes: The Meaning of the Double Helix*, World Scientific Publishing, 2000.
GRAS, A., 'Anthropologie et philosophie des techniques: le passé d'une illusion' in *Socio-anthropologie* (1998)3.
GRELON, A. and F. BIRK (eds.), *Des Ingénieurs pour la Lorraine, XIXe-XXe siècles*, Metz, Éd. Serpenoise, 1998.
GRIN, J. and H. DE GRAAF, 'Technology Assessment as Learning' in *Science, Technology and Human Values* 21(1996), pp. 72-99.
GRUBB, M., 'Technologies, Energy Systems, and the Timing of CO_2 Emissions Abatement: An Overview of Economic Issues' in *Energy Policy* 25(1996), pp. 159-172.
GRUNWALD, A., 'Ethik der Technik' in *Ethik und Sozialwissenschaften* 7(1996)2/3, pp. 191-204; 'Comments and Reply', pp. 205-81.
HABERMAS, J., *Der Philosophische Diskurs der Moderne*, Frankfurt a.M., Suhrkamp, 1988.
HALL, J., *How is SSADM4.2 Different from SSADMV4?: SSADM4+ for Academics*, CCTA and International SSADM User Group Workshop, Leeds Metropolitan University, 1995.
HAMELIN, F., 'L'École d'application de l'artillerie et du génie et les cours industriels de la ville de Metz ' in A. GRELON and F. BIRCK (eds.), *Des ingénieurs pour la Lorraine, XIXe-XXe siècles*, Metz, Éditions Serpenoise, 1998.
HARMAN, W., *Global Mind Change – The Promise of the Last Years of the Twentieth Century*, San Fransisco, Knowledge Systems Inc., Institute of Noetic Sciences, 1993.
HASTEDT, H., *Aufklärung und Technik*, Frankfurt a.M., Suhrkamp, 1991.
HAWK, S.R., 'The Effects of Computerised Performance Monitoring: An Ethical Perspective' in *Journal of Business Ethics* 13(1994), pp. 949-957.
HEIM, M., *The Metaphysics of Virtual Reality*, New York, Oxford University Press, 1993.
HELD, D., *Democracy and the Global Order*, Stanford, Stanford University Press, 1995.
HÉRIARD DUBREUIL, B., *Imaginaire technique et éthique sociale, essai sur le métier d'ingénieur*, Brussels, De Boeck Université, 1997.
HIRSCHMAN, A., *Exit, Voice and Loyalty: Responses to Declines in Firms, Organisations and States*, Cambridge, MA, Harvard University Press, 1970.
HIRST, P. and G. THOMPSON, *Globalization in Question*, Cambridge, Polity Press, 1996.
HMSO, *Report of the Working Party on the Practice of Genetic Manipulation*, Williams Working Party, Comnd. 6600, Her Majesty's Stationery Office, London, 1974.
HOFF, E.-H., 'Berufliche Verantwortung' in E.-H. HOFF and L. LAPPE (eds.), *Verantwortung im Arbeitsleben*, Heidelberg, Asanger, 1995, pp. 46-63.

HÖFFE, O., *Politische Gerechtigkeit*, Frankfurt a.M., Suhrkamp, 1989.
HOGENHUIS, C.T. and D.G.A. KOELEGA (eds.), *Technologie als levenskunst. Visies op instrumenten voor inclusieve technologie-ontwikkeling*, Kampen, Kok, 1996.
HOGENHUIS, C.T., 'Een nieuwe opening. Morele codes in de beroepspraktijk' in C.T. HOGENHUIS and D.G.A. KOELEGA (eds.), *Technologie als levenskunst. Visies op instrumenten voor inclusieve technologieontwikkeling*, Kampen, Kok, 1996, pp. 175-196.
HOGENHUIS, C.T., *Beroepscodes en morele verantwoordelijkheid in technische en natuurwetenschappelijke beroepen*, Driebergen/Zoetermeer, MCKS, 1993.
HOMANN, K. and F. BLOME-DREES, *Wirtschafts- und Unternehmensethik*, Göttingen, Vandenhoeck, 1992.
HUBIG, C., A. HUNING and G. ROPOHL (eds.), *Klassiker der Technikphilosophie*, Berlin, Sigma, 2000.
HUBIG, C., *Technik- und Wissenschaftsethik*, 2nd ed., Berlin/ Heidelberg/ New York, Springer, 1995.
HUGHES, T.P., *Networks of Power: Electrification in Western Society, 1880-1930*, Baltimore, John Hopkins University Press, 1983.
HUNING, A. (ed.), *Ingenieurausbildung und soziale Verantwortung*, Düsseldorf/ Pullach, VDI-Verlag, 1974.
HUSTED, B.W., 'Reliability and the Design of Ethical Organisations: A Rational Systems Approach' in *Journal of Business Ethics* 12(1993).
IHDE, D., *Technics and Praxis*, Dordrecht, Reidel, 1979.
Ingenieurinnen und Ingenieure für die Zukunft. Aktuelle Entwicklungen von Ingenieurarbeit und Ingenieurausbildung, Berlin, Neef/Pelz, 1997.
IPCC (Intergovernmental Panel on Climate Change), *Climate Change 1995: Impacts, Adaptations and Mitigation of Climate Change: Scientific-Technical Analysis*, Cambridge, Cambridge University Press, 1996.
IPCC (Intergovernmental Panel on Climate Change), *Climate Change 1995: Economic and Social Dimensions of Climate Change*, Cambridge, Cambridge University Press, 1996.
IPCC (Intergovernmental Panel on Climate Change), *Climate Change: The IPCC Scientific Assessment*, J.T. HOUGHTON, G.J. JENKINS and J.J. EPHRAUMS (eds.), Cambridge, Cambridge University Press, 1990.
IRWIN, A., *Citizen Science: A Study of People, Expertise and Sustainable Development*, London, Routledge, 1995.
JÄGER, J. and T. O'RIORDAN, 'The History of Climate Change Science and Politics' in T. O'RIORDAN and J. JAEGER (eds.), *Politics of Climate Change – A European Perspective*, London/ New York, Routledge, 1996, pp. 1-31.
JANSSEN, M.A. 'Optimization of a Non-linear Dynamical System for Global Climate Change' in *European Journal of Operational Research* 99(1997), pp. 322-335.

JANSSEN, M.A., *Modelling Global Change – The Art of Integrated Assessment Modelling*, Cheltenham, UK, Edward Elgar Publishers, 1998.
JANSSEN, M.A. and H.J.M. DE VRIES, 'The Battle of Perspectives: A Multi-Agents Model with Adaptive Responses to Climate Change' in *Ecological Economics* 26(1998), pp. 43-65.
JANTSCH, E., *The Self-Organizing Universe – Scientific and Human Implications of the Emerging Paradigm of Evolution*, Oxford, Pergamon Press, 1980.
JASANOFF, S. and B. WYNNE, *Science and Decisionmaking*, Vol. 1 of *Human Choice and Climate Change*, S. RAYNER and E.L. MALONE, (eds.), Columbus, Ohio, Battelle Press, 1998, pp. 1-87.
JAYARATNA, N., *Understanding and Evaluating Methodologies*, Maidenhead, McGraw-Hill, 1994.
JESPERSEN, J., 'Reconciling Environment and Employment by Switching from Goods to Services? A Review of Danish Experience' in *European Environment* 9(1999).
JOHNSON, D., *Ethical Issues in Engineering*, Englewood Cliffs, NJ, Prentice Hall, Englewood Cliffs, 1991.
JONAS, H., *Das Prinzip Verantwortung*, Frankfurt a.M., Suhrkamp, 1979.
JONAS, H., *The Imperative of Responsibility: In Search of an Ethic for the Technological Age*, Chicago, Chicago University Press, 1984.
JUFER, M. and F.L. PERRET, 'Swissmetro, une chance de renouveau pour l'industrie suisse' in *La Vie économique* (February 1994).
JUFER, M., *Swissmetro – Synthèse de l'étude préliminaire*, Report DFTCE-SET 201, Lausanne, LEME/EPFL, 1993.
JÜNGER, F.G., *Die Perfektion der Technik*, 1946, 6th ed., Frankfurt a.M., Klostermann, 1980.
JUSTMAN, M. and M. TEUBAL, 'Innovation Policy in an Open Economy: A Normative Framework for Strategic and Tactical Issues' *Research Policy* 15(1986), pp. 121-138.
KADRITZKE, U., 'Das berufliche Selbstverständnis von Ingenieuren – und die Realität auf die es trifft' in *Ingenieurinnen und Ingenieure für die Zukunft. Aktuelle Entwicklungen von Ingenieurarbeit und Ingenieurausbildung*, Berlin, Neef/ Pelz, 1995, pp. 71-84, Rpt. 1997.
KANT, I., *The Moral Law: Kant's 'Groundwork of the Metaphysics of Morals'*, ed. H. Paton, London, Hutchinson, 1948.
KAPP, E., *Grundlinien einer Philosophie der Technik*, Braunschweig, Westermann, 1877, reprint Düsseldorf, Stern, 1978.
KAPTEIN, M., *Ethics Management: Auditing and Developing the Ethical Content of Organisations*, Dordrecht, Kluwer, 1998.
KEMP, P., *L'Irremplaçable. Une éthique de la technologie*, Paris, Cerf, 1991.
KIELY, R. and P. MARFLET (eds.), *Globalisation and the Third World*, London/New York, Routledge, 1998.
KODAMA, F., 'Innovative Approach to Research Draws Conspiracy Cries From Abroad' in *Japan Economic Journal* (26 November 1988).

KOELEGA, D.G.A. (ed.), *De ingenieur buitenspel? Over maatschappelijke verantwoordelijkheid in technische en natuurwetenschappelijke beroepen*, Den Haag, Boekencentrum, 1989.
KOHLBERG, L., *Essays on Moral Development*, vol. 1: *The Philosophy of Moral Development*, San Francisco, Harper and Row, 1981.
KRANZBERG, M. (ed.), *Ethics in an Age of Pervasive Technology*, Boulder, CO, Westview Press, 1980.
KREBS, F. and H. BOSSEL, 'Emergent Value Orientation in Self-Organization of an Animat' in *Ecological Modelling* 96(1997), pp. 143-164.
KUHN, T.S., *The Structure of Scientific Revolutions*, 3rd ed., Chicago, University of Chicago Press, 1996.
KUSTERS B., C. LOBET-MARIS and N.T. NGUYEN, 'Some Methodological Issues in Information T.A. – Two Cases Studies' in *Third European Congress on Technology Assessment: Post Congress Workshop*, Copenhagen, 1992, reprint in Cahier de la CITA T.A. 2.
LANGE, H. and W. MÜLLER (eds.), *Kooperation in der Arbeits- und Technikgestaltung*, Hamburg, Munster, 1995.
LANGFORD, D. and J. WUSTEMAN, 'The Increasing Importance of Ethics in Computer Science' in *Business Ethics – A European Review* 3(1994)4, pp. 219-222.
LAROCHELLE, M., N. BEDNARZ and J. GARRISON (eds.), *Constructivism and Education*, Cambridge, Cambridge University Press, 1998.
LATOUR, B., *La science en action*, Paris, La Découverte, 1989.
LAW, J., 'Technology and Heterogeneous Engineering: The Case of Portugese Expansion' in W.E. BIJKER, T.P. HUGHES and T. PINCH (eds.), *The Social Construction of Technological Systems*, Cambridge, MA, The MIT Press, 1987, pp. 111-134.
LENK, H. and G. ROPOHL (eds.), *Technik und Ethik*, Stuttgart, Reclam, 1987, 2nd ed., 1993.
LENK, H. and G. ROPOHL, 'Toward an Interdisciplinary and Pragmatic Philosophy of Technology' in P. DURBIN (ed.), *Research in Philosophy and Technology*, vol. 2, Greenwich, CT, 1979.
LENK, H. and M. MARING (eds.), *Technikverantwortung*, Frankfurt/ New York, Campus, 1991.
LENK, H. and M. MARING, (eds.), *Wirtschaft und Ethik*, Stuttgart, Reclam, 1992.
LENK, H. and M. MARING, 'Verantwortung und soziale Fallen' in *Ethik und Sozialwissenschaften* 1(1990)1, pp. 49-57, 'Comments and Reply', pp. 57-105.
LEPORI, B., *L'avenir des transports en Europe. Résultats choisis de la recherche européenne*, Bern, Swiss Science Council, TA 5/1995.
LÉVY, P., *L'intelligence collective. Pour une anthropologie du cyber espace*, Paris, La Découverte, 1994.
LISBON GROUP, R. PETRELLA, *Limites à la Compétitivité*, Paris, La Découverte, 1995.

LLORY, M., 'Ce que nous apprennent les accidents industriels' in *Revue Générale Nucléaire* (January-February 1998)1, pp. 63-68.
LLORY, M., *Accidents industriels: Le coût du silence. Opérateurs privés de parole et cadres introuvables*, Paris, L'Harmattan,1996.
LLORY, M., *L'accident de la centrale nucléaire de Three Mile Island*, Paris, L'Harmattan, 1999.
LOCKE, J., *Two Treatises of Government*, P. LASLETT (ed.), Cambridge, Cambridge University Press, 1960.
LUNDGREEN, P., 'De l'école spéciale à l'université technique, étude sur l'histoire de l'école supérieure technique en Allemagne avant 1870' in *Culture technique* (mars 1984)12, pp. 305-311.
MACKELLAR, L.W., W. LUTZ, A.J. MCMICHAEL and A. SUKRHE, *Population and Climate Change*, vol. 1 of *Human Choice and Climate Change*, S. RAYNER AND E.L. MALONE (eds.), Columbus, OH, Battelle Press, 1998, pp. 1-87.
MAGLEV 98, *Proceedings of the 16th International Conference on Magnetically Levitated Systems*, Yamanashi, Japan, 10-14 April 1998.
MANER, W., 'Unique Ethical Problems in Information Technology' in *Science and Engineering Ethics* 2(1996)2, pp. 137-155.
MARKL, H., 'Wir Deutsche sind heute zu risikofürchtig', interview in *Die Welt* (1 Juli 1996), p. 9.
MARTIN, M.W. and R. SCHINZINGER, *Ethics in Engineering*, New York, McGraw-Hill, 1983.
MARX, K., *Das Kapital*, 1867, vol. 1. K. MARX and F. ENGELS, *Werke*, Berlin, Dietz, 1959 and later, vol. 23.
MARZ, L. 'Mensch, Maschine, Moderne. Zur diskursiven Karriere der posthumanen Vernunft', Paper des Wissenschaftszentrums Berlin für Sozialforschung, 1993, FSII 93-107.
MASLOW, A.M., *The Psychology of Science: A Reconnaissance*, New York, Harper and Row, 1966.
MCCARTHY, J., *Dynamics of Software Development*, Redmont, WA, Microsoft Press, 1996.
MEADOWS, D.H., D.L. MEADOWS, J. RANDERS and W.W. BERHENS III, *The Limits to Growth: A Report for The Club of Rome's Project on the Predicament of Mankind*, New York, Universe Books, 1972.
MENKES, J., 'The Role of Technology Assessment in Decision Making-Process' in *International Symposium on the Role of T.A. in Decision Making-Process*, Bonn, 1982.
MILLON-DELSOL, C., *L'Etat subsidiaire*, Paris, PUF, 1992.
MINKS, K.-H., *Beschäftigungs- und Weiterbildungssituation von Ingenieurinnen in den alten und neuen Ländern*, Hannover, HIS, 1994.
MITCHAM, C. and A. HUNING (eds.), *Philosophy and Technology*, Vol. 2: *Information Technology and Computers in Theory and Practice* (Vol. 90 in the Boston Series in the Philosophy of Science), Dordrecht, D. Reidel, 1986.

MITCHAM, C. and R. MACKEY (eds.), *Philosophy and Technology: Readings in the Philosophical Problems of Technology*, New York, Free Press, 1972.
MITCHAM, C., *Engineering Ethics Throughout the World: Introduction, Documentation, Commentary and Bibliography*, Pennsylvania State University Press, 1992.
MITCHAM, C., *Thinking Through Technology: The Pathway between Engineering and Philosophy*, Chicago, University of Chicago Press, 1994.
MOATTI, J.-P., 'L'expérience américaine de l'évaluation technologique aux U.S.A.' in *Culture Technique* (juin 1983)10.
MONOT, P. and M. SIMON, *Habiter le Cybermonde*, Paris, Éd. de l'Atelier, 1998.
MOOR, J.H., 'Reason, Relativity, and Responsibility in Computer Ethics' in *Computers and Society* 28(March 1998)1, pp. 14-21.
MOOR, J.H., 'What is Computer Ethics?' in *Metaphilosophy* 16(1985)4, pp. 266-279.
MORECROFT, J. and J. STERMAN, 'Modelling for Learning' in special issue of *European Journal of Operations Research* 59(1992)1.
MORONE, J.G. and E.J. WOODHOUSE, *Averting Catastrophe: Strategies for Regulating Risky Technologies*, Berkeley, University of California Press, 1987.
MOSER, S. and A. HUNING (eds.), *Werte und Wertordnungen in Technik und Gesellschaft*, Düsseldorf, VDI-Verlag, 1975.
MOSER, S. and A. HUNING (eds.), *Wertpräferenzen in Technik und Gesellschaft*, Düsseldorf, VDI-Verlag, 1976.
MOWERY, D. and N. ROSENBERG, *Technology and the Pursuit of Economic Growth*, Cambridge, Cambridge University Press, 1989.
MUMFORD, E., *Designing Participatively*, Manchester Business School, 1983.
MÜNCH, R., *Dialektik der Kommunikationsgesellschaft*, Frankfurt a.M., Suhrkamp, 1991.
NEEF, W., 'Paradigmenwechsel in Beruf und Ausbildung von Ingenieuren' in FRICKE, *Innovation in Technik, Wissenschaft und Gesellschaft*, Bonn, Forum Humane Technikgestaltung 19(1998).
NELSON, R. and S. WINTER, *An Evolutionary Theory of Economic Change*, Cambridge, MA, The Belknap Press of Harvard University Press, 1982.
NELSON, R., 'Government Support of Technical Progress: Lessons from History' in *Journal of Policy Analysis and Management* 2(1983)4, pp. 499-514.
NELSON, R., 'The Simple Economics of Basic Scientific Research' in *Journal of Political Economy* (1959)67, pp. 297-306.
NELSON, R., *High Technology Policy: A Five Nation Comparison*, Washington/London, American Enterprise Institute for Public Policy Research, 1984.

NELSON, R., M. PECK and E. KALACHEK, *Technology, Economic Growth and Public Policy*, Washington, DC, Brookings Institute, 1976.

NELSON, R., *Understanding Technical Change as an Evolutionary Process*, Amsterdam, North Holland, 1987.

NGUYEN, N.T., G. FOUREZ, D. DIENG, *La santé informatisée: Carte santé et questions éthiques*, Brussels, De Boeck Université, 1995.

NGUYEN, N.T., *Bio-éthique et Technology Assessment, (Contrôler la science?)*, Brussels, De Boeck-Wesmael, 10/1990.

NIDA-RÜMELIN, J. (ed.), *Angewandte Ethik*, Stuttgart, Kröner, 1996.

NISBET, R.J., *History of the Idea of Progress*, London, Heinemann; Dordrecht, D. Reidel, 1980.

NOZICK, R., *Anarchy, State, and Utopia*, New York, Basic Books, 1974.

O'CONNELL, F., *How To Run Successful Projects*, Englewood Cliffs, NJ, Prentice Hall, 1994.

OTT, K., 'Technik und Ethik' in J. NIDA-RÜMELIN (ed.), *Angewandte Ethik*, Stuttgart, Kröner, 1996, pp. 650-717.

OZ, E., 'Ethical Standards for Computer Professionals: A Comparative Analysis of Four Major Codes' in *Journal of Business Ethics* 12(1993)9, pp. 709-728.

PARSONS, T., *The System of Modern Society*, Englewood Cliffs, NJ, Prentice Hall, 1971.

PAVITT, K. and W. WALKER, 'Government Policies towards Industrial Innovation: An Overview' in *Research Policy* 5(1976), pp. 11-97.

PERROW, C., *Normal Accidents: Living with High Risk Technologies*, New York, Basic Books, 1984.

PETRELLA, R., *Le Bien Commun: Éloge de la solidarité*, Brussels, Labor, 1996; Lausanne, Presses de Page 2, 1996.

PIAGET, J., *The Moral Judgement of the Child*, trans. M. GABAIN, London, Routledge and Kegan Paul, 1960. First pub. in England, 1932.

PICHT, G., *Mut zur Utopie*, München, Piper, 1969.

PICON, A. and M. YVON, *L'ingénieur artiste*, Paris, Presses de l'École Nationale des Ponts et Chaussées, 1989.

PICON, A., *L'invention de l'ingénieur moderne. L'École des Ponts et Chaussées, 1747-1851*, Paris, Presses de l'École Nationale des Ponts et Chaussées, 1992.

PIEPER, A. and U. THURNHERR (eds.), *Angewandte Ethik*, München, Beck, 1998.

PIEPER, A., *Ethik und Moral*, München, Beck, 1985.

QVORTRUP, L. et al. (eds.), 'Social Experiments with Information Technology and the Challenges of Innovation', a report from the FAST Programme of the CEC, Dordrecht, Reidel, 1987.

RAE, H., *The World in 2020 – Power, Culture and Prosperity: A Vision of the Future*, London, Harper Collins Publishers, 1995.

Rahanu, R., J. Davies and S. Rogerson, 'Ethical Analysis of Software Failure Cases' in *Failure and Lessons Learned in Information Technology Management* 3(1990).
Raoul, J.C., 'How High-Speed Trains Make Tracks' in *Scientific American* (October 1997).
Rapp, F. (ed.), *Neue Ethik der Technik?*, Wiesbaden, Deutscher Universitäts-Verlag, 1993.
Rapp, F. (ed.), *Technik und Philosophie, Technik und Kultur*, vol. 1, Düsseldorf, VDI-Verlag, 1990.
Rapp, F., *Die Dynamik der modernen Welt. Eine Einführung in die Technikphilosophie*, Hamburg, Junius, 1994.
Rawls, J., *A Theory of Justice*, New York, Harvard University Press, 1971.
Rayner, S., 'A Cultural Perspective on the Structure and Implementation of Global Environmental Agreements' in *Evaluation Review* 15(1991)1, pp. 75-102.
Remmen, A., 'Constructive Technology Assessment', in T. Cronberg et al. (eds.), *Danish Experiment: Social Constructions of Technology*, Copenhagen, New Social Science Monographs, 1991, pp. 185-200.
Rhoads, S.E. (ed.), *Valuing Life: Public Policy Dilemmas*, Boulder, CO, Westview Press, 1980.
Ricoeur, P., 'Science et Idéologie' in *Revue Philosophique de Louvain* 72(1974)14, pp. 328-356.
Ricoeur, P., *Le juste*, Paris, Éd. Esprit, 1995.
Ricoeur, P., *Soi-même comme un autre*, Paris, Le Seuil, 1990.
Ringius, L., A. Torvanger and B. Holtsmark, 'Can Multi-Criteria Rules Fairly Distribute Climate Burdens? OECD Results from Three Burden Sharing Rules' in *Energy Policy* 26(1998)10, pp. 777-793.
Rip A., J. Misa and J. Schot, *Managing Technology in Society: The Approach of Constructive Technology Assessment*, London, Pinter Publishers, 1995.
Roberts, K., 'Some Characteristics of One Type of High Reliability Organisation' in *Organisation Science* 1(1990), pp. 160-176.
Rogers, M., 'The Pandora's Box Congress' in *Rolling Stone* (19 June 1975), p. 77.
Rogerson, S. and D. Gotterbarn, 'The Ethics of Software Project Management' in G. Collste (ed.), *Ethics and Information Technology*, Delhi, India, New Academic Publishers, 1998, pp. 137-154.
Rogerson, S. and T.W. Bynum, 'Cyberspace the Ethical Frontier, Multimedia pp iv' in *The Times Higher Education Supplement*, No. 1179, June 9, 1995.
Rogerson, S. and T.W. Bynum, *Information Ethics: The Second Generation*, UK Academy for Information Systems Conference, 1996.
Rogerson, S., 'Software Project Management Ethics' in C. Myers, T. Hall and D. Pitt (eds.), *The Responsible Software Engineer*, London, Springer-Verlag, 1996.

ROGERSON, S., J. WECKERT and C. SIMPSON, 'An Ethical Review of Information Systems Development: The Australian Computer Society's Code of Ethics and SSADM' in *Information Technology and People*, 1999.
ROOBEEK, A.J.M. and M.N.F. DE BRUIJNE, *Strategisch management van onderop*, Amsterdam, 1993.
ROPOHL, G., 'Das neue Paradigma in den Technikwissenschaften' in *Ingenieurinnen und Ingenieure für die Zukunft. Aktuelle Entwicklungen von Ingenieurarbeit und Ingenieurausbildung*, Berlin, Neef/Pelz, 1995, pp. 11-16, reprint, 1997.
ROPOHL, G., 'Das Risiko im Prinzip Verantwortung' in *Ethik und Sozialwissenschaften* 5(1994)1, pp. 109-120, Comments and Reply, pp. 121-194.
ROPOHL, G., 'Verantwortungskonflikte im technischen Handeln' in J. HOFFMANN (ed.), *Irrationale Technikadaptation als Herausforderung an Ethik, Recht und Kultur*, Frankfurt a.M., IKO, 1997, pp. 55-80.
ROPOHL, G., *Ethik und Technikbewertung*, Frankfurt a.M., Suhrkamp, 1996.
ROPOHL, G., *Technologische Aufklärung*, Frankfurt a.M., Suhrkamp, 1991, 2nd ed., 1999.
ROSENBERG, N., 'Why Do Firms Do Basic Research (With Their Own Money)?' in *Research Policy* (1989).
ROSENBERG, N., *Inside the Black-Box: Technology and Economics*, Cambridge, Cambridge University Press, 1982.
ROSENBERG, N., *Perspectives on Technology*, Cambridge, Cambridge University Press, (1976) 1985.
ROSSEL, P., 'Dimensions sociologiques de Swissmetro' in M. JUFER et al., *Rapport pour l'étude préliminaire*, Lausanne, IREC-EPFL, 1992.
ROSSEL, P., F. BOSSET, O. GLASSEY and R. MANTILLERI, *Les enjeux des transports à grande vitesse: Des méthodes pour l'évaluation des innovations technologiques, l'exemple de Swissmetro*, Bern, FNRS/PNR41, Rapport F3, 1999.
ROTBLAT, J., 'Valedictory Address to the 47th Annual Pugwash Conference' in *Pugwash Newsletter* 34(November 1997)3/4, pp. 245-250.
ROTBLAT, J., *Scientists in the Quest for Peace: A History of the Pugwash Conferences*, Cambridge, MA, MIT Press, 1972.
ROTMANS, J. and H.J.M. DE VRIES (eds.), *Perspectives on Global Change: The TARGETS Approach*, Cambridge, Cambridge University Press, 1997.
SACHSSE, H., *Technik und Verantwortung*, Freiburg, Rombach, 1972.
SASSEN, S., *Losing Control? Sovereignty in the Age of Globalisation*, New York, Columbia University Press, 1996.
SCHAUB, J.H. and K.R. PAVLOVIC, *Engineering Professionalism and Ethics*, New York, Wiley, 1983.
SCHIMEL, D., I.G. ENTING, M. HEIMANN, T.M.L. WIGLEY, D. RAYNAUD, D. ALVES and U. SIEGENTHALER, 'CO_2 and the Carbon Cycle' in *Climate Change 1994*, IPCC, Cambridge, Cambridge University Press, 1995.

SCHIMEL, D.S., 'The Carbon Equation' in *Nature* 393(1998), pp. 208-209.
SCHMIDHEINY, S., *Changing Course: A Global Business Perspective on Development and the Environment*, Cambridge, MA, MIT Press, 1992.
SCHOT, J. and A. RIP, 'The Past and Future of Constructive Technology Assessment' in *Technological Forecasting and Social Change* (1998)54, pp. 251-268.
SCHULER, M. and V. KAUFMANN, *Pendularité à longue distance: la vitesse comme facteur structurant l'urbain*, DISP 126, Zurich, ORL, ETH, 1966.
SCHULTZ, P.A. and J.F. KASTING, 'Optimal Reductions in CO_2 Emissions' in *Energy Policy* 25(1997), pp. 491-500.
SCHWARTZ, M. and M. THOMPSON, *Divided We Stand: Redefining Politics, Technology and Social Choice*, New York, Harvester Wheatsheaf, 1990.
SEELY, B., 'European Contributions to American Engineering Education: Blending Old and New' in *Quaderns d'Historia de l'Enginyeria* 3(1999), pp. 25-50.
SEN, A., *Poverty and Famines*, Oxford, Clarendon Press, 1981.
SENGHAAS-KNOBLOCH, E. and B. VOLMERG, *Technischer Fortschritt und Verantwortungsbewusstsein*, Opladen, Westdeutscher Verlag, 1990.
SENGHAAS-KNOBLOCH, E., 'Lust und Unlust am technischen Fortschritt' in *Informatikforum* 7(1993)1-2, pp. 15-22.
SENGHAAS-KNOBLOCH, E. and H. LANGE, *Konstructive Sozialwissenschaft*, Munster, 1997.
SÉRUSELAT, F., 'Vers une approche des définitions du concept de l'éthique' in *La lettre éthique* 5(1990).
SHAPIRO, S.P., 'The Social Control of Impersonal Trust' in *American Journal of Sociology* 93(1987)3, pp. 623-658.
SHINN, T., 'Pillars of French Engineering' in *Social Studies of Science* 29(February 99)1.
SHNEIDERMAN, B. and A. ROSE, 'Social Impact Statements: Engaging Public Participation in Information Technology Design', Technical Report of the Human Computer Interaction Laboratory, September, 1995, pp. 1-13.
SIMON, H.A., *The Sciences of the Artificial*, Cambridge, MA, The MIT Press, 1969.
SIMONDON, G., *Du mode d'existence des objets techniques*, Aubier, Paris, 1969.
SINGER, M.F. and D. SOLL, 'DNA Hybrid Molecules' in *Science* 181(1973), p. 1114.
SJÖBERG, L., 'Riskupplevelse' in *Risk och beslut: Individen inför samhällsriskerna*, L. SJÖBERG (ed.), Stockholm, Liber förlag, 1982.
SKIDMORE, S., R. FARMER and G. MILLS, *SSADM Version 4 Models and Methods*, 2nd. ed., Manchester, NCC Blackwell, 1994.
SKOLIMOWSKI, H., 'A New Social Philosophy as Technology Assessment' in P.T. DURBIN (ed.), in *Research in Philosophy and Technology*, Vol. 5, Greenwich, CT, JAI Press, 1982, pp. 130-141.

STIX, G., 'Maglev Racing to Oblivion?' in *Scientific American* (October 1997).
SWART, R., M. BERK, M. JANSSEN, E. KREILEMAN and R. LEEMANS, 'The Safe Landing Approach: Risks and Trade-Offs in Climate Change' in J. ALCAMO, R. LEEMANS and E. KREILEMAN (eds.), *Global Change Scenarios of the 21st Century – Results from the Image 2.1 Model*, Oxford, Elsevier Science Ltd., 1998, pp. 193-234.
TANGL, P., *A Short History of Expectations on Magnetically Levitated Trains*, Master's Thesis, ESST, University of East London, 1996.
THÉPOT, A., *Les ingénieurs des Mines au XIXe siècle. Histoire d'un corps technique d'Etat*, Paris, Eska, 1998.
THOMPSON, D.F., 'Moral Responsibility of Public Officials: The Problem of Many Hands' in *American Political Science Review* 74(1980), pp. 905-916.
THOMPSON, M., R. ELLIS and A. WILDAWSKY, *Cultural Theory*, Boulder, CO, Westview Press, 1990.
TOURAINE, A., Z. HEGEDUS, F. DUBET and M. WIEVIORKA, *La prophétie antinucléaire*, Paris, Seuil, 1980.
TURNER, B.A. and N.F. PIDGEON, *Man-made Disasters*, 2nd ed., Oxford, Butterworth Heinemann, U.K., 1997.
UNGER, S.H., *Controlling Technology: Ethics and the Responsible Engineer*, New York, Holt, Rinehart and Winston, 1982; 2nd enlarged ed., New York, Wiley, 1994.
VAN ASSELT, M.B.A. and J. ROTMANS, 'Uncertainty in Perspective: A Cultural Perspective Based Approach' in *Global Environmental Change* 6(1996)2, pp. 121-158.
VAN BOXSEL, J.A.M., 'The Relevance of Technology Dynamics for the Practice of Constructive Technology Assessment' in *Advanced Training Course on Technology Assessment Methodology*, Mol, 14-15 October 1993.
VAN DAALEN, E., W. THISSEN and M. BERK, 'The Delft Process: Experiences with a Dialogue between Policy Makers and Global Modellers' in J. ALCAMO, R. LEEMANS and E. KREILEMAN (eds.), *Global Change Scenarios of the 21st century – Results from the Image 2.1 Model*, Oxford, Elsevier Science Ltd., 1998, pp. 193-234.
VAN DER POT, J.J.H., *Die Bewehrtung des technischen Fortschritts. Eine systhematische Übersicht der Theorien*, Assen, Van Gorcum, 1985 (English edition: *Steward or Sorcerer's Apprentice? The Evaluation of Technical Progress: a Systematic Overview of Theories and Opinions*, Delft, 1994).
VAN DER SLUIJS, J., *Anchoring Amid Uncertainty: On the Management of Uncertainties in Risk Assessment of Anthropogenic Climate Change*, Ph.D. diss., Utrecht, University of Utrecht, 1997.

VAN DIEREN, W., *Taking Nature into Account – Towards a Sustainable National Income. A Report to the Club of Rome*, New York, Springer Verlag, 1995.

VAN DIJK, J.W.A. and N. VAN HULST, 'Grondslagen van het technologiebeleid' in *Economische Statistische Berichten* (September 1988)21.

VAN DIJK, P., 'Op zoek naar een verantwoorde technologie-ethiek. Een bijdrage vanuit oecumenisch gezichtspunt' in D.G.A. KOELEGA (ed.), *De ingenieur buitenspel? Over maatschappelijke verantwoordelijkheid in technische en natuurwetenschappelijke beroepen*, Den Haag, Boekencentrum, 1989, pp. 127-148.

VAN DIJK, P., *Op de grens van twee werelden. Een onderzoek naar het ethisch denken van de natuurwetenschapper C.J. Dippel*, Den Haag, Boekencentrum, 1985.

VAN EINDHOVEN, J. c.s., *Ethics: To Coerce or Counsel? Annual Report 1997*, Rathenau Institute, The Hague, 1998.

VAN HEUR, R.J.H.G. and A.H. MARINISSEN, 'Een goed ontwerp is het halve werk. De verantwoordelijkheid van een industrieel ontwerper' in C.T. HOGENHUIS and D.G.A. KOELEGA (eds.), *Technologie als levenskunst. Visies op instrumenten voor inclusieve technologie-ontwikkeling*, Kampen, Kok, 1996, pp. 80-98

VAN LUIJK, H., 'Business Ethics: The Field and Its Importance' in B. HARVEY (ed.), *Business Ethics: A European Approach*, New York, Prentice Hall, 1994.

VAN RIESSEN, H., *Filosofie en techniek*, Kampen, 1949.

VAUGHAN, D., *The Challenger Launch Decision: Risky Technology, Culture and Deviance at NASA*, Chicago/London, Chicago University Press, 1996.

VDE/ZVEI, *Auswirkungen des Strukturwandels in der Elektroindustrie auf die Ingenieurausbildung*, Frankfurt a.M., 1994.

VDI, 'Ideas for the Development of Training for Engineers in Germany', Düsseldorf, 1998a.

VDI, 'VDI Study on the Demand for Engineers', 1997.

VDI, *Ingenieurausbildung im Umbruch, Empfehlungen des VDI für eine zukunftsorientierte Ingenieurqualifikation*, Düsseldorf, 1995.

VDI-Richtlinie 3780, *Technikbewertung: Begriffe und Grundlagen*, Düsseldorf, VDI-Verlag, 1991. Reprint in F. RAPP (ed.), *Normative Technikbewertung*, Berlin, Sigma, 2000, pp. 221-250.

VELTZ, P., *Des territoires pour apprendre à innover*, Collection Territoires et Société, La Tour d'Aigues, Éditions de l'Aube, 1994.

VENDRYES, G., *Superphénix pourquoi?*, Paris, Nucléon, 1997.

VÉRIN, H., 'Le mot: ingénieur' in *Culture technique* (mars 1984)12, p. 22.

VIVERET, P., *Démocratie, passions et frontières*, Paris, Éditions Charles Léopold Mayer, 1995.

VOLMERG, B. 'Die Moral des Ingenieurs und die Ethik technischen Handelns – sozial-psychologische Überlegungen zur Gestaltbarkeit der Technik durch den Wandel kultureller Werte' (1990) in B. VOLMERG and E. SENGHAAS-KNOBLOCH (eds.), *Technikgestaltung und Veranwortung. Bausteine für eine neue Praxis*, Opladen, Westdeutscher Verlag, 1992, pp. 111-123.

VOLMERG, B. and E. SENGHAAS-KNOBLOCH (eds.), *Technikgestaltung und Veranwortung. Bausteine für eine neue Praxis*, Opladen, Westdeutscher Verlag, 1992.

VON KEMPSKI, J., *Brechungen*, Reinbek, Rowohlt, 1964.

VON NELL-BREUNING, O., *Philosophisches Wörterbuch*, W. BRUGGER (ed.), Freiburg, Herder, 1953.

VON WEIZSÄCKER, E.U., A.B. LOVINS and L.H. LOVINS, *Factor Four: Doubling Wealth, Halving Resource Use*, London, Earthscan, 1997.

VON WESTPHALEN, R. (ed.), *Technikfolgenabschätzung*, 3rd ed., München/Wien, Oldenbourg, 1997.

VORSTENBOSCH, J.M.G., 'Ethische commissies: instrumenten van verantwoordelijkheid, verantwoorde instrumenten?' in C.T. HOGENHUIS and D.G.A. KOELEGA (eds.), *Technologie als levenskunst. Visies op instrumenten voor inclusieve technologie-ontwikkeling*, Kampen, Kok, 1996, pp. 142-163.

VROON, P., 'Een oud brein in een nieuwe wereld' in *Informatie en Informatiebeleid* 7(1989)4, pp. 19-29.

WALSHAM, G., 'Ethical Theory, Codes of Ethics and IS Practice' in *Information Systems Journal* 6(1996), pp. 69-81.

WATERS, M., *Globalisation*, London/New York, Routledge, 1995.

WEARE, K.M., 'Engineering Ethics: History, Professionalism and Contemporary Cases', in *Louvain Studies*, 13(1988), pp. 252-271.

WEBER, M., 'Politik als Beruf. 1919' in J. WINCKELMANN (ed.), *Soziologie, Universalgeschichtliche Analysen, Politik*, Stuttgart, Kröner, 1973, pp. 167-85.

WEBER, M., *La ciencia como profesión*, Madrid, Espasa Calpe, 1992.

WEIL, V., *Beyond Whistleblowing: Defining Engineers' Responsibilities*, Chicago, Illinois Institute of Technology, 1983.

WEILER, R. and D. HOLEMANS (eds.), *De leefbaarheid op aarde. Global Change: voor welke toekomst?*, Leuven, Garant en TI-K VIV, 1997.

WEIZENBAUM, J., *Computer Power and Human Reason: from Judgment to Calculation*, San Francisco, Freeman, 1976.

WENDELING-SCHRÖDER, U., *Autonomie im Arbeitsrecht*, Frankfurt a.M., Klostermann, 1994.

WENK, E., *Tradeoffs: Imperatives of Choice in a High-Tech World*, Baltimore, The John Hopkins Press, 1986.

WESTRUM, R., *Technology and Society: The Shaping of People and Things*, Belmont, Wadsworth, 1991.

WHITE, L., 'Technikfolgen-Abschätzung aus der Sicht eines Historikers' in M. DIERKES, T. PETERMANN and V. VON THIENEN (eds.), *Technik und Parlament. Technikfolgen-Abschätzung: Konzepte, Erfahrungen, Chancen*, Berlin, Sigma, 1986, pp. 47-72.

WHITE, T.I., *Business Ethics: A Philosophical Reader*, New York, Macmillan, 1993.

WIDDIFIELD, R. and V. GROVER, 'Internet and the Implications of the Information Superhighway for Business' in *Journal of Systems Management* (May/June 1995), pp. 16-21, 65.

WINNER, L., *Autonomous Technology*, Cambridge, MA, MIT Press, 1977.

WINTER, M.C., D.H. BROWN and P.B. CHECKLAND, 'A Role for Soft Systems Methodology In Information Systems Development' in *European Journal of Information Systems* 4(1995)3, pp. 130-142.

WOOD-HARPER, A.T., S. CORDER and B. BYRNE, 'Ethically Situated Information Systems Development' in C.R. SIMPSON (ed.), *AICE99 Conference Proceedings*, Australian Institute of Computer Ethics, 1999.

WOOD-HARPER, A.T., S. CORDER, J.R.G. WOOD and H. WATSON, 'How We Profess: The Ethical Systems Analyst' in *Communications of the ACM* 39(March 1996)3, pp. 69-77.

WORLD COMMISSION ON ENVIRONMENT AND DEVELOPMENT (Chair: G.H. BRUNDTLAND), *Our Common Future*, Oxford, Oxford University Press, 1987.

WORLD COUNCIL OF CHURCHES, *Biotechnology: Its Challenges to the Church and to the World*, Report by WCC Subunit on Church and Society, WCC Geneva, 1989.

WRR (Wetenschappelijke Raad voor het Regeringsbeleid), 'Duurzame risico's: een blijvend gegeven', Den Haag: Report n°4, Staatsuitgeverij, 1994.

WYATT, S., *Technology's Arrow: Developing Information Networks for Public Administration in Britan and the United States*, Maastricht, Universitaire Pers, Maastricht, 1998.

ZIMAN, J., 'The Restructuring of the Links Between Fundamental and Applied Research', paper prepared for the TEP Conference on Technology and Competitiveness, Paris, OECD, June, 1990.

ZIMMERLI, W.C. and V.M. BRENNECKE (eds.), *Technikverantwortung und Unternehmenskultur*, Stuttgart, Poeschel, 1993.

ZIMMERLI, W.C., 'Wandelt sich die Verantwortung mit dem technischen Wandel?' in H. LENK and G. ROPOHL (eds.), *Technik und Ethik*, Stuttgart, Reclam, 1987, pp. 92-111.

ZWICKY, F., *Entdecken, Erfinden, Forschen im Morphologischen Weltbild*, München/Zürich, Knaur, 1971.

CONTRIBUTORS

Madeleine AKRICH is a researcher at the Centre for Sociology of Innovation (École des Mine de Paris). Her work is mainly devoted to the sociology of technology. She has been working on the transfer of energy technologies in less developed countries and on the way users are represented during the innovation process. She is currently interested in the way actual uses of technologies transform activities, social relationships, experiences and self definition. She is involved in several teaching activities mostly directed at future engineers.

Brenda ALMOND is a professor of moral and social philosophy at the University of Hull in England. She is a graduate of London University and holds an honorary doctorate (*doc. honoris causa*) from the University of Utrecht. She is president of the Society for Applied Philosophy, and joint editor of the *Journal of Applied Philosophy*. As well as many articles and reviews on a wide range of philosophical topics, she is the author of several books including: *Exploring Ethics: A Traveller's Tale*, Oxford, Blackwell, 1998; *Exploring Philosophy: Second Edition, The Philosophical Quest*, Oxford, Blackwell, 1995 and *Introducing Applied Ethics* (ed.), Oxford, Blackwell, 1995.

Boel BERNER is a sociologist and professor in the multidisciplinary Department of Technology and Social Change at Linköping University, Sweden. Her main research interest is the social and gendered formation of technical expertise, historically and in relation to contemporary education, politics and work. Among her recent international publications are: *Gendered Practices. Feminist Studies of Technology and Society*, Stockholm, Almqvist & Wiksell International, 1997; 'Professional or Wage Worker? Engineers and Economic Transformation in Sweden,' in Peter Meiksins & Chris Smith (eds.) *Engineering Labour: Technical Workers in Comparative Perspective*, London, Verso, 1996; 'Explaining Exclusion: Women and Swedish Engineering Education from the 1890s to the 1920s,' in *History and Technology*, vol 14, 1997.

Jacques BERLEUR has been a professor in the Computer Science Faculty at the University of Namur since 1972 (See: http://www.info.fundp.ac.be/~jbl). His teaching involves 'Computers and Rationality' as well as 'Computers and Society,' and 'Ethics of Computing.' He has been: president (rector) of his University from 1984-1993; Belgian representative to IFIP-TC9 'Relationships between Computers and Society' since its creation (1976); chair of IFIP-WG9.2 'Social Accountability' (1990-96); chair of IFIP-Ethics Task Group, set up by the 1992 IFIP General Assembly (1992-94); and now chair of the Special Interest Group 'IFIP Framework on Ethics'(1994-present). He has been Belgian expert at the Commission of the European Communities for several programmes, including the COST A4 programme and the FAST programme (Forecasting and Assessment for Science and Technology) and co-author of the first 'Science and Technology Assessment Report' to the European Parliament. He is editor or author of 12 books, including: *Des rôles et missions de l'université*, Namur, Namur University Press, 1994; *Ethics of Computing: Codes, Spaces for Discussion and Law* (ed.) London,Chapman and Hall, 1996; *An Ethical Global Information Society: Culture and Democracy Revisited* (ed.), London, Chapman & Hall, 1997.

Pierre CALAME originally trained as a polytechnician and became a Civil Engineer in 1968. He was very interested in issues related to town planning at a time when the Ministère de l'Equipement (Ministry of Public Works) had considerable responsibilities in this field. From 1980 to 1983, Calame was appointed deputy Director of the Department of Town Planning and was involved in the debates regarding decentralisation. From 1983 to 1985, he joined the Department of International Relations of the Ministère de l'Equipement where he was in charge of relations with the North African region. In the meantime, he was Secretary General of Usinor, a public-owned company and since 1986, he has been Chairman of an international foundation under Swiss Law, the Charles Leopold Mayer Foundation. He is also a consultant on public management. He has published with André Talmant: *L'État au Cœur; Le meccano de la gouvernance*, Paris, Desclée De Brouwer, 1997.

Jose Angel CEBALLOS-AMANDI teaches engineering ethics in the Department of Industrial Organisation at the School of Higher Engineering (ICAI) and business ethics in the Department of Economics and Business (ICADE) at Comillas Pontifical University, Madrid.

Göran COLLSTE is a professor of ethics, doctor of theology and director of the Centre for Applied Ethics at Linköping University, Sweden. He has written numerous books and articles on ethics. Among these are: *Is Human Life Special? Philosophical Perspectives*, Studies in Applied Ethics, 3, Linköping, 1998; *Is Human Life Special? Religious Perspectives*, Studies in Applied Ethics, 5, Linköping, 1999; *Ethics and Information Technology*, New Delhi, New Academic Publisher, 1998.

Peter W.F DAVIES is visiting professor of business ethics at the Business School of Buckinghamshire Chilterns University College, and otherwise a freelance academic/writer. He was a founding member of COPE (Centre for Organisational and Professional Ethics) with colleagues at Brunel University. Formerly a mining and production engineer, he completed a PhD in the philosophy of technology in 1992. He has edited: *Current Issues in Business Ethics*, New York, Routledge, 1997, and co-edited *Ethics and Empowerment* (with John Quinn), London, Macmillan, 1999. He has also authored: 'Technology and Business Ethics Theory,' in *Business Ethics: A European Review*, 6(April 1997)2, pp. 76-80. He is a fellow of the Royal Society.

Stanislas DEMBOUR has studied philosophy and theology. He was employed as a worker for ten years in a glass factory and in the steel industry. He then worked twenty years for the state employment agency as a job counsellor for the unemployed. He is now retired. As an activist, he was a trade union delegate for many years in the companies and institutions in which he was employed.

H. (Bert) J.M. DE VRIES holds a degree in Theoretical Chemistry. He is one of the founders and coordinators of the Centre for Energy and Environmental Studies at the University of Groningen and has written his doctoral thesis on sustainable resource use. Since 1984 he has been a member of the Balaton Group. Since 1990 he has been a senior researcher at the Bureau for Environmental Assessment of the Dutch National Institute of Public Health and the Environment (RIVM) in Bilthoven. He has published in the fields of energy policy, technology and economics; environmental economics; and simulation gaming. His publications include: 'Energy and Environmental Consequences and Prospects – Taiwan's Miracle,' in *Energy Policy* 18(1990)4, pp. 949-961; 'Baseline Scenarios of Global Environmental Change,' (with J Alcamo, et. al.) in *Global Environmental Change*, 6(1996)4, pp. 261-303; *Perspectives on Global Change: the TARGETS Approach* (with J. Rotmans), Cambridge University Press, 1997; 'SUSCLIME - a Simulation Game on Population and Development in a Resource- and Climate-constrained Two-country World,' in *Simulation & Games*, 29(1998)2, pp. 216-37.

Christelle DIDIER first studied engineering, then turned to sociology. She has been working in the Centre d'éthique technologique at the Catholic University of Lille, and has been teaching ethics, sociology and the history of technology to engineering students for six years. She is co-editor of the first French text-book on engineering ethics, *Éthique industrielle* (1998) and author of *Pour un questionnement éthique des choix techniques:*

une ouverture dans la formation des ingénieurs (1999). She is currently completing her doctoral dissertation on French engineers and ethics at the École des Hautes Études en Sciences Sociales (EHESS).

Dominique DIENG holds a Master's Degree in business law, management control option (University of Law and Medicine, Lille) and a Certificate of Aptitude for business administration (University of Sciences and Techniques, Lille). She works as a researcher at the Facultés Universitaires Notre-Dame de la Paix in Namur in a multi-disciplinary research unit, the Cellule Interfacultaire de Technology Assessment (CITA) and part-time for the Centre de Recherche Informatique et Droit (CRID - IT and Law Research Centre). Her Research chiefly concerns analysing the social impact of new technologies. One of her research areas covers medical telematics, particularly health cards. In this context, she has taken over the coordination of a health card experiment, known as the Hemacard project, and has also assisted with the work of D.G.XIII, in the context of the concerted 'Eurocards' action. She has also contributed towards a work on health card and technology assessment: Nguyen N.T., Fourez G., Dieng D., *La santé informatisée: carte santé et questions éthiques*, Brussels, De Boeck Université, 1995. Her other research areas concern public administration in the context of the information society and electronic commerce. Before joining the CITA and CRID, Dominique Dieng worked in the private sector (market surveys and marketing consultancy, especially in the pharmaceutical industry).

José Luis FERNÁNDEZ FERNÁNDEZ, holds a PhD in philosophy and a Masters in business administration (MBA) and is co-ordinator of Professional Ethics at Comillas Pontifical University, Madrid. He is also a professor of business ethics at the Department of Economics and Business (ICADE) of the same university. He is a member of the editorial board of the journal *Ethical Perspectives* and president of Ethics, Economics and Management (EBEN-Spain). Among his numerous publications on subjects in his field of specialty are more than 10 books, written by himself or together with others, and more than 50 press and other widely disseminated articles.

Gerard M. FOUREZ holds a PhD in physics, and other degrees in mathematics, philosophy and theology. He is a professor at the University of Namur in the Department of Sciences, Philosophy and Society, which focuses its research on the relations between science, technology and society. He is the coordinator of the research cell EMSTES (Enseignement des Mathématiques et des Sciences, Technologies, Éthiques et Sociétés) and has also been involved in research on Technology Assessment. He is author of several books including: *Eduquer, Ecoles, Ethiques, Société*, Brussels, De Boeck Univ., 1990; *Alphabétisation scientifique et technique, Essai sur les finalités de l'enseignement scientifique*, Brussels, De Boeck Univ., 1994; *La construction des sciences*, De Boeck Univ., 1996; *Nos savoirs sur nos savoirs*, Brussels, De Boeck Univ, 1997; and some fifteen other books. He has also written numerous articles in periodicals ranging from *Physical Review* to *Social Studies of Sciences, La revue philosophique de Louvain, Esprit*. etc.

Annie GIREAU-GENEAUX is pedagogical head of the Social and Human Relations Department at the Institut supérieur d'Electronique du Nord (ISEN) where she teaches epistomology, ethics and management to engineering students. Originally trained in philosophy, she has carried out research in the history of the physical sciences in 18th century France. She is member of the team in Engineering Ethics which is attached to the Centre d'éthique technologique at the Catholic University of Lille. In addition, she is co-editor of the first engineering ethics textbook in France: *Ethique industrielle, Textes pour un Débat*, Brussels, De Boeck Univ., 1998.

Philippe GOUJON received his PhD from the University of Burgundy (1993) with a doctorate entitled *'Along the Paths of Information: From Communication to Complexity'*. He is author of articles and books on artificial life, self-organisation, thermodynamics, the complexity concept and also the connection between science, techniques, education, culture and ethics. One aspect of his research centres on history and the socio-economic impact of the European genome project and in a more general way biotechnology and the international genome programme. He is now addressing the epistemological and ethical problems of this new scientific field which is named genomics. At the same time, he is studying the epistemology of complexity as well as the interaction between science, technology and culture. Since 1997, he has been a member of the Centre d'éthique technologique and teaches various courses at the Catholic University of Lille. His publications include: S.M. Thomas, P.B. Joly and P. Goujon, *The Industrial Use Of Genome Resources In Europe* - Report for the European Commission - Contract N°BIO4-CT96-0686; B. Feltz, A. Crommelinck, Ph. Goujon (eds.), *Auto-Organisation et Émergence dans les Sciences de la Vie*, OUSIA, 1999; P. Goujon, *From Biotechnology to Genomes: the Meaning of the Double Helix*, World Scientific Publishing, 2001.

André GRELON is director of studies and head of the Sociology, Psychology and Social Anthropology Department at the École des Hautes

Etudes en Sciences Sociales (EHESS-Paris). He holds a doctorate in sociology from the Université de Paris VII (1983). He is also a researcher at the 'Laboratoires d'analyses secondaires et des méthodes appliquées à la sociologie' and at the 'Institut de recherche sur les sociétés contemporaines' (IRESCO)-CNRS. He is currently working on the history of higher technical training in France in the XIXth and XXth centuries and on the history and sociology of the Christian managers' movement (MCC). Among his numerous publications, some of the most recent include: *Des Ingénieurs pour la Lorraine, XIXe-XXe siècles*, (with F. Birck, ed.), Metz, Éd. Serpenoise, 1998; *La naissance de l'ingénieur électricien. Origine et développement des formations nationales électrotechniques* (with Girolamo Ramunni, ed.), Paris, PUF, 1997; 'La naissance de l'enseignement supérieur industriel en France,' in *Quaderns d'historia de l'Enginyera*, Barcelona, ESTEIB, vol.1(1996), pp. 53-81.

Bertrand HÉRIARD DUBREUIL practiced and taught "workshop designing" at the *Institut Catholique d'Arts et Métiers* (ICAM-Lille). After graduate studies in philosophy and theology, he has taught in the STS department at ICAM and is currently in charge of the Centre d'éthique technologique, which is part of the Catholic University of Lille. He has published an essay on the engineering profession entitled : *Imaginaire technique et éthique sociale, essai sur le métier d'ingénieur*, Brussels, De Boeck Univ., 1997 and is co-editor with Christelle Didier and Annie Gireau-Genaux of the first French handbook in engineering ethics entitled *Ethique industrielle, textes pour un débat*. Brussels, De Boeck Univ., 1998.

Christiaan T. HOGENHUIS studied physics at the University of Utrecht in the Netherlands, with specialties in both meteorology and environmental research and social philosophy. Since 1989 he has been a staff member and researcher at the Multidisciplinary Centre for Church and Society, where he is responsible for research and debate on topics related to technology, ethics and religion. He has written several books and articles on topics like 'tools for engineering ethics', 'the responsibility of the engineer', 'the value of professional codes of ethics for engineers', 'sustainability and technology' and 'sustainability, lifestyle and spirituality'. With Dick Koelega, he has written *Technologie als levenskunst. Visies op instrumenten woor inclusieve Technologie-ontwillelig*, Kampen, 1996.

Dick G.A. KOELEGA studied theology at the University of Leiden, with a special interest in the philosophy of religion, ethics and peace research. Since 1985 he has been a staff member of the multidisciplinary Centre for Church and Society. From 1990 to 1996, he also worked as a part-time researcher at the Institute for Ethics of the Free University, Amsterdam. He has written several books and articles on topics such as engineering ethics (together with Christiaan Hogenhuis), public morality, political philosophy and theology of technology. With Christiaan Hogenhuis, he has written *Technologie als levenkunst. Visies op instrumenten woor inclusieve Technologie-ontwillelig*, Kampen, 1996.

Hellmuth LANGE has been a professor at the University of Bremen since 1973. He teaches on problems concerning occupational sociology, sociology of science and technology, political science and political technology, in the scientific-technical courses of the University of Bremen. In addition, he carries out research on problems concerning occupational sociology (physicists and engineers), the theory and philosophy of technology, as well as the sociology and politics of science. He is the author and editor of several books and articles on questions of the sociology of engineering and on questions of the organisation of labour and technology, including: *Kooperation in der Arbeits- und Technilgestaltung* (ed. with W. Müller), Hamburg/Munster, Lit-Verlag, 1995; H. Lange, E. Senghaas-Knobloch, *Springen aus dem Stand. Akteure der Arbeist-und Technikgestalatung in der Transformation*, Hamburg/Munster, Lit-Verlag, 1994.

Michel LLORY was an engineer and researcher at Electricité de France for 27 years. He successively held the posts of head of probability studies (safety) and head of the department of safety and reliability studies, where he worked from 1986 to 1993. He has directed studies of security and safety, risk analysis of complex industrial systems, and in particular studies of human factors and ergonomics. He has taken part in field research and commissions of enquiry. In his work and his writings, he has contributed to introducing the social sciences into high-risk socio-technical systems. Since 1993, he has been active as a consulting engineer for the Institut du Travail Humain, which he founded. He has assisted in developing a new, organisational approach to accidents and security. He is intent on showing the importance of ergonomics, work organisation, input from personnel in the field, and the changing position of engineers in discussions about risks. At the same time, he carries out teaching and training activities in these areas as an associate professor at the University of Angers. He has authored various articles for specialist journals and conferences, and has written two books.

Martin MEGANCK, is a civil engineer (Ghent State University, 1982) and doctor in applied sciences (Catholic University of Leuven, 1987). From 1982 till 1987, he was a research engineer in the Central Laboratory of the Lyonnaise des Eaux in Le Pecq, France. After training in philosophy and moral theology in the Dominican Institutional Study Centre and at the Faculty of Theology in Leuven, he now teaches philosophy and professional ethics to engineering students at the Katholieke Hogeschool Sint-Lieven in Ghent and Aalst.

Carl MITCHHAM is professor of liberal arts and international studies at the Colorado School of Mines in Golden, Colorado, where he also serves as coordinator of 'ethics across the curriculum'. He has published extensively on the philosophy and ethics of technology. Among his more recent books are: *Thinking through Technology: The Path between Engineering and Philosophy*, Chicago, University of Chicago Press, 1994; *Engineer's Toolkit: Engineering Ethics*, Englewood Cliffs, NJ, Prentice Hall, 2000. He is also general editor of the annual series *Research in Philosophy and Technology* published by JAI-Elsevier Press.

Göran MÖLLER has an MSc in civil engineering and a ThD. He is associate professor in ethics in the Department of Theology at Uppsala University. As an engineer, Göran Möller has been engaged in traffic safety issues at the Swedish National Road Administration. He studied theology at Uppsala University. In his doctoral dissertation he critically examined a number of proposed methods of determining an acceptable level of risk of accidental death. Möller then worked with environmental issues at the Swedish National Road Administration. He returned to University and his research has resulted in the monography entitled *Ethics and the Life of Faith; a Christian Moral Perspective*. In this book Möller situates ethics within a wider and more comprehensive context of a philosophy of life and human nature. His current research concerns virtue ethics. He is the author of the following books: *Ethics and the Life of Faith; a Christian Moral Perspective*, Leuven, Peeters, 1998. *Etikens landskap; etik och kristen livstolkning*, Stockholm, Arena, 1995.

Riccardo PETRELLA is a professor at the Catholic University of Louvain-la-Neuve, advisor to the European Commission and president of the so-called Lisbon Group, which has published the report *Limits to Competitivity* in eleven languages. His latest book is *Le Bien Commun: Éloge de la Solidarité*, Brussels, Labor, 1996 and Lausanne, presses de Lausanne, 1996.

Bernard REBER holds a doctorate in philosophy and is currently a researcher at CNRS. From 1995-97 he was lecturer in ethics at the Swiss Federal Institute of Technology at Lausanne. Since 1997, he has been coordinator of public scientific controversies at the École Nationale

Supérieure des Mines de Paris. He also gave lectures in fundamental moral theology at the Institut Catholique de Paris (Rouen) in 1999. He is the author of various articles, including: 'L'éthique du virtuel: Apocalypse contre Platon,' in *Virtual Worlds 98*, first international colloquium on synthetic image and virtuality, 1-3.07.98, published by the Moscow Academy of Science, 1999; 'Croyances, éthique économique transversale et interreligieuse,' in *Ethical Perspectives, Journal of the European Ethics Network*, October 1999. 'L'irremplaçable. L'ethique technologique de Peter Kemp,' in *Esprit*, January 1998. He is currently preparing *Responsabilités et nouvelles Technologies. Archimède étranglé* to be published by L'Harmattan.

Simon ROGERSON is the director of the Centre for Computing and Social Responsibility at De Montfort University. Following a successful industrial career, he now combines research, lecturing and consultancy in the ethical and management aspects of computing. He has published over 140 papers and books and given presentations in many countries and at many international conferences. He conceived of and co-directs the ETHICOMP conference series on the ethical impacts of ICT. He is a member of the Parliamentary IT Committee in the UK, a fellow of the Institute for the Management of Information Systems and a fellow of the Royal Society for the encouragement of Arts, Manufactures and Commerce.

Günter ROPOHL is a professor of general technology in the Institute of Polytechnic and Work Studies at the University of Frankfurt on Main (Germany). He studied mechanical engineering and received a PhD in engineering with an investigation on 'Flexible Manufacturing Systems' (1971). Then he turned to the philosophy and sociology of technology and obtained the lecturer's qualification (habilitation) in this field (1978). He has been professor of philosophy and sociology of technology (1979-1981) and director of the Institute for General Studies at the University of Karlsruhe (1979-1987), and, since 1981, he has been working with Frankfurt University. He has written widely on systems engineering, technology assessment and the philosophy of technology. Among other works, his publications include: *Eine Systemtheorie der Technik* (1979), *Interdiszipliäre Technikforschung* (ed. 1981), *Die unvolkommene Technik* (1985), *Technik und Ethik* (co-ed., 1987, 2nd ed. 1993), *Schlüsseltexte zur Technikbewertung* (co-ed., 1990), *Technologische Aufklärung* (1991), *Ethik und Technikbewertung* (1996), *Wie die Technik zur Vernunft kommt* (1998) and *Allgemeine Technologie* (1999).

Pierre ROSSEL studied at the universities of Neuchatel and Geneva and received a PhD in cultural anthropology. Since 1985, he has been carrying out research in the area of science, technology and society, in particular on innovation dynamics and more recently on the technology assessment of information and communication technologies and on new transport systems such as magnetically-levitated trains. Since 1992, he has been in charge of the ESST European Master Programme at the Swiss Federal Institute of Technology at Lausanne.

Sir Joseph ROTBLAT was born in Warsaw and went to Liverpool, England in 1939. He then participated in the Manhattan Project, the Allied effort to develop nuclear weapons during the Second World War. He returned to England early in 1945, as soon as he learned that the Germans had abandoned their atom bomb work, and devoted himself to the development of peaceful applications of nuclear physics, especially medical applications. He was one of the founders of Pugwash, its secretary-general between 1957 and 1973, and its president between 1988 and 1997. The Pugwash Conferences on Science and World Affairs received the Nobel Peace Prize in 1995. Sir Joseph Rotblat continues to promote the ideals and activities of Pugwash.

Johan SCHOT is a professor of the history of technology at the Eindhoven University of Technology and the University of Twente in the Netherlands. He has been working on a number of related issues: technology assessment, environmental management, sociology of technology, history of technology. He is co-founder of the Greening of Industry Network, project leader of the Dutch national project on the history of technology in the 20th century and co-editor of various books, among others: *Managing Technology in Society* (with A. Rip and Th. J. Misa) and *Environmental Strategies for Industry* (with K. Fischer).

Eva SENGHAAS-KNOBLOCH is a professor of sociology of work and the humanisation of work at the Research Centre for Work and Technology at the University of Bremen. She graduated in sociology from the Free University of Berlin and holds a PhD from the University of Frankfurt, as well as a habilitation in political science from the University of Bremen. She has co-authored: *Konstructive Sozialwissenschaft* (with H. Lange), Munster, 1997; *Springen aus dem Stand. Akteure der Arbeits- und Technikgestaltung in der Transformation* (with H. Lange), Munster, 1994; and *Zukunft der industriellen Arbeitskultur* (with B. Nagler and A. Dohms), Munster, 1996.

Luc SOETE is a professor of international economics in the Faculty of Economics and Business Administration at Maastricht University. He completed his first degrees in economics and development economics at the University of Ghent, then obtained his PhD in economics at the University of Sussex. Before coming to Maastricht in 1986, he worked in the Science Policy Unit at the University of Sussex and in the Department of Economics at Stanford University. His research interests cover a broad range of theoretical and empirical studies of the impact of technological change on employment, economic growth and international trade and investment, as well as the related policy and measurement issues. In 1988 he set up MERIT, of which he has since been the director. In 1995, he was co-ordinator for the OECD's Directorate for Science, Technology and Industry, the G-7 project 'Technology, Productivity and Job Creation,' submitted to the G-7 Jobs Summit in Lille, April 1996. He chaired the high Level Expert Group on 'Social and Societal Aspects of the Information Society' for the European Commission which was set up in May 1995. More recently he has been engaged in discussions on taxing electronic transactions.

André STAEDLER holds a degree in political science and is currently a scientific assistant at the Research Center for Labour and Technology at the University of Bremen.

André TALMANT has been consecutively a civil engineer in charge of a district in the Valenciennois region, manager of the waterways at the regional level, regional manager of Public Works and last but not least general chief inspector at the Ministry of Public Works. He has contributed to: *Ethique industrielle, textes pour un débat*, Brussels, De Boeck Univ., 1997 and has published: *L'État au Cœur; Le meccano de la gouvernance* (with Pierre Calame), Paris, Desclée De Brouwer, 1997.

Marie D'UDEKEM-GEVERS has a doctorate in Anthropology from the University of Liege and a doctorate in Sciences (Zoologie) from the Catholic University of Louvain-la-Neuve, as well as a Degree and a Masters in Computer Sciences from the University of Namur (FUNDP). She is currently a researcher in the 'Cellule Interfacultaire de Technology Assessment' (CITA) of the FUNDP. Since 1994 her research has focused mainly on Information Highways and the Information Society. She has analysed notably codes of ethics, digital cities and Internet filtering. She is a member of the International Federation of Information Processing (IFIP) Working Group 9.2 on Social Accountability of Computing [IFIP WG 9.2] (1995-) and the Special Interest Group 9.2.2 on Ethics [IFIP SIG 9.2.2]. (The list of her papers and a longer bibliography can be found at http://www.info.fundp.ac.be/~mge/)

Johann VERSTRAETEN is a professor of ethics at the Catholic University of Leuven, director of the Centre for Catholic Social Thought, and chairman of the board of directors of the European Ethics Network. He teaches business ethics, peace ethics, engineering ethics, Catholic social thought. His research and writing focuses on professional ethics, social ethics, issues of narrative and imagination in ethics. His recent publications include: *Matter of Breath, Foundations for Professional Ethics* (ed. with G.de Stexhe), 2000; *Business Ethics, Broadening the Perspectives* (ed.), 2000. *Catholic Social Thought: Twilight or Transcendence* (ed. with J.Boswell and F.McHugh), 2001.

Raoul WEILER is a professor in societal aspects of technology at the Centre of Ethics on Agriculture, Life Sciences and Environment in the Faculty of Agricultural and Applied Biological Sciences at the Catholic University of Leuven (KUL). He is a bio-engineer and holds a doctorate in Applied Biological Sciences (KUL). He spent several years in different research institutions and universities in France and the US. Between 1970 and 1996 he worked in the chemical industry. He presently holds teaching positions at the Catholic University of Louvain-la-Neuve, the University of Antwerp and the Catholic University of Leuven. He is the editor of four books on the Philosophy of Technology, Sustainable Development and Global Change. He is a member of the European Academy of Sciences and Arts, Salzburg.

Sally WYATT holds a doctorate and works at the Department of Innovation Studies, University of East London, and with the 'sociaal-wetenschappelijke informatica' group, University of Amsterdam. She has published widely in the areas of technology policy and technology studies. She is currently working on a project funded by the British Economic and Social Research Council entitled, 'From the Net to the Web and Beyond: Actors and Interests in the Construction of the Internet'. Between 1996-99, she was the International Coordinator of the MA programme, 'Society, Science and Technology in Europe', involving fifteen European universities. She is author or co-author of the following articles and books: 'Shaping Cyberspace - Interpreting and Transforming the Internet' (with Graham Thomas), in *Research Policy* 28(1999)6; *Technology's Arrow: Developing Information Networks for Public Administration in Britain and the United States*, Maastricht, Universitaire Pers Maastricht, 1998; *Multinationals and Industrial Property, The Control of the World's Technology* (with Gilles Bertin), London, Harvester-Wheatsheaf, 1988 (first published in French as *Multinationales et propriété industrielle, Le contrôle de la technologie mondiale*, Paris, PUF, 1986).

PRINTED ON PERMANENT PAPER • IMPRIME SUR PAPIER PERMANENT • GEDRUKT OP DUURZAAM PAPIER - ISO 9706

N.V. PEETERS S.A., KLEIN DALENSTRAAT 42, B-3020 HERENT